Trends in Industrial Engineering Applications to Manufacturing Process

Jorge Luis García-Alcaraz ·
Arturo Realyvásquez-Vargas · Emigdio Z-Flores
Editors

Trends in Industrial Engineering Applications to Manufacturing Process

Springer

Editors
Jorge Luis García-Alcaraz
Industrial Engineering and Manufacturing
Universidad Autónoma de Ciudad Juárez
Ciudad Juárez, Chihuahua, Mexico

Division of Research and Postgraduate Studies
Tecnológico Nacional de México/IT Ciudad Juárez
Ciudad Juárez, Chihuahua, Mexico

Emigdio Z-Flores
Instituto Tecnológico de Tijuana
Tecnológico Nacional de México
Tijuana, Baja California, Mexico

Arturo Realyvásquez-Vargas
Instituto Tecnológico de Tijuana
Tecnológico Nacional de México
Tijuana, Baja California, Mexico

ISBN 978-3-030-71578-6 ISBN 978-3-030-71579-3 (eBook)
https://doi.org/10.1007/978-3-030-71579-3

This Springer imprint is published by the registered company Springer Nature Switzerland AG
The registered company address is: Gewerbestrasse 11, 6330 Cham, Switzerland

Preface

Industrial engineering started in the late eighteenth century and has been evolving gradually from a series of scattering ideas to a combination of tools, ideologies, policies, and methodologies. All of them are aimed to improve industrial processes by using adequate resources, either human or manufactured infrastructure, and the environment where companies are operating cycles impact more. The most direct benefit is the competitive advantage among other companies, regardless of the labor sector type and size. Nevertheless, this broad application of knowledge has profound repercussions globally, socially, economically, and environmentally.

In order to use this set of tools and methods, industrial engineers selected a project from which they measure and analyze specific data. They then develop, present, and implement a new working method aimed to improve company processes. Later, they collect and analyze data again with the new method and compare it with the data from the previous method. Suppose the latter method indeed improves the previous condition. In that case, it is established at the current process by a standardization technique, controlling its execution through monitoring and supervision until a new improvement stage is defined. Then, the cycle continues shifting through specific rules and objectives defined at a high-level decision-making operation. The project mentioned above can be originated from a wide range of different approaches being logistics, productivity, quality, or ergonomics, some of the most applied, to name a few.

In the case of the logistics approach, a key member is the supply chain (SC). That SC comprises different functions, processes, and activities that allow raw materials, finished products, and services to be transformed, delivered, and consumed by the end customer. The supply chain's functions, processes, and activities can be affected by political, technological, or environmental issues. A Bad SC can cause raw materials, finished products, or services to reach the consumer at the wrong time. In this situation, the application of industrial engineering can help improve SC activities.

Regarding the productivity approach applied to both services and manufacturing activities, it is vital to meet customer expectations regarding quality and quantity. Nowadays, there are cases where the companies do not meet these expectations, even with the well-established methodologies, so the need to explore new research areas is in constant demand. At the production level of goods, the decision-making, control of architectures, embedded and emergent intelligence are active and promising research fields. Tools and technologies, such as the Internet of Things, artificial intelligence, agile work environments, and big data analytics enable inert raw material and machines to become active, endowed with greater capabilities and intelligent decision-making. Therefore, this puts pressure on organizations to take advantage of existing and cutting-edge technologies to improve performance, thus minimizing the disruptions, risks, and disturbances within the company structure and maximizing sustainability at the same time.

Finally, concerning the ergonomics approach, whenever human beings carry out some activity, industrial engineering can help to improve it, whether it is a manufacturing or service activity, providing comfort and safety. Ergonomics analyzes the workers' risks by applying methods such as Rapid Upper limb Assessment (RULA), Rapid Entire Body Assessment (REBA), the National Institute for Occupational Safety and Health (NIOSH) model, and the NASA Task Load Index (NASA-TLX). Subsequently, it seeks to reduce or eliminate risks by redesigning workstations or work methods, thus improving workers' performance in the first instance and that of the company in the second instance.

This book is divided into three parts. Part I, called supply chain and logistics in industrial engineering, comprises Chaps. 1–9. In Chap. 1, a conceptual framework for a supply chain with a green approach is presented, the factors that influence it, green practices, and performance. Chapter 2 presents a study of the methodological and statistical treatments of the integration arcs of Frohlich and Westbrook, and De la Calle Vicente, considered as good diagnostic tools to measure the level of integration, both internal and external, of companies in the food industry. In Chap. 3, supply centers are reorganized based on economic and time improvements, compared with the company's previous model. Through a case study, in Chap. 4, a new methodology is employed to plan the production layout, using a multi-tier approach. In Chap. 5, using lean methodologies, the distribution process is improved and validated by a discrete simulation stage, finalizing with the implementation at a factory. Using a structural equation model integrating manufacturing technology, in Chap. 6, a supply chain is configured and statistically validated using surveys. Using the Factory Physics approach, in Chap. 7, the effects of variability on the optimal flow of goods are studied through analytic methods. In Chap. 8, a mathematical model is presented to optimally locate a set of street markets in an arid region, providing an adequate budget, schedule, and required quantities. Finally, in Chap. 9, a study is presented where characterization, evaluation, and improvement of the procurement process directly impact the organization's strategy for decision making.

In Part II, Chaps. 10–17 compile different trends for manufacturing and quality applications. Chapter 10 summarizes the latest approaches of manufacturing execution systems, their implications in the manufacturing processes, and their transformation as a dynamic and productive system. Through an exploratory factor analysis based on information collected from different companies in the manufacturing sector, a quality management system is synthesized, exposing barriers and benefits related to the implementation of quality standards, presented in Chap. 11. Applying the principles of coherence and cohesion, a model design to develop genetic information in an organization is shown in Chap. 12, an essential element for company survival. In Chap. 13, three elements within a company are studied using a structural equation modeling scheme: employer branding, servant leadership, and work engagement. Using a DMAIC methodology, a process is improved, as shown in Chap. 14 and validated using the first-time yield as a metric. In Chap. 15, a working environment is improved by adapting Lean Sigma methodology allowing a multi-culture workplace, exhibiting clear benefits. Chapter 16 focuses on using discrete simulation to provide an adequate solution for layout plans in a factory, using a real-case scenario, and presenting economic analysis based on the implementation results. In Chapter 17, a new design of operation strategy is presented to increase companies' competitiveness capability, providing good theoretical and practical implications in the overall research area of productivity and quality in industrial manufacturing.

Finally, Part III groups Chaps. 18–21, covering topics related to human factors and ergonomics. In Chap. 18, a review of the latest workplace design is presented considering health risks and regulations and further provides new perspectives in technology, human context, and safety. By assessing different factors that could potentially produce risks within a workplace environment, in Chap. 19, a study is presented through the analysis of a case study and provides insights at several levels of ergonomic design. Chapter 20 describes an analysis of the occurrence of electrical failures in industrial equipment and machinery and workers' discomfort, influenced by climatic variations and air pollutants indoors of the electronics industry. In Chap. 21, an investigation is performed to study if human factors impact manufacturing companies' loss of productivity and quality. The authors report a case study using different approaches, both for quantitative and for qualitative analyses.

Ciudad Juárez, Mexico — Jorge Luis García-Alcaraz
Tijuana, Mexico — Arturo Realyvásquez-Vargas
Tijuana, Mexico — Emigdio Z-Flores

Acknowledgements

The editors thank all the authors for their support in our editorial project. We know this year 2020 was challenging due to the COVID-19 pandemic and, even so, they continued to write their chapters. Thank you very much to all of them.

Likewise, we also wish to thank all the anonymous reviewers who gave their assessment to chapters initially proposed, without a doubt that this editorial project would not have the expected quality without their collaboration and wise judgment. Thanks to all of them for that invaluable collaboration.

Finally, the editors would like to thank our families for the support provided throughout this editorial project; since much of that time was invested in editing this book, we had to spend with them in harmony. Without a doubt, we are in debt with our families.

Ciudad Juárez, Mexico — Jorge Luis García-Alcaraz
Tijuana, Mexico — Arturo Realyvásquez-Vargas
Tijuana, Mexico — Emigdio Z-Flores

Contents

Part I Supply Chain and Logistics

1 **A Conceptual Framework of Green Supply Chain Management: Influential Factors, Green Practices, and Performance** 3
Mariana Trujillo-Gallego and William Sarache

2 **Arcs of Integration: Methodological and Statistical Analysis** 35
Bertha Molina-Quintana, Antonio Vaamonde-Liste, Jaime Apolinar Martínez-Arroyo, and María Berta Quintana-León

3 **A Proposal to Redesign the Distribution Networks of Steel Manufacturing and Distribution Companies** 59
Alexandra Ferrer, Yndira Guevara, Yereth Romero, and Mario Chong

4 **Layout Planning Approach at a Plumbing Department in a Manufacturing Industry: A Case Study** 77
Guadalupe Hernández-Escobedo, Cristian Omar González-Higuera, Karina Cecilia Arredondo-Soto, and Arturo Realyvásquez-Vargas

5 **Improving Distribution Process Using Lean Manufacturing and Simulation: A Seafood Packer Company Case** 103
Julian I. Aguilar-Duque, Jorge Luis García-Alcaraz, Juan L. Hernández-Arellano, and Guillermo Amaya-Parra

6 **Effect of Advanced Manufacturing Technology on Responsive Supply Chain Strategy, Pull System and Responsiveness to Market** 133
José Roberto Díaz-Reza, Adrián Salvador Morales-García, and Jorge Luis García-Alcaraz

7 Effect of Variability on the Optimal Flow of Goods in Supply Chains Using the Factory Physics Approach 157
Yamil Mejía-Hernández, Salvador Hernández-González, Aarón Guerrero-Campanur, and José Alfredo Jiménez-García

8 Mathematical Approach for Street Markets Location in Food Desert Regions: A Case Study in Valparaíso Chile Region 181
Julián Alberto Espejo-Díaz, Daniela Granados-Rivera, and Gonzalo Mejía

9 Optimization of Supply Management in a Fishing Sector Company 199
Carlos Gonzales, Natalí Zavala, Yuri Diaz, Manuel Peña, Ana Luna, and Mario Chong

10 Decision Support Model for Solid Waste Management in a Closed-Loop Supply Chain 215
Vivian Lorena Chud-Pantoja, Claudia Cecilia Peña-Montoya, and Juan Carlos Osorio-Gómez

Part II Manufacturing and Quality

11 Manufacturing Execution System State-Of-The-Art: Its Evolution and Dynamism Focused on Industry 4.0 245
Saúl Manuel Favela-Camacho, Javier Molina-Salazar, and Lázaro Rico-Pérez

12 Enablers and Barriers for a Quality Management System Implementation in Mexico: An Exploratory Analysis 263
Marcos Alberto Sanchez-Lizarraga, Jorge Limon-Romero, Diego Tlapa, Yolanda Baez-Lopez, Lizbeth Puerta-Sierra, and Marco Maciel-Monteon

13 Gestation of the Genetics of a Company 289
Jorge Vera Jiménez

14 Work Engagement as Result of Leadership and Employer Branding: A Case of a Mexican Manufacturing Plant 311
María Teresa de la Garza-Carranza, Eugenio Guzmán-Soria, Jorge Armando López-Lemus, and Quetzalli Atlatenco-Ibarra

15 The DMAIC Methodology as a Tool for Process Improvement: The Case of a Mexican Manufacturing Company 335
Arturo Realyvásquez-Vargas, Jorge Luis García-Alcaraz, Guadalupe Hernández-Escobedo, Karina Cecilia Arredondo-Soto, Joel Eduardo García-Ortíz, Julio Blanco-Fernández, and Emilio Jiménez-Macías

16 Adaptability of the Lean-Sigma Methodology for Operations in a Multicultural Workplace ... 365
Noé Gaudencio Alba-Baena and Aldo Salcido-Delgado

17 Using Simulation for Facility Layout Problems: A Case Study of an Ecuadorian Handcraft Chocolate Company ... 389
Diego Viteri-Viteri, Andreé Rodríguez-Luna, Sonia Avilés-Sacoto, and Galo Mosquera-Recalde

18 A Framework for the Formulation of an Operations Strategy in Manufacturing Systems ... 435
Jorge A. Vivares, William Sarache, and Jorge E. Hurtado

Part III Human Factors and Ergonomics

19 The Human Factor as a Central Element in the Design of the Workplace. A Systematic Review ... 465
Gabriela Jacobo-Galicia, Carlos Raúl Navarro-González, Mildrend Montoya-Reyes, Ismael Mendoza-Muñoz, and Eusebio Jiménez-López

20 Ergonomic Study of Construction Workers in Odisha (India): A Case Study in Construction Sites ... 507
Debesh Mishra and Suchismita Satapathy

21 Are Productivity and Quality in Electronics Manufacturing Industry Affected by Human Factors? A Quantitative Analysis Using Statistical Tools ... 529
Berónica Botello-Lara, Margarita Gil-Samaniego Ramos, Juan Ceballos-Corral, and Arturo Sinue Ontiveros-Zepeda

Index ... 561

Editors and Contributors

About the Editors

Jorge Luis García-Alcaraz is Full-time Professor with adscription to the Department of Industrial Engineering at the Autonomous University of Ciudad Juárez and Division of Research and Postgraduate Studies/Tecnológico Nacional de México/IT Ciudad Juárez. He received a bachelor's degree and a M.Sc. in Industrial Engineering from the Instituto Tecnológico de Colima (Mexico), a Ph.D. in Industrial Engineering from Instituto Tecnológico de Ciudad Juárez (Mexico), a Ph.D. in Innovation in Product Engineering and Industrial Process from the University of La Rioja (Spain), a Ph.D. in Engineering and industrial technologies and a Postdoc in Manufacturing Process from University of La Rioja (Spain). His main research areas are related to multicriteria decision making applied to lean manufacturing, production process modeling and statistical inference. He is Founding Member of the Mexican Society of Operation Research and active member in the Mexican Academy of Industrial Engineering. Currently, Dr. Garcia is National Researcher level III recognized by the National Council of Science and Technology of Mexico (CONACYT) and is working with research groups from Colombia, Spain, Chile, and Dominican Republic. Dr. Garcia is author/coauthor of around 150 papers publisher in journals indexed in the Journal Citation Reports, more than 200 international conferences and congress around the world. Dr. Garcia is author of 12 books published by international publishers as Springer and IGI Global, all related to lean manufacturing, the main tools and techniques.

ORCID: http://orcid.org/0000-0002-7092-6963
Scopus Author ID: 55616966800
Department of Industrial Engineering and Manufacturing
Autonomous University of Ciudad Juárez
Av. del Charro 450 Norte, Ciudad Juárez, Chihuahua, Mexico, Z.P. 32310
Phone: +52 656 6884843 ext. 5433 and 4249 Fax: +52 656 6884841
Email: jorge.garcia@uacj.mx

Arturo Realyvásquez-Vargas is Full-time Professor in the Department of Industrial Engineering at Tecnológico Nacional de Mexico/Instituto Tecnológico de Tijuana in Mexico. He received master's degree in Industrial Engineering, a Ph.D. in Engineering Sciences from the Autonomous University of Ciudad Juarez in Mexico and a Ph.D. in Innovation in Product Engineering and Industrial Process at the University of La Rioja (Spain). Specifically, his main research areas are related to optimization of industrial processes, lean manufacturing and ergonomics. Also, he is Active Member of the Society of Ergonomists of Mexico Civil Association (Sociedad de Ergonomistas de México, SEMAC) and the Network of Optimization of Industrial Processes (Red de Optimización de Procesos Industriales, ROPRIN). Currently, Dr. Realyvásquez is National Researcher recognized by the National Council of Science and Technology of Mexico (CONACYT) as candidate. Dr. Realyvásquez is an author/coauthor of around 12 papers published in journals indexed in the Journal Citation Reports, and also, he has attended international conferences and congress in Mexico as well as in the USA. Nowadays, Dr. Realyvásquez has supervised more than 20 bachelor theses and five master theses. Dr. Realyvásquez is author of one book published by the international publisher Springer, related to ergonomics. Also, Dr. Realyvásquez has edited two books in IGI Global, all of them related to ergonomics.

ORCID: https://orcid.org/0000-0003-2825-2595

Scopus Author ID: 56167726800

Emigdio Z-Flores received a bachelor's degree in Electrical Engineer and a doctoral degree in Engineering Sciences at Instituto Tecnológico de Tijuana and his master's degree in Digital Systems at Centro de Investigación y Desarrollo de Tecnología Digital. He worked in the industry sector for nearly 8 years as product designer, developing several projects in different key areas. He currently works as Full-time Professor in Tecnológico Nacional de Mexico, campus Tijuana, at the Industrial Engineering Department, both in undergraduate and graduate programs. He is author/coauthor of several publications in indexed journals and international conferences and holds one patent. His current interests involve applied statistics, computer science, neuroscience and industrial real-world applications.

ORCID: https://orcid.org/0000-0002-1442-5320

Industrial Engineering Department

Tecnológico Nacional de México, Campus Tijuana

Calzada Del Tecnológico S/N, Fraccionamiento Tomas Aquino. Tijuana, Baja California, Mexico, 22414

Email: emigdio.zflores@tectijuana.edu.mx

Contributors

Julian I. Aguilar-Duque Facultad de Ingeniería, Arquitectura y Diseño. Universidad Autónoma de Baja California, Ensenada, B.C., C.P., México

Noé Gaudencio Alba-Baena Department of Industrial Engineering and Manufacturing, Universidad Autónoma de Ciudad Juárez, Ciudad Juárez, Chihuahua, Mexico

Guillermo Amaya-Parra Facultad de Ingeniería, Arquitectura y Diseño. Universidad Autónoma de Baja California, Ensenada, B.C., C.P., México

Karina Cecilia Arredondo-Soto Faculty of Chemical Sciences and Engineering, Universidad Autónoma de Baja California, Tijuana, Baja California, Mexico; Chemical Sciences and Engineering Department, Universidad Autónoma de Baja California, Tijuana, México

Quetzalli Atlatenco-Ibarra Departamento de Estudios Multidisciplinarios, Universidad de Guanajuato, Yuriria, Guanajuato, Mexico

Sonia Avilés-Sacoto Departamento de Ingeniería Industrial, Colegio de Ciencias e Ingenierías, Instituto de Innovación en Productividad y Logística CATENA-USFQ, Universidad San Francisco de Quito (USFQ), Quito, Pichincha, Ecuador

Yolanda Baez-Lopez Facultad de Ingeniería, Arquitectura y Diseño, Universidad Autónoma de Baja California, Ensenada, Baja California, Mexico

Julio Blanco-Fernández Department of Mechanical Engineering, University of La Rioja. Edificio Departamental—C/San José de Calasanz, 31, Logroño, La Rioja, Spain

Berónica Botello-Lara Department of Industrial Engineering, Universidad Autónoma de Baja California, Mexicali, Baja California, México

Juan Ceballos-Corral Department of Industrial Engineering, Universidad Autónoma de Baja California, Mexicali, Baja California, México

Mario Chong Facultad de Ingeniería, Universidad del Pacífico, Lima, Perú; Universidad del Pacífico, Jesús María, Perú

Vivian Lorena Chud-Pantoja Escuela de Ingeniería Industrial, Programa de Ingeniería Industrial, Universidad del Valle, Sede Zarzal, Zarzal, Valle del Cauca, Colombia

María Teresa de la Garza-Carranza Departamento de Ciencias Económico Administrativas, Instituto Tecnológico de Celaya. Av, Celaya, Guanajuato, Mexico

Yuri Diaz Universidad del Pacífico, Jesús María, Perú

José Roberto Díaz-Reza Department of Electric Engineering and Computation, Autonomous University of Ciudad Juárez, Chihuahua, México

Julián Alberto Espejo-Díaz Faculty of Engineering, Universidad de La Sabana, Autopista Norte de Bogotá, Chía, Colombia

Saúl Manuel Favela-Camacho Department of Industrial Engineering and Manufacturing, Universidad Autónoma de Ciudad Juárez. Av. Del Charro 450 Norte. Col. Partido Romero. Juárez, Chihuahua, México

Alexandra Ferrer Pacífico Business School, Universidad del Pacífico, Lima, Perú

Jorge Luis García-Alcaraz Department of Industrial Engineering and Manufacturing, Universidad Autónoma de Ciudad Juárez, Juárez, Chihuahua, Mexico; Division of Research and Postgraduate Studies, Tecnológico Nacional de México/IT Ciudad Juárez, Ciudad Juárez, Chihuahua, Mexico

Joel Eduardo García-Ortíz Departamento de Ingeniería Industrial, Tecnológico Nacional de México/IT Tijuana, Calzada del Tecnológico S/N, Col. Tomás Aquino, Tijuana, Baja California, Mexico

Carlos Gonzales Universidad del Pacífico, Jesús María, Perú

Cristian Omar González-Higuera Departamento de Ingeniería Industrial, Tecnológico Nacional de México/I. T. de Tijuana, Calzada Tecnológico s/n, Tijuana, México

Daniela Granados-Rivera Faculty of Engineering, Universidad de La Sabana, Autopista Norte de Bogotá, Chía, Colombia

Aarón Guerrero-Campanur Tecnológico Nacional de México, Instituto Tecnológco de Uruapan, Uruapan, Mexico

Yndira Guevara Facultad de Ingeniería, Universidad del Pacífico, Lima, Perú

Eugenio Guzmán-Soria Departamento de Ciencias Económico Administrativas, Instituto Tecnológico de Celaya. Av, Celaya, Guanajuato, Mexico

Juan L. Hernández-Arellano Instituto de Arquitectura Diseño y Arte, Universidad Autónoma de Ciudad Juárez, Ciudad Juárez, Chihuahua, C.P., México

Guadalupe Hernández-Escobedo Departamento de Ingeniería Industrial, Tecnológico Nacional de México/IT Tijuana, Calzada del Tecnológico S/N, Col. Tomás Aquino, Tijuana, Baja California, Mexico; Tecnológico Nacional de México, Instituto Tecnológico de Celaya, Celaya, Mexico

Jorge E. Hurtado Departamento de Ingeniería Civil, Facultad de Ingeniería y Arquitectura, Universidad Nacional de Colombia, Manizales, Colombia

Gabriela Jacobo-Galicia Department of Industrial Engineering, Universidad Autónoma de Baja California, Mexicali, Baja California, México

José Alfredo Jiménez-García Tecnológico Nacional de México, Instituto Tecnológico de Celaya, Celaya, Mexico

Eusebio Jiménez-López CIAAM, Southern Sonora Technological University—Northwest La Salle University-IIMM, Cd. Obregón, Sonora, México

Emilio Jiménez-Macías Department of Electrical Engineering, University of La Rioja, Edificio Departamental—C/San José de Calasanz, 31, Logroño, La Rioja, Spain

Jorge Vera Jiménez Department of Industrial Engineering, TecNM Campus Oaxaca, Oaxaca, Mexico

Jorge Limon-Romero Facultad de Ingeniería, Arquitectura y Diseño, Universidad Autónoma de Baja California, Ensenada, Baja California, Mexico

Ana Luna Universidad del Pacífico, Jesús María, Perú

Jorge Armando López-Lemus Departamento de Estudios Multidisciplinarios, Universidad de Guanajuato, Yuriria, Guanajuato, Mexico

Marco Maciel-Monteon Unidad Académica San Luis, Universidad Estatal de Sonora, Carretera Sonoyta-San Luis Rio Colorado Km, Sonora, Mexico

Jaime Apolinar Martínez-Arroyo Faculty of Accounting and Administrative Sciences, Universidad Michoacana de San Nicolas de Hidalgo, Gral. Francisco J. Múgica S/N, Morelia, Michoacán, México

Yamil Mejía-Hernández Tecnológico Nacional de México, Instituto Tecnológico de Celaya, Celaya, Mexico

Gonzalo Mejía Faculty of Engineering, Universidad de La Sabana, Autopista Norte de Bogotá, Chía, Colombia

Ismael Mendoza-Muñoz Department of Industrial Engineering, Universidad Autónoma de Baja California, Mexicali, Baja California, México

Debesh Mishra KIIT Deemed to be University, Bhubaneswar, Odisha, India

Bertha Molina-Quintana Faculty of Accounting and Administrative Sciences, Universidad Michoacana de San Nicolas de Hidalgo, Gral. Francisco J. Múgica S/N, Morelia, Michoacán, México

Javier Molina-Salazar Department of Industrial Engineering and Manufacturing, Universidad Autónoma de Ciudad Juárez. Av. Del Charro 450 Norte. Col. Partido Romero. Juárez, Chihuahua, México

Mildrend Montoya-Reyes Department of Industrial Engineering, Universidad Autónoma de Baja California, Mexicali, Baja California, México

Adrián Salvador Morales-García Department of Industrial Engineering and Manufacturing, Autonomous University of Ciudad Juárez, Chihuahua, México

Galo Mosquera-Recalde Departamento de Ingeniería Industrial, Colegio de Ciencias e Ingenierías, Universidad San Francisco de Quito (USFQ), Quito, Ecuador

Carlos Raúl Navarro-González Department of Industrial Engineering, Universidad Autónoma de Baja California, Mexicali, Baja California, México

Arturo Sinue Ontiveros-Zepeda Department of Industrial Engineering, Universidad Autónoma de Baja California, San Fernando, Tecate, Baja California, Mexico

Juan Carlos Osorio-Gómez Escuela de Ingeniería Industrial, Universidad del Valle, Cali, Valle del Cauca, Colombia

Claudia Cecilia Peña-Montoya Departamento de Operaciones y Sistemas, Facultad de Ingeniería, Universidad Autónoma de Occidente, Cali, Valle del Cauca, Colombia

Manuel Peña Universidad Nacional Mayor de San Marcos, Lima, Perú

Lizbeth Puerta-Sierra Facultad de Economía y Negocios, Universidad Anáhuac México, Colonia Lomas Anáhuac Huixquilucan, Estado de México, Mexico

María Berta Quintana-León Faculty of Accounting and Administrative Sciences, Universidad Michoacana de San Nicolas de Hidalgo, Gral. Francisco J. Múgica S/N, Morelia, Michoacán, México

Margarita Gil-Samaniego Ramos Department of Industrial Engineering, Universidad Autónoma de Baja California, Mexicali, Baja California, México

Arturo Realyvásquez-Vargas Departamento de Ingeniería Industrial, Tecnológico Nacional de México/IT Tijuana, Calzada del Tecnológico S/N, Col. Tomás Aquino, Tijuana, Baja California, Mexico

Lázaro Rico-Pérez Department of Industrial Engineering and Manufacturing, Universidad Autónoma de Ciudad Juárez. Av. Del Charro 450 Norte. Col. Partido Romero. Juárez, Chihuahua, México

Andreé Rodríguez-Luna Departamento de Ingeniería Industrial, Colegio de Ciencias e Ingenierías, Universidad San Francisco de Quito (USFQ), Quito, Ecuador

Yereth Romero Facultad de Ingeniería, Universidad del Pacífico, Lima, Perú

Aldo Salcido-Delgado Department of Industrial Engineering and Manufacturing, Universidad Autónoma de Ciudad Juárez, Ciudad Juárez, Chihuahua, Mexico

Marcos Alberto Sanchez-Lizarraga Facultad de Ingeniería, Arquitectura y Diseño, Universidad Autónoma de Baja California, Ensenada, Baja California, Mexico

William Sarache Departamento de Ingeniería Industrial, Facultad de Ingeniería y Arquitectura, Universidad Nacional de Colombia Sede Manizales, Manizales, Caldas, Colombia

Suchismita Satapathy KIIT Deemed to be University, Bhubaneswar, Odisha, India

Diego Tlapa Facultad de Ingeniería, Arquitectura y Diseño, Universidad Autónoma de Baja California, Ensenada, Baja California, Mexico

Mariana Trujillo-Gallego Departamento de Ingeniería Industrial, Universidad Nacional de Colombia Sede Manizales, Manizales, Caldas, Colombia

Antonio Vaamonde-Liste Department of Statistics and Operational Research, University School of Business Studies of the University of Vigo, Vigo, Spain

Diego Viteri-Viteri Departamento de Ingeniería Industrial, Colegio de Ciencias e Ingenierías, Universidad San Francisco de Quito (USFQ), Quito, Ecuador

Jorge A. Vivares Escuela de Ciencias Básicas, Tecnología e Ingeniería, Ingeniería Industrial, Universidad Nacional Abierta y a Distancia, Dosquebradas, Colombia

Natalí Zavala Universidad del Pacífico, Jesús María, Perú

List of Figures

Fig. 1.1	Publications distribution over time	6
Fig. 1.2	Contributions by region	8
Fig. 1.3	GSCM and sustainable classification themes	10
Fig. 1.4	Proposed conceptual framework	11
Fig. 3.1	AS-IS model. *Source* Own elaboration	64
Fig. 3.2	Sets and index configuration. *Source* Own elaboration	66
Fig. 3.3	TO-BE model. *Source* Own elaboration	69
Fig. 4.1	Initial layout	88
Fig. 4.2	Proposed locations	89
Fig. 4.3	Spaghetti diagram	92
Fig. 4.4	Layout draft	94
Fig. 4.5	Operational needs	95
Fig. 4.6	Last layout draft and final layout	96
Fig. 4.7	Comparative of initial and final layouts	97
Fig. 5.1	Trends in simulation paper reports	105
Fig. 5.2	VSM according to Thomas et al. (2018), Mei et al. (2018)	108
Fig. 5.3	Types of models	109
Fig. 5.4	Physical layout of the Seafood Packer Company, Ensenada	111
Fig. 5.5	Original physical layout of the Seafood Packer Company	112
Fig. 5.6	VSM of EZR	114
Fig. 5.7	VSM of EZP	114
Fig. 5.8	New physical layout of the Seafood Packer Company, Ensenada	117
Fig. 5.9	EZR and EZM process flow in the Seafood Packer Company, Ensenada	119
Fig. 5.10	Configuration system for online operation restricted by no mobile shared	128
Fig. 5.11	Configuration system for online operation restricted by mobile shared and packing one product at a time	129
Fig. 5.12	Comparison of simulated lead times versus real lead times	130
Fig. 5.13	Comparison of simulated value added versus real value added	130

Fig. 6.1 Proposed model . . . 142
Fig. 6.2 Evaluated model . . . 147
Fig. 7.1 Generic supply chain network . . . 158
Fig. 7.2 Example of variability. Service time = 0.583 and different variability . . . 165
Fig. 7.3 Merge, queue, and split flow in a Jackson network . . . 166
Fig. 7.4 Supply chain case . . . 167
Fig. 7.5 Deterministic solution . . . 168
Fig. 7.6 Probability distributions of service time in the three scenarios of variability . . . 170
Fig. 7.7 Probability distributions of time between arrivals in the three scenarios of variability . . . 171
Fig. 7.8 Cycle time trend . . . 176
Fig. 8.1 Current street market locations . . . 188
Fig. 8.2 F&V local small retailers' location in Valparaíso region . . . 190
Fig. 8.3 Percentage of coverage quality and covered demand varying the budget . . . 191
Fig. 8.4 Proposed street markets location with 76% of customers demand coverage . . . 193
Fig. 8.5 Proposed street market locations in Valparaíso food desert zones . . . 194
Fig. 8.6 Proposed street markets location with 90.78% of customers demand coverage . . . 195
Fig. 9.1 Supply chain. *Source* Own elaboration (2019) . . . 204
Fig. 9.2 Level process map. *Source* Own elaboration (2019) . . . 204
Fig. 9.3 Maturity model. *Source* Own elaboration . . . 207
Fig. 9.4 Current functional strategy map. *Source* Own elaboration (2019) . . . 208
Fig. 9.5 Problem statement. *Source* Own elaboration (2019) . . . 209
Fig. 10.1 Generic reverse supply chain (Pochampally et al. 2009b) . . . 218
Fig. 10.2 Types of supply chains in a reverse logistics process (Monroy and Ahumada 2006) . . . 219
Fig. 10.3 Structure of a generic supply chain (Pochampally et al. 2009b) . . . 220
Fig. 10.4 Return sources in a closed-cycle supply chain (Stindt and Sahamie 2014) . . . 221
Fig. 10.5 Proposed methodology (Chud et al. 2018) . . . 226
Fig. 10.6 CLSC for glass container adapted from (Chud et al. 2018) . . . 227
Fig. 10.7 Hierarchical structure for glass container manufacturer decisions (Chud et al. 2018) . . . 231
Fig. 10.8 Result of multi-criteria priorities for the manufacturer . . . 231
Fig. 10.9 Preferences of alternatives in manufacturer's multi-criteria model with expert choice . . . 233
Fig. 10.10 Initial status of manufacturer's alternative preferences considering the priority of criteria . . . 233

Fig. 11.1 Control recipe model based on the ISA-S88 (ISA 2006) 250
Fig. 11.2 Multi-level functional hierarchy of activities based on the ISA-S95 (ISA 2005) 251
Fig. 11.3 Manufacturing operations management model based on ISA-S95 (ISA 2005) 251
Fig. 11.4 MESA model based on business strategic initiatives (MESA 2019) 252
Fig. 12.1 Organizations with ISO 9001 certification worldwide over time. *Source* ISO survey (2017) 267
Fig. 12.2 Eight quality management principles in the ISO 9001:2000 standard 269
Fig. 12.3 Seven quality management principles in ISO 9001:2015 standard 273
Fig. 13.1 Interrelation of the leadership process with the other processes 301
Fig. 13.2 Interrelation of the organization's system processes 301
Fig. 13.3 Concatenation of values 306
Fig. 13.4 Genetic information of the organization 307
Fig. 14.1 Hypothetical model 319
Fig. 14.2 Results of the model 1 for work engagement (dedication). *Note* The model shows the standardized regression weights ... 325
Fig. 14.3 Model 2: Results of the model for work engagement (absorption). *Note* The model shows the standardized regression weights 326
Fig. 14.4 Model 3: Results of the model for work engagement (vigor). *Note* The model shows the standardized regression weights 326
Fig. 15.1 FTY values 346
Fig. 15.2 Defects and frequency of occurrence 347
Fig. 15.3 Damaged T3 (defect) vs T3 in good condition 347
Fig. 15.4 Damaged/polluted optical fiber (defect) vs optical fiber in good condition 348
Fig. 15.5 Ishikawa diagram for damaged T3s defect 349
Fig. 15.6 Cause–effect diagram of the damaged optical fibers defect 352
Fig. 15.7 Visual inspection of fiber 354
Fig. 15.8 Original alignment fixture with narrow cavity 355
Fig. 15.9 Improvement action implemented for T3 defect 355
Fig. 15.10 Sample requested to suppliers 356
Fig. 15.11 *Gain* results with 30 samples 356
Fig. 15.12 *Slope* results with 30 samples 357
Fig. 15.13 *Flatness* results with 30 samples 357
Fig. 15.14 Evolution of FTY in the AR subassembly area 358
Fig. 15.15 Evolution of defect occurrence in 2019 for damaged T3 and damaged optical fiber 359
Fig. 15.16 Visual aid for T3 electrical testing 359

Fig. 15.17 Visual aid for optical fiber inspection 360
Fig. 16.1 Failure rates after implementing WCMS [1] Six-Sigma and [2] Lean Manufacturing (Roser, 2017) 367
Fig. 16.2 Comparisons between the scores of Mexico, Italy, France, and the USA against the scores obtained in cases 1 and 2 for the dimensions **a** distance to power, **b** resistance to uncertainty, **c** femininity, **d** masculinity, **e** individualism, and **f** collectivism respectively 370
Fig. 16.3 Graph comparing means of data values from the tools used in the analyzed projects 371
Fig. 16.4 Twenty-one activities chosen for adapting Lean-Sigma tools to the workplace culture (Alba-Baena unpublished data 2020) 373
Fig. 16.5 Schematic of a tampography process 377
Fig. 16.6 **a** Initial setup of the tampography process, **b** Processing method used for stamping the product 378
Fig. 16.7 Examples of cosmetic defects due to **a** lack of paint, and **b** excess of paint. **c** Cpk and data of the edge to the printed stamp 379
Fig. 16.8 Results for the gage R&R test on the edge distance for the stamping process 380
Fig. 16.9 Partially filled cause and effect (Ishikawa) diagram used in this case 381
Fig. 16.10 Schematic of the selected inkjet stamping process proposed in the solution step 382
Fig. 16.11 Process capability study results after implementing the solution 383
Fig. 16.12 Two-sample t statistical comparison for cycle times using the pad printing and inkjet printing processes 384
Fig. 16.13 Statistical comparison study for printing variation before and after the solution 384
Fig. 17.1 Regular and irregular shapes (authors elaboration) 392
Fig. 17.2 Material handling equation (Tompkins et al. 2010) 396
Fig. 17.3 Market distribution 2019 (authors elaboration) 398
Fig. 17.4 Production planning 2018 (authors elaboration) 401
Fig. 17.5 Arrangement of production departments inside the facility for each scenario (authors elaboration) 408
Fig. 17.6 Final alternatives (authors elaboration) 410
Fig. 18.1 Proposed framework for OS formulation 439
Fig. 18.2 Histograms obtained in the four investment scenarios 448
Fig. 18.3 Probability curves by scenario 450
Fig. 18.4 Implementation schedule for strategic projects that support OS 450
Fig. 18.5 Current and expected (MI_k) maturity indices (MI_{k+1}) 450

Fig. 19.1 Manufacturing process design stages. *Source* Own creation based on Grewal (2011) 467
Fig. 19.2 Workplace design stages. *Source* Adapted from Solorzano (n.d.) retrieved from Kuka (2018) 477
Fig. 20.1 Steps followed 514
Fig. 20.2 Worker doing wood works 514
Fig. 20.3 REBA input for neck, trunk and legs for worker doing wood works 515
Fig. 20.4 REBA input for load during wood works 515
Fig. 20.5 REBA input for upper arms, lower arms and wrists during wood works 516
Fig. 20.6 REBA input for type of coupling during wood works 516
Fig. 20.7 REBA input for type of activity during wood works 517
Fig. 20.8 REBA score for worker doing woodwork 517
Fig. 20.9 Laborer doing load-carrying work 518
Fig. 20.10 REBA score for laborer doing load-carrying work 518
Fig. 20.11 QEC input from observer's point of view during load-carrying works 519
Fig. 20.12 QEC input from worker's point of view during load-carrying works 520
Fig. 20.13 Result sheet of QEC output 521
Fig. 20.14 Worker doing iron and steel work 522
Fig. 20.15 REBA score for worker doing iron and steel work 522
Fig. 20.16 Worker doing water supply work 523
Fig. 20.17 REBA score for worker doing water supply work 523

List of Tables

Table 1.1	Journal publication distribution	7
Table 1.2	Influential authors in terms of publications and citations	8
Table 1.3	Research methodology distribution	9
Table 1.4	Pressures and drivers identified in literature review	14
Table 1.5	Barriers identified in the literature review	16
Table 1.6	Success factors identified in the literature review	17
Table 1.7	Sustainable performance dimensions and measures	24
Table 2.1	Operationalization of arcs of integration	39
Table 2.2	Operationalization of the description of the level of internal integration	40
Table 2.3	Statistical summary for integrative activities	41
Table 2.4	Analysis of factorial saturations and index of adaptation to the factorial model of integrative activities	42
Table 2.5	External integration factor scores (suppliers and clients) and normalized scores	44
Table 2.6	Recodification to obtain the arcs of integration	47
Table 2.7	Number of companies in each combination of levels for suppliers and customers	47
Table 2.8	Percentages of companies in each level of combination for suppliers and customers	47
Table 2.9	Arcs of integration	48
Table 2.10	Number of cases for each arc of integration	50
Table 2.11	Classification matrix with discriminant analysis column percentages	51
Table 2.12	Description of the level of internal integration	53
Table 2.13	Relationship between integration arcs and the level of internal integration (absolute frequencies)	55
Table 2.14	Relationship between integration arcs and the level of internal integration (relative frequencies)	56
Table 3.1	Sets of the model	65
Table 3.2	Variables of the model	65
Table 3.3	Parameters of the model	66

Table 3.4 Cost saving . . . 70
Table 3.5 Variation in days . . . 70
Table 3.6 Project charter . . . 71
Table 4.1 Proposed locations . . . 89
Table 4.2 Proposed locations and distance to warehouse . . . 89
Table 4.3 Proposed locations and quantity of corridors . . . 90
Table 4.4 Proposed locations and distance to the shipping warehouse . . . 90
Table 4.5 Summary of the process flowchart . . . 91
Table 4.6 Summary . . . 93
Table 4.7 Summary of transport vehicles . . . 94
Table 4.8 Summary of the final results . . . 98
Table 4.9 Summary of other results . . . 98
Table 5.1 Distance cover by employees during the packing activities of red and purple sea urchin . . . 113
Table 5.2 Capacity versus % of workstation utilization red sea urchin and purple sea urchin . . . 113
Table 5.3 Productivity indicators for EZR and EZP . . . 115
Table 5.4 Evaluation criteria for system experimentation selection . . . 115
Table 5.5 Distances between nodes, for the EZR network segments . . . 116
Table 5.6 Distances between nodes, for the new network . . . 118
Table 5.7 Entity summary . . . 120
Table 5.8 Proportion of times for each entity . . . 121
Table 5.9 Location summary . . . 122
Table 5.10 Resources summary . . . 123
Table 5.11 Entity summary . . . 124
Table 5.12 Proportion of times for each entity . . . 125
Table 5.13 Location summary . . . 126
Table 5.14 Resources summary . . . 127
Table 5.15 Lead time and value added, for the scenarios considered . . . 129
Table 6.1 Industrial sector versus position . . . 146
Table 6.2 Latent variables validation . . . 146
Table 6.3 Model fit and quality indexes . . . 147
Table 6.4 Hypotheses test . . . 148
Table 6.5 Sum of indirect effects . . . 148
Table 6.6 Total effects . . . 148
Table 6.7 Sensitivity analysis . . . 149
Table 7.1 Optimal flow . . . 168
Table 7.2 Mean time between arrivals, coefficient of variation and parameters of gamma distribution, low variability . . . 171
Table 7.3 Mean service time, coefficient of variation, and parameters of gamma distribution, low variability . . . 172
Table 7.4 Performance measures of the supply chain as a network of queues . . . 172
Table 7.5 Global measures of the supply chain . . . 172

Table 7.6 Mean time between arrivals, coefficient of variation, and parameters of gamma distribution, moderate variability ... 173
Table 7.7 Mean service time, coefficient of variation, and parameters of gamma distribution, moderate variability ... 173
Table 7.8 Performance measures of the supply chain as a network of queues, moderate variability ... 174
Table 7.9 Global measures of the supply chain, moderate variability ... 174
Table 7.10 Mean time between arrivals, coefficient of variation, and parameters of gamma distribution, high variability ... 174
Table 7.11 Mean service time, coefficient of variation, and parameters of gamma distribution, high variability ... 175
Table 7.12 Performance measures of the supply chain as a network of queues, high variability ... 175
Table 7.13 Global measures of the supply chain, high variability ... 175
Table 8.1 Valparaíso city street markets current configuration, source (Saavedra Manríquez 2016) ... 188
Table 8.2 Street markets' coverage level ... 191
Table 8.3 Number of street markets in the planning horizon ... 192
Table 9.1 Strategic matrix ... 205
Table 9.2 Strategic objectives matrix ... 205
Table 9.3 Synergic between objectives matrix ... 206
Table 9.4 Confrontation matrix ... 207
Table 9.5 Operationalization matrix ... 210
Table 9.6 Projects proposal ... 211
Table 9.7 Amount of the project ... 211
Table 10.1 Use of multi-criteria tools in waste management Osorio et al. (2013) ... 222
Table 10.2 Multi-criteria tool application in waste management ... 223
Table 10.3 Multi-criteria methods related to solid waste management problems ... 224
Table 10.4 Categories of criteria involved in waste management ... 228
Table 10.5 Subcriteria identification ... 229
Table 10.6 Priorities of subcriteria in the AHP multi-criteria model for the manufacturer ... 232
Table 10.7 Summary of sensitivity analysis scenarios for the manufacturer's multi-criteria model (%) ... 234
Table 12.1 Countries with the highest number of certified organizations with ISO 9001:2015 standard ... 266
Table 12.2 Major differences in terminology between ISO 9001:2008 and ISO 9001:2015 ... 271
Table 12.3 HLS of the ISO 9001:2015 standard ... 275

Table 12.4	Organizations certified with standards based on ISO 9001 standard worldwide	277
Table 12.5	QMPs in ISO 9001:2015 standard	278
Table 12.6	Results of the EFA	281
Table 13.1	Genetic information of an organization	291
Table 13.2	Approaches to genetic aspects	291
Table 13.3	Beneficiaries of operating a commercial organization	293
Table 13.4	Structure of the organizational purpose	293
Table 13.5	Mission structure	295
Table 13.6	Functions to develop	297
Table 13.7	Policy	299
Table 13.8	Relationship between mission and policy	300
Table 13.9	Institutional vision	303
Table 13.10	Individual vision (one for each person)	304
Table 14.1	Age of participants of the study	320
Table 14.2	Descriptive statistics work engagement scale	321
Table 14.3	Descriptive statistics employer branding scale	321
Table 14.4	Descriptive statistics servant leadership scale	322
Table 14.5	Validity and reliability analysis	323
Table 14.6	Pearson correlation of the latent variables	324
Table 14.7	Goodness-of-fit parameters of the model	325
Table 14.8	Weights of the items of the latent variables	327
Table 14.9	Hypothesis testing	328
Table 15.1	Literature review: success cases of DMAIC implementation	338
Table 15.2	Use of Pareto charts: success cases	342
Table 15.3	Root cause analysis for damaged T3 defect	350
Table 15.4	Root cause analysis for damaged optical fiber defect	353
Table 15.5	Internal fiber inspection parameters in Aurora	354
Table 15.6	Specifications of the IEC 61300-3-35 standard	354
Table 15.7	Parameters of the Arris simplex profile	354
Table 16.1	Characteristic parameters of the personnel surveyed in two locations	369
Table 16.2	List of tools suggested for use in each stage for solving problems (EC0264)	372
Table 16.3	Relationship between problem-solving tools and dimensions of work-culture	374
Table 16.4	Options for solving the pad printing problem	381
Table 17.1	Company production 2018–2019 [units]	400
Table 17.2	Tukey's procedure *P*-value	413
Table 17.3	Tukey's procedure Alternative 1	413
Table 17.4	Tukey's procedure Alternative 2	414
Table 17.5	Monthly profitability for each alternative	414
Table 18.1	Maturity levels for POS evaluation (Phase 1)	440
Table 18.2	POS evaluation levels (Phase 1)	440

Table 18.3 Criteria for company selection 443
Table 18.4 General Phase 1 (POS evaluation) results 444
Table 18.5 General information about the project portfolio (PP) 445
Table 18.6 Solutions found for OS formulation in the different investment scenarios 449
Table 18.7 Result evaluation by company practitioners 452
Table 19.1 Taxonomy of psychosocial hazards 476
Table 19.2 Differences between original process design and process redesign stages 479
Table 19.3 Health and safety proposals per workstation design stages 481
Table 19.4 Health workplace elements at design stages 482
Table 19.5 ISO standards regarding ergonomics 486
Table 19.6 ISO standards regarding anthropometry and biomechanics 487
Table 19.7 Digital technology used on workplace design 490
Table 19.8 Methodologies that include human factors on workplace design 493
Table 19.9 Workplace design and health and safety considerations 496
Table 20.1 Interpretation of QEC scores 521

Part I
Supply Chain and Logistics

Chapter 1
A Conceptual Framework of Green Supply Chain Management: Influential Factors, Green Practices, and Performance

Mariana Trujillo-Gallego and William Sarache

Abstract The green supply chain management (GSCM) approach has emerged as a response to pressures derived from global interest in environmental protection and sustainability, in order to satisfy the present generation's needs without overlooking those of future generations. Organizations must, consequently, identify those factors that influence GSCM adoption to achieve sustainable performance. Despite the increase in the number of studies that have examined the various factors that influence GSCM adoption, there is not a clear, defined classification thereof. In addition, notwithstanding the importance of GSCM as a means for sustainability achievement, these factors have been established as separate branches. In order to address this gap, the aim of the present chapter is to propose a conceptual framework that integrates those factors that influence GSCM adoption, GSCM practices, and sustainable performance. The framework includes a classification of influential factors, GSCM practices, and vital measures that would enable sustainable performance to support effective GSCM implementation.

Keywords GSCM · Influential factors · Green practices · Performance · Sustainability

1.1 Introduction

A major source of ecosystemic imbalance has been attributed to manufacturing operations (Olugu et al. 2011; Mumtaz et al. 2018), which constitute a threat to human, animal, and plant health and life, as well as to future generations (Savita et al. 2012).

As such, increased pressure has been exerted upon manufacturers, as polluters and resource consumers (Manimaran et al. 2018). Undoubtedly, the impact of manufacturing organizations on the environment is a growing concern, which has led to demands for sustainable practices that fulfill environmental, economic, and social

M. Trujillo-Gallego (✉) · W. Sarache
Departamento de Ingeniería Industrial, Universidad Nacional de Colombia Sede Manizales, Campus La Nubia, Km 9 vía al Aeropuerto La Nubia, Manizales, Caldas, Colombia
e-mail: matrujilloga@unal.edu.co

J. L. Garcia Alcaraz et al. (eds.), *Trends in Industrial Engineering Applications to Manufacturing Process*, https://doi.org/10.1007/978-3-030-71579-3_1

needs alike (Zaid et al. 2018). Manufacturing companies are thus now obliged to make more concerted efforts to balance their economic growth with environmental and social concerns (Choudhary and Sangwan 2019b).

These pressures, which initially targeted focal firms, later expanded to the supply chain (Seuring and Müller 2008; Tuni et al. 2018). The World Resource Institute (2009) has concluded that companies apart from focal firms are responsible for up to 80% of overall supply chain emissions. The need to incorporate a green dimension into supply chain management (SCM), from which the green supply chain management (GSCM) concept emerged, has arisen (Saeed et al. 2018).

Carter and Rogers (2008) and Hong et al. (2019) define GSCM as the strategic, transparent, and systematic integration of key interorganizational business processes, throughout the supply chain, for the achievement of both company and supply chain social, environmental, and economic goals. In accordance with Diabat et al. (2013) and Saeed et al. (2018), GSCM may be performed through a set of green practices (GP) which ranges from idea generation to green manufacturing (Mafini and Loury-Okoumba 2018), green distribution (Vachon and Klassen 2006; Saeed et al. 2018), and reverse logistics (Diabat et al. 2013; Rao and Holt 2005; Srivastava 2007).

On the other hand, sustainability has become a dominant trend in both human society and economic development (Yongge et al. 2009). The three principal pillars of sustainable development are as follows: the economy, society, and environment (Yildiz Çankaya and Sezen 2019). In addition, Seuring and Müller (2008), and Zaid et al. (2018) all affirm that, to measure progress toward being truly sustainable, the triple bottom line (TBL) (Elkington 1998), which provides a practical framework for sustainability, based on the measurement of business performance and the success of an organization using economic, social, and environmental pillars, may be employed (Pinto et al. 2018).

GSCM has emerged as a feasible, logical management framework that permits balance between ecological, financial, and social benefits in an enterprise facing competitive, regulatory, and societal pressures (Zhu et al. 2005; Bon et al. 2018; Chin et al. 2015). According to Muduli and Barve (2013), GSCM can lead to cost savings, in terms of reductions in material and energy usage, regulatory fines, waste disposal expenses, and costly environmental accidents, thus increasing profitability. Moreover GSCM enables challenges regarding worker and community safety, associated with accidents and occupational health hazards to be addressed, thus improving the social dimension of sustainable development (Eltayeb et al. 2011).

However, despite the increase in the number of studies that have examined different factors that influence GSCM adoption, there is no clear, defined classification for them (Ali et al. 2017; Choudhary and Sangwan 2019a, b; Yildiz Çankaya and Sezen 2019). Additionally, notwithstanding the importance attributed by scholars and practitioners of the positive effect of GSCM on sustainability, these factors have been established as separate branches (Gong et al. 2019; Tseng et al. 2019). As such, several authors have highlighted the importance for managers and practitioners to understand the driving forces in GSCM practice implementation and its subsequent effect on performance (Ahmed et al. 2019; Sharma and Gandhi 2016; Reddy Maditati et al. 2018).

In this regard, Björklund et al. (2012), Vachon and Klassen (2008), Govindan et al. (2015), and Tseng et al. (2019) add that a significant number of influential papers in the GSCM literature lack sound theoretical support, when it comes to the assessment of the relationship between drivers, practices, and performance. In addition, several authors have mentioned the lack of holistic framework available to represent organizations' practical roadmaps to environmental activities and sustainable performance measurement (Laosirihongthong et al. 2013; Dubey et al. 2017; Yildiz Çankaya and Sezen 2019), as many existing studies consider isolated factors (Laosirihongthong et al. 2013; Sharma et al. 2015; Al-Sheyadi et al. 2019). Therefore, the design of an appropriate framework, that includes the main external factors that influence GSCM, as well as main GSCM practices and performance measurement, provides vital support in strategy development, decision making, and performance improvement (Dey and Cheffi 2013).

Based on the above, the present study aims to answer two primary research questions:

RQ1 What are the key topics (influential factors, green practices, and performance measures) and trends related to GSCM?

RQ2 Is it possible to propose a comprehensive framework for GSCM and its related factors?

In an attempt to answer these questions, this study included a systematic literature review. It will thus be able to address the longstanding dissatisfaction of previous researchers. To this end, the remainder of this paper is organized as follows: In the following section, the research methodology and bibliometric analysis are presented. Section 1.3 discusses the proposed conceptual framework, and finally, the most relevant conclusions and future research lines are presented in Sect. 1.4.

1.2 Research Methodology

A three-stage systematic literature review process occurred in the present study. In Stage 1, two databases were selected for article obtention: Scopus, the largest peer-reviewed journal database in the management and engineering fields (Ahi and Searcy 2013), and ISI Web of Science, which focuses specifically on management (Taticchi et al. 2013). The keywords used for data collection included "GSCM" or "GREEN SUPPLY CHAIN MANAGEMENT" or "GREEN SUPPLY CHAIN" and "PERFORMANCE". Using the "title, abstract, keywords" search in the Scopus and Web of Science databases, journal articles were collected and stored (conference papers, books, and book chapters excluded) for the defined search terms. The initial search attempts resulted in a total of 1,239 articles.

Considering the objective of the present contribution, in Stage 2, titles and abstracts were analyzed, so as to establish the papers' suitability. Peer-reviewed publications addressing topics directly related to GSCM and performance were included in the sample. Gray literature (company/industry reports, market reports, editorials, and news) and those addressing topics in sectors unrelated to manufacturing were excluded. Further, duplicates found in Scopus and ISI Web of Science were removed. As a result, 376 papers were selected. In Stage 3, all selected papers were analyzed, and those with alternate scopes were removed. Finally, 354 relevant papers were identified to aid in responding to the research questions.

1.2.1 Bibliometric Analysis: GSCM Trends

The third stage of the review procedure was paper classification. The content of said documents was further assessed by means of descriptive analysis. Three main questions were addressed in said section as follows: (1) How are the publications distributed throughout time periods, journals, geographic regions, and influential authors? (2) What are the study trends in GSCM and sustainable performance fields? (3) What are the factors studied regarding the relationship between GSCM and sustainability?

1.2.1.1 Distribution Throughout Time

Paper temporal distribution, as included in the analysis, is depicted in Fig. 1.1. The initiation of the GSCM debate was traced to the 1990s (Seuring and Müller 2008; Srivastava 2007). Thus, it came as no surprise that the oldest paper in the search dated back to 1995 (Hart 1995) and that steady growth ensued until 2010. From 2010, it was clear from Fig. 1.1 that there has been exponential growth, up to the present day. Moreover, the trend line also indicated a pattern increase, which implies that the GSCM literature is still increasing, with its peak publication number reached in 2018, at 63 papers. This indicates increasing concern for and interest in the GSCM topic.

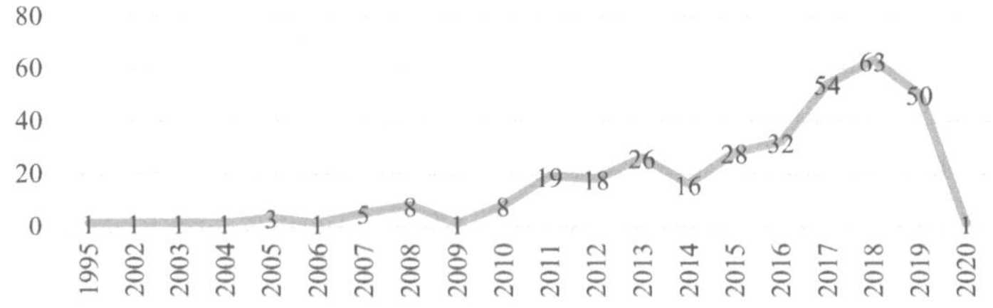

Fig. 1.1 Publications distribution over time

Table 1.1 Journal publication distribution

Journal	Publications
Journal of Cleaner Production	35
Resources, Conservation and Recycling	13
Benchmarking: An International Journal	12
International Journal of Production Economics	11
International Journal of Supply Chain Management	11
International Journal of Production Research	9
Sustainability (Switzerland)	9
International Journal of Operations and Production Management	8
Supply Chain Management	8
Business Strategy and the Environment	7

1.2.1.2 Publication by Journal

Table 1.1 shows the publication distribution among the top 10 journals that published most often regarding GSCM and sustainable performance. It should be noted that the top 20 journals published 173 out of 314 articles, accounting for nearly 56% of all articles yielded. The remaining 44% had published three papers or fewer. Also, the Journal of Cleaner Production published the highest number of papers (35), which constitutes approximately 11% of the total (314 papers). Said journal also has the second highest impact factor (6.395). Resources, Conservation and Recycling Journal was the second most popular journal, with 13 papers published (4.14%); however, it also had the highest impact factor (7.044). Therefore, it may be stated that, although Journal of Cleaner Production is ranked as number one, due to its popularity, the Resources, Conservation and Recycling Journal may be ranked number one, due to its impact. Finally, 79 journals appear only once in the sample, showing the multidisciplinary nature of the field: These journals encompass various disciplines, including mathematics, energy, earth and the environment, science, and computer science.

1.2.1.3 Publications by Geographic Region

Figure 1.2 shows the contribution of each region to the green supply chain management literature. It may be observed that Asia, Europe, and South America are the regions with the highest number of contributions. Note that countries with large numbers of publications include China, India, and Brazil. Compared to other countries in Asia, China accounts for the greatest number of publications (45), representing 16% of the total. It is important to mention that in South America, only Brazilian and Colombian researchers have made contributions to the GSCM approach.

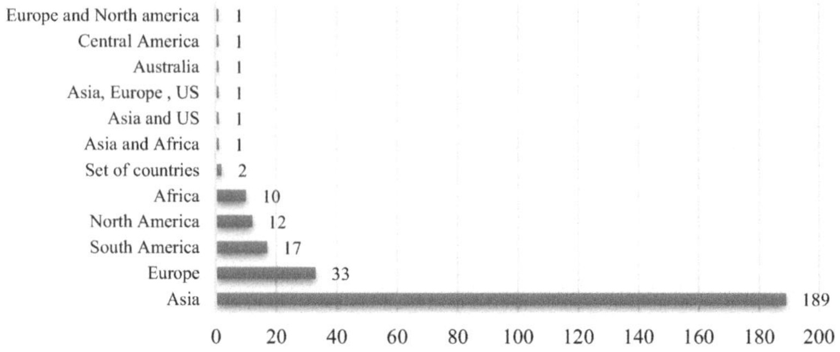

Fig. 1.2 Contributions by region

1.2.1.4 Influential Authors

To a certain extent, the devoted efforts of a researcher may be reflected by their number of publications. Similarly, the extent to which a researcher's publications have been cited by other studies can also represent their influence (Zhu and Wang 2018). The ratio of citations/publications shows the average influence or popularity of each article published. As Table 1.2 shows, in terms of the number of publications, Joseph Sarkis, who published 18 papers on this topic, comes first, followed by Zhu, Q. H. and Tseng, M.-L. From the perspective of citation frequency, Sarkis, J., who has been cited 6468 times, remained the most influential researcher, followed by Zhu, Q. H. with 4627 citations.

Table 1.2 Influential authors in terms of publications and citations

Rk	Author	Publications	Author	Citations	Author	Citations/Publications
1	Sarkis, J.	18	Sarkis, J.	6468	Seuring	1428
2	Zhu, Q.	13	Zhu, Q.	4627	Vachon, S.	610
3	Tseng, M.-L.	7	Lai, K.-H.	2062	Srivastava	545
4	Jabbour, C. J. C.	7	Klassen, R. D.	1505	Lai, K.-H.	515.5
5	Geng, Y.	6	Seuring	1428	Klassen, R. D.	501.7
6	Govindan, K.	6	Rao	1361	Sarkis, J.	359.3
7	Green, K.	6	Vachon, S.	1220	Zhu, Q.	355.9
8	Jabbour A. B. L. D	6	Srivastava	1090	Rao	340.3
9	Lai, K.-H.	4	Geng, Y.	998	Lee, S.-Y.	289.0
10	Rao	4	Govindan, K.	583	Geng, Y.	166.3

Rk Ranking

Table 1.3 Research methodology distribution

Research methodology	Number	Percent of total (%)
Case of study	28	8
Theoretical and conceptual paper	37	11
Literature review	30	9
Mathematical/analytical modeling	76	22
Survey	164	48
Other	7	2

Note Math includes model and evaluation

1.2.1.5 Research Methodologies

Table 1.3 shows the methodology distribution among the 342 selected articles. Five research methodologies were differentiated as follows: (1) theoretical and conceptual papers, (2) case studies, (3) surveys, (4) mathematical/analytical modeling, and (5) literature reviews. It is clear that survey and mathematical/analytical modeling were the most frequently adopted methods, accounting 70% of the total, followed by theoretical and conceptual papers and case studies. The articles that utilized the survey method usually designed a questionnaire based on the proposed research framework and then collected data to confirm or validate research hypotheses (Zhu and Wang 2018). The mathematical/analytical modeling included both optimal programming and specific evaluation methods, such as multi-criteria decision-making methods (MCDM). In addition, 30 papers were literature reviews, conducted to consolidate the existing pool of knowledge. Additionally, 28 articles carried out case studies, which are suitable for exploring newly emergent topics, such as GSCM/SSCM and supply chain risk management. Theoretical and conceptual papers usually aim to develop frameworks and propositions for future empirical tests (Zhu and Wang 2018). Seven articles adopted multiple methods, such as "questionnaire-based surveys + expert interview."

1.2.1.6 Research Topic Analysis

Figure 1.3 summarizes various categories used for the classification of the 336 papers, based on research topics. Seven categories were proposed, namely (i) drivers, pressures, barriers, and success factors, (ii) pressures, practices, and performances, (iii) GSCM assessment and performance evaluation, (iv) environmental management practices, (v) GSCM and other approaches (SSCM, supply chain risk management, circular economy, lean manufacturing), (vi) sustainability, and (vii) theories and challenges.

In accordance with the above, pressures, practices, and performances are currently the most studied category in the literature, with 172 articles (51%), followed

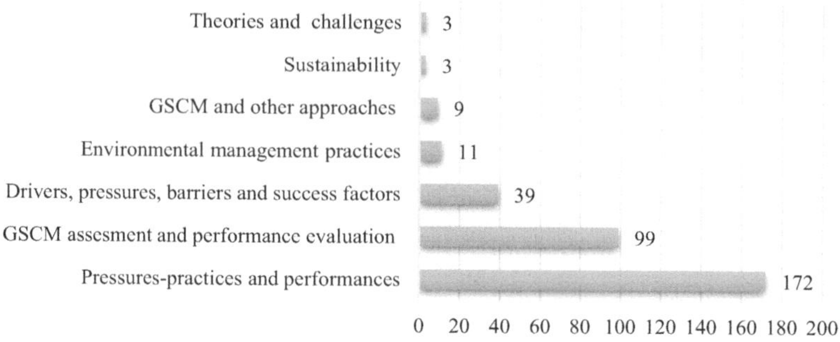

Fig. 1.3 GSCM and sustainable classification themes

by GSCM assessment and performance evaluation, with 99 articles (29%). The pressures, practices, and performances category includes empirical studies that examine the influence of drivers/pressures on the implementation of different environmental practices and, in turn, the effect of environmental practices on organizational performance (economic, operational, export, social, competitive, organizational, and marketing performance). The main methodologies used with this approach were structural equations, modeling, and multiple regression analysis. The GSCM assessment and performance category aims to investigate GSCM implementation and propose performance measurement systems, models, methods, frameworks, and methodologies. The methodologies used in this category are multi-criteria decision methods combined with fuzzy logic.

The third most studied category focused on the identification, classification, and prioritization of drivers (D), pressures (P), barriers (B), and success factors (S) in the GSCM and sustainability implementation process. Also, it has been recognized that the main barriers to GSCM implementation include a lack of support from both regulatory authorities and top management. The main data analysis technique used was interpretative structural modeling (ISM). The environmental management practices category includes studies focused on the exploration of new GSCM practices and the importance thereof. The GSCM and other approach category includes studies that investigate the relationship between the GSCM approach and other new approaches, such as sustainable supply chain management and the circular economy. The sustainability category includes research on new concepts, practices, theories, and challenges. Finally, it should be mentioned that the content analysis reveals that the papers related to theories, conceptual development, and challenges to do with GSCM are the least studied category, which indicates that authors are still adjusting the GSCM concept.

1.3 The Proposed Conceptual Framework

The foundation of the theoretical framework is composed of three elements as follows: (1) pressures/drivers, barriers, and success factors, (2) GSCM practices, and (3) sustainable performance (Fig. 1.4). Pressures are the external determinants for GSCM adoption (Famiyeh and Kwarteng 2018), and drivers are the internal factors that motivate focal firms to adopt the GP (Tachizawa et al. 2015). According to Zhu et al. (2005), Zhu et al. (2013), a stronger presence of pressures and drivers results in more rapid adoption of GSCM practices, given the possibility that companies, being unable to implement GP, see their very existence threatened. In contrast, there are barriers which hinder the GSCM implementation process (Dube and Gawande 2016). Success factors, conversely, are enablers that allow GSCM implementation and ensure success and competitiveness (Chiappetta Jabbour et al. 2017).

According to Tseng et al. (2019), GSCM is conceptual, whereas GSCM practices (GP) are actions. As shown in Fig. 1.4, GP are classified into three types as follows: first, practices that form the supply chain of a given product, from green purchasing and green manufacturing to green distribution, green marketing, use and return of products, components, and materials to focal firms for disposal (reverse logistics) (Diabat et al. 2013; Srivastava 2007). A second type includes those practices such as eco-design, environmental collaboration, and internal environmental management, which, according to Zhu and Sarkis (2004), Jawaad and Zafar (2019), Khan and Qianli (2017) and Bae and Grant (2018), are difficult to copy from competitors, as they involve the development of skills that are difficult to replicate and are essential for implementing other practices at lower levels (Green et al. 2012a, b; Reddy Maditati et al. 2018). Finally, a third type includes those practices that support the

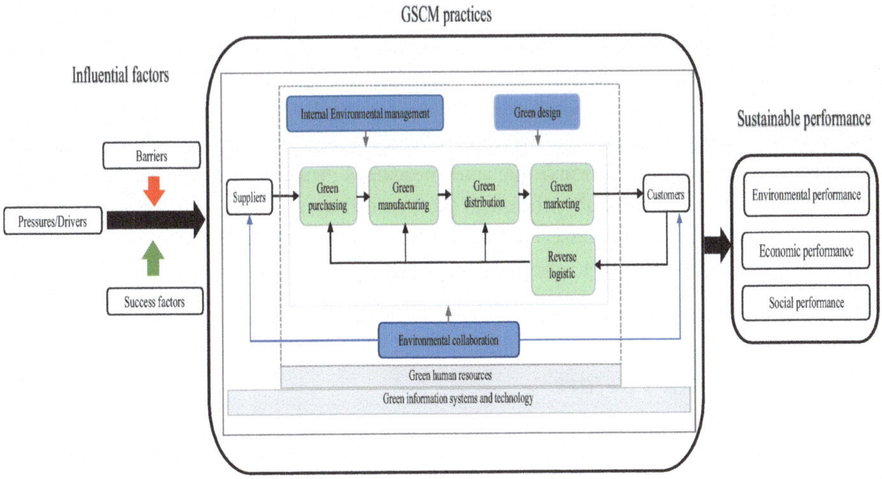

Fig. 1.4 Proposed conceptual framework

organization, allowing for the development and improvement of the remaining practices (Kusi-Sarpong et al. 2015). These include green human resource management (Longoni et al. 2018; Zaid et al. 2018; Singh and El-Kassar 2019) and green information systems and technologies (Green et al. 2012a, b; Bae and Grant 2018; Yang et al. 2018).

Once the firms successfully implement GP, it can expect superior performance (Green et al. 2012a, b). According to the natural resource-based view (NRBV) (Hart 1995), GSCM practices can be regarded as a unique organizational capability that improves firm performance. De Giovanni (2012) and Setyadi et al. (2019) support the idea that GSCM is not just a tool to minimize the environmental impacts of products and operations, but also a unique strategy to provide economic benefits and enhance social welfare. Therefore, enhanced firm performance can be considered improved environmental performance, improved economic performance, and improved social performance (Tseng et al. 2019).

According to the above, Fig. 1.4 shows that pressures/drivers, barriers, and success factors influence focal firms to adopt GSCM. Black arrows represent the positive relationship between pressures/drivers and GP. Similarly, the successful implementation of GP lead to firms improved sustainable performance, as represented in economic, environmental, and social performance, which is indicated by another black arrow. The red arrow indicates the adverse impact of barriers on GP implementation, while the green arrow represents the facilitating role of success factors for GSCM adoption. The smaller black arrows represent the flow of raw materials and products throughout the supply chain, from purchasers to suppliers to distribution to consumers and their return to the company for disposal. The gray arrows indicate the strategic role of internal environmental management, eco-design, and environmental collaboration for the other practices in the supply chain. The dashed lines indicate the need for external collaboration with suppliers and customers to help firms adopt GP (Zhu et al. 2008). Finally, the position of the gray boxes represents the transversal and supportive role of green human resources and green information systems and technology, in the focal firm and its supply chain.

In order to keep the framework simple, the figure does not incorporate all items. However, the lists of pressures, drivers, barriers, success factors, practice aspects, sustainable performance dimensions, and measures employed in the framework are presented below. All of the GP, influential factors and sustainable performance dimensions have been discussed based on different aspects of the literature. Thus, this framework is a contribution to the GSCM literature.

1.3.1 Influential Factors

The first component of the proposed conceptual framework is influential factors for GSCM implementation. Some organizational theories, such as the resource-based view, transaction cost economics, agency network, resource dependence theory (RDT), natural resource-based view (NRBV), and institutional theory, have been used to understand how companies adopt, assimilate, and develop GSCM operations (Zhu et al. 2005; Laosirihongthong et al. 2013; Lee et al. 2013; Ali et al. 2017). Many authors have identified different internal and external factors from stakeholders that influence GSCM adoption, such as pressures/drivers (Mathiyazhagan et al. 2018; Singh et al. 2018), barriers (Kaur et al. 2019; Majumdar and Sinha 2018), and success factors (Chiappetta Jabbour et al. 2017; Garg et al. 2017; Mauricio and De Sousa Jabbour 2017).

1.3.1.1 Pressures/Drivers

In the literature, the determinants of GSCM adoption and enterprise environmental performance improvement can be broadly divided into: "external factors," mostly linked to stakeholder pressure (Zhu et al. 2005; Zhu et al. 2012) and "internal factors" or "internal drivers" (Testa and Iraldo 2010; Lee et al. 2013; Ali et al. 2017). Institutional theory highlights three kinds of isomorphic pressures as follows: normative, coercive, and mimetic (DiMaggio and Powell 1983). Coercive pressures are exerted by powerful entities by applying imposition and inducement mechanisms (Famiyeh and Kwarteng 2018). Normative isomorphism occurs as a result of professionalization, which is defined as shared norms and standards, formalized by the environment, from a cultural expectation of that environment (Saeed et al. 2018). Finally, mimetic isomorphism occurs when an organization imitates the actions of successful industry competitors (Manimaran et al. 2018; Saeed et al.2018).

Limiting the analysis to only external pressures does not allow for a complete understanding of why organizations operating within the same context (market or sector) pursue different strategies, despite experiencing similar pressures (Testa and Iraldo 2010). According to this, Tseng et al. (2019) define drivers as factors that motivate firms to engage in GSCM initiatives. Pressures and drivers identified in the literature review are shown in Table 1.4.

1.3.1.2 Barriers

A barrier is an obstacle that hinders the implementation GSCM processes (Dube and Gawande 2016). It can be stated that the stronger the presence of said barriers, the poorer the GSCM implementation level is (Tseng et al. 2019). Different authors have classified barriers into governmental, cultural, structural, and contextual categories. The barriers identified in the literature review are shown in Table 1.5.

1.3.1.3 Success Factors

Success factors (SFs) are the organizational actions necessary to ensure successful competitive performance, thus supporting a company's organizational change process (Mauricio and De Sousa Jabbour 2017; Chiappetta Jabbour et al. 2017).

Table 1.4 Pressures and drivers identified in literature review

Coercive pressures	National environmental regulations (such as waste emission, cleaner production)	Zhu et al. (2005, 2007b, 2013), Sharma et al. (2017), Miras-Rodríguez et al. (2018), Famiyeh and Kwarteng (2018) Zhang et al. (2020), Dubey et al. (2015a, b)
	National resource saving and conservation regulations	Zhu et al. (2005, 2007b, 2013), Sharma et al. (2017), Famiyeh and Kwarteng (2018), Zhang et al. (2020), Dubey et al. (2015a, b)
	Regional environmental regulations (such as waste emissions, cleaner production)	Zhu et al. (2005, 2007b, 2013), Sharma et al. (2017), Famiyeh and Kwarteng (2018), Zhang et al. (2020), Dubey et al. (2015a, b)
	Regional resource saving and conservation regulations	Zhu et al. (2005, 2007b, Zhu 2013), Sharma et al. (2017), Famiyeh and Kwarteng (2018), Zhang et al. (2020), Dubey et al. (2015a, b)
	Export countries' environmental regulations	Zhu et al. (2005, 2013)
	Products potentially conflict with laws (such as circular economy, EPR, EHS)	Zhu et al. (2005, 2007b, 2013)
Normative pressures	Export market	Zhu et al. (2007b, 2013), Vanalle et al. (2017), Choi et al. (2018), Zhang et al. (2020)
	Sales to foreign customers	Zhu et al. (2007a, 2013), Vanalle et al. (2017), Zhan et al. (2020)
	Customer awareness	Walker et al. (2008), Miras-Rodríguez et al. (2018), Singh et al. (2018), Zhang et al. (2020)
	The industry associations requirement to plan and implement environmental management practices	Zhu et al. (2005, 2013), Famiyeh and Kwarteng (2018)
	Pressure as of non-government organization (NGO) to put into practice GSCM	Sharma et al. (2017)
	Increased awareness of supply chain partners including suppliers and logistics service providers	Zhu et al. (2007a), Vanalle et al. (2017), Sharma et al. (2017), Tseng et al. (2019)
	Media tracking the industry	Zhang et al. (2020)

(continued)

Table 1.4 (continued)

	Public environmental protection awareness	Zhang et al. (2020)
	Internal multinational policies (subsidiaries or divisions of a multinational firm)	Zhang et al. (2020)
Mimetic pressures	Effect of competitor's green strategies to implement GSCM	Zhang and Yang (2016), Sharma et al. (2017), Famiyeh and Kwarteng (2018), Choi et al. (2018)
	Green strategy of substitute product producers	Zhu et al. (2013)
Internal drivers	A corporate image strategy (reputation-led)	Testa and Iraldo (2010), Vanalle et al. (2017), Sharma et al. (2017), Singh et al. (2018)
	Company's environmental mission	Zhu et al. (2007b)
	Presence of ethical leadership	Tseng et al. (2019)
	ISO 14000 compliance requirement	Tseng et al. (2019)
	Environmental awareness among members of the organization	Tseng et al. (2019), Miras-Rodríguez et al. (2018)
	Cost-saving strategy (efficiency-led)	Testa and Iraldo (2010), Tseng et al. (2019)
	Economic benefits	Singh et al. (2018)
	Top management support	Singh et al. (2018), Miras-Rodríguez et al. (2018)
	A product and/or process development strategy (innovation-led)	Testa and Iraldo (2010)

Besides that, CSFs are mechanisms used to plan and identify goals in an organization, assess threats and opportunities, evaluate organizational strengths and weaknesses, assist managers in improving performance, and define a manager information needs (Ab Talib and Muniandy 2013; Mauricio and De Sousa Jabbour 2017). Several authors have pointed out that, when CSFs are not properly managed, they tend to become barriers to organizational success (Luthra et al. 2014). Therefore, CSFs must be effectively identified, assessed, and managed by organizations to facilitate GSCM implementation (Chiappetta Jabbour et al. 2017; Mauricio and De Sousa Jabbour 2017), in order to achieve sustainability (Luthra et al. 2014). The success factors identified in the literature review are shown in Table 1.6.

Table 1.5 Barriers identified in the literature review

Governmental	Inefficient/lack of national and regional government policies and regulations that support GSCM	Walker et al. (2008), Mitra and Datta (2014), Dube and Gawande (2016), Wang et al. (2016), Ghadge et al. (2017), Agyemang et al. (2018)
Structural	Lack of environmental knowledge and understanding	Walker et al. (2008), Kaur et al. (2018), Ghadge et al. (2017), Tseng et al. (2019)
	Cost implications/financial constraints	Muduli and Barve (2013), Dube and Gawande (2016), Wang et al. (2016), Tseng et al. (2019)
	Lack of eco-technology	Tseng et al. (2019)
	Lack of training in GSCM	Walker et al. (2008), Dube and Gawande (2016)
	Lack of technical expertise to implement GSCM	Muduli and Barve (2013), Dube and Gawande (2016), Kaur et al. (2018)
Cultural	Lack of environmental awareness	Muduli and Barve (2013), Ghadge et al. (2017), Tseng et al. (2019)
	Lack of top-level management commitment	Muduli and Barve (2013), Dube and Gawande (2016), Agyemang et al. (2018), Tseng et al. (2019), Majumdar and Sinha (2018)
	Lack of employee commitment	Muduli and Barve (2013), Walker et al. (2008), Tseng et al. (2019)
	Fear of failure	Muduli and Barve (2013), Dube and Gawande (2016), Tseng et al. (2019)
	Focus on cost reductions at expense of green practices	Walker et al. (2008)
	Lack of legitimacy	Walker et al. (2008)
	Focus on short-term strategic goals	Ghadge et al. (2017)
	Resistance to change and adoption	Muduli and Barve (2013), Dube and Gawande (2016)
Contextual	Market competition and uncertainty	Walker et al. (2008), Dube and Gawande (2016), Ghadge et al. (2017), Kaur et al. (2018)
	Poor commitment from partners	Walker et al. (2008)
	Limited supplier capabilities and resources	Ghadge et al. (2017)
	Lack of integrated management information and traceability system	Agyemang et al. (2018), Wang et al. (2016)
	Lack of understanding among supply chain stakeholders	Walker et al. (2008), Dube and Gawande (2016), Agyemang et al. (2018)
	Lack of customer awareness toward GSCM	Dube and Gawande (2016), Wang et al. (2016), Agyemang et al. (2018)

(continued)

Table 1.5 (continued)

	Inadequate support and guidance from industry bodies, NGOs, and development agencies	Dube and Gawande (2016), Agyemang et al. (2018)

Table 1.6 Success factors identified in the literature review

Governmental	Government support	Ab Talib and Muniandy (2013), Gandhi et al. (2016), Luthra et al. (2014), Garg et al. (2017)
	Environmental management certifications	Gandhi et al. (2016)
	International environment agreements	Luthra et al. (2014)
External factors	Supplier management and integration	Gandhi et al. (2016), Garg et al. (2017), Mauricio and De Sousa Jabbour (2017)
	Supply chain members' awareness and literacy	Luthra et al. (2014)
Internal factors	Corporate goal	Gandhi et al. (2016)
	Total involvement of employees	Gandhi et al. (2016), Mauricio and De Sousa Jabbour (2017)
	Organization's values	Gandhi et al. (2016)
	Benchmarking	Gandhi et al. (2016)
	Proper workplace management	Luthra et al. (2014)
	Technology advancement and adaption	Luthra et al. (2014)
	Performance measure	Mauricio and De Sousa Jabbour (2017), Chiappetta Jabbour et al. (2017)
	Information management	Mauricio and De Sousa Jabbour (2017), Chiappetta Jabbour et al. (2017)
	Top management commitment	Mauricio and De Sousa Jabbour (2017), Chiappetta Jabbour et al. (2017)
	Training	Mauricio and De Sousa Jabbour (2017), Chiappetta Jabbour et al. (2017)
	Competencies for greener products and processes	Mauricio and De Sousa Jabbour (2017), Chiappetta Jabbour et al. (2017)

1.3.2 GSCM Practices

GPs are environmentally friendly cooperative activities throughout the supply chain that can reduce firms' impact on the natural environment because of their industrial activities (Balakrishnan and Suresh 2018; Choudhary and Sangwan 2019b) and then capture added value that may emerge from these activities (Ryoo and Koo 2013) without affecting quality, productivity, or operating costs (Vilchez et al. 2017). According to Srivastava (2007) and Diabat et al. (2013), the green supply chain concept covers all phases of a product's life cycle, from the extraction of raw materials through the design, purchasing, production, and green logistics phases, as well as marketing, to the use of the product by consumers and its disposal at the end of the product's life cycle. In addition, Diabat et al. (2013), Mafini and Loury-Okoumba (2018), and Al-Ghwayeen and Abdallah (2018) argue that GSCM encompasses other GPs, such as reverse logistics, environmental collaboration with suppliers, and internal environmental management.

Eco-design, also called design for the environment (Eltayeb et al. 2011; Rostamzadeh et al. 2015), includes activities that aim to minimize negative impacts caused during products' entire life cycles (Rostamzadeh et al. 2015; de Sousa Jabbour et al. 2015), such as minimizing the consumption of materials, water, and energy, as well as promoting positive environmental practices, such as the reuse, recycling, and recovery of component materials and parts (Choi et al. 2018). The above is done without compromising other essential criteria, such as functionality, quality, or cost (Green et al. 2012a, b; Saeed et al. 2018). This practice systematically involves life cycle analysis (Govindan et al. 2015), environmental risk management, product safety, pollution prevention (Diabat et al. 2013), waste and resource management, and conservation (Srivastava 2007). The main benefit of adopting eco-design is its ability to offer advanced and proactive solutions to environmental problems and increase product value and usefulness (Hartmann and Germain 2015). Likewise, companies can reduce ecological risks and operational costs (Mumtaz et al. 2018), increase profitability (Khan and Qianli 2017), and improve innovation capabilities (Saeed et al. 2018).

Internal environmental management is the practice of incorporating GSCM into an organization's strategy and showing their commitment through environmental policies, environmental management systems, environmental impact assessments, quantifiable environmental objectives, action plans, training activities, responsibilities, and environmental auditing, in order to monitor, control, and evaluate the effect of the organization's actions on the environment (Zhu et al. 2008; de Sousa Jabbour et al. 2015). The main measure of this practice, as stated by Zhu and Sarkis (2006) and Green et al. (2012a, b), is commitment and support from top management, since without this, environmental programs are destined to fails and are much less likely to be initiated. Other measures include the middle management involvement and support and the establishment of cross-functional teams (Zhu et al. 2005; Saeed et al. 2018). Also, this practice includes the efforts to create an organizational culture, in which organization members share a set of values and beliefs related to environmental protection (Fraj et al. 2013).

Environmental collaboration refers to the involvement of an organization with its suppliers and customers in joint planning for environmental solutions, including shared environmental knowledge (Kusi-Sarpong et al. 2015), joint product development, and green innovations (Govindan et al. 2015), the exchange of information (de Soussa Jabbour et al. 2015), diffusion of new capabilities (Vachon and Klassen 2008; Bae and Grant 2018), and willingness to learn about each other's operations in order to plan and set environmental improvement objectives (Diabat et al. 2013). Likewise, the main objective of this practice is to increase mutual trust, commitment among members of the supply chain, and work in synergy to resolve conflicts and obtain mutual benefit (Vachon and Klassen 2006; Vachon and Klassen 2007). In this way, environmental collaboration promotes direct long-term benefits, such as rapid development of environmental products, savings in company inventory management costs, and increases in customer satisfaction (Kusi-Sarpong et al. 2015; Dubey et al. 2015a, b).

Green purchasing includes eco-conscious activities, based on the purchase of inputs, parts, products, materials, and equipment that meet environmental needs, in terms of reduction of waste, recycling, reuse, resource reduction, and material substitution (Diabat et al. 2013; Lin 2013; Eltayeb et al. 2011; Younis et al. 2016). Since green purchases are located at the beginning of the flow of materials within an organization, they play a key role in the ecological transformation of products and activities (Eltayeb et al. 2011). Thus, company ecological performance can be affected by the selection, evaluation, and auditing of suppliers (Liu et al. 2018; Younis et al. 2016). Therefore, among the factors that suppliers have evaluated are environmental certifications (Zhu et al. 2008), the products purchased do not contain environmentally undesirable items like lead and other harmful or noxious materials (Laosirihongthong et al. 2013; Younis et al. 2016; Zaid et al. 2018), and likewise, are easy to recycle, remanufacture, or reuse (Eltayeb et al. 2011; Rostamzadeh et al. 2015).

Green manufacturing is also called green production and is associated with production practices mixed with advanced energy-efficient technologies (Sangode and Metre 2019), in order to provide maximum output, using more environmentally friendly resources (Dubey et al. 2015a, b; Mafini and Loury-Okoumba 2018), less energy and resources, and at the same time, causing the least possible amount of environmental pollution and waste during the production process (Yildiz Çankaya and Sezen 2019). The implementation of the above activities promotes the reduced use of raw materials (Rao and Holt 2005), lower environmental, and occupational safety expenses (Rostamzadeh et al. 2015) and, finally, efficiency, quality, and production performance increases, at a minimum cost, with little or no waste (Dubey et al. 2015a, b).

Green distribution is defined as the integration of green packaging practices and the use of energy-efficient transportation methods that minimize energy usage, carbon emissions, and transportation costs (Rao and Holt 2005; Mutingi et al. 2014; Yildiz Çankaya and Sezen 2019; Hutomo et al. 2018). According to Chin et al. (2015), Rao and Holt (2005) and Mutingi et al. (2014), green packaging includes activities such as the use of environmentally friendly materials, recycled packaging, saving energy in warehouses, cooperating with vendors to standardize packaging, minimize material uses, time to unpack, and the adoption of returnable containers. Green transportation and logistics to reduce CO2 emissions include activities such as the consolidation of deliveries and route optimization and rearranged loading patterns. Yongge et al. (2009) state that green distribution provides benefits such as reducing the use of materials, increasing the use of warehouse and container space, reducing the amount of handling required, and reducing logistics costs.

Green marketing is the initiative that an organization takes to involve the incorporation of environmental criteria into product and services marketing and promotion so as to reduce negative environmental impacts (Deng et al. 2018; Malviya and Kant 2017). Green marketing activities include voluntary eco-labeling, green alliances, participation in community activities to learn its point of view in the development of green products, franchising, licensing, pricing activities, and the use of ecological issues in the company's global message (Fraj et al. 2013; Deng et al. 2018; Abu Seman et al. 2019).

Reverse logistics is defined as the process of planning, execution, and efficient control of the flow of raw materials, inventory in process, finished products, and related information, from the point of consumption to manufacturing sites, in order to recover product and material value (Sarkis 2003; Srivastava 2007; Mafini and Loury-Okoumba 2018). Thus, reverse logistics redesigns the traditional ssupply chain (Olugu and Wong 2012), by including the collection of end user products and returning them to the factory for inspection and classification (Pourjavad and Shahin 2018; Srivastava 2007), followed by disposal activities such as reuse, remanufacturing, and recycling (Diabat et al. 2013; Eltayeb et al. 2011).

Green human resources are defined as the management of work and people toward desired environmental targets (Longoni et al. 2018). It includes activities such as green hiring, green training, involvement, green performance, and compensation (Zaid et al. 2018). Sarkis et al. (2010) suggest that in order to adopt GSCM initiatives, it is important to offer necessary organizational knowledge to all employees in the organization. For this, it is necessary to develop effective environmental awareness policies in the hiring processes (Bon et al. 2018), performance evaluation (Jabbour et al. 2013), and training and education programs (Zaid et al. 2018). These initiatives can facilitate the evolution of knowledge capabilities, skills, and competencies among employees and their "green empowerment" (Sarkis et al. 2010).

Green information systems and technologies have been primarily associated with hardware and software to address energy conservation, pollution control, and waste reduction. Examples of green IT include green data centers, virtualization software, server consolidations, and energy-saving devices (Kusi-Sarpong et al. 2016; Yang et al. 2018; Ryoo and Koo 2013). They are also used to summarize and communicate information about the environmental impact of the materials, parts, and components of the products and their available alternatives (Kusi-Sarpong et al. 2016). Green information systems, "Green IS," concern the establishment of systems and practices to support and promote organizational environmentally friendly operations for sustainable development and green innovation (Kusi-Sarpong et al. 2016; Yang et al.

2018) and monitor the results of their application (Green et al. 2012a, b). Examples of "Green IS" include environmental information systems, groupware, teleconferencing, environmental auditing systems, monitoring and control systems for orders, material flows, transport, and routes, among others (Kusi-Sarpong et al. 2016; Yang et al. 2018).

1.3.3 Sustainable Performance

Sustainability or sustainable development has received increasing attention since the release of *Our Common Future* by Brundtland (1987), where it was defined as the use of resources, progress, and advancements to meet the needs of the present generations without compromising the ability of future generations to meet their own needs (Seuring and Müller 2008). However, due to the ambiguity and vagueness of this concept, Elkington (1998) proposed a more specific and measurable conceptualization of sustainability, which differentiates social, environmental, and economic performance dimensions, commonly referred to as the triple bottom line (TBL). According to Malviya et al. (2018), organizations should endeavor to perform well on all three dimensions of the TBL.

In recent decades, organizations have realized the need to integrate the principles of sustainability and TBL into their strategy and decision-making process, as well as the importance of the relationship between sustainability and performance (Dubey et al. 2015a, b; Ahi and Searcy 2015). In the world of business, the term "performance" generally relates to the accomplishment of a given task, as measured against predetermined standards of accuracy, cost, speed, and completeness (Mafini and Loury-Okoumba 2018). In addition, performance measurement is the process of collecting, analyzing and/or reporting information, monitoring, controlling, and achieving specific objectives and outcomes (Sahu et al. 2015; Kazancoglu et al. 2018). It provides a directive for the company to implement the desired practices, evaluate their execution, reveal the effects of the implementation, and determine future actions or corrective plans (Savita et al. 2016). Also, performance measures are tools to help us to understand, manage, and improve that which organizations do (Sahu et al. 2018).

However, scholars assert that focusing only on profits or economic performance fosters short-term thinking within the organization (Sundram et al. 2018; Ghadimi et al. 2019). Therefore, firms should identify sustainable performance measures classified into the three groups as follows: economic, environmental, and social performance, using the TBL approach (Aital and Vijai 2015; Reddy Maditati et al. 2018). For sustainability, in order to achieve a balance in economic, social, and environmental processes, all businesses must succeed in these performance dimensions. However, due to the complexity thereof and their relationship to each other, it is complex to successfully achieve this balance (Yildiz Çankaya and Sezen 2019).

Environmental performance (EP), according to Dubey et al. (2015a, b), is the relationship between the organization and the environment. Specifically, it refers to the impact of organizational activities on the environment (Choudhary and Sangwan 2019a). This relationship then includes the environmental effects of resources consumed, the environmental impacts of organizational process, the environmental implications of its products and services, the recovery and processing of products, and meeting legal environmental requirements (Dubey et al. 2015a, b). In addition, some authors (Green et al. 2015; Zhu et al. 2008; Jawaad and Zafar 2019; Luthra et al. 2014; Bae and Grant 2018) agree with the statement that EP reflects attempts by manufacturing organizations to reduce the environmental impacts of their manufacturing processes and the effectiveness of the ecological initiatives of a firms' commitment to social concerns about the natural environmental preservation.

On the other hand, Govindan et al. (2015) and Abu Seman et al. (2019) affirm that EP describes and evaluates the positive effects of GSCM practice implementation on the natural environment. According to Hervani et al. (2005) and Al-Ghwayeen and Abdallah (2018), although EP indicators are essential requirements for GSCM, EP measurement is difficult and complex because of the multiple EP indicators. Some of the widely used EP indicators in the literature include reduced air, carbon dioxide, sulfur dioxide (SO_2), and nitrogen oxide (NO) emissions, decreased solid and effluent waste, less use of water and fuel, reduction of water and air pollution, lessening energy/resource utilization, reduction of hazardous and toxic materials, improved employee and community health, as well as lowering the frequency of environmental accidents, which reduces organizational environmental risk (Zhu et al. 2008; Kazancoglu et al. 2018; Namagembe et al. 2019; Eltayeb et al. 2011; Zhu and Sarkis 2004; Setyadi et al. 2019; Vijayvargy et al. 2017). In addition, environmental performance is also measured through management performance indicators, such as the approval rate of management systems (Diabat et al. 2013), environmental plans, actions, policies and measures (Zhu and Sarkis 2004; Diabat et al. 2013), community relation enhancement, improved organizational image (Lin 2013), and reports of firm environmental performance (Saeed et al. 2018).

Economic performance is an outcome that influences an organization to be environmentally oriented (Ahmed et al. 2019). Academics agree that, in order to be competitive and to improve business benefits, firms need to address environmental problems (Saeed et al. 2018; Ahmed et al. 2019). In this sense, economic performance outcomes expected financial benefits that result from GSCM practice adoption (Eltayeb et al. 2011; Luthra et al. 2014). Studies on the economic advantages of green supply chain initiatives recognized different results. According to Choudhary and Sangwan (2019a), the economic benefits gained by companies as a result of GSCM practice implementation are called positive economic performance. However, negative GSCM adoption results are called negative economic performance (Zhu and Sarkis 2004).

Positive economic performance is based on the assessment of firm cost reduction in material purchasing, energy consumption, waste treatment and discharge, and accident occurrence (Lin 2013; Kazancoglu et al. 2018), revenue growth (Eltayeb et al. 2011), increase in market share contribution (Zhu and Sarkis 2004, Green et al. 2012a, b), profit maximization (Vijayvargy et al. 2017), increase in return on assets (Setyadi et al. 2019), productivity improvement (Govindan et al. 2015), increased firm competitiveness (Luthra et al. 2014), and additional market opportunities (Diabat et al. 2013). In contrast, negative economic performance is reflected in operational cost, the cost of buying environmentally friendly materials, and investment costs (Zhu and Sarkis 2004; Lin 2013; Govindan et al. 2015).

Social performance has not received much attention as an outcome of GSCM (Eltayeb et al. 2011; Laosirihongthong et al. 2013), unlike economic and environmental performance, because it is relatively new to the field of supply chain management (Seuring and Muller 2008). In addition, since social aspects depend on the perceptions, preferences, and values of the different people involved, the quantity, variety and diversity of social features complicate the identification, classification, and measurement process (Guang Shi et al. 2012). However, it is a fact that social performance measurement has become an obligation for organization to guarantee business durability (Yildiz Çankaya and Sezen 2019).

In this regard, social performance can be defined as firms' capability to contribute positively to society, creating health and welfare for people by translating institutional social goals into action, in line with accepted social values (de Giovanni 2012). Then, social performance includes employee health and safety, the improvement of occupational health and safety in the workplace, developing health and safety performance measurement systems (Guang Shi et al. 2012), social responsibility toward clients, employees, and the community (Setyadi et al. 2019), education, training, and human resource advancement (Kazancoglu et al. 2018), improving the economic and social conditions of clients (de Giovanni 2012). Also, it refers to the real effects of green practices on social aspects, such as enhanced product image and company image (Luthra et al. 2014), improved stakeholder relationships (Laosirihongthong et al. 2013), perceived firm goodwill by stakeholders (suppliers customers, employees, and community) (Eltayeb et al. 2011; Setyadi et al. 2019), employee satisfaction and development (Younis et al. 2016), customer loyalty and satisfaction (Laosirihongthong et al. 2013), and improved firm acceptance by local communities (Silva et al. 2019).

The sustainable performance dimensions and measures are shown in Table 1.7.

1.4 Conclusions

GSCM has emerged as a unique organizational strategy for the reduction of environmental impacts that provide both economic benefits and social welfare. In this sense, organizations need to be able to identify the pressures and drivers that influence GSCM adoption, remove the major barriers, and identify factors for success that facilitate GSCM implementation, hence the improvement of sustainable performance.

The bibliometric analysis revealed that The Journal of Cleaner Production was most popular in the knowledge field. However, the Resources, Conservation and Recycling Journal had the greatest impact. Also, the study showed that Joseph Sarkis, Zhu, Q.H and Seuring, and Muller were the most influential authors in GSCM,

Table 1.7 Sustainable performance dimensions and measures

Variable	Measures	Authors
Environmental performance	Reduction of air emission Reduction of water and/or solid Decrease of consumption for hazardous/harmful/toxic materials Reduction of amount of energy used Lower frequency of environmental accidents and health hazards Improve enterprise's environmental situation	Zhu et al. (2005, 2007, 2013), De Giovanni and Esposito Vinzi (2012), Diabat et al. (2013), Dubey et al. (2015a, b), Al-Ghwayeen and Abdallah (2018), Yildiz Çankaya and Sezen (2019), Choudhary and Sangwan (2019a, b)
Social performance	Improvement in customer satisfaction Improvement in its image in the eyes of its customers, suppliers, employees Improvement in investments on social projects (education, culture, sports) Improvement in relations with community stakeholders, e.g., non-governmental organizations (NGOs) and community activists Improved awareness and protection of the claims and rights of people in community served Improvement in employee satisfaction Improvement in occupational health and safety of employees Improvement in overall stakeholder welfare or betterment	Eltayeb et al. (2011), De Giovanni (2012), Diabat et al. (2013), Laosirihongthong et al. (2013), Ahi and Searcy (2015), Yildiz Çankaya and Sezen (2019), Zhang et al. (2020)
Economic performance	Decrease of cost for materials purchasing Decrease of cost for energy consumption Decrease of fee for treatment and waste discharge Decrease of fine for environmental accidents Average return on sales and investment over the three years Average profit and profit growth over the past three years Average growth in market share over the past three years	Zhu et al. (2005, 2007b, 2013), Diabat et al. (2013), Younis et al. (2016), Balakrishnan and Suresh (2018), Silva et al. (2019), Yildiz Çankaya and Sezen (2019), Zhang et al. (2020)

in terms of numbers of publications and citations. The countries with the largest numbers of publications included China, India, and Brazil, and the industrial sectors with the most studies were the automotive industry and electrical and electronics industry, due to their strict regulations. Moreover, survey and mathematical/ analytical modeling were the most frequently adopted investigative methods. The present study also identified four collaborative network clusters. Surprisingly, theoretical and conceptual papers category is the category furthest behind, which indicates that GSCM is still an open field, from this research perspective.

In accordance with the findings derived from the literature review, the proposed conceptual framework encompasses the relationship between three main components as follows: influential factors, green practices, and performance. Regarding influential factors, three aspects were identified as follow: pressures/drivers, barriers, and success factors. Pressures/drivers are related to external determinants and firms' internal motivators for GSCM adoption. Said relationship is influenced by barriers, which may hinder GSCM adoption, and their counterbalance, success factors, which act as enablers for GSCM practice implementation.

Based on these triggers, companies can implement GSCM practices from three perspectives as follows: strategic, operational, and supportive practices. Strategic practices refer to the unique capabilities that differentiate an organization from its competitors. Operational practices encompass the main activities involved in supply chain echelons (sourcing, manufacturing, and distribution). Finally, supportive practices allow for the development and improvement of strategic and operational practices. The literature review provided several insights into the positive effect of green practice adoption on environmental, economic, and social performance. Along these lines, the proposed conceptual framework is a contribution that may be useful to support firm strategy development, decision making, and performance improvement.

In terms of future research directions, the proposed conceptual framework could be used to study the relationship between influential factors, GSCM practices, and sustainable performance. Also, future studies could examine aspects related to the conceptual framework in emerging economies, such as those of countries in South America and Africa. The proposed framework may further be extended, considering that environmental and social metrics are topics that require additional investigation. Finally, from content analysis, a lack of theoretical and conceptual background is evident. This should be considered in future research into the GSCM approach.

References

Ab Talib MS, Muniandy S (2013) Green supply chain initiatives in Malaysia: a conceptual critical success factors framework. World Appl Sci J 26:276–281. https://doi.org/10.5829/idosi.wasj.2013.26.02.1479

Abu Seman NA, Govindan K, Mardani A et al (2019) The mediating effect of green innovation on the relationship between green supply chain management and environmental performance. J Clean Prod 229:115–127. https://doi.org/10.1016/j.jclepro.2019.03.211

Agyemang M, Zhu Q, Adzanyo M, et al (2018) Evaluating barriers to green supply chain redesign and implementation of related practices in the West Africa cashew industry. Resour Conserv Recycl 136:209–222. https://doi.org/10.1016/j.resconrec.2018.04.011

Ahi P, Searcy C (2013) A comparative literature analysis of definitions for green and sustainable supply chain management. J Clean Prod 52:329–341. https://doi.org/10.1016/j.jclepro.2013.02.018

Ahi P, Searcy C (2015) An analysis of metrics used to measure performance in green and sustainable supply chains. J Clean Prod 86:360–377. https://doi.org/10.1016/j.jclepro.2014.08.005

Ahmed W, Najmi A, Arif M, Younus M (2019) Exploring firm performance by institutional pressures driven green supply chain management practices. Smart Sustain Built Environ 8:415–437. https://doi.org/10.1108/sasbe-04-2018-0022

Aital P, Vijai JP (2015) Operational practices and performances of green supply chain management in Indian firms. Int J Process Manag Benchmarking 5:352–374. https://doi.org/10.1504/IJPMB.2015.70819

Al-Ghwayeen WS, Abdallah AB (2018) Green supply chain management and export performance. J Manuf Technol Manag 29:1233–1252. https://doi.org/10.1108/JMTM-03-2018-0079

Al-Sheyadi A, Muyldermans L, Kauppi K (2019) The complementarity of green supply chain management practices and the impact on environmental performance. J Environ Manage 242:186–198. https://doi.org/10.1016/j.jenvman.2019.04.078

Ali A, Bentley Y, Cao G, Habib F (2017) Green supply chain management–food for thought? Int J Logist Res Appl 20:22–38. https://doi.org/10.1080/13675567.2016.1226788

Bae H-SS, Grant DB (2018) Investigating effects of organisational culture and learning on environmental collaboration and performance of Korean exporting firms. Int J Logist Res Appl 21:614–630. https://doi.org/10.1080/13675567.2018.1470232

Balakrishnan AS, Suresh J (2018) Green supply chain management in Indian automotive sector. Int J Logist Syst Manag 29:502. https://doi.org/10.1504/ijlsm.2018.090476

Björklund M, Martinsen U, Abrahamsson M (2012) Performance measurements in the greening of supply chains. Supply Chain Manag 17:29–39. https://doi.org/10.1108/13598541211212186

Bon AT, Zaid AA, Jaaron A (2018) Green human resource management, Green supply chain management practices and Sustainable performance. Proc Int Conf Ind Eng Oper Manag Bandung, Indones 6–8 March 2018

Brundtland GH (1987) Our common future: report of the 1987 World Commission on Environment and Development, United Nations, Oslo, 1–59

Carter CR, Rogers DS (2008) A framework of sustainable supply chain management: Moving toward new theory. Int J Phys Distrib Logist Manag 38:360–387. https://doi.org/10.1108/09600030810882816

Chiappetta Jabbour CJ, Mauricio AL, Jabbour ABL de S (2017) Critical success factors and green supply chain management proactivity: shedding light on the human aspects of this relationship based on cases from the Brazilian industry. Prod Plan Control 28:671–683. https://doi.org/10.1080/09537287.2017.1309705

Chin TA, Tat HH, Sulaiman Z, Muhamad Zainon SNL (2015) Green supply chain management practices and sustainability performance. Adv Sci Lett 21:1359–1362. https://doi.org/10.1166/asl.2015.6029

Choi S-B, Min H, Joo H-Y (2018) Examining the inter-relationship among competitive market environments, green supply chain practices, and firm performance. Int J Logist Manag IJLM-02-2017-0050. https://doi.org/10.1108/IJLM-02-2017-0050

Choudhary K, Sangwan KS (2019a) Multiple case study analysis and development of an interpretive structural model for greening of supply chains in Indian ceramic enterprises. Manag Environ Qual an Int J 30:1279–1296. https://doi.org/10.1108/MEQ-11-2018-0196

Choudhary K, Sangwan KS (2019b) Adoption of green practices throughout the supply chain: an empirical investigation. Benchmarking 26:1650–1675. https://doi.org/10.1108/BIJ-09-2018-0293

Chuang Y-C, Hu S-K, Liou JJH, Lo H-W (2018) Building a decision dashboard for improving green supply chain management. Int J Inf Technol Decis Mak. https://doi.org/10.1142/S0219622018500281

de Giovanni P (2012) Do internal and external environmental management contribute to the triple bottom line?

de Giovanni P, Esposito Vinzi V (2012) Covariance versus component-based estimations of performance in green supply chain management. Int J Prod Econ 135:907–916. https://doi.org/10.1016/j.ijpe.2011.11.001

de Sousa Jabbour SAB, Jabbour AFA, Chiappetta CJ (2013) Green supply chain management : mapping the territory Ana Beatriz Lopes de Sousa Jabbour, Ariana Fernandes Arantes and Charbel José Chiappetta Jabbour. Int J Environ Sustain Dev 12:145–167

Deng H, Luo F, Wibowo S (2018) Multi-criteria group decision making for green supply chain management under uncertainty. Sustain 10:1–13. https://doi.org/10.3390/su10093150

Dey PK, Cheffi W (2013) Green supply chain performance measurement using the analytic hierarchy process: a comparative analysis of manufacturing organisations. Prod Plan Control 24:702–720. https://doi.org/10.1080/09537287.2012.666859

Diabat A, Govindan K (2011) An analysis of the drivers affecting the implementation of green supply chain management. Resour Conserv Recycl 55:659–667. https://doi.org/10.1016/j.resconrec.2010.12.002

Diabat A, Khodaverdi R, Olfat L (2013) An exploration of green supply chain practices and performances in an automotive industry. Int J Adv Manuf Technol 68:949–961. https://doi.org/10.1007/s00170-013-4955-4

Dube AS, Gawande RS (2016) Analysis of green supply chain barriers using integrated ISM-fuzzy MICMAC approach. Benchmarking 23:1558–1578. https://doi.org/10.1108/BIJ-06-2015-0057

Dubey R, Gunasekaran A, Papadopoulos T (2017) Green supply chain management: theoretical framework and further research directions. Benchmarking 24:184–218. https://doi.org/10.1108/BIJ-01-2016-0011

Dubey R, Gunasekaran A, Papadopoulos T, Childe SJ (2015) Green supply chain management enablers: mixed methods research. Sustain Prod Consum 4:72–88. https://doi.org/10.1016/j.spc.2015.07.001

Dubey R, Gunasekaran A, Samar Ali S (2015) Exploring the relationship between leadership, operational practices, institutional pressures and environmental performance: A framework for green supply chain. Int J Prod Econ 160:120–132. https://doi.org/10.1016/j.ijpe.2014.10.001

Elkington J (1998) Partnerships from cannibals with forks: the triple bottom line of 21st-century business. Environ Qual Manag 8:37–51. https://doi.org/10.1002/tqem.3310080106

Eltayeb TK, Zailani S, Ramayah T (2011) Green supply chain initiatives among certified companies in Malaysia and environmental sustainability: investigating the outcomes. Resour Conserv Recycl 55:495–506. https://doi.org/10.1016/j.resconrec.2010.09.003

Famiyeh S, Kwarteng A (2018) Implementation of environmental management practices in the Ghanaian mining and manufacturing supply chains. Int J Product Perform Manag 67:1091–1112. https://doi.org/10.1108/IJPPM-04-2017-0095

Fraj E, Martínez E, Matute J (2013) Green marketing in B2B organisations: an empirical analysis from the natural-resource-based view of the firm. J Bus Ind Mark 28:396–410. https://doi.org/10.1108/08858621311330245

Gandhi S, Mangla SK, Kumar P, Kumar D (2016) A combined approach using AHP and DEMATEL for evaluating success factors in implementation of green supply chain management in Indian manufacturing industries. Int J Logist Res Appl 19:537–561. https://doi.org/10.1080/13675567.2016.1164126

Garg CP, Sharma A, Goyal G (2017) A hybrid decision model to evaluate critical factors for successful adoption of GSCM practices under fuzzy environment. Uncertain Supply Chain Manag 5:59–70. https://doi.org/10.5267/j.uscm.2016.7.002

Ghadge A, Kaklamanou M, Choudhary S, Bourlakis M (2017) Implementing environmental practices within the greek dairy supply chain drivers and barriers for SMEs. Ind Manag Data Syst 117:1995–2014. https://doi.org/10.1108/IMDS-07-2016-0270

Ghadimi P, Wang C, Lim MK (2019) Sustainable supply chain modeling and analysis: past debate, present problems and future challenges. Resour Conserv Recycl 140:72–84. https://doi.org/10.1016/j.resconrec.2018.09.005

Gong R, Xue J, Zhao L et al (2019) A bibliometric analysis of green supply chain management based on the web of science (wos) platform. Sustainability 11:1–18. https://doi.org/10.3390/su11123459

Govindan K, Khodaverdi R, Vafadarnikjoo A (2015) Intuitionistic fuzzy based DEMATEL method for developing green practices and performances in a green supply chain. Expert Syst Appl 42:7207–7220. https://doi.org/10.1016/j.eswa.2015.04.030

Green KW, Toms LC, Clark J (2015) Impact of market orientation on environmental sustainability strategy. Manag Res Rev 38:217–238. https://doi.org/10.1108/MRR-10-2013-0240

Green KW, Zelbst PJ, Bhadauria VS, Meacham J (2012a) Do environmental collaboration and monitoring enhance organizational performance? Ind Manag Data Syst 112:186–205. https://doi.org/10.1108/02635571211204254

Green KW, Zelbst PJ, Meacham J, Bhadauria VS (2012b) Green supply chain management practices: Impact on performance. Supply Chain Manag 17:290–305. https://doi.org/10.1108/13598541211227126

Guang Shi V, Lenny Koh SC, Baldwin J, Cucchiella F (2012) Natural resource based green supply chain management. Supply Chain Manag Int J 17:54–67. https://doi.org/10.1108/13598541211212203

Hart SL (1995) A natural-resource-based view of the firm. Acad Manag Rev 20:986–1014

Hartmann J, Germain R (2015) Understanding the relationships of integration capabilities, ecological product design, and manufacturing performance. J Clean Prod 92:196–205. https://doi.org/10.1016/j.jclepro.2014.12.079

Hervani AA, Helms MM, Sarkis J (2005) Performance measurement for green supply chain management. Benchmarking 12:330–353. https://doi.org/10.1108/14635770510609015

Hong J, Zheng R, Deng H, Zhou Y (2019) Green supply chain collaborative innovation, absorptive capacity and innovation performance: evidence from China. J Clean Prod 241:118377. https://doi.org/10.1016/j.jclepro.2019.118377

Hutomo A, Haizam M, Sinaga O (2018) The mediating role of organizational learning capability on green distribution and green packaging towards sustainability performance as a function environmental dynamism: Indonesia and Malaysia fishery industries. IOP Conf Ser Earth Environ Sci 164:1–11. https://doi.org/10.1088/1755-1315/164/1/012018

Jabbour ABLDS, Frascareli FCDO, Jabbour CJC et al (2015) Green supply chain management and firms' performance: understanding potential relationships and the role of green sourcing and some other green practices. Resour Conserv Recycl 104:366–374. https://doi.org/10.1016/j.resconrec.2015.07.017

Jawaad M, Zafar S (2019) Improving sustainable development and firm performance in emerging economies by implementing green supply chain activities. Sustain Dev 1–14. https://doi.org/10.1002/sd.1962

Kaur J, Sidhu R, Awasthi A, et al (2018) A DEMATEL based approach for investigating barriers in green supply chain management in Canadian manufacturing firms. Int J Prod Res 56:312–332. https://doi.org/10.1080/00207543.2017.1395522

Kaur J, Sidhu R, Awasthi A, Srivastava SK (2019) A Pareto investigation on critical barriers in green supply chain management. Int J Manag Sci Eng Manag 14:113–123. https://doi.org/10.1080/17509653.2018.1504237

Kazancoglu Y, Kazancoglu I, Sagnak M (2018) A new holistic conceptual framework for green supply chain management performance assessment based on circular economy. J Clean Prod 195:1282–1299. https://doi.org/10.1016/j.jclepro.2018.06.015

Khan SAR, Qianli D (2017) Impact of green supply chain management practices on firms' performance: an empirical study from the perspective of Pakistan. Environ Sci Pollut Res 24:16829–16844. https://doi.org/10.1007/s11356-017-9172-5

Kusi-Sarpong S, Bai C, Sarkis J, Wang X (2015) Green supply chain practices evaluation in the mining industry using a joint rough sets and fuzzy TOPSIS methodology. Resour Policy 46:86–100. https://doi.org/10.1016/j.resourpol.2014.10.011

Kusi-Sarpong S, Sarkis J, Wang X (2016) Assessing green supply chain practices in the Ghanaian mining industry: a framework and evaluation. Int J Prod Econ 181:325–341. https://doi.org/10.1016/j.ijpe.2016.04.002

Lee SM, Rha JS, Choi D, Noh Y (2013) Pressures affecting green supply chain performance. Manag Decis 51:1753–1768. https://doi.org/10.1108/MD-12-2012-0841

Laosirihongthong T, Adebanjo D, Choon Tan K (2013) Green supply chain management practices and performance. Ind Manag Data Syst 113:1088–1109. https://doi.org/10.1108/IMDS-04-2013-0164

Lin RJ (2013) Using fuzzy DEMATEL to evaluate the green supply chain management practices. J Clean Prod 40:32–39. https://doi.org/10.1016/j.jclepro.2011.06.010

Liu J, Yuan C, Hafeez M, Yuan Q (2018) The relationship between environment and logistics performance: evidence from Asian countries. J Clean Prod 204:282–291. https://doi.org/10.1016/j.jclepro.2018.08.310

Longoni A, Luzzini D, Guerci M (2018) Deploying environmental management across functions: the relationship between green human resource management and green supply chain management. J Bus Ethics 151:1081–1095. https://doi.org/10.1007/s10551-016-3228-1

Luthra S, Garg D, Haleem A (2014) Empirical analysis of green supply chain management practices in Indian automobile industry. J Inst Eng Ser C 95:119–126. https://doi.org/10.1007/s40032-014-0112-6

Mafini C, Loury-Okoumba WV (2018) Extending green supply chain management activities to manufacturing small and medium enterprises in a developing economy. South African J Econ Manag Sci 21:1–12. https://doi.org/10.4102/sajems.v21i1.1996

Majumdar A, Sinha S (2018) Modeling the barriers of green supply chain management in small and medium enterprises: a case of Indian clothing industry. Manag Environ Qual an Int J 29:1110–1122. https://doi.org/10.1108/MEQ-12-2017-0176

Malviya RK, Kant R (2017) Modeling the enablers of green supply chain management: an integrated ISM—fuzzy MICMAC approach. Benchmarking 24:536–568. https://doi.org/10.1108/BIJ-08-2015-0082

Malviya RK, Kant R, Gupta AD (2018) Evaluation and selection of sustainable strategy for green supply chain management implementation. Bus Strateg Environ 27:475–502. https://doi.org/10.1002/bse.2016

Manimaran A, Muthuraman V, Jayakumar V (2018) Implementation of tolerance evaluation framework for green supply chain management in a gear manufacturing industry : a case study. 118:345–354

Mathiyazhagan K, Datta U, Bhadauria R et al (2018) Identification and prioritization of motivational factors for the green supply chain management adoption: case from Indian construction industries. Opsearch 55:202–219. https://doi.org/10.1007/s12597-017-0316-7

Mauricio AL, De Sousa Jabbour ABL (2017) Critical success factors for GSCM adoption: case studies in the automotive battery industry. Gest E Prod 24:78–94. https://doi.org/10.1590/0104-530x2267-16

Miras-Rodríguez M del M, Machuca JAD, Escobar-Pérez B (2018) Drivers that encourage environmental practices in manufacturing plants: a comparison of cultural environments. J Clean Prod 179:690–703. https://doi.org/10.1016/j.jclepro.2017.11.029

Mitra S, Datta PP (2014) Adoption of green supply chain management practices and their impact on performance: An exploratory study of Indian manufacturing firms. Int J Prod Res 52:2085–2107. https://doi.org/10.1080/00207543.2013.849014

Muduli K, Barve A (2013) Establishment of a sustainable development framework in small scale mining supply chains in India. Int J Intell Enterp 2:84–100. https://doi.org/10.1504/IJIE.2013.057340

Mumtaz U, Ali Y, Petrillo A, De Felice F (2018) Identifying the critical factors of green supply chain management: environmental benefits in Pakistan. Sci Total Environ 640–641:144–152. https://doi.org/10.1016/j.scitotenv.2018.05.231

Mutingi M, Mapfaira H, Monageng R (2014) Developing performance management systems for the green supply chain. J Remanufacturing 4:1–20. https://doi.org/10.1186/s13243-014-0006-z

Namagembe S, Ryan S, Sridharan R (2019) Green supply chain practice adoption and firm performance: manufacturing SMEs in Uganda. Manag Environ Qual an Int J 30:5–35. https://doi.org/10.1108/MEQ-10-2017-0119

Nejati M, Rabiei S, Chiappetta Jabbour CJ (2017) Envisioning the invisible: understanding the synergy between green human resource management and green supply chain management in manufacturing firms in Iran in light of the moderating effect of employees' resistance to change. J Clean Prod 168:163–172. https://doi.org/10.1016/j.jclepro.2017.08.213

Olugu EU, Wong KY, Shaharoun AM (2011) Development of key performance measures for the automobile green supply chain. Resour Conserv Recycl 55:567–579. https://doi.org/10.1016/j.resconrec.2010.06.003

Olugu EU, Wong KY (2012) An expert fuzzy rule-based system for closed-loop supply chain performance assessment in the automotive industry. Expert Syst Appl 39:375–384. https://doi.org/10.1016/j.eswa.2011.07.026

Pinto GMC, Pedroso B, Moraes J et al (2018) Environmental management practices in industries of Brazil, Russia, India, China and South Africa (BRICS) from 2011 to 2015. J Clean Prod 198:1251–1261. https://doi.org/10.1016/j.jclepro.2018.07.046

Pourjavad E, Shahin A (2018) The application of Mamdani fuzzy inference system in evaluating green supply chain management performance. Int J Fuzzy Syst 20:901–912. https://doi.org/10.1007/s40815-017-0378-y

Rao P, Holt D (2005) Do green supply chains lead to competitiveness and economic performance? Int J Oper Prod Manag 25:898–916. https://doi.org/10.1108/01443570510613956

Reddy Maditati D, Munim ZH, Schramm H-J, Kummer S (2018) A review of green supply chain management: from bibliometric analysis to a conceptual framework and future research directions. Resour Conserv Recycl 139:150–162. https://doi.org/10.1016/j.resconrec.2018.08.004

Rostamzadeh R, Govindan K, Esmaeili A, Sabaghi M (2015) Application of fuzzy VIKOR for evaluation of green supply chain management practices. Ecol Indic 49:188–203. https://doi.org/10.1016/j.ecolind.2014.09.045

Ryoo SY, Koo C (2013) Green practices-IS alignment and environmental performance: the mediating effects of coordination. Inf Syst Front 15:799–814. https://doi.org/10.1007/s10796-013-9422-0

Saeed A, Jun Y, Nubuor SA et al (2018) Institutional pressures, green supply chain management practices on environmental and economic performance: a two theory view. Sustain 10:1–24. https://doi.org/10.3390/su10051517

Sahu AK, Narang HK, Rajput MS, Sahu NK, Sahu AK (2018) Performance modeling and benchmarking of green supply chain management: an integrated fuzzy approach. Benchmarking Int J 25:2248–2271. https://doi.org/10.1108/BIJ-02-2017-0032

Sahu AK, Datta S, Mahapatra SS (2015) Green supply chain performance appraisement and benchmarking using fuzzy grey relation method. Int J Bus Inf Syst 20:157–194. https://doi.org/10.1504/IJBIS.2015.071533

Sangode PB, Metre SG (2019) Green supply chain practices for environmental sustainability: a proposed framework for manufacturing firms. Int J Mech Prod Eng Res Dev 9:287–298. https://doi.org/10.24247/ijmperdapr201928

Sarkis J, Gonzalez-Torre P, Adenso-Diaz B (2010) Stakeholder pressure and the adoption of environmental practices: the mediating effect of training. J Oper Manag 28:163–176. https://doi.org/10.1016/j.jom.2009.10.001

Savita KS, Dominic PDD, Ramayah T (2012) Eco-design strategy among ISO 14001 certified manufacturing firms in Malaysia: green drivers and its relationship to performance outcomes. In: 2012 International Conference on Computer and Information Science, ICCIS 2012 - A Conference of World Engineering, Science and Technology Congress, ESTCON 2012 - Conference Proceedings. IEEE, pp 154–159

Savita KS, Ramayah T, Lumpur K (2016) The drivers, practices and outcomes of green supply chain management: insights from ISO14001 manufacturing firms in Malaysia. Int J Inf Syst Supply Chain Manag 9:35–60. https://doi.org/10.4018/IJISSCM.2016040103

Setyadi A (2019) Does green supply chain integration contribute towards sustainable performance? Uncertain Supply Chain Manag 7:121–132. https://doi.org/10.5267/j.uscm.2018.10.012

Seuring S, Müller M (2008) From a literature review to a conceptual framework for sustainable supply chain management. J Clean Prod 16:1699–1710. https://doi.org/10.1016/j.jclepro.2008.04.020

Sharma S, Gandhi MA (2016) Exploring correlations in components of green supply chain practices and green supply chain performance. Compet Rev 26:332–368. https://doi.org/10.1108/CR-04-2015-0027

Sharma VK, Chandana P, Bhardwaj A (2015) Critical factors analysis and its ranking for implementation of GSCM in Indian dairy industry. J Manuf Technol Manag 26:911–922. https://doi.org/10.1108/JMTM-03-2014-0023

Silva GM, Gomes PJ, Sarkis J (2019) The role of innovation in the implementation of green supply chain management practices. Bus Strateg Environ 28:819–832. https://doi.org/10.1002/bse.2283

Sharma VK, Chandna P, Bhardwaj A (2017) Green supply chain management related performance indicators in agro industry: a review. J Clean Prod 141:1194–1208. https://doi.org/10.1016/j.jclepro.2016.09.103

Singh M, Jawalkar CS, Kant S (2018) Analysis of drivers for green supply chain management adaptation in a fertilizer industry of Punjab (India). Int J Environ Sci Technol 1–12. https://doi.org/10.1007/s13762-018-1759-y

Singh SK, El-Kassar AN (2019) Role of big data analytics in developing sustainable capabilities. J Clean Prod 213:1264–1273. https://doi.org/10.1016/j.jclepro.2018.12.199

Srivastava SK (2007) Green supply-chain management: a state-of-the-art literature review. Int J Manag Rev 9:53–80. https://doi.org/10.1111/j.1468-2370.2007.00202.x

Sundram VPK, Premkumar R, Atikah SB, Geetha S (2018) The role of supply chain integration on green practices and performance in a supply chain context: a conceptual approach to future research. Int J Supply Chain Manag 7:95–104

Tachizawa EM, Gimenez C, Sierra V (2015) Green supply chain management approaches: drivers and performance implications. Int J Oper Prod Manag 35:1546–1566. https://doi.org/10.1108/IJOPM-01-2015-0023

Taticchi P, Tonelli F, Pasqualino R (2013) Performance measurement of sustainable supply chains. Int J Product Perform Manag 62:782–804. https://doi.org/10.1108/IJPPM-03-2013-0037

The World Resource Institute (2009) The greenhouse gas protocol initiative: scope 3 accounting and reporting standard

Testa F, Iraldo F (2010) Shadows and lights of GSCM (green supply chain management): determinants and effects of these practices based on a multi-national study. J Clean Prod 18:953–962. https://doi.org/10.1016/j.jclepro.2010.03.005

Tseng M-L, Islam MS, Karia N et al (2019) A literature review on green supply chain management: trends and future challenges. Resour Conserv Recycl 141:145–162. https://doi.org/10.1016/j.resconrec.2018.10.009

Tuni A, Rentizelas A, Duffy A (2018) Environmental performance measurement for green supply chains. A systematic analysis and review of quantitative methods. Int J Phys Distrib Logist Manag 48:765–793. https://doi.org/10.1108/EUM0000000001124

Vachon S, Klassen RD (2006) Extending green practices across the supply chain: the impact of upstream and downstream integration

Vachon S, Klassen RD (2007) Supply chain management and environmental technologies: the role of integration. Int J Prod Res 45:401–423. https://doi.org/10.1080/00207540600597781

Vachon S, Klassen RD (2008) Environmental management and manufacturing performance: the role of collaboration in the supply chain. Int J Prod Econ 111:299–315. https://doi.org/10.1016/j.ijpe.2006.11.030

Vijayvargy L, Thakkar J, Agarwal G (2017) Green supply chain management practices and performance: the role of firm-size for emerging economies. J Manuf Technol Manag 28:299–323. https://doi.org/10.1108/JMTM-09-2016-0123

Vilchez VF, Darnall N, Aragón Correa JA, Correa JAA (2017) Stakeholder influences on the design of firms' environmental practices. J Clean Prod 142:3370–3381. https://doi.org/10.1016/j.jclepro.2016.10.129

Walker H, Di Sisto L, McBain D (2008) Drivers and barriers to environmental supply chain management practices: lessons from the public and private sectors. J Purch Supply Manag 14:69–85. https://doi.org/10.1016/j.pursup.2008.01.007

Wang Z, Mathiyazhagan K, Xu L, Diabat A (2016) A decision making trial and evaluation laboratory approach to analyze the barriers to Green Supply Chain Management adoption in a food packaging company. J Clean Prod 117:19–28. https://doi.org/10.1016/j.jclepro.2015.09.142

Yang Z, Sun J, Zhang Y, Wang Y (2018) Synergy between green supply chain management and green information systems on corporate sustainability: an informal alignment perspective. Environ Dev Sustain. https://doi.org/10.1007/s10668-018-0241-9

Yildiz Çankaya S, Sezen B (2019) Effects of green supply chain management practices on sustainability performance. J Manuf Technol Manag 30:98–121. https://doi.org/10.1108/JMTM-03-2018-0099

Yongge G, Jiyong L, Yunfeng S (2009) Performance evaluation of green supply chain management based on membership conversion algorithm. In: 2009 Second ISECS international colloquium on computing, communication, control, and management, CCCM 2009, pp 237–240

Younis H, Sundarakani B, Vel P (2016) The impact of implementing green supply chain management practices on corporate performance. Compet Rev 26:216–245. https://doi.org/10.1108/CR-04-2015-0024

Yu W, Chavez R, Feng M, Wiengarten F (2014) Integrated green supply chain management and operational performance. Supply Chain Manag 19:683–696. https://doi.org/10.1108/SCM-07-2013-0225

Zaid AA, Jaaron AAM, Bon AT (2018) The impact of green human resource management and green supply chain management practices on sustainable performance: an empirical study. J Clean Prod 204:965–979. https://doi.org/10.1016/j.jclepro.2018.09.062

Zhang H, Yang F (2016) On the drivers and performance outcomes of green practices adoption an empirical study in China. Ind Manag Data Syst 116:2011–2034. https://doi.org/10.1108/IMDS-06-2015-0263

Zhang J, Zhang X, Wang Q, Ma Z (2020) Relationship between institutional pressures, green supply chain management practices and business performance: an empirical research on automobile industry. Springer International Publishing

Zhu Q, Sarkis J (2004) Relationships between operational practices and performance among early adopters of green supply chain management practices in Chinese manufacturing enterprises. J Oper Manag 22:265–289. https://doi.org/10.1016/j.jom.2004.01.005

Zhu Q, Sarkis J (2006) An inter-sectoral comparison of green supply chain management in China: drivers and practices. J Clean Prod 14:472–486. https://doi.org/10.1016/j.jclepro.2005.01.003

Zhu Q, Sarkis J, Geng Y (2005) Green supply chain management in China: pressures, practices and performance. Int J Oper Prod Manag 25:449–468. https://doi.org/10.1108/01443570510593148

Zhu Q, Sarkis J, Lai K, Hung (2007a) Initiatives and outcomes of green supply chain management implementation by Chinese manufacturers. J Environ Manage 85:179–189. https://doi.org/10.1016/j.jenvman.2006.09.003

Zhu Q, Sarkis J, Lai K, Hung (2007b) Green supply chain management: pressures, practices and performance within the Chinese automobile industry. J Clean Prod 15:1041–1052. https://doi.org/10.1016/j.jclepro.2006.05.021

Zhu Q, Sarkis J, Hung LK (2008) Confirmation of a measurement model for green supply chain management practices implementation. Int J Prod Econ 111:261–273. https://doi.org/10.1016/j.ijpe.2006.11.029

Zhu Q, Sarkis J, Lai K, Hung (2013) Institutional-based antecedents and performance outcomes of internal and external green supply chain management practices. J Purch Supply Manag 19:106–117. https://doi.org/10.1016/j.pursup.2012.12.001

Zhu Q, Tian Y, Sarkis J (2012) Diffusion of selected green supply chain management practices: an assessment of Chinese enterprises. Prod Plan Control 23:837–850. https://doi.org/10.1080/09537287.2011.642188

Zhu W, Wang Z (2018) The collaborative networks and thematic trends of research on purchasing and supply management for environmental sustainability: a bibliometric review. Sustain 10:1–28. https://doi.org/10.3390/su10051510

Chapter 2
Arcs of Integration: Methodological and Statistical Analysis

Bertha Molina-Quintana, Antonio Vaamonde-Liste, Jaime Apolinar Martínez-Arroyo, and María Berta Quintana-León

Abstract There are several research works that use the methodology of Arcs of Integration of (Frohlich and Westbrook in J Oper Manag 19:185–200, 2001), based on the level of integration of the external supply chain with the key dimensions to represent a strategic position as the direction (toward customers and/or suppliers) and the measure (degree of integration), graphically illustrated through an arc. As well as the arcs of integration of (De la Calle Vicente in La Integración de la Cadena de Suministro como Herramienta Competitiva: El Caso de la Industria Manufacturera del País Vasco, Universidad de Deusto, Bilbao, España, 2015) describing the level of internal integration, to measure the integration of the supply chain, however, none of these works explain a detail how to develop the methodology or the statistical process for the use of arcs of integration. Is for this reason that the objective of this work is to study in depth the methodological and statistical treatment of arcs of integration, since it is considered a good diagnostic tool to measure the level of integration, both internal and external of companies. The measurement instrument was applied to 93 food manufacturing companies in Michoacán, Mexico, and different statistical techniques were used to analyze the data and obtain valid results, such as factor analysis, discriminant analysis, linear and ordinal correlations, "t" tests., nonparametric tests, or independence tests, among others, using the R statistical programs (R Core Team V 3.6.1) and SPSS (IBM SPSS Statistics V 24).

Keywords Arcs of integration · Supply chain integration · Methodological and statistical treatment

B. Molina-Quintana (✉) · J. A. Martínez-Arroyo · M. B. Quintana-León
Faculty of Accounting and Administrative Sciences, Universidad Michoacana de San Nicolas de Hidalgo, Gral. Francisco J. Múgica S/N, Felícitas del Río, 58030 Morelia, Michoacán, México
e-mail: bertha.molina@umich.mx

A. Vaamonde-Liste
Department of Statistics and Operational Research, University School of Business Studies of the University of Vigo, Rúa Conde de Torrecedeira, 105, Vigo, Spain

J. L. Garcia Alcaraz et al. (eds.), *Trends in Industrial Engineering Applications to Manufacturing Process*, https://doi.org/10.1007/978-3-030-71579-3_2

2.1 Introduction

Due to intense global competition, organizations create a cooperative and mutually beneficial relationship between supply chain partners, as supply chain activities are increasingly dispersed among customers, suppliers, and service providers (Leuschner et al. 2013), and there is a greater need for customers and suppliers to work together more closely. Effective supply chain management has become a potentially valuable way to gain competitive advantage and enhance organizational development, as competition no longer occurs between organizations, but between supply chains (Li et al. 2006; Christopher 2011).

Supply chain integration is an organizational model guideline, which considers the dynamic supply chain alliance, seeks to improve competitive performance by closely integrating internal cross-functions within a company and effectively linking them with the external operations of providers, customers, and other channel members (Otchere et al. 2013a, b; Kumar et al. 2017), by establishing and maintaining a long-term strategic partnership based on information integration, business and function reengineering, integration of organizations, cultural adaptation, and reorganization of strategic resources.

The benefits of integration are often translated into customer-valued aspects such as product quality, delivery reliability, process flexibility, and cost leadership (Rosenzweig et al. 2003). However, as the market has increasingly demanding customers and companies, they can no longer survive on their own. To achieve integration, chain members must see each other as partners, working together to develop strategic planning, demand, forecasting, and goal setting (Kim and Lee 2010). However, to achieve the integration of the supply chain, some characteristics that must be identified in the relationship between members are as follows: trust, information exchange, association, cooperation, collaboration, and coordination (Arantes et al. 2014, 2018).

Supply chain integration is divided into internal integration and external integration, while external integration is divided into supplier integration and customer integration (Tolu Feyissa et al. 2018; Lin and Lin 2018; Huo et al. 2016; Lii and Kuo 2016; Baofeng et al. 2016; Saleh 2015; Chang et al. 2015; Zhao et al. 2015; Ramesh et al. 2014; Otchere et al. 2013a, b; Lotfi et al. 2013; Afshan 2013; Alfalla-Luque et al. 2012; Mahmood et al. 2012; Schoenherr and Swink 2012; Kotcharin et al. 2012; Wong et al. 2011; Zhao et al. 2011; Flynn et al. 2010; Koufteros et al. 2010; Narasimhan and Kim 2002). Supply chain integration is generally recognized as critical to achieving performance and competitive advantage (Flynn et al. 2010; Frohlich and Westbrook 2001; Vickery et al. 2003; Zailani and Rajagopal 2005).

It is very complex to replicate the ability of companies to integrate their own functional areas to customers and to suppliers, since they are due to a social nature, tacit knowledge, and a specific context. For this reason, these are some of the reasons why it can be said that the integration of the supply chain is an element that generates sustainable competitive advantages over time (Krasnikov and Jayachandran 2008).

For Ireland and Webb (2007), strategic supply chains serve as a potential source of competitive advantage for companies.

Rosenzweig et al. (2003) investigated the role of manufacturing capabilities (quality, flexibility, delivery, and cost) between supply chain integration intensity (a proxy variable for supply chain integration) and business performance. The result supported a positive relationship between the intensity of supply chain integration and manufacturing capabilities. In addition, the outcome that relates to the intensity of integration and sales growth and customer satisfaction was specifically mediated by manufacturing capabilities.

The strategy that a company may have regarding the integration of its supply chain is unique, that is, it depends on the context, its capabilities, and the objective it pursues. Businesses can design what type of integration they should focus on, what action should be taken when there are various types of integration overlap, and what procedures should be followed (Van Nguyen and Thanh Nguyen 2017). The level of supply chain integration will be able to generate a competitive advantage for the firm (Mentzer et al. 2001); for this reason, it is important for companies to know their level of integration.

The integration with suppliers affects cost, quality, flexibility, and delivery performance in a positive way, while integration with customers has no effect according to Devaraj et al. (2007). Wong et al. (2011) investigated the moderating role of environmental uncertainty in the relationships between supply chain integration and operational performance. The result suggested that supplier integration, customer integration, and internal integration have a positive impact on delivery, cost of production, product quality, and production flexibility. Furthermore, the result showed that under high environmental uncertainty, internal integration will have a greater impact on production cost and product quality. External integration will have a greater impact on delivery and production flexibility. However, the moderating effect of environmental uncertainty was not found to be significant in the relationship between customer/supplier integration and product quality and cost of production.

Schoenherr and Swink (2012) investigated the impact of supply chain integration strategies on quality, delivery, flexibility, and cost performance. The result showed that the relationship between the external integration strategies is moderated by internal integration only for flexibility and delivery, but not for quality and cost.

There is a methodology to identify the level of integration of the supply chain presented for Frohlich y Westbrook in 2001, based on the level of integration of the external supply chain with the key dimensions to represent a strategic position as the direction (toward customers and/or suppliers) and the measure (degree of integration), graphically illustrated through an arc, with the direction of the segment that shows if the company is inclined by the supplier or the customer and the degree of arc that indicates the extent of the integration, this methodology is named arcs of supply chain integration (Frohlich and Westbrook 2001).

Scales are developed to measure supply chain integration, and five different strategies are identified. Each of these strategies is characterized by a different integration arc, representing the direction and degree of integration activity. There is consistent evidence that the widest degree of arc of integration with suppliers and customers has the strongest association with performance improvement.

The vision of the supply chain reduced to the external dimension, which loses an essential part of the sense of supply chain integration and internal integration. For this reason, completing the proposal by Frohlich and Westbrook (2001), De la Calle Vicente (2015) proposes the incorporation of internal integration in the measurement instrument. To determine their categories, the same criteria are followed as that used in the arcs of integration of Frohlich and Westbrook.

Nevertheless, none of the papers who works with these methodologies show or specify how to develop the methodology or the statistical process for the use of arcs of integration that is why the objective of this work is to try to accomplish this according to the work we did at food manufacturing companies.

2.2 Methodological Analysis of Arcs of Integration

2.2.1 External Integration

According to Frohlich and Westbrook (2001), companies can be classified according to the level of external integration into five categories: (1) Inward-Facing, (2) Periphery-Facing, (3) Supplier-Facing, (4) Customer-Facing, and (5) Outward-Facing. The internal, peripheral, and external arcs are considered symmetrical arcs, while the arc toward suppliers and the arc toward customers are considered asymmetric.

The factor score for supplier integration is used to rank each manufacturer in the top, middle, or bottom quartiles. Similarly, the factor score for customer integration is used to rank each manufacturer in the correct quartile. Companies are classified according to the quartile in which they are located according to their level of integration with customers and suppliers. Specifically, three levels are differentiated: the level below the Q1 quartile, the level above the Q3 quartile, and the intermediate level between both quartiles.

The operationalization of the arcs of integration that measures external integration is presented in Table 2.1. External integration influences competitive capabilities, product innovation performance, and quality performance (Koufteros et al. 2005).

2.2.2 Internal Integration

The vision of the supply chain reduced to the external dimension, which loses an essential part of the sense of supply chain integration and internal integration. For this reason, complementing the proposal of Frohlich and Westbrook (2001), De la Calle

Table 2.1 Operationalization of arcs of integration

Inward-facing	
Classified as inward-facing if the answer is as follows • In lower quartile for suppliers, and • In lower quartile for customers	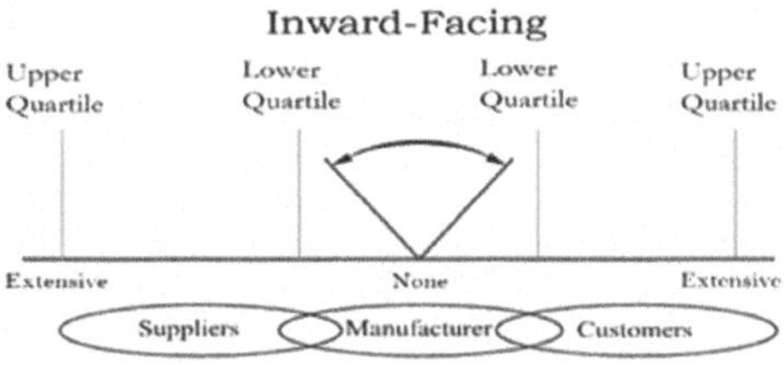
Periphery-facing	
Classified as periphery-facing if the answer is as follows • Above lower quartile for suppliers or customers, but • Below upper quartile for suppliers and customers	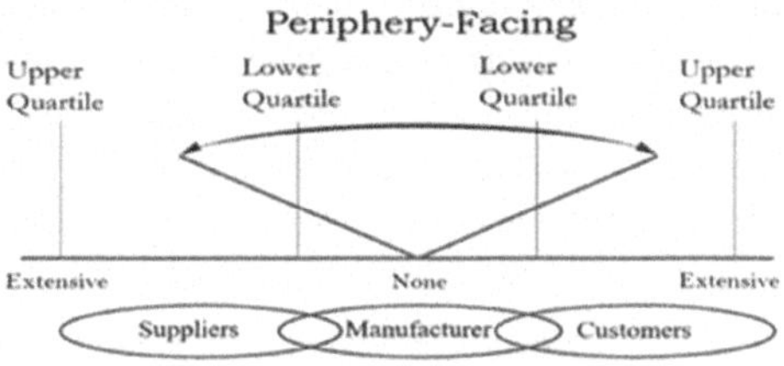
Supplier-facing	
Classified as supplier-facing if the answer is as follows • In upper quartile for suppliers, and • Below upper quartile for customers	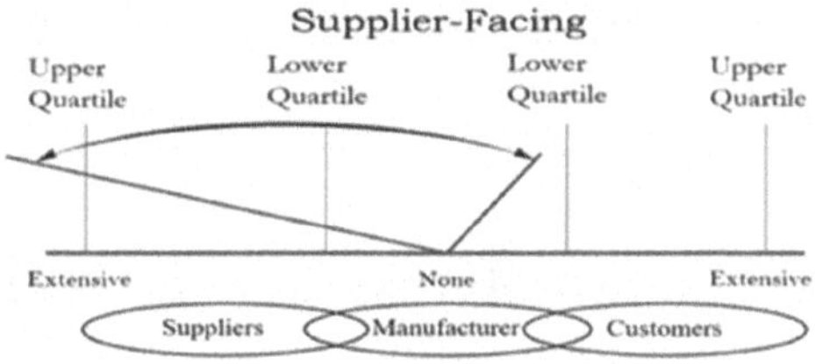
Customer-facing	
Classified as customer-facing if the answer is as follows • In upper quartile for customers, and • Below upper quartile for suppliers	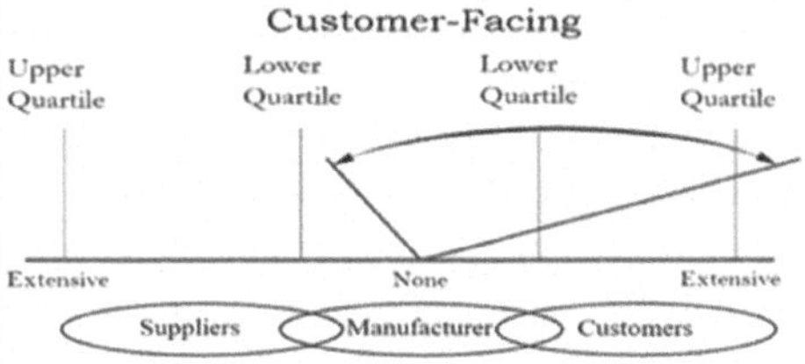
Outward-facing	
Classified as outward-facing if the answer is as follows • In upper quartile for suppliers, and • In upper quartile for customers	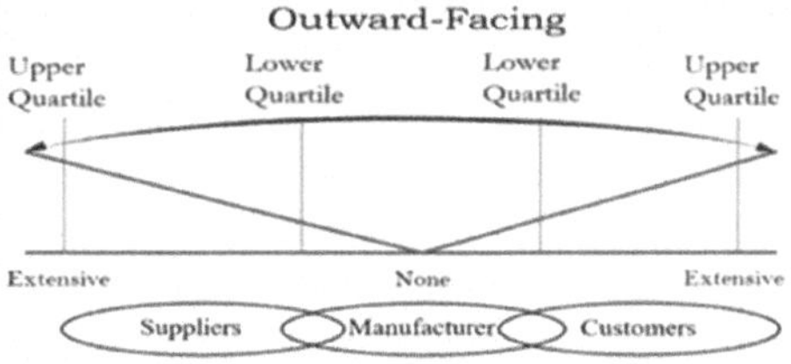

Source Frohlich and Westbrook (2001)

Table 2.2 Operationalization of the description of the level of internal integration

Poorly integrated	Integrated	Very integrated
The companies that are classified in the poorly integrated level are because they are below the quartile 25	The companies that are classified in the integrated level are because their factorial score is between the quartiles 25 and 75	The highly integrated level shows the set of companies whose factorial score is higher than quartile 75

Source De la Calle Vicente (2015)

Vicente (2015) proposes the incorporation of the internal integration, to measure the supply chain integration.

To determine their categories, the same criteria is followed as that used in the arcs of integration.

According to De la Calle Vicente (2015), companies can be classified according to the level of internal integration into three categories: (1) poorly internally integrated, (2) integrated, and (3) very internally integrated, taking as reference the quartile 1 (bottom quartile) and quartile 3 (top quartile).

Table 2.2 shows the operationalization of the description of the level of internal integration. Internal integration is defined by de Abreu and Chicarelli Alcantara (2017) as the quality of the state of collaboration between the departments is necessary to develop the unit of effort in accordance with environmental demands.

2.3 Statistical Analysis of the Methodology of Arcs of Integration

Our measurement instrument regarding integration toward suppliers and customers coincides in nine different types of activities that manufacturers commonly use to integrate their operations with suppliers and customers. They were measured on five-point Likert-type scales from 1 to 5 from (1) never to (5) always. Summary statistics for integrative supplier and customer integration activities are shown in Table 2.3.

Negative t-values indicate that the difference is always in favor of clients (all variables take higher mean values for clients than for suppliers), although this difference is only statistically significant when the *p*-value of the significance test is less than 0.05 (6 of the nine comparisons).

Factor analysis is a data reduction technique (dimensionality reduction) that serves to find homogeneous groups of variables from a large set of variables that aims to simplify the information provided by a correlation matrix to make it easily interpretable and allows to express the original variables as a linear function of a small set of implicit (unobservable) variables called common factors.

A factorial model has been constructed with a single factor for external integration relative to suppliers and another model for external integration relative to customers, after analyzing the correlations.

Table 2.3 Statistical summary for integrative activities

Suppliers		Integrative activities	Customers		Paired samples t-test	
Mean	Standard deviation		Mean	Standard deviation	*t*	*p* Value
2.882	1.552	Share your demand forecasts	3.527	1.372	−2.844	0.006
2.419	1.432	Share your production plans	2.957	1.474	−2.459	0.017
4.226	1.190	Establish long-term relationships	4.516	0.892	−1.388	0.170
2.140	1.426	Joint planning to anticipate and solve problems	3.161	1.454	−1.248	0.217
2.376	1.481	Share information through information technology	3.065	1.436	−3.536	0.001
2.871	1.408	Involvement in product development processes	3.591	1.369	−6.727	0.000
2.323	1.423	Participation in product design processes	2.591	1.393	−2.200	0.032
2.312	1.459	Set joint goals	2.473	1.388	−2.079	0.042
2.707	1.264	Develop joint responsibilities	3.640	1.199	−1.342	0.185

Source Own elaboration based on the collected data

The saturations or factor loads are presented in Table 2.4. Because it is pertinent, from a statistical point of view, to carry out a series of related tests that indicate how appropriate it is to apply factor analysis with the data and samples available, that is why we annex the adequacy index to the KMO factorial model (Kaiser–Meyer–Olkin), which measures the adequacy of the sample and its relevance.

The saturations that indicate the importance of the factor in the explanation of the variability of each of the original variables are also shown. The saturations indicate for each variable or question of the questionnaire the weight that it has in the construction of the factor. The variables used are integrative activities referring to demand forecasts, production plans, long-term relationships, joint planning, information through information technologies, involvement in product development processes, participation in design processes, production, joint objectives, and joint responsibilities.

Table 2.4 Analysis of factorial saturations and index of adaptation to the factorial model of integrative activities

Suppliers		Integrative activities	Customers	
Factor saturations	Factor model adequacy Index (KMO)		Factor saturations	Factor model adequacy Index (KMO)
0.705	0.80	Share your demand forecasts	0.531	0.83
0.563	0.74	Share your production plans	0.523	0.92
0.252	0.61	Establish long-term relationships	0.260	0.79
0.532	0.84	Joint planning to anticipate and solve problems	0.629	0.85
0.722	0.89	Share information through information technology	0.272	0.75
0.805	0.84	Involvement in product development processes	0.515	0.86
0.732	0.81	Participation in product design processes	0.511	0.77
0.831	0.78	Set joint goals	0.928	0.67
0.745	0.79	Develop joint responsibilities	0.927	0.67
Sample adequacy measure (KMO) = 0.81			Sample adequacy measure (KMO) = 0.79	

Source Own elaboration based on the collected data

The activities are the same for both suppliers and clients, which allows us to build indicators based on factor scores that will allow the two concepts to be related.

In general, we observe high saturations, which indicates a clear relationship between each variable and the common factor. Although some such as long-term relationships and information sharing through information technology in dealing with customers are not explained by the factor model in equal measure, this could be explained by the low loyalty of buyers of food products (unlike banking or insurance customers).

Factorial loads can have a maximum value of 1; therefore, the maximum value that the eigenvalue can reach is equal to the number of variables.

KMO values below 0.5 are not acceptable, considering the data inadequate to a factor analysis model. For KMO values >0.5, the adequacy of the data to a factor analysis model is considered acceptable. The closer the KMO values are to 1, the better the data fits to a factor model, considering that it is excellent for values greater than 0.8. Because the KMO indexes are values greater than 0.6, in all cases they indicate that it is applicable to apply it.

KMO always takes values between 0 and 1, and the authors themselves established the following interpretation guide:

<0.5 Unacceptable

$0.5-0.6$ Bad

$0.6-0.7$ Regular

$0.7-0.8$ Good

>0.8 Excellent

So it is considered excellent in the two previous applications for suppliers and customers. There is also an individual sample adequacy measure for each of the variables, in addition to the global one. The index for suppliers is 0.81 and for clients 0.77.

Table 2.5 shows the factor scores for all the companies in the sample, both direct and normalized scores. Normalization consists of a linear transformation that results in a variable that takes values between 0 and 10 and that allows a better interpretation, whereby a value close to zero, for example less than 2, indicates a low degree of integration and a high value close to 10, for example greater than 8, would indicate a high degree of integration.

Some companies have a high degree of integration with suppliers and a low degree of integration with customers and vice versa.

The next step is to categorize the companies according to both integration factors using the methodology of the integration arcs; for this, quartiles 1 and 3 (Q1 and Q3) are used, which indicate for each factor the value below which is a quarter or three quarters of the total sample, which allows establishing three levels (low, medium, and high) of integration for each factor.

To obtain the integration arcs, we recode the following in Table 2.6; on the customers' side: Low (B) = 10, Medium (M) = 20, High (A) = 30; on the suppliers´ side: Low (B) = 1, Medium (M) = 2, High (A) = 3.

We use same Table 2.6 in Table 2.7 to exemplify the combinations that occur with both suppliers and customers. In Table 2.7, the frequencies obtained in the sample data corresponding to each combination of levels (B = Low, M = Medium, A = High) of both factors are shown, which will serve to find out what is the arc of integration for each business.

Table 2.7 is presented below, in Table 2.8, but expressed as percentages of the total number of companies for each combination of levels for suppliers and customers.

Table 2.5 External integration factor scores (suppliers and clients) and normalized scores

Suppliers		Company	Customers	
Punctuation factorial	Normalized score		Punctuation factorial	Normalized score
0.076	3.9326	Company 1	−1.1355	0.8745
1.1397	6.9858	Company 2	−0.5434	2.6829
2.1328	9.8366	Company 3	1.7381	9.6511
0.6124	5.4722	Company 4	0.2973	5.2505
0.0921	3.9788	Company 5	−0.4358	3.0117
1.8357	8.9838	Company 6	−0.1239	3.9641
−0.5173	2.2294	Company 7	−1.2445	0.5417
−0.2344	3.0417	Company 8	−0.2802	3.4867
0.0001	3.7148	Company 9	−1.1677	0.7761
−0.4695	2.3668	Company 10	−1.0294	1.1986
0.5139	5.1896	Company 11	−0.2992	3.4287
−0.0904	3.4549	Company 12	0.2211	5.0176
1.1803	7.1023	Company 13	0.348	5.4055
−0.596	2.0036	Company 14	−1.0222	1.2205
0.5028	5.1576	Company 15	0.2939	5.24
0.4766	5.0823	Company 16	−0.7284	2.1179
0.2242	4.3581	Company 17	1.8524	10
−1.294	0	Company 18	−0.0316	4.2461
0.2291	4.3719	Company 19	−1.241	0.5522
0.3947	4.8473	Company 20	−0.04	4.2203
−0.2604	2.9669	Company 21	0.344	5.393
−0.9759	0.9133	Company 22	1.7196	9.5945
−0.699	1.7079	Company 23	−0.4409	2.9959
−0.6514	1.8448	Company 24	1.3949	8.6027
−1.1014	0.5529	Company 25	−0.6747	2.2818
−0.2008	3.138	Company 26	0.2814	5.2018
−0.3717	2.6475	Company 27	−0.5138	2.7732
1.0837	6.8252	Company 28	0.6199	6.2359
0.8531	6.1631	Company 29	1.0287	7.4845
2.1395	9.8557	Company 30	1.8524	10
0.604	5.4481	Company 31	−0.4951	2.8302
−0.8986	1.1352	Company 32	−0.0086	4.3162
−1.2077	0.2479	Company 33	1.8524	10
0.8868	6.2599	Company 34	1.0133	7.4373

(continued)

Table 2.5 (continued)

Suppliers		Company	Customers	
Punctuation factorial	Normalized score		Punctuation factorial	Normalized score
−1.2293	0.1859	Company 35	0.4849	5.8234
1.2265	7.235	Company 36	0.4626	5.7554
−1.2077	0.2479	Company 37	−1.1883	0.7132
−0.9514	0.9834	Company 38	−0.3538	3.262
0.0774	3.9365	Company 39	−1.0494	1.1375
0.8555	6.1702	Company 40	0.756	6.6515
0.28	4.5183	Company 41	−0.5014	2.811
2.1898	10	Company 42	0.9748	7.3198
2.1898	10	Company 43	1.8524	10
0.3833	4.8147	Company 44	0.2633	5.1467
−0.665	1.8057	Company 45	−1.2628	0.4857
−0.6057	1.9759	Company 46	−0.03	4.2509
−0.9258	1.0568	Company 47	1.1341	7.8061
0.0388	3.8257	Company 48	0.4823	5.8157
−1.1507	0.4114	Company 49	−1.2272	0.5946
−0.6379	1.8834	Company 50	−1.0527	1.1275
−1.2077	0.2479	Company 51	−0.9661	1.3919
−1.2077	0.2479	Company 52	−0.9817	1.3443
−1.0938	0.5748	Company 53	1.0827	7.6492
−0.2573	2.9757	Company 54	−0.8723	1.6782
−0.5473	2.1435	Company 55	−0.3355	3.318
−0.7429	1.5821	Company 56	−0.8317	1.8023
−1.2077	0.2479	Company 57	−1.0579	1.1114
0.4285	4.9443	Company 58	1.6889	9.5006
−0.7674	1.5118	Company 59	0.971	7.3081
−0.7674	1.5118	Company 60	0.971	7.3081
−1.2508	0.124	Company 61	−1.194	0.6958
1.3083	7.4699	Company 62	−0.2948	3.4422
1.2335	7.2552	Company 63	0.7066	6.5005
0.0331	3.8093	Company 63	−0.1264	3.9563
0.026	3.7889	Company 65	1.8524	10
0.8698	6.2112	Company 66	−1.1921	0.7015
−0.3494	2.7114	Company 67	−1.4218	0

(continued)

Table 2.5 (continued)

Suppliers		Company	Customers	
Punctuation factorial	Normalized score		Punctuation factorial	Normalized score
−0.3494	2.7114	Company 68	−1.4218	0
0.2867	4.5374	Company 69	−1.0611	1.1017
−0.3493	2.7117	Company 70	0.1755	4.8786
−0.5134	2.2408	Company 71	1.163	7.8946
−0.8028	1.4102	Company 72	−0.204	3.7195
0.9819	6.533	Company 73	0.9795	7.3341
1.894	9.151	Company 74	0.1847	4.9066
1.7328	8.6884	Company 75	1.7841	9.7915
0.2517	4.437	Company 76	−0.8338	1.7959
1.2496	7.3012	Company 77	1.0106	7.4289
0.7488	5.8637	Company 78	−0.9991	1.291
−0.1508	3.2816	Company 79	−0.6332	2.4088
−1.2077	0.2479	Company 80	−1.1957	0.6906
0.0836	3.9544	Company 81	−0.3011	3.4229
0.1142	4.0423	Company 82	1.597	9.22
−0.5511	2.1326	Company 83	−0.6693	2.2985
−1.2077	0.2479	Company 84	−0.7278	2.1196
−0.775	1.4897	Company 85	−0.6674	2.3042
1.2195	7.2148	Company 86	0.4088	5.5909
−0.5369	2.1732	Company 87	−0.4763	2.8879
−1.1935	0.2886	Company 88	0.3653	5.4582
1.0187	6.6385	Company 89	1.1481	7.8489
−0.0811	3.4818	Company 90	0.491	5.8421
0.7617	5.9008	Company 91	0.411	5.5978
−1.294	0	Company 92	−1.0382	1.1718
−1.1071	0.5365	Company 93	−0.6341	2.4058

Source Own elaboration based on the collected data

These percentages can be used as estimators of the distribution of levels in the population from which the sample comes.

The level of integration with suppliers and the level of integration with customers are shown for each company, so once analyzed and according to the combination, the integration arc is decided, as shown in Table 2.9.

Table 2.10 shows the frequencies of the different integration arcs observed in the sample.

Table 2.6 Recodification to obtain the arcs of integration

	Customers			
Suppliers		L 10	M 20	H 30
	L 1	11 Inward-facing	21 Inward-facing	31 Customer-facing
	M 2	12 Inward-facing	22 Periphery-Facing	32 Customer-facing
	H 3	13 Supplier-facing	23 Supplier-facing	33 Outward-facing

Source Own elaboration based on the collected data

Table 2.7 Number of companies in each combination of levels for suppliers and customers

	Customers				Inward-facing arc
Suppliers		L	M	H	Periphery-facing arc
	L	8	10	6	Supplier-facing arc
	M	14	26	6	Customer-facing arc
	H	2	10	11	Outward-facing arc

Source Own elaboration based on the collected data

Table 2.8 Percentages of companies in each level of combination for suppliers and customers

	Customers				
Suppliers		H	L	M	Total
	H	11.8	2.2	10.8	24.7
	L	6.5	8.6	10.8	25.8
	M	6.5	15.1	28.0	49.5
	Total	24.7	25.8	49.5	100

Source Own elaboration based on the collected data

The statistical technique of discriminant analysis consists in the construction of one or more functions of the original variables, called discriminant functions, which allow deciding in which class each element should be, using the proximity (or similarity) of each element to the different existing classes or groups. The construction of the discriminant functions, linear combinations of the original variables, is performed looking for those that maximize the separation between groups, which is equivalent to minimizing variability within the classes. The obtained functions allow to discriminate in the best possible way between the belonging elements.

The discriminant analysis uses as data a set of quantitative explanatory variables, in this case all the indicators of competitive advantages and a classification factor in this case the integration arcs. Discriminant analysis constructs a set of functions

Table 2.9 Arcs of integration

Company	Supplier level	Customer level	Arc of integration
Company 1	M	L	Inward-facing
Company 2	H	M	Supplier-facing
Company 3	H	H	Outward-facing
Company 4	M	M	Periphery-facing
Company 5	M	M	Periphery-facing
Company 6	H	M	Supplier-facing
Company 7	M	L	Inward-facing
Company 8	M	M	Periphery-facing
Company 9	M	L	Inward-facing
Company 10	M	L	Inward-facing
Company 11	M	M	Periphery-facing
Company 12	M	M	Periphery-facing
Company 13	H	M	Supplier-facing
Company 14	M	L	Inward-facing
Company 15	M	M	Periphery-facing
Company 16	M	M	Periphery-facing
Company 17	M	H	Customer-facing
Company 18	L	M	Inward-facing
Company 19	M	L	Inward-facing
Company 20	M	M	Periphery-facing
Company 21	M	M	Periphery-facing
Company 22	L	H	Customer-facing
Company 23	M	M	Periphery-facing
Company 24	M	H	Customer-facing
Company 25	L	M	Inward-facing
Company 26	M	M	Periphery-facing
Company 27	M	M	Periphery-facing
Company 28	H	M	Supplier-facing
Company 29	H	H	Outward-facing
Company 30	H	H	Outward-facing
Company 31	M	M	Periphery-facing
Company 32	L	M	Inward-facing
Company 33	L	H	Customer-facing
Company 34	H	H	Outward-facing
Company 35	L	M	Inward-facing
Company 36	H	M	Supplier-facing

(continued)

Table 2.9 (continued)

Company	Supplier level	Customer level	Arc of integration
Company 37	L	L	Inward-facing
Company 38	L	M	Inward-facing
Company 39	M	L	Inward-facing
Company 40	H	H	Outward-facing
Company 41	M	M	Periphery-facing
Company 42	H	H	Outward-facing
Company 43	H	H	Outward-facing
Company 44	M	M	Periphery-facing
Company 45	M	L	Inward-facing
Company 46	M	M	Periphery-facing
Company 47	L	H	Customer-facing
Company 48	M	M	Periphery-facing
Company 49	L	L	Inward-facing
Company 50	M	L	Inward-facing
Company 51	L	L	Inward-facing
Company 52	L	L	Inward-facing
Company 53	L	H	Customer-facing
Company 54	M	L	Inward-facing
Company 55	M	M	Periphery-facing
Company 56	M	M	Periphery-facing
Company 57	L	L	Inward-facing
Company 58	M	H	Customer-facing
Company 59	L	H	Customer-facing
Company 60	L	H	Customer-facing
Company 61	L	L	Inward-facing
Company 62	H	M	Supplier-facing
Company 63	H	M	Supplier-facing
Company 63	M	M	Periphery-facing
Company 65	M	H	Customer-facing
Company 66	H	L	Supplier-facing
Company 67	M	L	Inward-facing
Company 68	M	L	Inward-facing
Company 69	M	L	Inward-facing
Company 70	M	M	Periphery-facing
Company 71	M	H	Customer-facing
Company 72	L	M	Inward-facing

(continued)

Table 2.9 (continued)

Company	Supplier level	Customer level	Arc of integration
Company 73	H	H	Outward-facing
Company 74	H	M	Supplier-facing
Company 75	H	H	Outward-facing
Company 76	M	L	Inward-facing
Company 77	H	H	Outward-facing
Company 78	H	L	Supplier-facing
Company 79	M	M	Periphery-facing
Company 80	L	L	Inward-facing
Company 81	M	M	Periphery-facing
Company 82	M	H	Customer-facing
Company 83	M	M	Periphery-facing
Company 84	L	M	Inward-facing
Company 85	L	M	Inward-facing
Company 86	H	M	Supplier-facing
Company 87	M	M	Periphery-facing
Company 88	L	M	Inward-facing
Company 89	H	H	Outward-facing
Company 90	M	M	Periphery-facing
Company 91	H	M	Supplier-facing
Company 92	L	L	Inward-facing
Company 93	L	M	Inward-facing

Source Own elaboration based on the collected data

Table 2.10 Number of cases for each arc of integration

Arc of integration	Number of Companies
Inward-facing arc	32
Periphery-facing arc	26
Supplier-facing arc	12
Customer-facing arc	12
Outward-facing arc	11

Source Own elaboration based on the collected data

called discriminant functions that are linear combinations of the explanatory variables that allow us to predict the class of membership.

Table 2.11 shows the class predicted by the rows in rows and the real class in columns so that the percentages of each original prediction/class combination appear in the body of this table. It is observed how most companies are correctly identified: 100% of the companies of Outward-Facing Arc, 75% of Customers-Facing Arc

Table 2.11 Classification matrix with discriminant analysis column percentages

	Outward-facing arc	Customer-facing arc	Suppliers-facing arc	Inward-facing arc	Periphery-facing arc
Outward-facing arc	100.00	8.33	0.00	3.12	7.69
Customer-facing arc	0.00	75.00	0.00	6.25	7.69
Suppliers-facing arc	0.00	8.33	75.00	6.25	0.00
Inward-facing arc	0.00	8.33	16.67	81.25	3.85
Periphery-facing arc	0.00	0.00	8.33	3.12	80.77
Global hit percentage 81.72043					

Source Own elaboration based on the collected data

and Suppliers-Facing Arc, 81.25% of Inward-Facing Arc and 80.77% of Periphery-Facing Arc, with a total degree of success of 81.72%.

This high degree of success indicates that competitive advantages are clearly related to the arc of integration, since they allow to reasonably predict the arc of integration that corresponds to each company.

A factorial analysis was carried out with a single factor to obtain an aggregate indicator (internal integration factor), showed in Table 2.12 formed by the scores of the common factor. The following have been used as indicators of internal integration: (1) knowledge and information exchange, (2) support from senior management, (3) multifunctional teams, (4) support of information systems, (5) planning of objectives and joint problem solving, (6) organizational structure, (7) alignment metrics and reward systems, (8) organizational culture, and (9) alignment between organizational strategy and functional objectives.

With this factorial analysis, the descriptions of the level of internal integration are shown for each company, so once analyzed and according to the score, the integration level is decided, as shown in Table 2.12.

The relationship between the arcs of integration and the internal integration aggregate is analyzed in Table 2.13 that shows the number of companies for each relationship. The external integration (rows) includes the orientation of the company toward clients or suppliers (arcs of integration), and the internal integration (columns) includes the elements of description of the level of internal integration.

Below is the same relationship from previous Table 2.13 but expressed in relative frequencies or percentages, which are presented in Table 2.14.

2.4 Conclusions

This research contributes to the study of supply chain integration by providing empirical evidence to explain the existence of different arcs or configurations of supply chain integration in the industry. The treatment of this methodology is presented in a more complete way which can help understand the statistical process for any company in any sector could obtain a diagnosis of their level of external and internal integration.

It considers also an achievement having included in this research two types of methodologies, since one is to measure the level of external integration and the other to describe the level of internal integration of the company, which complements and fulfills the level of integration, both external and internal.

Also as relatively the supply chain integration is a current topic, and it is really not so deep, this research accomplished concepts and statements that in the literature were not very well defined.

The contribution of this research work could help the academic world as it provides guidance on the integration of the supply chain, and the business environment because it allows a diagnosis or analysis of the performance of their supply chains, that is, this study has, considerable application for professionals by providing guidance on the

Table 2.12 Description of the level of internal integration

Company	Factorial score	Integration level
Company 1	0.0290	Integrated
Company 2	0.2054	Integrated
Company 3	0.5666	Integrated
Company 4	−0.3534	Integrated
Company 5	−0.9310	Poor integrated
Company 6	−0.0441	Integrated
Company 7	0.0192	Integrated
Company 8	0.3372	Integrated
Company 9	−0.2798	Integrated
Company 10	1.0293	Very integrated
Company 11	−0.2526	Integrated
Company 12	−0.3478	Integrated
Company 13	−0.4081	Integrated
Company 14	−0.2264	Integrated
Company 15	−1.2880	Poor integrated
Company 16	−0.5933	Poor integrated
Company 17	0.7782	Integrated
Company 18	−4.1763	Poor integrated
Company 19	−2.0305	Poor integrated
Company 20	−0.7945	Poor integrated
Company 21	−0.4247	Poor integrated
Company 22	−0.0581	Integrated
Company 23	−0.8051	Poor integrated
Company 24	0.1329	Integrated
Company 25	−1.2282	Poor integrated
Company 26	−0.0964	Integrated
Company 27	−0.1711	Integrated
Company 28	−0.6431	Poor integrated
Company 29	−0.3793	Integrated
Company 30	1.0649	Very integrated
Company 31	−1.4657	Poor integrated
Company 32	−0.1068	Integrated
Company 33	0.8574	Very integrated
Company 34	0.3601	Integrated
Company 35	0.9072	Very integrated
Company 36	1.1198	Very integrated
Company 37	−0.8053	Poor integrated

(continued)

Table 2.12 (continued)

Company	Factorial score	Integration level
Company 38	−0.0753	Integrated
Company 39	0.5889	Integrated
Company 40	0.8564	Very integrated
Company 41	−1.1050	Poor integrated
Company 42	1.1394	Very integrated
Company 43	1.1748	Very integrated
Company 44	1.1748	Very integrated
Company 45	0.5370	Integrated
Company 46	0.4315	Integrated
Company 47	0.0768	Integrated
Company 48	0.3284	Integrated
Company 49	0.8167	Very integrated
Company 50	0.8279	Very integrated
Company 51	−0.1879	Integrated
Company 52	0.7306	Integrated
Company 53	1.0898	Very integrated
Company 54	0.6210	Integrated
Company 55	1.0709	Very integrated
Company 56	0.9797	Very integrated
Company 57	0.5860	Integrated
Company 58	0.8201	Very integrated
Company 59	0.3070	Integrated
Company 60	0.3070	Integrated
Company 61	0.0953	Integrated
Company 62	−0.5484	Poor integrated
Company 63	0.4320	Integrated
Company 63	−0.2290	Integrated
Company 65	1.0977	Very integrated
Company 66	−1.8067	Poor integrated
Company 67	−1.5976	Poor integrated
Company 68	−1.5976	Poor integrated
Company 69	0.2719	Integrated
Company 70	−0.9448	Poor integrated
Company 71	0.8959	Very integrated
Company 72	0.8587	Very integrated
Company 73	−0.8347	Poor integrated
Company 74	−0.3392	Integrated

(continued)

Table 2.12 (continued)

Company	Factorial score	Integration level
Company 75	1.1571	Very integrated
Company 76	−0.9004	Poor integrated
Company 77	0.1746	Integrated
Company 78	−0.7112	Poor integrated
Company 79	−1.9294	Poor integrated
Company 80	−2.0748	Poor integrated
Company 81	0.1464	Integrated
Company 82	0.7001	Integrated
Company 83	0.8985	Very integrated
Company 84	0.5786	Integrated
Company 85	0.8349	Very integrated
Company 86	1.1394	Very integrated
Company 87	−1.8505	Poor integrated
Company 88	0.8127	Integrated
Company 89	0.6093	Integrated
Company 90	0.1923	Integrated
Company 91	0.4230	Integrated
Company 92	1.0063	Very integrated
Company 93	0.4454	Integrated

Source Own elaboration based on the collected data

Table 2.13 Relationship between integration arcs and the level of internal integration (absolute frequencies)

	Poor integrated	Integrated	Very integrated
Outward-facing arc	1	5	5
Customer-facing arc	0	7	5
Suppliers-facing arc	4	6	2
Inward-facing arc	8	17	7
Periphery-facing arc	11	11	4

Source Own elaboration based on the collected data

conditions that must exist for a company to implement supply chain management with its suppliers and customers to include all typical business functions in supply chain management; planning, organization, and processes. Without such cross-functional coordination, supply chains cannot reach their full potential. Businesses should stay focused on the final goals of supply chain management: lower costs, higher value and customer satisfaction, and finally obtain competitive advantages.

Table 2.14 Relationship between integration arcs and the level of internal integration (relative frequencies)

	Poor integrated	Integrated	Very integrated
Outward-facing arc	10.9	21.7	4.2
Customer-facing arc	15.2	21.7	0.0
Suppliers-facing arc	13.0	8.7	16.7
Inward-facing arc	37.0	30.4	33.3
Periphery-facing arc	23.9	17.4	45.8
Total	100.0	99.9	100

Source Own elaboration based on the collected data

References

Afshan N (2013) The performance outcomes of dimensions of supply chain integration: a conceptual framework. Verslas: Teorija ir Praktika Business: Theor Pract 14(4):323–331. https://doi.org/10.3846/btp.2013.34

Alfalla-Luque R, Medina-López C, Kumar Dey P (2012) Supply chain integration framework using literature review. Prod Plann Control 1–18. https://doi.org/10.1080/09537287.2012.666870

Arantes FP, Leite MS, Maciel J. L. (2014). SCMI: a study to improve the integration of the SCs. In: 2014 Proceeding of the industrial and systems engineering research conference, Montreal

Arantes FP, Leite MS, Bornia AC, Barbetta PA (2018) Multidimensionality evaluation of supply chain management integration. Independent J Manag Prod 9(1):170–193. https://doi.org/10.14807/ijmp.v9i1.664

Baofeng H, Yuxiao Y, Xiande Z, Yongyi S (2016) The impact of human capital on supply chain integration and competitive performance. Int J Prod E 178:132–143. https://doi.org/10.1016/j.ijpe.2016.05.009

Chang W, Ellinger AE, Kim KK, Franke GR (2015) Supply chain integration and firm financial performance: a meta-analysis of positional advantage mediation and moderating factors. Eur Manag J 1-14. https://doi.org/10.1016/j.emj.2015.11.008

Christopher M (2011) Logistics and supply chain management. Prentice Hall, Great Britain

de Abreu A, Chicarelli Alcântara RL (2017) Internal integration in supply chain integration: a systematic literature review. Reuna 22(4):40–64. https://doi.org/10.21714/2179-8834/2017v22n4p40-64

De la Calle Vicente A (2015) La Integración de la Cadena de Suministro como Herramienta Competitiva: El Caso de la Industria Manufacturera del País Vasco. Universidad de Deusto, Bilbao, España

Devaraj S, Krajewski L, Wei JC (2007) Impact of eBusiness technologies on operational performance: the role of production information integration in the supply chain. J Oper Manag 25:1199–1216. https://doi.org/10.1016/j.jom.2007.01.002

Flynn BB, Huo B, Zhao X (2010) the impact of supply chain integration on performance: a contingency and configuration approach. J Oper Manag 28:58–71. https://doi.org/10.1016/j.jom.2009.06.001

Frohlich MT, Westbrook R (2001) Arcs of integration: an international study of supply chain strategies. J Oper Manag 19:185–200

Huo B, Ye Y, Zhao X, Shou Y (2016) The impact of human capital on supply chain integration and competitive performance. Int J Prod Econ 178:132–143. https://doi.org/10.1016/j.ijpe.2016.05.009

Ireland DR, Webb JW (2007) Strategic entrepreneurship: creating competitive advantage through streams of innovation. Bus Horiz 50(1):49–59. https://doi.org/10.1016/j.bushor.2006.06.002

Kim D, Lee RP (2010) Systems collaboration and strategic collaboration: their impacts on supply chain responsiveness and market performance. Decis Sci J 41(4):955–981. https://doi.org/10.1111/j.1540-5915.2010.00289.x

Kotcharin S, Eldridge S, Freeman J (2012) Investigating the relationships between internal integration and external integration and their impact on combinative competitive capabilities. In: Proceedings of the 17th international working seminar on production economics, pp 1–12

Koufteros XA, Rawski GE, Rupak R (2010) Organizational integration for product development: the effects on glitches, on-time execution of engineering change orders, and market sucess. Decis Sci 41(1):49–80. https://doi.org/10.1111/j.1540-5915.2009.00259.x

Koufteros X, Vonderembse M, Jayaram J (2005) Internal and external integration for product development: the contingency effects of uncertainty, equivocality, and platform strategy. Decis Sci 36(1):97–133

Krasnikov A, Jayachandran S (2008) The relative impact of marketing, research-and-development, and operations capabilities on firm performance. J Mark 72(4):1–11. https://doi.org/10.1509/jmkg.72.4.1

Kumar V, Nwakama Chibuzo E, Garza Reyes J, Kumari A, Rocha Lona L, López Torres GC (2017) The impact of supply chain integration on performance: evidence from the UK food sector. Procedia Manuf 11:814–821. https://doi.org/10.1016/j.promfg.2017.07.183

Leuschner R, Rogers DS, Charvet FF (2013) A meta-analysis of supply chain integration and firm performance. J Supply Chain Manag 49(2):34–57

Li S, Ragu-Nathan B, Ragu-Nathan T, Subba Rao S (2006) The impact of supply chain management practices on competitive advantage and organizational performance. Int J Manag Sci 34:107–124. https://doi.org/10.1016/j.omega.2004.08.002

Lii P, Kuo F-I (2016) Innovation-oriented supply chain integration for combined competitiveness and firm performance. Int J Prod Econ 174:142–155. https://doi.org/10.1016/j.ijpe.2016.01.018

Lin H, Lin Q (2018) Research on the impact of supply chain integration of startups: service supply chain perspective. Open J Soc Sci 6:255–274. https://doi.org/10.4236/jss.2018.64022

Lotfi Z, Sahran S, Mukhtar M, Zadeh A (2013) The relationships between supply chain integration and product quality. Procedia Technol 11:471–478. https://doi.org/10.1016/j.protcy.2013.12.217

Mahmood HS, Azizi S, Sheikhi N (2012) An investigation on the effect of supply chain integration on competitive capability: an empirical analysis of Iranian food industry. Int J Bus Manag 7(5):73–90. https://doi.org/10.5539/ijbm.v7n5p73

Mentzer JT, DeWitt W, Keebler JS, Min S, Nix NW, Smith CD, Zacharia ZG (2001) Defining supply chain management. J Bus Logist 22(2):1–25. https://doi.org/10.1002/j.2158-1592.2001.tb00001.x

Narasimhan R, Kim SW (2002) Effect of supply chain integration on the relationship between diversification and performance: evidence from Japanese and Korean firms. J Oper Manag 20:303–323. https://doi.org/10.1016/S0272-6963(02)00008-6

Otchere AF, Annan J, Anin EK (2013a) Achieving competitive advantage through supply chain integration in the cocoa industry: a case study of Olam Ghana Limited and produce buying company limited. Int J Bus Soc Res 3(2):131–145

Otchere AF, Annan J, Quansah E (2013b) Assessing the challenges and implementation of supply chain integration in the cocoa industry: a factor of cocoa farmers in Ashanti region of Ghana. Int J Bus Soc Sci 4(5):112–123

Ramesh V, Kumar YV, Sindhu S (2014) A conceptual framework of supply chain integration for competitive advantage. Ushus J Bus Manag 13(4):63–81. https://doi.org/10.12725/ujbm.29.5

Rosenzweig ED, Roth AV, Dean JW Jr (2003) The influence of an integration strategy on competitive capabilities and business performance: an exploratory study of consumer products manufacturers. J Oper Manag 21:437–456. https://doi.org/10.1016/S0272-6963(03)00037-8

Saleh H (2015) The impact of supply chain integration on operational performance at Jordanian pharmaceutical manufacturing organizations. Middle East University, Amman, Jordan

Schoenherr T, Swink M (2012) Revisiting the arcs of integration: cross-validations and extensions. J Oper Manag 30:99–115. https://doi.org/10.1016/j.jom.2011.09.001

Tolu Feyissa T, Kumar Sharma R, Kuei-Kuei L (2018) The impact of the core company's strategy on the dimensions of supply chain integration. Int J Logistics Manag. https://doi.org/10.1108/IJLM-03-2017-0080

Van Nguyen P, Thanh Nguyen L (2017) Performance measurement of supply chain integration in manufacturing firms' of southeast Vietnam. Eur J Econ Manag 6(2):23–32

Vickery SK, Jayaram J, Droge C, Calantone R (2003) The effects of an integrative supply chain strategy on customer service and financial performance: an analysis of direct versus indirect relationships. J Oper Manag 21:523–539. https://doi.org/10.1016/j.jom.2003.02.002

Wong CY, Boon-itt S, Wong CW (2011) The contingency effects of environmental uncertainty on the relationship between supply chain integration and operational performance. J Oper Manag 29:604–615. https://doi.org/10.1016/j.jom.2011.01.003

Zailani S, Rajagopal P (2005) Supply chain integration and performance: US versus East Asian companies. Supply Chain Manag Int J 10(5):379–393. https://doi.org/10.1108/13598540510624205

Zhao X, Huo B, Selen W, Yeung JH (2011) The impact of internal integration and relationship commitment on external integration. J Oper Manag 29:17–32. https://doi.org/10.1016/j.jom.2010.04.004

Zhao G, Feng T, Wang D (2015) Is more supply chain integration always beneficial to financial performance? Ind Mark Manage 45:162–172. https://doi.org/10.1016/j.indmarman.2015.02.015

Chapter 3
A Proposal to Redesign the Distribution Networks of Steel Manufacturing and Distribution Companies

Alexandra Ferrer, Yndira Guevara, Yereth Romero, and Mario Chong

Abstract Steel product manufacturing and distribution companies use diverse optimization tools to improve its operations. This study seeks to compare the current model (AS-IS) of a company, in order to design a new distribution network and location of its current supply centers, allowing economic and time improvements. Using this methodology, companies will optimize the customer's service level without neglecting its quality proposal; the distribution centers will enhance their capacity as well as the use of their resources, and the decision makers will simulate their models using operations research to define optimal policies. Moreover, we will present a proposed model (TO-BE) to the company. The comparison between the AS-IS and TO-BE models indicates a reduction in the delivery times and costs, currently being the main problems of the company. The contribution of this work lies in the optimization process of the current model and its proposals.

Keywords Optimization · Manufacturing companies · Distribution network · Operations research

3.1 Introduction

Globally, there is trust in the steel sector infrastructure investments (Felipe and Carvajal 2011), despite the latent problem, the global economic instability, and the Asian manufacturing industry recession (World Bank 2019). Concerning Latin America, its growth will come from the improvement of the global market, the raw materials prices, and the affordable exchange in recent years as well as the public and private investment in this sector (Madias and Madías 2014). Even though the expected

A. Ferrer
Pacífico Business School, Universidad del Pacífico, Jr. Gral. Luis Sánchez Cerro 2141, Jesús María, Lima, Perú

Y. Guevara · Y. Romero · M. Chong (✉)
Facultad de Ingeniería, Universidad del Pacífico, Jr. Gral. Luis Sánchez Cerro 2141, Jesús María, Lima, Perú
e-mail: m.chong@up.edu.pe

J. L. Garcia Alcaraz et al. (eds.), *Trends in Industrial Engineering Applications to Manufacturing Process*, https://doi.org/10.1007/978-3-030-71579-3_3

economic growth, in the main Latin American countries, the political environment could be unfavorable.

China maintains its growth tendency, not at the same level as previous years, but superior to other countries (Guide 2017). Many countries find China´s competition very complicated due to its dumping prices and its manufacturing overcapacity. China exports products surplus to the world, reducing the steel prices and forcing companies to be competitive (Le and Tong 2009); if China increases its production, it decreases worldwide steel production (Le and Tong 2009; Felipe and Carvajal 2011).

Peru has had two private steel manufacturing and distribution companies for decades. In the face of global dependency, these companies made an improvement in increasing their efficiency and maintaining its competitive advantage with the optimization of the working capital, by improving their production processes, reducing the operating costs and expenses as well as strengthening their functional plans of sales and operations.

In this work, we will focus on one of these companies, which was founded in 1966. This company began with a factory that produced rolling products, angles, plates, and profiles. In 1982, they inaugurated the second factory of steel billet. In 1996, they started the direct reduction of steels and produced fine steels in the third factory. Finally, in 2013, they built the second rolling products factory.

Despite the investments and the processes improvement, the company has two main problems. The first is the non-competitive delivery times; the competitor strategy is based on distribution centers close to its customers and shorter delivery time. The second problem is the non-competitive prices, consequence of their production cost and unnecessary expenses.

Our proposal has three phases. The first phase is the AS-IS model evaluation of the company. The current distribution network has production plants, distribution centers, and local customers. The second phase is the simulation of the proposals, which uses operations research tools to eliminate or reduce the problems and have better efficiency in delivery time and costs. The third phase is the formulation of the proposal, which is the TO-BE model.

This chapter is organized as follows. We revised the literature review in Sect. 3.2. We defined the methodology using a deterministic model in Sect. 3.3. We presented the results in Sect. 3.4 and finally the conclusion and final remarks.

3.2 Literature Review

A supply chain can be defined as the set of all the activities related to the movement of goods, from the state of raw material to the end user state (Mintzberg et al. 1997); this includes all decisions related to generating value based on the competitive advantage in the cross-functional logistics drivers: facilities, inventory, information, sourcing, pricing, and transportation (Chopra and Meindl 2020) as well as the logistics components: supply resources for the production and distribution of goods; which represents 40% of the total logistics cost (Frazelle 2002).

The supply chain strategy must be focused on responding adequately to high distribution costs, the variability of inventory levels (Stevens 1989), the lack of data and information (Chopra and Meindl 2020), the behavior of the push and pull demand, the constraints of allocations, among others. Sitek and Wikarek (2012) presented the importance of cost optimization in a supply chain, which is focused on a problem in a multimodal logistics provider, with a linear programming including discrete and continuous variables as costs, volume, capacity, and mode of transportation for the development of the model.

3.2.1 Operations Research as a Tool for Improvement

The operations research is one of the approaches for decision makers to model and find an optimal solution of the systems (Ziegel and Winston 1988; Hillier and Lieberman 2020), under different constraints, such as, cost, time, transportation, allocation, and others (Ziegel and Winston 1988; Hillier and Lieberman 2020). The phases of the operations research are the following: present the problem, observe the system, formulate the model, verify the model, predict the outcomes, select an appropriate option, analyze the results, and finally implement and evaluate the recommendations (Ziegel and Winston 1988).

The optimal solutions to the problems can be found using linear programming or nonlinear programming, with deterministic or stochastics models (Hillier and Liebermann 2020) including mathematical and computational fundamentals (Anderson and Sweeney 2016). According to Ng (2020), linear programming is the recommended approach to start solving problems, based on the Simplex algorithm, developed by Dantzig and Wolfe (1960) in 1947 (Hillier and Lieberman 2020) and supported by multiple optimization cases such as Camm et al. (1997), which integrates distribution and product problems with a linear programming model in Procter and Gamble; Kumanan et al. (2007), which developed techniques to minimize production and total distribution cost in a supply chain network (Ziegel and Winston 1988). Another algorithm is the Hungarian; it converts the original cost table of the problem into a series of equivalent tables until reaching the optimal one (Hillier and Lieberman 2020).

3.2.2 Machine Learning as a Tool for the Management of Data and Information

Machine learning is a tool based on different algorithms allowing the process of large amounts of data using computers (Gollapudi 2016). There are several definitions for machine learning; according to Samuel (2000), it allows computer systems to learn; Mitchell (2006) defines it as a computer system that learns from experience with respect to the task and the measure of performance (Ng 2020). Some software allows the implementation of learning algorithms such as Octave, Matlab, Phyton, and R (Ng 2020). The machine learning algorithm types can be divided into two categories: supervised and unsupervised learning.

The supervised learning refers to a prediction technique based upon training sets of data (Ng 2020), which includes linear regression, logistic regression, and others, that allow to recover a variable in function of independent attributes in the testing sets of data (Palacios-Cruz et al. 2013). For example, when the prediction of the dependent variable has a relation with a set of independent linear variables in the interval or ratio (Molinero 2001), it is known as a multiple linear regression (Grégoire 2015; Kolter 2008).

Other supervised learning algorithms are neural networks (Çoban 2016) which are used to imitate the activities of the brain and machine learning problems with a complex nonlinear hypothesis (Ng 2020). They function as a vital data mining tool (Yang and Yang 2014) for the classification and grouping of data, specifically for learning (Cilimkovic 2010). Mathematically, an artificial neural network is defined as a function that maps the input values and its parameters (Ng 2020). This function should be minimized with the back-propagation method (Cilimkovic 2010), which evaluates the neural network outputs against the expected output (Buscema 1998). The third algorithm is the support vector machine, which is considered as the most powerful supervised learning algorithm (Ng 2020). This is defined as a mathematical entity which maximizes a function with a certain dataset (Noble 2006), using binary, multi-nominal, or continuous variables (Cortes and Vapnik 1995; Vapnik 2013). The support vector machine allows to generate results (Maroco et al. 2011) (Hastie 2003; Verplancke et al. 2008) and convert a nonlinear problem into a linear one in a large dimensional space (Ng 2020).

In the actual state of the literature, machine learning and deep learning techniques have found little adoption in operations research and business analytics (Kraus et al. 2020). Nevertheless, supervised machine learning and deep learning could contribute to business strategies and supply chain management very well (Fairley et al. 2018). For the case of hotel revenue management, machine learning allowed pricing to be optimal, and it proved itself as a cost-effective tool for the business core (Zhang et al. 2019). Wherever, or whenever, a decision, both total or partially, can be automated, machine learning is a good alternative (Basso et al. 2018; Memeti et al. 2018).

The unsupervised learning algorithms objective is to find a dataset structure that has no labels. One of the algorithms is clustering, which groups the data into coherent subsets, depending on the patterns of the data (Ng 2020) and grouping tasks in "K-Means" (Boehmke and Greenwell 2020). Its goal is to divide the points in dimensions into groups (Hartigan and Wong 1979). The most used algorithm for making dimensional reductions is the principal components analysis whose objective is to find a vector on which all the data is projected and to reduce the projected error (Ng 2020).

We have ample reason to conclude that according to the objectives of the project, which are the optimization of the delivery time and the reduction of the processing costs, we will use supervised algorithms to summarize the information based on the data results and the interrelationships between variables. In this work, the aim will still be the operation research, which will be conducted to optimize the whole production process.

3.3 Methodology and Methods

During decades, several approaches such as operations research, analytical models, and simulations have been used to formulate the optimal solution for manufacturing and distribution models (McKenney and Rosenbloom 1969; Karlsson 2016; García et al. 2019), where the objective function and constraints are presented using linear and nonlinear programming. However, recently, the machine learning approach has improved the modeling of the systems and the decision-making process.

This chapter proposes three phases: present the AS-IS model, simulate the proposals and formulate the TO-BE model. The original multilevel transportation and allocation problem is presented in Fig. 3.1. The objective is to align the regions' demand to the production system as much as possible and minimize the cost of production and distribution. The simulation of the different scenarios in a steel manufacturing and distribution company faces its main problems, such as non-competitive delivery times and non-competitive prices. There are strategic decisions (closing or opening distribution centers), tactical decisions (the flow of raw materials, products, information, and money), and operational decisions (adequate the decision to the customer reality), considering an acceptable degree of deviation or error, in order to guarantee a reliable result (Hillier and Lieberman 2020).

3.3.1 The Model Formulation

In this model formulation, we defined three initial components: sets, variables, and parameters. The sets are groups of related objects with one or more characteristics associated (Schrage 2009), and they describe the model. In this case, the sets are production source, distribution center, and point of distribution. The variables are elements of the model that change during simulation or time, and they represent the sets' characteristics and store the results of the simulation (Brailsford et al. 2014) (Grigoryev 2016). The parameters describe characteristics of the objects in a static manner, and they are constant during the simulation or time and change when we need to adjust the model (Andres et al. 2016). The variables and parameters of the model are grouped according to the sets (Chong et al. 2019).

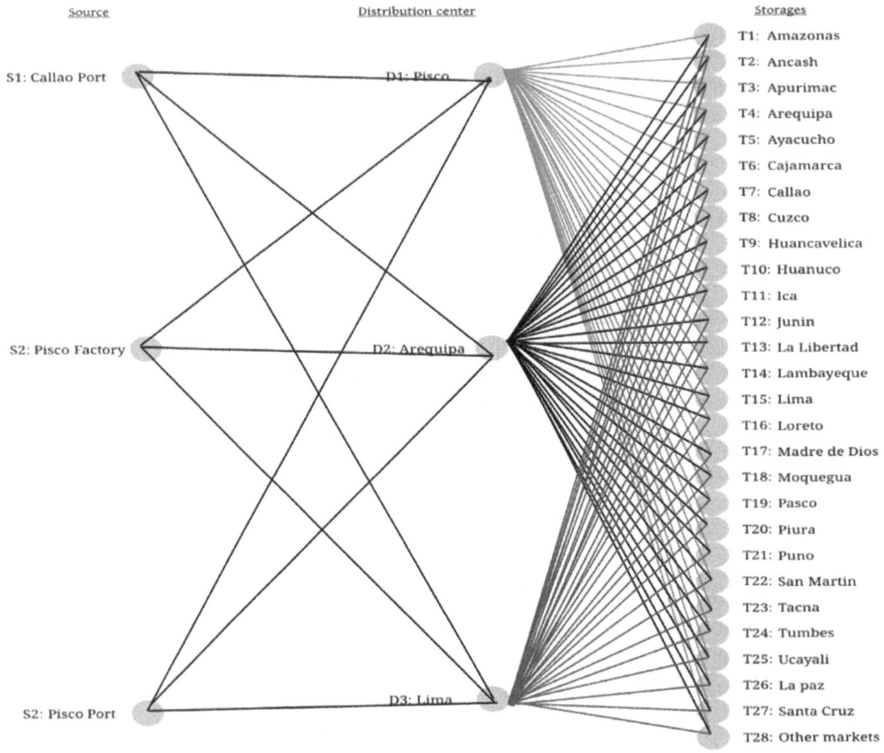

Fig. 3.1 AS-IS model. *Source* Own elaboration

3.3.2 Define the Sets, Variables, and Parameters

The production set (S) is composed of the products' sources. The variables are the quantity of steel product managed in tons (QS) and the control of operation in binary (BinQS). The parameters of the production set are the fixed cost of operation in US$ (cFijoS), the variable cost of sourcing in US$/ton (cVariableS), and the capacity in ton (capacidadS). The distribution center set (D) is composed of the products transferred from the sources to the point of distribution (P). The variables are the quantity of steel product managed in tons (QD) and the control of operation in binary (BinQD). The parameters of the distribution center set are the fixed cost of operation in US$ (cFijoD), the variable cost of distribution in US$/ton (cVariableD), and the capacity in ton(capacidadD). The point of distribution set (P) is composed of the products distributed to the customers. The parameter of the point of distribution center is the demand in tons (QP).

The relation of the production, distribution center, and point of distribution sets generates the secondary or derived sets. These sets are composed of the relation

Table 3.1 Sets of the model

Type of set	Set	Characteristic
Primary	S	Production source
Primary	D	Distribution center
Primary	P	Point of distribution
Secondary	SxD	Relation between production source and distribution center
Secondary	DxP	Relation between distribution center and point of distribution

Source Own elaboration

Table 3.2 Variables of the model

Variable	Set	Characteristic	Unit
QS	S	Amount of steel managed by S	Tons
BinQS	S	Binary of S	
QD	D	Amount of steel managed by D	Tons
BinQD	D	Binary of D	
QSD	SxD	Amount of steel transported from S to D	Tons
QDP	DxP	Amount of steel transported from D to P	Tons

Source Own elaboration

between two or more sets (Schrage 2009), and they bring more variables and parameters to represent the virtual reality (Chong et al. 2019) for the simulation model. The derived sets are SxD and DxP. SxD is comprised of the production source and the distribution center. The variables of SxD are the amount of steel transported from the production source to the distribution center in tons (QSD). The parameter is the transportation cost from the production source to the distribution center in US$/ton (cTransporteSD). The derived sets are DxP is comprised of the distribution center and the point of distribution. The variables of DxP are the amount of steel transported from the distribution center to the point of distribution in tons (QDP). The parameter is the transportation cost from the distribution center to the point of distribution in US$/ton (cTransporteDP). Respectively, Tables 3.1, 3.2, and 3.3 define the sets, the variables, and the parameters of the model.

3.3.3 Formulate the Model

Figure 3.2 shows the configuration of the main sets of the model:

The objective function in this case is to minimize the costs (Z): fixed cost, variable cost, and transportation cost. In Eq. (3.1), the objective function is shown

Table 3.3 Parameters of the model

Parameter	Set	Characteristic	Unit
cFijoS	S	Fixed cost of the production source	US$
cVariableS	S	Variable cost of the production source	US$/tons
capacidadS	S	Capacity of the production source	Tons
cFijoD	D	Fixed cost of the distribution center	US$
cVariableD	D	Variable cost of the distribution center	US$/tons
capacidadD	D	Capacity of the distribution center	Tons
qT	P	Demand of the point of distribution	Tons
minCapacidadSD	SxD	Minimum amount of products from the production source to the distribution center	Tons
maxCapacidadSD	SxD	Maximum amount of products from the production source to the distribution center	Tons
minCapacidadDP	DxP	Minimum amount of products from the distribution center to the point of distribution	Tons
maxCapacidadDP	DxP	Maximum amount of products from the distribution center to the point of distribution	Tons
cTransporteSD	SxD	Transportation cost from production source to distribution center	US$/tons
cTransporteDP	DxP	Transportation cost from distribution center to point of distribution	US$/tons

Source Own elaboration

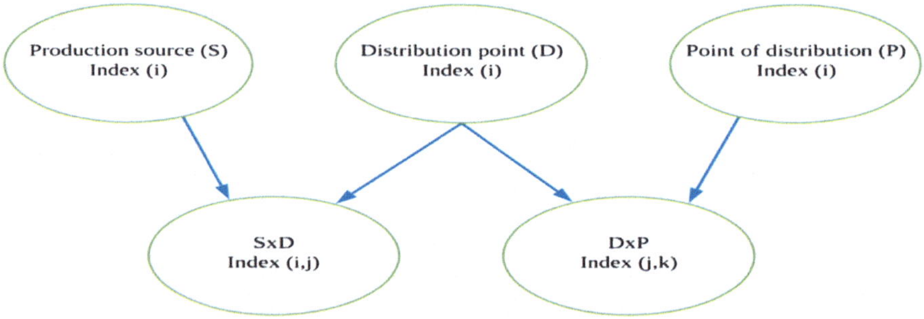

Fig. 3.2 Sets and index configuration. *Source* Own elaboration

$$\min Z = \sum_{i\epsilon S} BinQS(i) * CFijoS(i) + \sum_{j\epsilon D} BinQD(j) * CFijoD(j) + \sum_{i\epsilon S} QS(i) * CVariableS(i) + \sum_{j\epsilon D} QD(j) * CVariableoD(i) + \sum_{i\epsilon S, j\epsilon D} QSD(i,j) * CTransporteSD(i,j)$$

$$+\sum_{j\epsilon D,k\epsilon P} QPD(j,k)*CTransporteDP(j,k) \tag{3.1}$$

The model group of constraints are the following:

(a) The first group represents the amount of steel managed by the production source (QS) (Eq. 3.2).

$$QS(i)=\sum_{j\epsilon D} QSD(i,j)\quad \forall(i)\epsilon S \tag{3.2}$$

(b) The second group represents the amount of steel managed by the distribution center (QD) (Eqs. 3.3 and 3.4):

$$QD(j)=\sum_{i\epsilon S} QSD(i,j)\quad \forall(j)\epsilon D \tag{3.3}$$

$$QD(j)=\sum_{k\epsilon P} QDP(j,k)\quad \forall(j)\epsilon D \tag{3.4}$$

(c) The third group represents the amount of steel demanded by the point of distribution (QP) (Eq. 3.5):

$$QP(k)=\sum_{j\epsilon D} QDP(j,k)\quad \forall(k)\epsilon P \tag{3.5}$$

(d) The fourth group represents the binary and capacity restrictions (*QS*, *QD*, *BinQD*) (Eqs. 3.6, 3.7, 3.8 and 3.9):

$$QS(i) <= BinQS(i)*CapacidadS(i)\quad \forall(i)\epsilon P \tag{3.6}$$

$$QD(j) <= BinQD(j)*CapaciadD(j)\quad \forall(j)\epsilon D \tag{3.7}$$

$$\sum_{j\epsilon D} BinQD(j)\geq 3 \tag{3.8}$$

$$BinQD(j)=1\quad \forall(j=1)\epsilon D \tag{3.9}$$

(e) The fifth group represents the capacity from the production source to the distribution center (Eqs. 3.10 and 3.11):

$$QSD(i,j) <= maxCapacidadSD(i,j) \quad \forall(i)\epsilon P\forall(j)\epsilon D \tag{3.10}$$

$$QSD(i,j) >= minCapacidadSD(i,j) \quad \forall(i)\epsilon P\forall(j)\epsilon D \tag{3.11}$$

(f) The sixth group represents the capacity from the distribution center to the point of distribution (Eqs. 3.12 and 3.13):

$$QDP(j,k) <= maxCapacidadDP(j,k) \quad \forall(j)\epsilon D\forall(k)\epsilon P \tag{3.12}$$

$$QDP(j,k) >= minCapacidadDP(j,k) \quad \forall(j)\epsilon D\forall(k)\epsilon P \tag{3.13}$$

(g) The seventh group represents specifics constraints (Eqs. 3.14 and 3.15):

$$\sum_{i=1,3\epsilon S} QS(i) = 0.08 * \sum_{j\epsilon D} QD(j) \tag{3.14}$$

$$QS(i) <= 115{,}500 \quad \forall(s=2)\epsilon P \tag{3.15}$$

The first step of the procedure is a deterministic analysis (Chong et al. 2019), without uncertainties. The second step of the procedure is stochastics, and the level of deviation and variability of the demand was obtained using the historical data by machine learning tools with a supervised algorithm.

3.3.4 Results

In this section, the application of the proposed model is presented. According to the results of the optimization, the products must enter through the Port of Callao since it is cheaper than Pisco's port. Similarly, the distribution center in Lima and Arequipa should be closed, as there is a double freight incurred only in Lima, which is unnecessary. During the first three years, the company will work with the distribution centers of Pisco, Lambayeque, La Paz, and Santa Cruz. In the fourth year, the company could work with three distribution centers, while closing the distribution center of Santa Cruz, since the distribution center in La Paz would take in the demand from Santa Cruz.

Finally, the TO-BE model can be observed in Fig. 3.3, the annual savings in Table 3.4, and the improvement in lead times in days in Table 3.5.

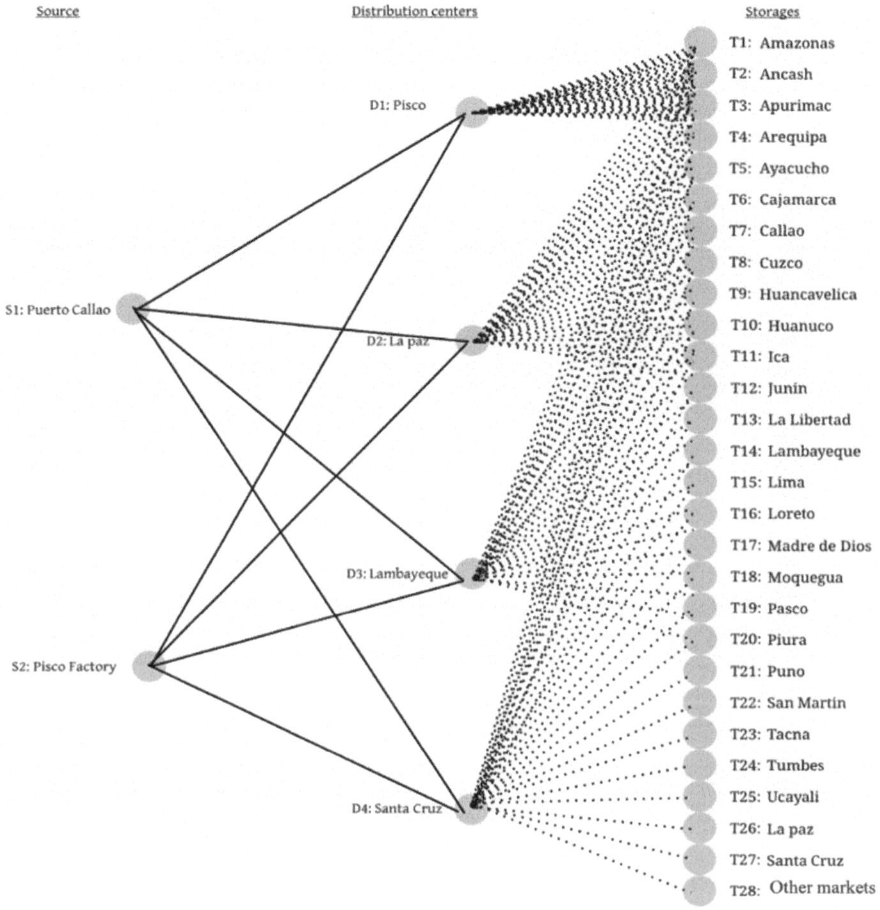

Fig. 3.3 TO-BE model. *Source* Own elaboration

An economic evaluation of the new centers opening was carried out, in which the NPV of the project would be $15,289,968.7, the 74% of IRR and the recovery period of 1.3 years. Table 3.6 presents the complete and final project that the company should implement in its business model if it wants to reduce delivery times and improve cost efficiency.

Table 3.4 Cost saving

Year	Situation	Fixed cost of the distribution centers (US$)	Variable cost (US$)	Variable cost of transport 1 (US$)	Variable cost of transport 2 (US$)	Annual savings (US$)
X	Proposal	9,047,607.76	486,267.16	7,023,994.93	22,612,358.81	
	Current	12,542,651.94	2,043,001.19	10,054,336.12	28,559,242.09	
	Cost saving (US$)	3,495,044.18	1,556,734.03	3,030,341.19	5,946,883.28	14,029,002.69
X + 1	Proposal	10,404,748.96	500,979.40	6,906,096.72	23,850,262.39	
	Current	14,424,049.85	2,085,940.90	10,304,122.69	29,377,165.37	
	Cost saving (US$)	4,019,300.90	1,584,961.49	3,398,025.97	5,526,902.99	14,529,191.64
X + 2	Proposal	11,965,461.19	528,701.49	7,388,066.87	26,232,646.27	
	Current	16,587,657.31	2,223,322.39	10,988,201.19	31,701,790.45	
	Cost saving (US$)	4,622,196.12	1,694,620.90	3,600,134.33	5,469,144.18	15,386,095.52
X + 3	Proposal	13,760,280.30	559,830.45	7,822,998.81	27,777,127.46	
	Current	19,075,805.67	2,354,222.99	11,635,148.66	33,568,206.57	
	Cost saving (US$)	5,315,525.37	1,794,392.84	3,812,149.85	5,791,079.10	16,713,147.16
X + 4	Proposal	15,824,322.39	591,151.94	8,260,840.90	29,331,264.78	
	Current	21,937,176.42	2,485,943.58	12,286,151.64	35,446,517.31	
	Cost saving ($)	6,112,854.03	1,894,791.64	4,025,310.75	6,115,252.54	18,148,209.25

Source Own elaboration

Table 3.5 Variation in days

Year	Proposal time (days)	Current time (days)	Variation (%)
X	244.24	249.83	−2.24
X + 1	253.39	257.66	−1.66
X + 2	270.6	275.56	−1.80
X + 3	286.54	291.78	−1.80
X + 4	302.56	308.12	−1.80

Source Own elaboration

Table 3.6 Project charter

Project charter		
Project name: Redesign the distribution network		
Project description		
The project aims to modify the actual distribution network thus producing a higher advantage among the services offered to clients, reducing the delivery time and establish a competitive production cost. The company will be in charge of the project's development: the project leader and eight team members; and		
the project's costs: installations (including opening and closure), distribution centers, freight, demand, material flow and frequency		
The project will take place for over a year. The project's management will be carried out in the facilities		
Project definition		
Opening of three distribution centers, including Bolivia; and the closure of the same in Lima and Arequipa		
Project requirements		
Distribution centers located in the north, center and south of Peru, including Bolivia		
Quality requirements: Achieve quality certifications in the new distribution centers		
Project objectives		
Concept	*Objectives*	*Success criteria*
1. Source	Elaborate: contracts and reports	Approve by the board
2.Time	Complete the project on time	Complete the project on time
3. Cost	Meet the project's budget	Do not exceed the project's budget
Project purpose		
Optimize the current distribution network. Delivery time and cost		
Project justification		
Qualitative justification		*Quantitative justification*
Reduce logistic costs		Outflows
Expand the demand		Income
Economic impact		NPV, IRR
Project manager designation		
Role	Project leader	*Authority*
Reports to	Sponsor	Require compliance
Supervises	Team member	
Schedule		
Milestone or significant event		*Scheduled date*
Project's kick-off		01.01.XX
General inventory cost: Lima and Arequipa		28.02.XX

(continued)

Table 3.6 (continued)

Transportation and equipment	15.12.XX
Sale of buildings and infrastructure	21.12.XX
Acquisition of distribution centers	30.04.XX
Steel distribution	15.06.XX
Operations implementation	15.10.XX
Inventory disposal	15.12.XX
Project milestone schedule	
Milestone or significant event	*Scheduled date*
Review of material and lead times for sale, replenishment, imports, and reverse logistics	15.08.XX
End of project	31.12.XX
Project main threats	
Commodities' price drop	
Conflicts with union workers	
Competition effective distribution and its influence in the demand	
Project main opportunities	
Increase in demand	
Reconstruction in the north of Peru and the economic growth of Bolivia	
Civil constructions	
Project budget	
Concept	*Amount $*
1. Infrastructure	$5,681,811.94
2. Materials	$2,600,431.04
3. Personal	$268,656.72
Total base line	$8,550,899.70
4. Contingency reserve	$855,089.85
5. Management reserve	$855,089.85
Total budget	$10,261,079.70
Project sponsor	

Source Own elaboration

3.4 Conclusions

This chapter develops the optimal allocation of the production sources, the distribution centers, and the point of distributions in a steel manufacturing and distribution company. As such, it developed three phases, the AS-IS model evaluation, the simulation of the proposal, and the formulation of the TO-BE model proposal.

The objective of the project is to reduce the delivery time and the processing costs, using two approaches: operations research and machine learning. The literature review contributes to the existing literature; since, not only an improvement in costs is sought, but also a reduction in delivery times is to maintain the quality of the company's service. This methodology increases the applicability of these approaches for the strategic, tactical, and operational levels in this manufacturing industry. In fact, the combination of these approaches in the modeling process has been useful and presents a real model to decision makers. This project presents a practical contribution, optimizing the entities and flows of this supply chain and focusing on chain sources and distribution centers through the proposed methodology. The results benefit the company under the external threat, which is the delivery time and the internal weakness, which is the production cost, facing the pressure of external competitors. The results presented are also limited by the data available in the formal and informal systems of the company. Nonetheless, the current results have an economic impact in the organization.

Despite the limitations in the study, the uncertainty of the steel price and the competition reaction, this work could contribute to a more efficient use of the resources. It is possible that all the investment would not have good economic results, especially during times when the price of commodities continues to fall and the competition increases their distribution effectiveness. Therefore, it is necessary to consider the correct execution of the project, the adequate risk management, and the team in charge of the project.

Finally, another beneficial suggestion to this study would be to identify and evaluate import channels in ports near the new distribution centers, in order to further reduce delivery times and costs.

References

Anderson D, Sweeney D (2016) Métodos cuantitativos para los negocios, 13th edn. Cengage Learning, México D. F.

Andres B, Sanchis R, Poler R (2016) Modelado y simulación de la cadena de suministro con AnyLogic®. Universitat Politècnica de València

Basso S, Ceselli A, Tettamanzi A (2018) Random sampling and machine learning to understand good decompositions. Ann Oper Res 284(2):501–526. https://doi.org/10.1007/s10479-018-3067-9

Boehmke B, Greenwell B (2020) Hands-on machine learning with R. https://doi.org/10.1201/9780367816377-20

Brailsford, S, Churilov L, Dangerfield B (2014) Discrete-event simulation and system dynamics for management decision making

Buscema M (1998) Back propagation neural networks. Subst Use Misuse 33(2):233–270. https://doi.org/10.3109/10826089809115863

Camm JD, Chorman TE, Dill FA, Evans JR, Sweeney DJ, Wegryn GW (1997) Blending OR/MS, judgment and GIS: restructuring P&G's supply chain. Interfaces 27(1):128–142. https://doi.org/10.1287/inte.27.1.128

Chong M, Lazo J, Pereda M, Machuca J (2019) Goal programming optimization model under uncertainty and the critical areas characterization in humanitarian logistics management. Emerald Group Publishing https://doi.org/10.1108/JHLSCM-04-2018-0027

Chopra S, Meindl P (2020) Supply chain management: strategy, planning, and operation, 7th edn. Pearson
Cilimkovic M (2010) Neural networks and back propagation algorithm. Institute of Technology Blanchardstown
Çoban E (2016) Neural networks and their applications. Istanbul Sehir University. https://doi.org/10.13140/RG.2.1.5176.0404
Cortes C, Vapnik V (1995) Support-vector networks. Mach Learn 20(3):273–297. https://doi.org/10.1023/A:1022627411411
Dantzig GB, Wolfe P (1960) Decomposition principle for linear programs. Oper Res 8(1):101–111. https://doi.org/10.1287/opre.8.1.101
Felipe L, Carvajal C (2011) Reflexión sobre la industria del acero en el mercado globalizado, vol 30, N°. 51 Primer Semestre 2011 51(30):165–182
Frazelle E (2002) Supply chain strategy: the logistic of supply chain management. McGraw-Hill Companies Inc., Estados Unidos
García J, Rivera L, González-Ramírez R, Leal G, Chong M (2019) Best practices in manufacturing processes: experiences from Latin America
Gollapudi S (2016) Practical machine learning, 6a edn. Packt Publishing Limited
Grégoire G (2015) Multiple linear regression. EAS Publ Ser 66:45–72. https://doi.org/10.1051/eas/1466005
Grigoryev I (2016) Anylogic 7 in three days: A quick course in simulation modeling
Guide P (2017) Trade and commodity finance solutions 2017
Hartigan A, Wong MA (1979) A K-means clustering algorithm. J Roy Stat Soc 28(1):100–108. https://doi.org/10.2307/2346830
Hastie J (2003) Observer data analysis and bycatch modeling status report. Northwest Fisheries Science Center
Hillier FS, Lieberman GJ (2020) Introduction to operations research, 11 edn. McGraw
Karlsson C (2016) Research methods for operations management 162–192:265–304
Kolter Z (2008) Linear algebra review and reference
Kraus M, Feuerriegel S, Oztekin A (2020) Deep learning in business analytics and operations research: models, applications and managerial implications. Eur J Oper Res 281(3):628–641. https://doi.org/10.1016/j.ejor.2019.09.018
Kumanan S, Venkatesan S, Kumar J (2007) Optimization of supply chain logistics network using random search techniques. Int J Logistics Syst Manag 3(2):252–266. https://doi.org/10.1504/IJLSM.2007.011824
Le T, Tong S (2009) China and anti-dumping: regulations, practices and responses. East Asian Institute, Singapore
Madias J, Madías P (2014) Productos de acero para la minería (Aug 2012), 42–49
Maroco J, Silva D, Rodrigues A, Guerreiro M, Santana I, de Mendonça A (2011) Data mining methods in the prediction of Dementia: A real-data comparison of the accuracy, sensitivity and specificity of linear discriminant analysis, logistic regression, neural networks, support vector machines, classification trees and random forests. BMC Res Notes 4.https://doi.org/10.1186/1756-0500-4-299
McKenney J, Rosenbloom R (1969) Cases in operations management. Wiley, New York
Memeti S, Pllana S, Binotto A, Kołodziej J, Brandic I (2018) Using meta-heuristics and machine learning for software optimization of parallel computing systems: a systematic literature review. Computing 101(8):893–936. https://doi.org/10.1007/s00607-018-0614-9
Mintzberg H, Quinn J, Voyer J (1997) El proceso estratégico: conceptos, contextos y casos. Pearson
Mitchell TM (2006) The discipline of machine learning. Carnegie Mellon University, School of Computer Science, Machine Learning Dept, Pittsburgh, Pa
Molinero L (2001) La regresión logística. Asociación De La Sociedad Española De Hipertensión 1(2):1–6
Ng A (2020) Machine Learning [Tipo de video]. Retraived from https://www.coursera.org/learn/machine-learning/home/welcome

Noble WS (2006) What is a support vector machine? Nat Biotechnol. https://doi.org/10.1038/nbt1206-1565
Palacios-Cruz L, Rivas-Ruiz R, Talavera J (2013) Modelo de regresión lineal. Rev Med Inst Mex Seguro Soc 51(6):656–661
Samuel AL (2000, Jan 01) Some studies in machine learning using the game of checkers. Ibm J Res Dev 44(1):206–227
Schrage L (2009) Optimization modeling with LINGO. Lindo Systems, Chicago, Illinois
Sitek P, Wikarek J (2012) Cost optimization of supply chain with multimodal transport. In: 2012 Federated conference on computer science and information systems, FedCSIS 2012, pp 1111–1118
Stevens G (1989) Integrating the supply chain. Int J Phys Distrib Logist Manag 46(1):19–42
Vapnik V (2013) The nature of statistical learning theory. Springer science and business media
Verplancke T, Van Looy S, Benoit D, Vansteelandt S, Depuydt P, De Turck F, Decruyenaere J (2008) Support vector machine versus logistic regression modeling for prediction of hospital mortality in critically ill patients with haematological malignancies. BMC Med Inf Deci Making 8. https://doi.org/10.1186/1472-6947-8-56
World Bank (2019) Opinion: China's value to the global economy. https://www.caixinglobal.com/2019-09-12/opinion-chinas-value-to-the-global-economy-101461912.html.
Yang ZR, Yang Z (2014) Artificial neural networks. Compr Biomed Phys 6(1):1–17. https://doi.org/10.1016/B978-0-444-53632-7.01101-1
Zhang Q, Qiu L, Wu H, Wang J, Luo H (2019, Nov) Deep learning based dynamic pricing model for hotel revenue management. In: 2019 international conference on data mining workshops (ICDMW). https://doi.org/10.1109/icdmw.2019.00061
Ziegel ER, Winston W (1988) Operations research: applications and algorithms. Technometrics 30(3):361. https://doi.org/10.2307/1270107

Chapter 4
Layout Planning Approach at a Plumbing Department in a Manufacturing Industry: A Case Study

Guadalupe Hernández-Escobedo, Cristian Omar González-Higuera, Karina Cecilia Arredondo-Soto, and Arturo Realyvásquez-Vargas

Abstract The chapter presents the use of the layout planning approach in the design and installation of the production line. This was carried out in a company dedicated to the manufacture of hydromassage tubs. Diverse problematic situations in an American plant served to make the decision to transfer that production line to a Mexican plant. As a case of study, the systematic layout planning methodology was employed to design the layout of that production line. Diverse phases gave the opportunity to include other methods and tools enriching its process. This iterative and multi-criteria approach permitted to include personnel converging from various departments exhibiting their operational needs. A definitive layout emerged after diverse drafts. It was consequently installed permitting pilot tests to make final adjustments. For 12 months, diverse data were collected to visualize its results, which were associated with diverse criteria. For instance, the layout was installed occupying less area than in previous plant. Moreover, there were reductions in the distance used in material handling, the number of personnel handling the materials and finished goods, and the costs relating to these activities. Similarly, there were an increase in the productivity index and an efficient use of available area. In addition, it allowed the implementation of lean manufacturing.

Keywords Layout planning · Plumbing department · Lean manufacturing · Systematic layout planning

G. Hernández-Escobedo (✉) · C. O. González-Higuera · A. Realyvásquez-Vargas
Departamento de Ingeniería Industrial, Tecnológico Nacional de México/I. T. de Tijuana, Calzada Tecnológico s/n, Fraccionamiento Tomás Aquino, 22414 Tijuana, México
e-mail: ghernan@tectijuana.edu.mx

K. C. Arredondo-Soto
Chemical Sciences and Engineering Department, Universidad Autónoma de Baja California, Calzada Universidad #14418, Parque Industrial Internacional, 22390 Tijuana, México

J. L. Garcia Alcaraz et al. (eds.), *Trends in Industrial Engineering Applications to Manufacturing Process*, https://doi.org/10.1007/978-3-030-71579-3_4

4.1 Introduction

The project was carried out at the company TP or WW (from here to onwards and for confidentiality reasons, its name and products have been changed). The company was founded in 1977 and has become one of the production leaders of hydro-massage tubs. Currently, it makes an average of 350 units per day. One of its production plants is located in the city of Tijuana, Mexico. Specifically, this plant put the basis to redesign its layout making possible the project. To put on perspective, the production line of tubs involves diverse activities and phases, which starts in the mentioned plant and the final phase and its activities are done in a production plant located in the United States of America (USA). In other words, this means that some processes were done in the Tijuana´s plant and the rest, in the USA´s plant. Within Tijuana's plant, there are different areas that contain activities and processes required in the production of the diverse types of tubs. Particularly, the research was done within the HL area or production line, used indistinctly from here to onwards, and involved the manufacturing engineering and plumbing departments of the plant. As a manner of detail, the area includes different types of plumbing assemblies, which are used in the final phase of the production process. It is important to remark that the company produces at least 50 different models, and the HL area was chosen because it assembles nine models representing the 30% of the total sales. Moreover, 25% of this production initiates in Mexico and is finalized in the USA. In addition, its personnel of both plants is recognized as excellent providers of after sales service to internal and external clients.

Consequently, some substantial changes in the area of the USA's plant have been done in order to satisfy both services, the after sales service and production demands. For instance, the installation of new equipment and uses of diverse materials within the production process have facilitated the satisfaction of both services. Substantial adaptations were also done to meet them. The methods and techniques were changed, and the personnel were trained, among other practices in order to improve the services. However, they do not fully satisfy new challenges consequently of increasing the production demands. So, the managers suggested to attend the production area but to leave outside distribution and managerial issues. In other words, internal aspects of the production line should be considered but external elements should not be left outside (Ning et al. 2010, 2011). For instance, it was discovered that, in general terms, some spaces required in the production line were not adequate for the activities and processes. Additionally, some of these activities and processes pertained to other models diminishing the resources used. Furthermore, the production line included some corridors, shelves, and other warehouses used in other production lines. In a few words, a mixture of activities, processes, and other production elements was uncovered stopping the increase of production rates demanded. In order to make feasible that increase, the use of a holistic viewpoint in this type of internal problematic situations was suggested (Ning et al. 2011). In this line, the managers uncovered diverse problematic situations as a follows:

- Unnecessary transports
- High transportation costs
- Incomplete use of the space
- Waste of materials
- Lack of techniques for designing layouts
- Absenteeism and lack of personnel.

Having this as a background, the managers made the decision to transfer the total production of the HL area, including the majority of the plumbing department personnel of the American plant, to Mexican plant. This decision was supported in statistical information, but it was not considered in this document by ethical reasons. In addition, the decision of transferring that area was another source, the existence of available areas to increase the size of the HL area. There were discovered four potential areas that could shelter it. Previously, the plant was expanded adding diverse spaces and areas to other production lines. Similarly, the relocation of the area could represent the opportunity to reduce its costs. In fact, the company had experience on transference of production lines with this objective in mind. For example, two production lines were transferred in the past from the USA to Mexico. Besides, a new product was developed and the entire production process was established in that plant. However, the managers uncovered the need of use a technique that helps them to design and install efficiently production layouts. It was clear that this type of projects arose from the need to systematize the process of design and installation of production lines using various approaches in the process (Tommelein and Zouein 1993). Specifically, varied tools of industrial engineering and lean manufacturing could be considered within this systematization and could provide differentiators in that process. Nevertheless, it was discovered that some Mexican governmental regulations banned the completely transference of the production process to Mexico. So, the managers made the decision to increase the size of the HL area. This means that some activities were realized in the Mexican plant, and the rest, in the American plant. It is important to remark that, from here to onwards, the term finished goods is related to the products that represent work-in-process products but for the Mexican plant are finalized. This uncovered the necessity of redesign the layout of both areas. Referring to the American plant, the project was made by personnel from that plant. Relating to the Mexican plant, the assignment was done by personnel from varied departments as explained in the next paragraphs and methodology section.

The diverse considerations mentioned before were taken into account in the definition of the objective of the layout planning. This was defined as the discovery of various elements required in the production process within the HL area using various industrial engineering and lean manufacturing tools in order to reduce the travel distances of parts and assemblies and the finished product. Moreover, it included the increase in the productivity of the production line, as a nested objective. In addition, it was expected that the available area should be used effectively in order to meet the production requirements. Although these objectives could be seen as ambitious, there are various tools that should provide what was necessary to achieve them (Xu and Li 2012). Also, it was expected to reduce the costs of distributing materials on the production line and to reduce the areas dedicated to the storage of work-in-process and finished products. This was from the viewpoint of taking advantage of existing resources in the best way as expected (Yang et al. 2000).

Based on these ideas, the managers empathized additional considerations. They were the reduction the costs in the transference, the redesign of the production areas, the reduction of the transportation costs within the production lines, including the reduction of the transportation distance, and finally, the efficient use of the space. These mandatory issues were put on practice in both plants. It was expected, in a similar way, further benefits in terms of increases of productivity, optimization of production resources and efficient use of equipment. However, some of these considerations can cause confusions in the personnel. So, they decided to revise literature relate to layout planning and diverse themes of modern techniques of manufacture. The result of this task is presented in the following section. Similarly, another consideration was the time in which the project should be finalized. The managers suggested 18 months initiating on June 2017. Having this time, the involved personnel added another consideration. It was the redesign of new tools and equipment that will be used in the transport of raw materials and work-in-process and finished products. They saw the opportunity to achieve diverse objectives at the same time in a holistic view point.

Nevertheless, the achievement of those objectives uncovered several limitations in the project, that were combined with the components of the production system (Jo and Gero 1998). For instance, the redesign of the layout and the installation of the production line should be initially approved by the Engineering Department. This was done by exploring various layouts until the final arrangement was found (Hegazy and Elbeltagi 1999). Once the approval was received, the layout design was presented to other departments to receive feedback and their acceptance. The departments were Maintenance, Production, Human Resources, Finance, among others. Another limitation was the validation of the production line. This served to verify the consistency of the routine activities in relation with the planned activities. Furthermore, this activity aided to engage the internal departments and external companies with the production process. They uncovered the relevance of their activities in that production process. In other words, the validation should serve to create strong relationships between internal and external elements within the production process. One expected consequence should be redesign and fabrication of the diverse equipment and tools that should be employed in the process. The pilot tests could be help on this matter. The recruitment and selection of personnel were other aspects that could be considered within the redesign of the production layout.

The next section presents a review of some topics that were considered into this project. It is followed by the employed methodology. Subsequently, the results are listed emphasizing the objectives achieved. After, the conclusions and recommendations are presented. Finally, the discussion section is exhibited.

4.2 Background

This part presents the main concepts and theoretical foundations creating the foundations of this project.

4.2.1 Layout Planning

The planning of facilities, distribution plant or layout, used interchangeably throughout this document, can be viewed in various ways. It could represent the satisfaction of the production needs of people or organizations that require diverse elements to produce services and goods. Similarly, it could exhibit the opportunity to reorganize the elements in the production process. Moreover, it could show the location of production elements within limited areas, among others (Naqvi et al. 2016). Consequently, in the current literature, there are various definitions of layout planning uncovering the diverse viewpoints in how it could be seen in strong relation with its context of application (Muther and Hales 2015). Consequently, layout planning is considered as the main reason for the plant arrangement in order to optimize the distribution of machines, resources, materials and auxiliary services, so that the value created by the production system is maximized (Martínez Carbajal 2004). Therefore, the transference of the production process from one place to another, it should be used to make changes that could be complicated to do in the initial plant. It creates an opportunity to make changes to correct detected areas of opportunity. One of its objectives is to make efficient the production system. As a manner of detail, it allows the opportunity to eliminate diverse structural aspects and conditions of the production systems (Martínez Carbajal 2004), such as:

- Materials: long distances in transport activities; excessive movements; damaged materials; congested areas for work-in-process products; lost materials; high amount of raw material for delivery; lack of materials, among others.
- Production: unnecessary movements of personnel; materials on the floor (work-in-process products); congestions in corridors; inadequate arrangements of the workplace, among others.
- Production context: accidents; personnel turnover, among others.
- General: disorganized production program; little interest from staff, among others.

Consequently, some of the benefits of having a good distribution plan are the increase of the production rates, the reduction of delays, the saving of areas dedicated to productive activities, the improvement of the materials handling, the reduction of the work-in-process inventories, among other benefits. Specifically, Muther and Rabada (1981) pointed out some advantages of a layout design and installation according to the production needs:

- Reduction of health risks and rise of safety for workers.
- Increased morale and job satisfaction.

- Increase in production rates.
- Reduction of production delays.
- Efficient use of available areas.
- Reduction of material handling.
- Decrement of the production time.
- Markdown of congestion and confusion.
- Among other advantages.

4.2.2 Lean Manufacturing

The term Lean Manufacturing (LM) was coined by a study group from the Massachusetts Institute of Technology (MIT). This group studied and analyzed the evolution of the manufacturing methods of companies in the automotive industry worldwide. They found differences between diverse companies of that industry in terms of how they made their practices more efficient (Reyes Aguilar 2002). The group highlighted the advantages of the manufacturing systems used in the Toyota company. They also uncovered that the company was the best manufacturer in the automotive industry and they had a remarkable leadership in their production practices (Allen 2018). So, this type of manufacturing was called Lean Manufacturing and it represented exceptional manufacturing methods used since 1960. Despite, these techniques exposed good practices, they were improved by Taiichi Onho and Shigeo Shingo (Allen 2018). Specifically, the improvements were done from the viewpoint of minimizing the use of resources in the entire production system and the delivery of goods to the clients. In other words, this means a holistic posture on reducing the costs in the entire system and satisfying the client needs. The reduction of the defects within the production process and the decrement of the delivery time of goods to clients were the principal aims. As a result, the diverse production systems of multiple sectors were aligned to these in order to improve their practices.

In the same way, these considerations have been mandatory in actual companies so that they implement techniques and methods to make possible the achievement of mentioned objectives. In this line, Pineda (2004) suggested that it is important to meet them at operational and tactical levels involving personnel from similar levels using various techniques and tools of LM. Eventually it is expected to obtain benefits, which could be: the reduction of production costs, inventory and lead time; the increase of quality; the reduction of waste, among others.

4.2.3 Kanban as a Visual Tool

The term *Kanban* is related to the administration of visual information within the diverse production areas and processes to reduce confusion on the material handling (Hernández Matías and Vizán Idoipe 2013). On the other hand, Reyes Aguilar (2002) pointed out that Japanese coined the term *Kanban* and has been translated as "signal card". Its use in the production systems suggested the change from normal practices to the visual practice in order to reduce confusion within the processes. The real significance of its intrinsic objective is to improve the practices reducing the errors on material handling. For these reasons, it is common the use of signs within the production processes. The signs take the form of charts, cards, colored lights, colored containers, level lines on walls, etc. The idea is to make possible the easy identification of those signs, which have strong relation with some activities required in the processes. For instance, a sign could indicate that some activities should be done in order to continue with the process. Eventually, the lack of materials in the process could be discovered in the signs employed so that personnel could bring the necessary materials, once they discovered this lack within the production areas. In addition, these activities should be done without the consultation of supervisors expecting the elimination of paperwork and reducing conflicts among personnel, inventories of work-in-process products, among other benefits.

In this line, the *Kanban* provided signals that should suggest the needs for material handling by collecting and transporting raw materials and finished products. Similarly, it is suggested that *Kanban* might avoid overproduction by exposing the necessity of occupying determined equipment to make certain quantity of finished goods. It implied the production of fixed quantities in strong relation with the sales of products. In addition, the use of the signs within the processes served to prevent the production of defective goods by controlling the transportation of quality products as well as to control the inventory in the production areas. For example, the *Kanban* boxes could be used among operations of manufacturing cells or manufacturing cells and processes. This is principally done to regulate differences in the production speeds in order to achieve a constant production flow. In general terms, the process begins with the customer's order; after, it continues with the preparation of the tooling and materials used to produce the goods; and subsequently, a card is generated by the finished product warehouse. This card is the start of the production pointing out the necessity to produce demanded goods. This process is iterative and intuitive (Gutiérrez Garza 2000) and its consideration on layout planning is crucial. It should add production elements so that the manufacturing efficiency should be improved.

4.2.4 Kaizen as Continuous Improvement

The term *Kaizen* is referred to the opportunity to make changes for the improvement of the practices in production. It is also an important part of Lean Manufacturing and represents how organizations employ diverse human aspects for continuous improvement. It focuses on the people and the standardization of production processes (Pineda 2004), in order to recognize people as an important asset of the companies by bringing out as the production element. They add value to the goods, on the one hand, and propose actions and activities to improve the processes, on the other. Consequently, common practices in the current companies are to continuously calibrate their operation involving personnel from all organizational levels employing the point and system approaches. The idea is to put attention to the personnel done their activities and operations so that they look for better practices and make possible them. It represents the organizational culture that reigns in the company. Consequently, there are benefits for the organization and its personnel. Some of them are the increase of the productivity, the control of the production, the reduction of the production cycles, the standardization of the quality criteria and production methods, among others. Hence, it is seen normal that personnel participate in the relevant decision made processes within the companies. The layout planning is one of them and embraces personnel from diverse areas of the organizations. It is common that the teams are created of personnel converging from the majority of areas of the companies. Expected results are layouts in which residues are minimized or eliminated. It is personnel should participate in the design, installation and operation of the production process so that wastes are reduced or eliminated in all phases of the design and installation of the layouts. The benefits could be observed in different periods of time. This could be possible because the majority of the personnel involved in the processes were acknowledged (Gutiérrez Garza 2000).

4.2.5 Systematic Layout Planning (SLP)

In order to carry out the layout planning, diverse aspects in its design were considered. Initially, a revision of the current literature was done and various methodologies were found (Monga and Khurana 2015). Each of them provided different approaches to design layouts. This means the diverse necessities of layout planning, which have strong relationships with the diverse contexts of utilization (Naik and Kallurkar 2016). Each company needs particular layouts that can give them technological advantages. For instance, companies from similar sectors usually have different layouts. They permitted to discover how their objectives might be achieved. In other words, the disposition of the diverse production elements might be a clear signal in how those elements are connected to achieve the organizational goals. However, the majority of the approaches converge in looking for the best layout to achieve organizational goals. So, it was found that the Systematic Layout Planning Plant (SLP) could be useful to achieve the aim of layout planning in the considered context.

Moreover, this methodology has multiple benefits in its application providing flexibility in the design and installation of layouts. In other words, it has practicality when it is applied and gives advantages over others. Moreover, it suggests a systemic procedure that includes multiple criteria, and the same time, it also delineates activities for layout design in new facilities and/or layout redesign in utilized facilities (Muther and Rabada 1981). The basis of the methodology is to put on perspective and emphasize that layouts are prepared to achieve organizational goals. Specifically, it incorporates as triggers the flow of materials, information, human needs, among other elements within the companies. Consequently, the layout planning process establishes diverse phases and techniques that should help on organizing the production elements. This could be seen in the disposition of the facilities to achieve the organizational goals. To do so, it allows the identification, valuation and visualization of all elements involved in the processes and their relationships (Muther and Rabada 1981). Furthermore, it provides details facilitating the adjustment of those elements to minimize the transportation of information, materials, and products in process as well as to foresee the use of the areas of the layouts, the risks and comfort within the production areas. In other words, the methodology helps on taking care of the relevant elements of the systems so that it permits the achievement of organizational goals (Muther and Hales 2015). Thus, it was found that this methodology could aid with the achievement of an efficient layout. Then, it was found that it is required to implement the methodology in certain phases as follows:

- Phase I: Localization. In this phase, the location of the plant that should be distributed is decided. Here, it was suggested to find a place that has a competitive geographic position satisfying those relevant factors, which were weighted as crucial to the organization. Moreover, in the case of a redistribution, the phase is aligned to use the existing area or assigned area with this objective in mind. For instance, it could be normal the redesign of layouts on existing production areas in order to open new production areas. The companies move their processes to other areas within their buildings in other type of situations. The efficient use of the areas is important to the organizations.
- Phase II: General Distribution Plan. This phase is distinguished by finding patterns of the transportation of information, materials and products in process so that it permits the creation of the basis of the physical location of the production elements. The quantification of the area required by each element, the relationships between those elements, and the particular configuration of them are indicated in this phase. These indications are in general terms and particular questions related to the disposition of the elements are not answered. Its result could be sketches o diagrams presenting the needs of the diverse areas in the organizations and the relationships between them.
- Phase III: Detailed Distribution Plan. This phase is characterized by the study and preparation in detail of a distribution plan. It is a continuation of the previous phase and it is suggested to develop the layout with personnel from the diverse areas of the organization. Specifically, it includes the analysis, definition and planning of the areas where the jobs, machinery or equipment should be installed or placed. The consideration of requirements and facilities to operate those areas are estimated to satisfy the discovered needs.
- Phase IV: Installation. The last phase is distinguished by the installation of the system elements into the designated areas. The physical movements and adjustments are made in order to collocate the elements. It included the installation of equipment, machines and facilities. The result of the phase is the materialization of the layout. Particular adjustments and details are made in order to achieve the solicited layout.

4.2.5.1 Process Mapping

This methodology allows the discovery of certain elements of the products adding value so that companies should guide and redefine their efforts to reinvent their products and production process in order to offer valuable goods (López Cuevas and Briceida Noemí 2013). It starts with the elaboration of the process map or Value Stream Mapping (VSM). This map allows the planning and identification of the inputs and outputs of the production elements to improve the design and operation of the process. Thus, the main objective is to establish coherent strategies to satisfy the needs of the internal and external clients. Similarly, the methodology allows to systematically measure the internal progress on accomplishing the client needs. Furthermore, it highlights the main obstacles and opportunities that might arise in the process considering all steps in the production process, delivery of goods and post-sales services (Cabrera Calva 2014). In other words, this methodology helps to understand in detail the need to improve the production process considering its elements and services provided to internal and external customers.

4.2.5.2 Spaghetti Diagram

This is a tool that helps on visualizing and defining different types of waste within the processes. Its emphasis is in the transportation activities of materials, products in process and finished goods (Cabrera Calva 2014). To do so, it is necessary to create a plant layout and draw the diverse transportations and trips that are done by diverse production elements in study. Each of these elements is drawn into the layout exhibiting the distances traveled within and/or between different areas. So each element used a distinctive color to facilitate the quantification of distances as well as to discover the mix of routes into the transportations. As this diagram is a graphical tool, the flow of the elements served to uncover the unproductive activities and put the grounds to redesign the process (Cabrera Calva 2014). Consequently, the name of the spaghetti diagram comes from the mix of the colored lines as a whole and exhibiting similar appearance as a plate of spaghetti. However, the color of the lines indicates the various routes that follow the diverse production elements exposing the risks and wastes into it. Consequently, the use of the diagram is primarily used to visualize the routes of materials, products in process and finished goods into the process; to detect diverse types of waste within the transportation activities, and to provide the basis for the optimization of transportation of production elements, the increase of the productivity and the reduction of waste (López Cuevas and Briceida Noemí 2013).

4.2.5.3 Flow Process Chart

Flow process charts are used to describe the transformation processes in production systems (Niebel and Freivalds 2004). Furthermore, they help to develop strategies to improve the effectiveness or efficiency of them since they allow the visualization of all required activities emphasizing those that aggregate value to products. Similarly, it permits to distinguish those activities that are unnecessary and constitute costs. For instance, transportations, storages, delays and inspections represent costs under this approach. Consequently, the analysis of processes should give an ample visualization of the effect of each these activities on the process. The chart offers a clear description of the activities required to make goods. Moreover, it graphically presents the entrance of materials, the use of machinery and equipment, the transformation processes, the order of the activities required in the manufacturing process and the relationships between the activities and the production elements (Niebel and Freivalds 2004).

4.3 Methodology

This part presents the techniques and tools used in the design and installation of the layout utilized in the production line. As mentioned in the last section, the SLP methodology was considered the basis to outline the activities that define the solicited layout in the production line. It is important to remark that its phases were reinforced with the utilization of other tools so that the design of the layout concentrates other tools from LM and Industrial Engineering. The administration suggested a holistic viewpoint in which these tools should be enrich the process of layout design. Other tools were exploited, but they were omitted. The company authorize their use in the project with the condition that were omitted in this report. These tools gave technological and competitive advantages. Some of these were qualitative and quantitative criteria granting a particular character to the project. This is in line the use of the SLP methodology, which can be utilized to make all kinds of plant layouts in deep relationship with the context of adoption (Muther and Hales 2015). It suggested the steps which were adapted to the observed needs. It is because some tools and designs were developed for the exclusive practices of this company. So, ethical and confidentiality aspects were considered.

4.3.1 Phase I: Localization

The start of project was the consideration of both HL areas in the American and Mexican plants. It was found that in the Mexican plant, the HL area can be closed and a new area will be destined to this production line. Four areas were offered and analyzed until one of them was chosen to install the production line. On the other hand, the HL area in the American plant served to create the foundations of that line. The occupied area of the line was 5061 square feet (sq. ft.). Figure 4.1 presents the area occupied by the HL production line, which contains diverse production elements as warehouses, corridors, workstations, etc. A brief description of these are disclosed

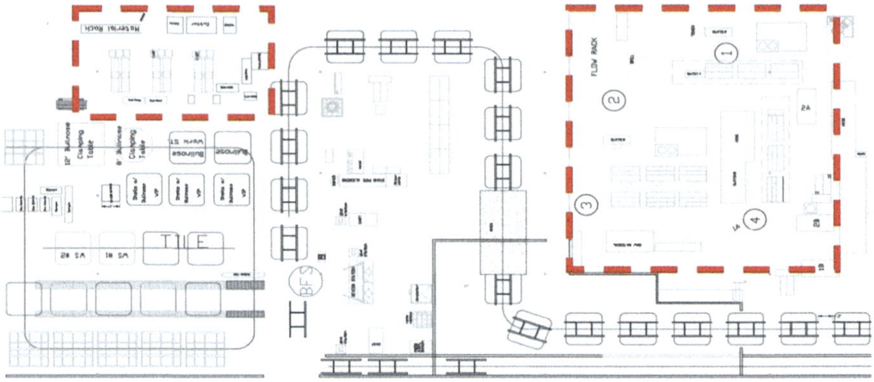

Fig. 4.1 Initial layout

after. Note: the figures used along this document are general representations of the developed technical drawings. The utilized drawings were a scale representation of all production elements involved in the production process and considered in this project.

The relevant data of the production elements in the HL area were:

- Number of workstations: 23; occupied area by each workstation: 32 sq. ft.; occupied area: 736 sq. ft.
- Number of temporary warehouses: 24; occupied area by each storage: 32 sq. ft.; occupied area: 768 sq. ft.
- Other elements as corridors, equipment, among others: 3557 sq. ft.
- Occupied area in the HL area: 5061 sq. ft.

Using these data as background, the offered areas in the Mexican plant were analyzed. In order to define a convenient area in the plan, diverse criteria were defined to help on the process of choosing it. In the process, diverse personnel converging from different departments were consulted. They listed four criteria that should be satisfied. The location of the production line, the distance from raw material warehouse to production areas, the appropriate area and number of corridors, and the distance from the production line to the shipping warehouse were the factors considered as relevant. They are explained below.

- **Location**: This criterion was associated with the use of the size of the area employed in the American plant, which was 5061 sq. ft., as an initial point of reference. The idea was to exploit less area in the new production line. As mentioned before, four areas were offered to house that line. Figure 4.2 presents the location of offered areas in the plant and the Table 4.1 exhibits the size of each area and the compliance of the criterion.

- **Distance to warehouse**: This criterion was associated with the distance between the production line and the raw materials warehouse. It was because the process

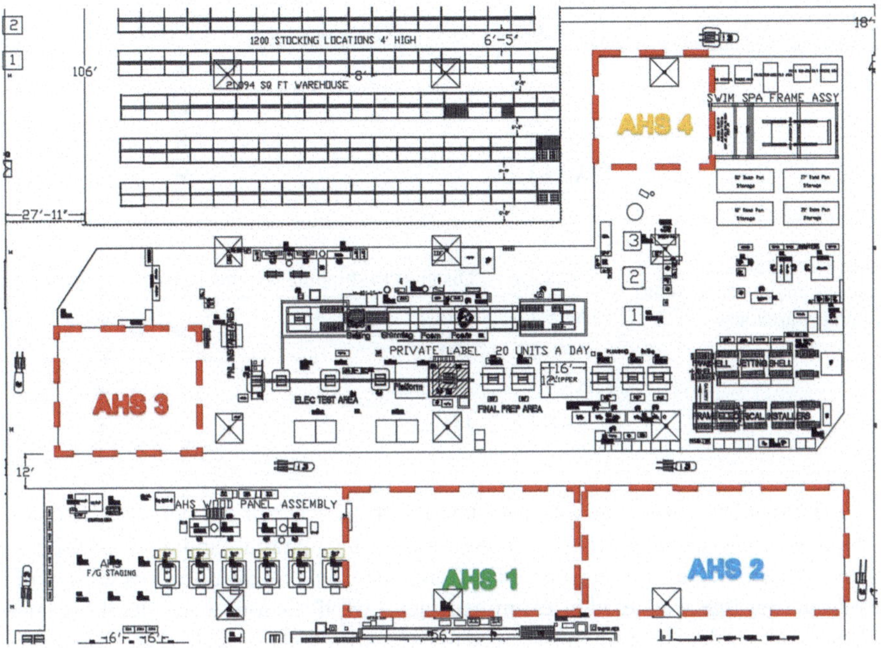

Fig. 4.2 Proposed locations

Table 4.1 Proposed locations

Location	Surface (sq. ft.)	Criterion
AHS 1	3557	✔
AHS 2	3516	✔
AHS 3	2064	✔
AHS 4	1638	✔

requires large quantities of raw material and was found that the distance between these locations was 920 feet (ft.). So, the aim was to reduce it so that the total distance traveled should also decrease. Table 4.2 presents the distances between these locations and the compliance of the criterion.

Table 4.2 Proposed locations and distance to warehouse

Location	Distance to warehouse (feet –ft.)	Criterion
AHS 1	397	✔
AHS 2	428	✔
AHS 3	245	✔
AHS 4	460	✔

Table 4.3 Proposed locations and quantity of corridors

Location	Corridors	Criterion
AHS 1	2	✖
AHS 2	3	✔
AHS 3	2	✖
AHS 4	2	✖

Table 4.4 Proposed locations and distance to the shipping warehouse

Location	Distance to shipping warehouse (ft.)	Criterion
AHS 1	430	✔
AHS 2	450	✔
AHS 3	278	✔
AHS 4	493	✖

- **Corridors**: The criterion was correlated with the size of the corridors because there was a necessity to transport continuously raw materials and work-in-process inventories. In addition, the consideration to use the existing corridors around the production line was another examined factor. Both factors were glued into this criterion to reflect the imperative need to have sufficient area to transport materials and to give access from diverse locations. In the American plant, three corridors with enough area satisfied these needs. Table 4.3 shows the number of corridors and the compliance with the criterion using the number of corridors as minimum condition.
- **Distance to the shipping warehouse**: This criterion was associated with the distance between the production line (specifically the final assembly station) and the shipping warehouse. It was because the product was voluminous and complex for transportation and was found that the distance between these locations was 920 feet (ft.). So, the aim was to reduce it so that the total distance traveled should also decrease. Table 4.4 presents the distances between these locations and the compliance of the criterion.

Based on the obtained results, it was decided to exploit the AHS 2 location. This location met all indicated criteria. That is, the existing area in AHS 2 location was less than the one used previously; the distances from this location to raw material and shipping warehouse were less that those traveled previously, and the appropriate area and number of corridors in this location were similar to the previous line.

4.3.2 *Phase II. General Distribution Plan*

Two diagrams were employed in this phase and were generated within the American plant. One was employed to describe the materials flow and the second one, to uncover the personnel movements. They put the basis to study the convenient materials flow and personnel movements. Particularly, the flow process chart exhibited all activities required in the production process. Table 4.5 shows a summary of these activities. With this chart, the designers carefully studied the entire process and discovered those activities that could be eliminated without affect the process. Furthermore, the diagram helped on in an intermediate level the transportation of materials and finished goods as well as the relationships between the activities permitting to realize a draft to the layout. The emphasis was in the traveled distance of the raw materials and finished products and certain difficulties on these transportations (Muther and Hales 2015).

On the other hand, the spaghetti diagram was used to visualize the personnel movements into the production line and to the diverse warehouses. Moreover, the discovery of the movements permitted to measure of their distances. The uncovered activities in the flow process chart aided with the creation of the spaghetti diagram. The purpose of each activity in the production process was uncovered allowing the planning of the reduction of the activities and when I was convenient, the cutback of the distances of movements. In other words, these activities put the basis to redesign the entire production process to propose a new one. An extraordinary discovery was that in the actual plant, the production was fragmented in two areas. This was one cause of the excessive distance in transportations and personnel movements. The average distance that personnel traveled each workday was 35,000 ft. Thus, additional problematic situations were exposed: lack of communication and some activities overlapped. Figure 4.3 presents the developed spaghetti diagram.

Another aspect contemplated in both diagrams was the design and construction of vehicles which will be adopted within the production process. On the one hand, the size and form of them allowed some changes in the methods employed in the diverse activities. On the other, the size and form granted specific characteristics that were implied onto the redesign of the size of corridors, warehouses, workstations, among other elements. It is important to remark that the transportation of raw materials and finished goods were manually transported. Their fabrication also boosts the implementation of LM tools.

During the design of the layout draft, personnel with strong relation with the production process were invited to participate. So, contributions from operators and supervisors representing the Production Department were valuable as well from

Table 4.5 Summary of the process flowchart

Activity	Quantity
Operations	3
Transport	6
Inspection	1
Delay	0
Warehouse	3
Total	13

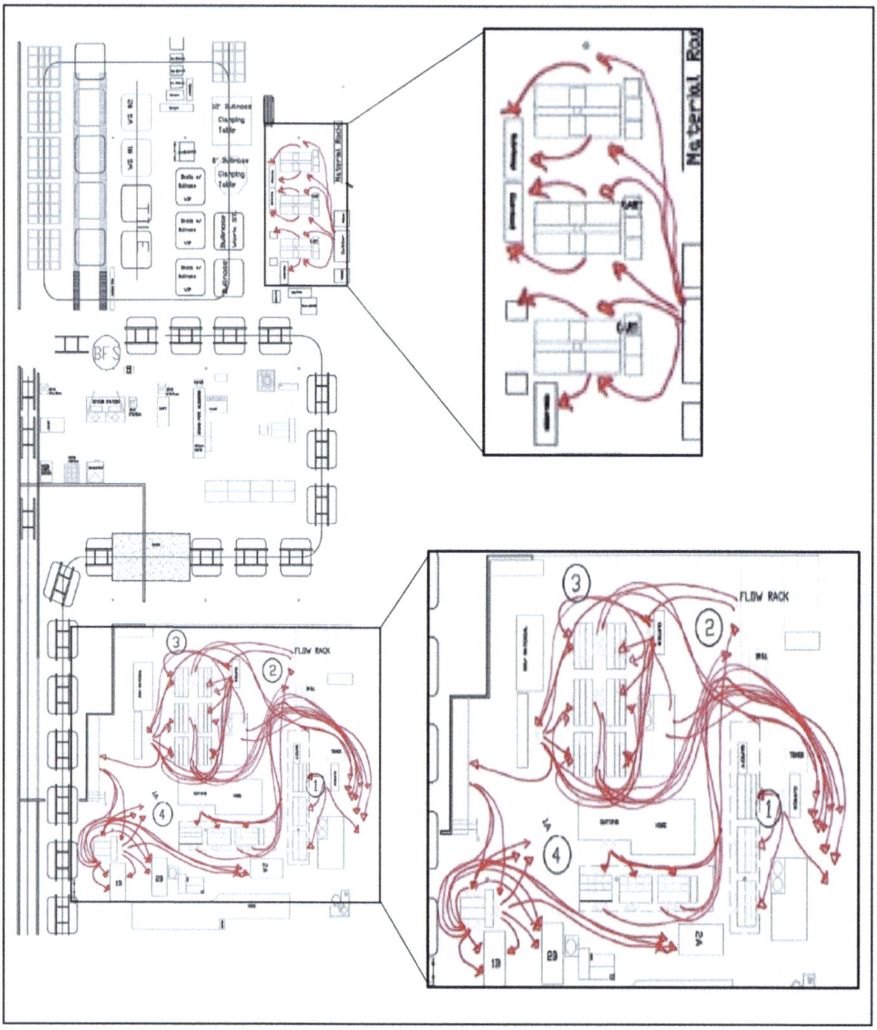

Fig. 4.3 Spaghetti diagram

personnel of the Maintenance, Hygiene and Safety, Engineering and Administration departments. In this particular task, various tools were adopted in order to catch their contributions. The *paper doll* tool was exploited because its flexible for proposing the initial and subsequent layout drafts until the final version of the layout was generated. The representation of the area and production elements were made using a commercial software of CAD. Their representations were made to scale for easy manipulation and two proposals were developed. This act was manually done because the company did not have appropriate software to do in a simulated environment. Moreover, Muther and Hales (2015) suggested the use of the activity-relationship

Table 4.6 Summary

Criterion	Proposal 1	Proposal 2
Available area (sq. ft.)	3516	3516
Number of employees	27	23
Cutter (piece)	5	4
Transport vehicle (piece)	4	15
Area for warehouse (sq. ft.)	564	462
Traveled distance (ft.)	22,000	17,000

and space-relationships diagrams in this phase; however, they were mixed with other tools presented before. This enriched the process and the layout and its implications were aligned to future tasks as the implementation of LM.

Since the process of layout design was challenging, diverse criteria were considered a quantitative and objective approach to choose an option that met the criteria. As mentioned before, in this phase a mix of diverse tools was crucial to develop a layout that satisfied the needs of the implied departments. The available area for install the production line, the number of required employees, the number of cutters as manual tool, the number of vehicles that fit within the corridors transporting raw materials and finished goods, the area assigned for warehouse and the average distance traveled per workday were the criteria scrutinized. The use of the spaghetti diagram was essential to quantify the traveled distance per each option. Table 4.6 presents the results obtained by each proposal and the compliance of the criteria. The proposal 2 was chosen by exhibiting better conditions for being a suitable layout, which is presented in Fig. 4.4 and was denominated layout draft.

4.3.3 Phase III. Detailed Distribution Plan

The third phase of the project consisted on defining in detail aspects related to the transportation of raw materials and the finished goods, the inventory systems to control their transportation, and the vehicles that should be utilized on it. Referring to the vehicles, these should be used to transport the raw materials in the plant and to put the basis in the design of the corridors and in the final arrangements of the production elements. Furthermore, they should be used to transport the finished goods from the Mexican plant to the American plant. As mentioned in the Introduction section, the finished goods are referred to the work-in-process inventories that have to be finalized in the American plant. For these reasons, the vehicles were designed and constructed having this goal in mind. Related to the inventory systems, the phase served to design and implement varied actions and tasks in conjunction with the design of the vehicles and how they should be exploited within the routine work. For instance, the inventory management and its control techniques were implemented, so that they similarly allowed the operation of diverse operational techniques to forecast the needs of materials and diverse resources in the production process. They tried to

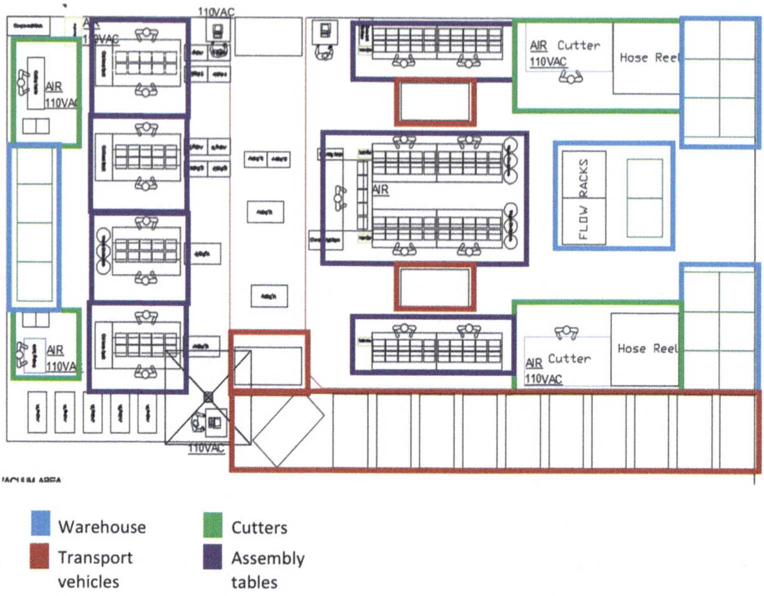

Fig. 4.4 Layout draft

make the production system efficient. Table 4.7 presents the number of vehicles used and considered in the plant layout in strong relationships with the expected production rates in both plants. The name of the models was changed for confidentiality reasons.

Another contemplated aspect related to the use of the vehicles was the area demanded by them. So, the desired size of the corridors was in connection with size of the vehicles and the considered area for them in the layout draft. This phase of the layout design went from the plane to the space. Although, it is preferable to design layouts in space, sometimes there are limitations such that plane design

Table 4.7 Summary of transport vehicles

Model	Capacity of operations	Required vehicles
A	15	8
B	13	9
C	18	6
D	18	4
E	24	4
F	24	4
G	32	3
H	32	3
I	20	2
Total	43	

is sufficient (Muther and Hales 2015). However, the designers explored the space using the constructed vehicles to have data from the context where they should be operated. Consequently, they helped on pondering barriers in risk areas, limitation of work areas, ramps, among other aspects of the production area. Once this process was completed, final adjustments were done to the layout draft and the installation plan was defined, which were the results of this phase.

4.3.4 Phase IV. Installation

The phase 4 was the materialization of the personnel needs from the diverse departments involved in the production process. The convergence of diverse personnel verifying that yours needs were satisfied was the particular characteristic of the stage. Layout designers and potential users were engaged on the installation of the production elements so that additional changes were done in the layout. This gave certainty that personnel were complete satisfied and the aim of the transference were achieved. Particularly, the operational needs were priority at the moment of install the equipment and tools. For instance, the installation of the electrical power, the compressed air intakes, air and steam extractors, among others were a challenge because they should be met technical requirements. Figure 4.5 presents the main operational needs within the production line.

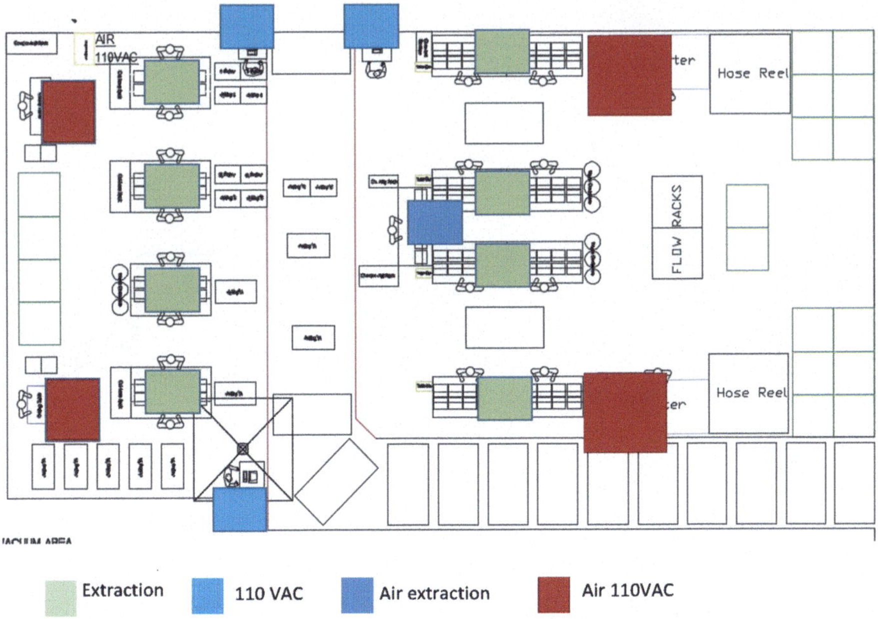

Fig. 4.5 Operational needs

Once the installation of the equipment, tools and the rest of the production elements finalized, the personnel considered that the production lines was ready to start the production. Nevertheless, diverse technical and pilot tests and observations were made so that they tuned the last details found in the production line. This means that with the obtained results were found many areas of opportunity or changes to make within the layout. The changes were not big into the layout but these had notable impact in the production flow. For example, some arrangements and dispositions of determined equipment were necessitated. Moreover, the size of temporary warehouses was increased and some depots were replaced. Figure 4.6 presents the last layout draft and the final layout.

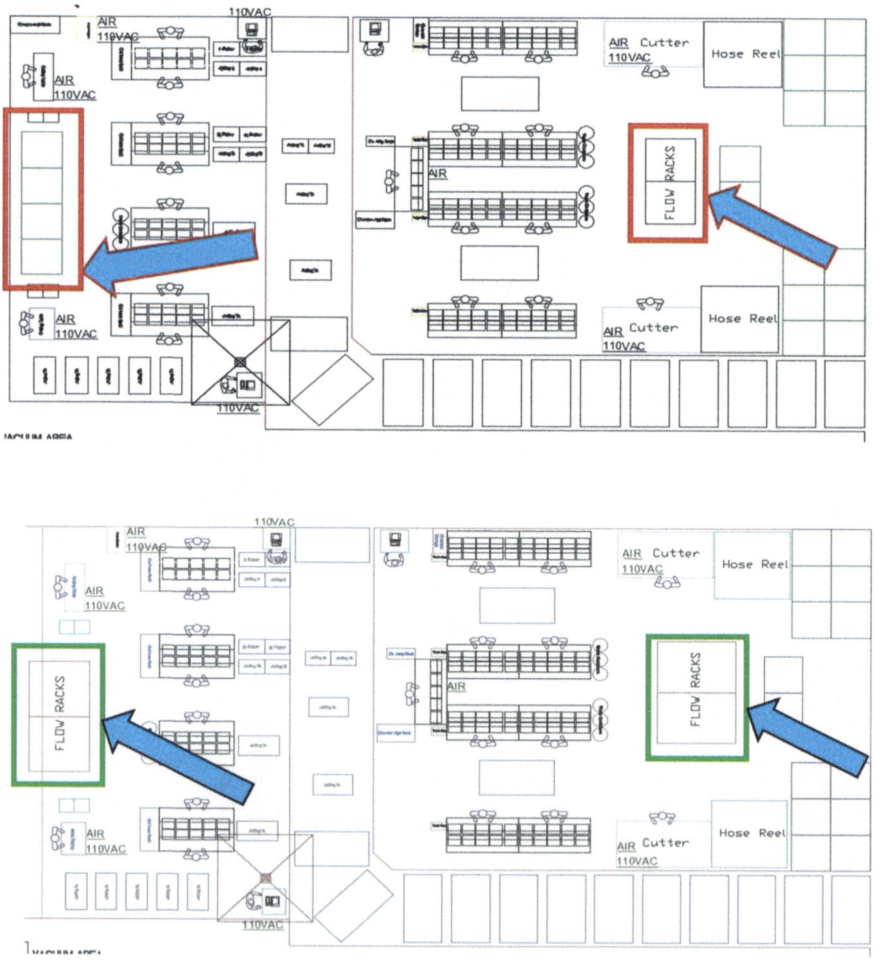

Fig. 4.6 Last layout draft and final layout

4.4 Results

This project was an iterative process in which the results of one phase were the inputs of the following phase. Each of the presented phases utilized the results of previous phases in order to advance in the achievement of its aim. In this form, the layout planning approach exhibited a multi-criteria path acceding similarly to multiple viewpoints coming from various origins. The idea was to start a production process in a designated place. Consequently, the proposed objectives were achieved. They converge in the design and installation of the HS production line. To do so, multiple tools were employed within the four phases included in the SLP methodology. Additionally, distinct tools were adopted to accomplish the multiple objectives indicated by the managers. In the following figure and tables, the representative results are disclosed exposing previous and actual status of the production line in order to compare quantitatively both layouts and the impact in performance and productivity. Initially, Fig. 4.7 presents previous (or initial) and actual layouts of the production lines. Both layouts exhibit the diverse routes following by the raw

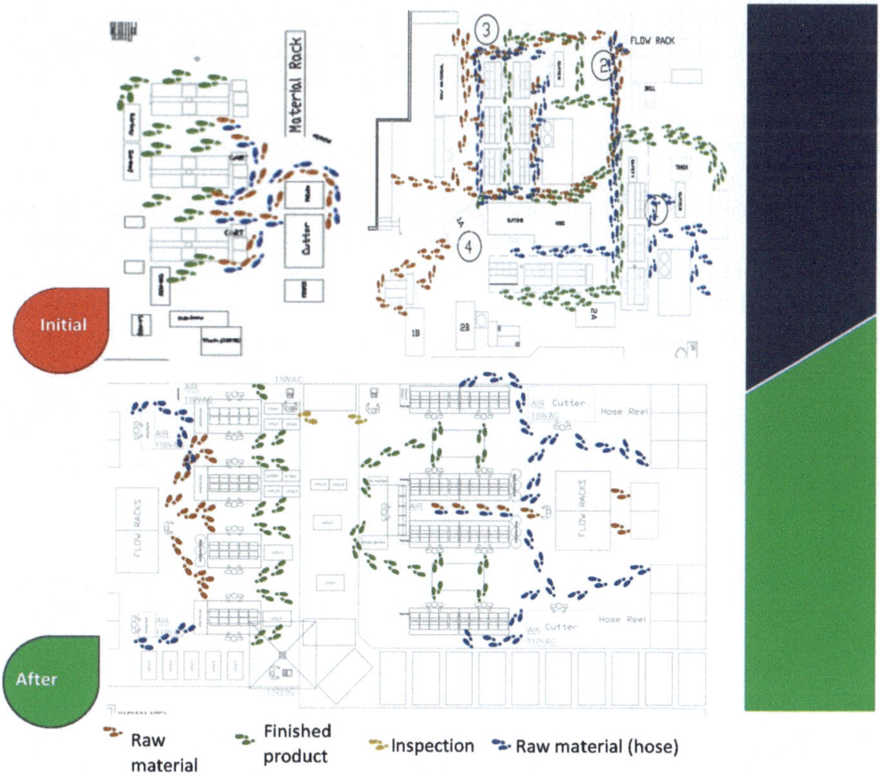

Fig. 4.7 Comparative of initial and final layouts

materials, the finished products, the inspection activities, and the approaching of raw materials to the production lines. This figure was requested by the administrative authorities to graphically and intuitively show differences between both layouts.

Moreover, various indicators were developed to be able to evaluate iteratively the layout through the phases of its design and installation. For instance, one employed index was the traveled distance so that it permitted to explore other indicators as the productivity and the transportation costs of raw materials and finished goods within the production line. It is important to notice that costs were defined and associated with the salary perceived by the personnel carrying out activities of transporting materials and finished products in the production line. Similarly, productivity was related to the result obtained in basis of the efficient utilization of resources within the production process. Another aspect was the entire duration of the project. Initially, managers considered 18 months to finish it and they started on June 2017 focusing in the general tasks of preparation the design and installation of the layout. However, it was delayed and phase 1 was activated on February 2018. And the final phase finished on December 2018. The pilot tests were done during the next two months. Subsequently, managers considered that the production line was in the condition to be exploited. During the following twelve months, diverse data were collected so that permitted to evaluate its performance uncovering a clear visualization of the implied trends on its routine use. The averages of the results are presented in Table 4.8.

Furthermore, other indicators helped to visualize other benefits as a result of the layout design and installation. These were closely related to SLP methodology. For example, as mentioned in the phase 2 section, the vehicles used to transport raw materials and finished goods were also redesigned and built to suit the actual layout. 11 different types of vehicles were constructed and adapted to the production process. Moreover, various fixtures were designed and constructed to aid the diverse operations within the process. Their aims were to reduce the damaged during the operations and to reduce the errors in the assembly operations. Table 4.9 shows the total occupied area, the number of used fixtures and the staff operating the process in both plants.

Table 4.8 Summary of the final results

Indexes	Initial	Final
Traveled distance (ft.)	35,000	17,000
Personnel transporting materials	4	2
Productivity (%)	74	104
Area of warehouses (sq. ft.)	629	458

Table 4.9 Summary of other results

Indexes	Initial	Final
Area used (sq. ft.)	5061	3516
Fixtures (pieces)	7	12
Personnel	27	23

4.5 Conclusions and Recommendations

The design and installation of a layout were the objective of this project and were solicited as a part of the production transfer from an American plant to a Mexican plant. The transfer was based on trying to solve various problems detected in the American plant. The initial goal of it was to transfer the entire process, but administrative regulations forced to the company to transfer diverse activities to the Mexican plant and the product should be finished in the American plant. For practical reasons, the managers considered the work-in-process inventories from the Mexican plant as finished goods. The process of the layout design and installation was a challenge, and the SLP methodology helped to direct efforts to achieve that objective. That methodology was an iterative process in which diverse tools were employed in order to enrich it. Additionally, this process aided in the implementation of LM, but it was not presented in this chapter. Referring to the methodology, the results of the diverse phases served as inputs of the subsequent phases. This allowed the involvement of personnel converging from the diverse areas that conform the company. It also boosted a holistic viewpoint along the entire process. The personnel agreed to share their needs in order to be added to the layout design, which subsequently impacted in the installation of the facilities.

Furthermore, the iterative path of the SLP methodology included a multi-criteria approach in which diverse goals were reached during that process. One of them was the use of the existing area in the Mexican plant, which met the criteria in reference to the necessary area to install the production line. In fact, one available location with less area met the criteria. The company saved approximately 30% of the area used initially. Another goal was related to the traveled distances of the transportation of raw materials and finished goods. The layout design should reduce these to minimal quantities. The obtained results exhibited that it was possible to reduce 50%. Similarly, this result impacted in other aims as presented after. For instance, the productivity in the production line increased approximately 30%. On the other hand, the costs were reduced in the same line and 50% of them were cut benefiting the company. The costs were associated with the personnel performing the transportation activities. So, the costs were reduced because fewer employees were hired to perform those activities. Furthermore, new types of vehicles were designed and constructed in consideration of the proposed layout. The 11 models met diverse criteria having as basis on the dimensions of the size of the corridors and the areas where they should be utilized. Moreover, the number of fixtures employed in the process was increased in order to give competitive advantages to the production process. These had effect on other benefits as reduction of scrap and human errors.

In short, the initial objectives were achieved and exceed the expectations that were initially outlined. The formed team who carried out the project was result of a mix of backgrounds, experiences, and knowledge that converged in it. The viewpoint was respected until the end of the project. Similarly, it worked well into the diverse phases of the endeavor involving in the same form personnel from other departments. This way helped on listen to individuals that expose their operational needs permitting to subsequently satisfying them in the form of an efficient layout. The results obtained in the diverse indicators uncovering the achievement of multiple objectives outlined. The majority of them was associated with the production flow so that productivity was increased and the costs were reduced.

However, it is important to comment that this process had many challenges along its diverse tasks. It was possible by exploring different sources of information and knowledge. This also provided an opportunity to probe the necessity to develop experiences, skills, and additional knowledge in order to achieve the proposed goals. These principally took the form of social skills, interpersonal communication, systematization of processes, and use of technological tools. Referring to the technological tools, the process was done using rudimentary tools employing considerable resources. For instance, the personnel were summoned every week in order to manifest their advances, on the one side; and these meetings were made in person, on the other. This represented the movement of personnel to the Mexican plant. In addition, only one software was utilized to develop the diverse proposed layouts. Thus, it is proposed to do financials studies in order to measure the benefits on buying software to conduct virtual meetings and to simulate the diverse proposed layouts in order to reduce the duration of future projects. The virtual and augmented reality technology would be options in this line.

Related to the systematization of processes, it was associated with the organization and ordering of the generated information along the project, which contributed with the awareness of all type of happened situations that, what factors were crucial, the results obtained and the lessons learnt along the project. This should result on the creation of a knowledge society in which the personnel can be exposed to the generation and sharing of knowledge so that it should improve these kinds of projects. It is relevant to remark that it could include the development of interpersonal communication and the social skills. For instance, the leadership adopted along the project can open favorable circumstances to increase the interpersonal communication and social skills. This can be visualized as contrary to the use of technological tools in the communication; nonetheless, these tools can be used in determined situations and not in crucial state of affairs. Here, the leadership can create the guidelines when the technology can be employed and when not. In conclusion, information, knowledge, and personnel can be the difference in the success of the project. These factors must be carefully monitored to see the performance within the project since their deviations can mean delays or non-compliances on it.

4.6 Discussion

The design and installation of a layout were the principal objective of this project. Initially, the motivation of the design and installation of the layout were to use the existing area in a Mexican plant. So, with this motivation in mind, it was inferred that the principal motivation to do those activities were based on the efficient use of the actual resources the company has. This was similar to other authors had exposed in the current literature (Naik and Kallukar 2016; Monga and Khurana 2015). Moreover, the SLP methodology was utilized to achieve the stated objective. To put on practice, many tasks were required to do. They were challenging and demanding in which many viewpoints were added. For instance, many researchers have found that other tools should be incorporated to the methodology in order to enrich the process (Fahad et al. 2017; Naqvi et al. 2016; Liao et al. 2015; Wiyaram and Watanapa 2010). It also allowed to include quantitative and qualitative approaches giving a holistic view. Furthermore, the methodology granted the utilization of a multi-criteria path where diverse aims were reached during the phases of the methodology. Thus, the process was iterative grating the convenience of exploiting the path to consummate various aims at the same time (Fahad et al. 2017), including pure quantitative approaches (Krajčovič et al. 2019; Li et al. 2018). It also reiterated the principal motivation of design and install layout, the efficient use of the resources. Finally, the results obtained were associated with the motivation, the methodology, the tools utilized, and the multi-criteria path. They can generally take quantitative forms in order to offer objectivity to the entire process (Naik and Kallukar 2016). To sum up, the efficient achievement in the design and installation of the layout had in basis its motivation, the employed methodology, the use of diverse tools within the process, and the inclusion of a multi-criteria approach. The results should exhibit them in strong relation with the context where the layout approach is applied.

Acknowledgements We would like to acknowledge to the Tecnológico Nacional de México/Instituto Tecnológico de Tijuana, Universidad Autónoma de Baja California, PRODEP and CONACYT for partially funding the study.

References

Allen TT (2018) Introduction to engineering statistics and lean six sigma: statistical quality control and design of experiments and systems. Springer, Berlin

Cabrera Calva RC (2014) Lean six sigma TOC. Simplificado. PYMES, Rafael Carlos Cabrera Calva

Fahad M, Naqvi SAA, Atir M, Zubair M, Shehzad MM (2017) Energy management in a manufacturing industry through layout design. Procedia Manuf 8:168–174

Gutiérrez Garza G (2000) Justo a tiempo y calidad total, principios y aplicaciones. Quinta edición. Ediciones Castillo, Monterrey

Hegazy T, Elbeltagi E (1999) EvoSite: evolution-based model for site layout planning. J Comput Civ Eng 13(3):198–206

Hernández Matías JC, Vizán Idoipe A (2013) Lean Manufacturing. Escuela de Organización Industrial, Madrid

Jo JH, Gero JS (1998) Space layout planning using an evolutionary approach. Artif Intell Eng 12(3):149–162

Krajčovič M, Hančinský V, Dulina Ľ, Grznár P, Gašo M, Vaculík J (2019) Parameter setting for a genetic algorithm layout planner as a toll of sustainable manufacturing. Sustainability 11(7):2083

Li J, Tan X, Li J (2018) Research on dynamic facility layout problem of manufacturing unit considering human factors. Math Probl Eng

Liao S, He W, Zhou H (2015) Study on the layout planning and optimization for an electronic product workshop based on cell manufacturing. In: 2015 international conference on automation, mechanical control and computational engineering. Atlantis Press

López Cuevas, Briceida Noemí (2013) Mapeo de la Cadena de Valor (VSM) como Estrategia de Reducción de Costos. Tesis de maestría, UABC. Tijuana

Martínez Carbajal A (2004) Planeación estratégica de la planta. Doctoral dissertation, Universidad Autónoma de Nuevo León. Monterrey

Monga R, Khurana V (2015) Facility layout planning: a review. Int J Innovative Res Sci Eng Technol IJIRSET 4(3):976–980

Muther R, Rabada CC (1981) Distribución en planta. Hispano Europea, Barcelona

Muther R, Hales L (2015) Systematic layout planning. Richard Muther and Associates, Marieta

Naik SB, Kallurkar S (2016) A literature review on efficient plant layout design. Int J Ind Eng Res Dev (IJIERD) 7(2)

Naqvi SAA, Fahad M, Atir M, Zubair M, Shehzad MM (2016) Productivity improvement of a manufacturing facility using systematic layout planning. Cogent Eng 3(1):1207296

Niebel BW, Freivalds A (2004) Ingeniería industrial: métodos, estándares y diseño del trabajo. Alfaomega, México

Ning X, Lam KC, Lam MCK (2010) Dynamic construction site layout planning using max-min ant system. Autom Constr 19(1):55–65

Ning X, Lam KC, Lam MCK (2011) A decision-making system for construction site layout planning. Autom Constr 20(4):459–473

Pineda K (2004) Manufactura esbelta. Manual y herramientas de aplicación. Recuperado de https://www.gestiopolis.com: https://www.gestiopolis.com/manufactura-esbelta-manual-y-herramientas-de-aplicacion/. (Recovered: 15 Jan 2020)

Reyes Aguilar P (2002) Manufactura Delgada (Lean) y Seis Sigma en empresas mexicanas: experiencias y reflexiones. Contaduría Y Administración 205:51–69

Tommelein ID, Zouein PP (1993) Interactive dynamic layout planning. J Constr Eng Manag 119(2):266–287

Wiyaratn W, Watanapa A (2010) 'Improvement plant layout using systematic layout planning (SLP) for increased productivity'. World Academy of science, engineering and technology, open science index 48. Int J Ind Manufac Eng 4(12):1382–1386

Xu J, Li Z (2012) Multi-objective dynamic construction site layout planning in fuzzy random environment. Autom Constr 27:155–169

Yang T, Su CT, Hsu YR (2000) Systematic layout planning: a study on semiconductor wafer fabrication facilities. Int J Oper Prod Manag 20(11):1359–1371

Chapter 5
Improving Distribution Process Using Lean Manufacturing and Simulation: A Seafood Packer Company Case

Julian I. Aguilar-Duque, Jorge Luis García-Alcaraz, Juan L. Hernández-Arellano, and Guillermo Amaya-Parra

Abstract During last decades, production systems have developed strategies to increase their competitiveness in a global market in manufacturing and services, and lean manufacturing and simulation methods have been consolidated as tools that support that. This paper reports a case study in a food packer company where a simulation model was applied to reduce waste time due to a poor layout in operations and transportation areas. The company has detected problems on layout that avoid fulfilling the market demand. The simulation model was aimed to test different scenarios and layout designs as alternatives for a better distribution without modifying its facilities and run in Promodel® simulation software. The production systems' scenarios were evaluated using the production system performance as percentage of used locations, percentage of resources utilization, number of finished products, and the level of work in process (WIP). Finally, the verification and validation stages were performed before running the scenarios in the real production area. After implementing the best simulated scenario, company reports an increase of 68% in production capacity as well as a reduction of 5% in the WIP.

Keywords Production system · Simulation manufacturing process · Simulation model · Work in process

J. I. Aguilar-Duque (✉) · G. Amaya-Parra
Facultad de Ingeniería, Arquitectura y Diseño. Universidad Autónoma de Baja California, Carretera Transpeninsular Tijuana-Ensenada 3917, Colonia Playitas, 22860 Ensenada, B.C., C.P., México
e-mail: julian.aguilar@uabc.edu.mx

J. L. García-Alcaraz
Department of Industrial Engineering and Manufacturing, Autonomous University of Ciudad Juárez., Av. Del Charro 450 Norte. Col. Partido Romero. Juárez, Chihuahua, México

Division of Research and Postgraduate Studies, Tecnológico Nacional de México/Instituto Tecnológico de Ciudad Juárez. Av. Tecnológico, 1340, Fuentes del Valle, 32500. Ciudad Juárez 32500, Chihuahua, México

J. L. Hernández-Arellano
Instituto de Arquitectura Diseño y Arte, Universidad Autónoma de Ciudad Juárez, Av. Del Charro 450. Colonia Partido Romero, 32310 Ciudad Juárez, Chihuahua, C.P., México

J. L. Garcia Alcaraz et al. (eds.), *Trends in Industrial Engineering Applications to Manufacturing Process*, https://doi.org/10.1007/978-3-030-71579-3_5

5.1 Introduction

Nowadays, globalization makes necessary to improve manufacturing and service systems, looking to prevail in a competitive market (Altinkemer et al. 2011), and the productive sector has been focused on cost reduction and resources optimization without neglecting quality and competitiveness (Deshkar et al. 2018; Mei et al. 2018). Currently, there are several alternatives for increasing the productivity indexes that were developed by transnational companies with complex systems and global presence (Group 2019). Those strategies have several purposes, for instance, Mei et al. (2018) presents a "Leadership" analysis focused on the appropriate employees training by developing their leader abilities, considering that employees are a key factor to improve competitiveness. Therefore, it is clear the importance to identify the potential in human resources, providing them the appropriate training, such as Toyota is doing it during the last decades (Group 2019; Stahl 2016).

Other strategies are focus in technological investment (Feltrinelli et al. 2017; Kavanagh and Johnson 2017; Ford and Despeisse 2016), product innovation (Garretson et al. 2016), supply chain management (Chen et al. 2010), among others. In the regional area, Jiménez-García et al. (2014) and Padilla-Pérez and Hernández (2010) have analyzed how strategies developed in Mexico allow to increase the presence of electronic manufacturers in South America, as well as indicate how the technological upgrading has faced the Asian competition based on the company localization. This strategy is focused on logistics for increasing the service level by reducing distance between the customer and supplier, creating alliances, and identifying business opportunities with micro, small, and middle companies that may produce the quantity demanded by customers.

Also, several companies have been adopting another strategies for increase their productivity, reduce their costs, and optimize their resources (Jiang and Qu 2020; Ibarra Cisneros et al. 2017; Feltrinelli et al. 2017), as for example, lean manufacturing, six sigma, automatization, additive manufacturing, new materials, among others. However, those strategies sometimes are complex, and companies do not have access to all of them, mainly because of a lack in financial resources, and then, they deploy improvement actions aimed to generate an appropriate adaptation for complex plans, strategies, or methodologies (Jansson and Söderman 2013). In this way, micro-, small-, and medium-size organizations have started to adopt strategies developed by international companies, making adaptations according to their own context, as simulation. Additionally, these companies want the success that has been achieved by bigger companies. Results from the integration of unique adaptations and improvements frequently are denominated "customized strategies" (Jansson and Söderman 2013; Swarnakar et al. 2020; Nithya and Haridass 2019; Thomas et al. 2018) 0.1.1 Simulation in the industry.

Dynamical changes in production systems have forced to improve strategies (López 2016; Ibarra Cisneros et al. 2017), where some national and international companies have implemented simulation as a tool to improve their capital flow, economic savings, and to facilitate the decision-making process according to their

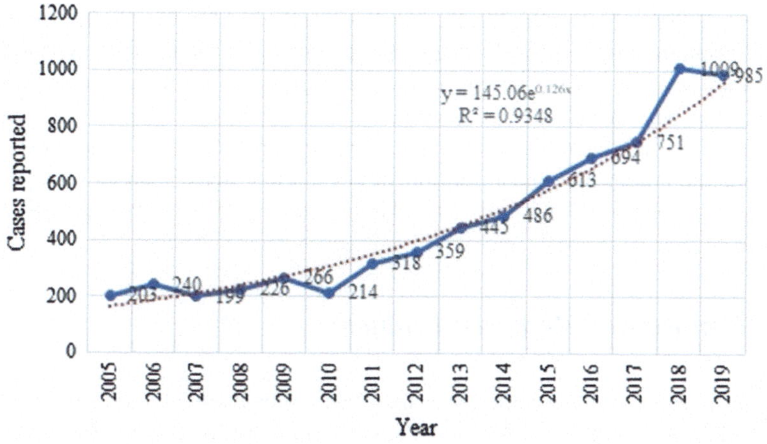

Fig. 5.1 Trends in simulation paper reports

own production process (Srivatsan and Sudarchan 2016). Since simulation was available for manufacturers, this technique has been considered as one of the most rentable ways to improve and optimize a manufacturing and service systems, because it makes possible to create different scenarios for decision making according to the company needs, which is always integrated as many variables as possible (Mourtzis et al. 2014).

Benefits gained from simulation in the industrial engineering field are many, which use has grown in an exponential way, as it is illustrated in Fig. 5.1; however, its correct use and implementation is still a challenge for micro and median industrial manufacturers (Omogbai and Salonitis 2016; Oleghe and Salonitis 2019; Nithya and Haridass 2019).

As soon as the simulation was available for manufacturers and education systems, software developers have created new alternatives for users, offering customized solutions depending on the user's needs (Esmaeilian et al. 2016). Also, using the simulation as a strategy of global competition, the construction of separate events that are implemented to reproduce some real manufacturing situations has been a common industrial application, and using simulation as an alternative for solving problems has generated three types of manufacturing simulation systems: system design, manufacturing system operation, and simulation language/package development (Negahban and Smith 2014).

Nevertheless, simulation has not been only used by manufacturers as a pure technique, which in several cases, lean manufacturing (LM) processes have accomplished their success where several cases are reported. For example, the development of new performance material flow indicators validated through simulation modeling (Liu et al. 2017), production flow analysis (Ye et al. 2020; Tanenbaum and Holstein 2019), and logic distribution simulated with multistage simulations (Mei et al. 2018;

Deshkar et al. 2018; Nagi et al. 2017), or the stock control levels with the creation of separate simulation models (Badurdeen and Jawahir 2017; Hong 2017; Abdul 2018).

As a matter of fact, LM has been used in continuous improvement systems in the past along with simulation as an economic optimization strategy that increased its power as a manufacturing tool, changing the perspective of many companies throughout the adaptation and imitation of simulation success cases on their own companies (Omogbai and Salonitis 2016; Helleno et al. 2017; Swarnakar et al. 2020).

5.1.1 Context and Research Problem

This paper reports the lean manufacturing (LM) improvement supported by simulation modeling applied to a Seafood Packer Company located in Ensenada City, Mexico. The fish and seafood packing process carried out by the company is leaded by the level of market demand, which represents 85% of the company's production. From 85% of the seafood packed in the company, 70% belong to the echinoderm or sea urchin, which can be either red or purple.

The arrival of containers with alive animals to the plant requires immediate processing action. The delay in the packaging process after killing the sea urchin is a potential generator of environments for harmful bacteria or microorganisms, therefore, the packing time is critical. The company currently uses two production lines, one for the red sea urchin and another for the purple urchin. In the packing lines, the process begins with the transportation of the sea urchin to the slaughterhouse table, one for each type of sea urchin, which must be handled carefully, because their spikes are highly sharp and may cause injuries to the operators. The opening of the sea urchin is a manual process which is done by cutting the lower central part, preventing the spines from ditching into the interior. This process consumes an average of 90 s for each red sea urchin, while 60 s for each purple sea urchin.

The next activity is called spooning, which is the extraction of the shellfish, and it is done manually, since it is a time-consuming activity (30 s per unit for each type of sea urchin) 30 to 40 pieces are concentrated on the slaughterhouse, before transporting the sea urchin to the spoon table. The average time consumed by transportation and storage in process between the two activities is 64 min for the red sea urchin and 95.9 min for the purple sea urchin. During this time, the seafood is exposed to the environment of the production area.

Once the seafood extraction process has been carried out, it is necessary to clean the product before sending it to the inspection and selection process. The cleaning process consists in eliminating the pieces of spines and debris that could contaminate the product during slaughter and extraction until leaving the yolks (seafood) completely clean. This process is critical, since the product must be completely clean. The units that are extracted from each sea urchin must be stored until they have enough product to justify the cleaning process. During this time, the product is exposed to the environment on average six minutes regarding the red sea urchin and 9.7 min for the purple sea urchin.

Once the sea urchin has been cleaned, a seafood selection process must be carried out according to size, since one of the most important characteristics for the standardization of product quality is the seafood dimension. For this reason, the average time required for the selection of the red sea urchin is about 118 s versus 155 s required by the selection activity of purple sea urchin. As in previous activities, seafood is exposed to the environment until an amount of product that justifies the time required to carry out the activity which is accumulated. In this process, the product is exposed to the environment 3.7 min regarding the red sea urchin and 5.56 min for the purple sea urchin.

Finally, the seafood is sent to the warehouse where the product is vacuum packed and placed in refrigeration chambers. The current distribution of the area assigned to the sea urchin, which is restricted by the characteristics of the building. Because of this and despite the presumption that is a process in which the activities are carried out in a line manner, the distribution of the equipment and the workstations are subject to the condition of the material being manufactured or by product. The current conditions of the building force the product flow from one station to another intermittently, thus, generating a path in a "S" shape. This displacement of the product is a dead time generating factor, since the time required for the displacement depends on the distance required by the workforce to carry out the activity.

The distance originally covered by the sea urchin during the packing process is about 130 m (38 m from warehouse to the slaughterhouse table, 22 m from the slaughterhouse table to the spoon table, 12 m from the spoon table to the cleaning area, 20 m from the cleaning area to the sort area, and 30 m from the sort area to the warehouse. Considering the average speeds generated by the employees during the transportation of the product, it is estimated that 92 min are consumed by the transportation of the red sea urchin and 139 min by the transportation of the purple sea urchin. Specifically, it is important to highlight that during this time; the product is held due to the risk of contamination by elements in the air or reproduction of microorganisms.

Moreover, regarding to the lead time, the red sea urchin consumes 100 min where 6.5 min are assigned to activities that add value to the product, the rest of the time is consumed in transportation between locations; for the purple sea urchin, the lead times is about 146 min, where 7 min are use in activities that add value to the product and 139 min are used to transport the product inside production line. It is seen that the building characteristics force the process to have sections for routes that directly affect the flow of the product by increasing the process time. As it is mentioned, the exposure of the product to humid environment is a risk, due to harmful bacteria. Due to this situation, the quality department identified during 2018 that 30% of the production of red sea urchin and 27% of the production of purple sea urchin were affected by microorganisms. This product contamination is an important indicator since the product must be discarded.

The current situation of the company forces optimization strategies to be explored with the least amount of economic resources. In this sense, it has been decided to use simulation as an optimization tool due to the low cost that allows to generate several distribution scenarios of the production lines, which are focused on minimizing the routes made by the product, consequently, the time that the product is exposed to the environment.

The main objective was to develop a simulation model to test different layout distributions scenarios and validate the best alternative to increase the company productivity. During the layout re-design, the model looks for reducing covered distances by operators in production lines during the seafood package process, trying to keep the process as linear as possible.

5.2 Materials and Methods

5.2.1 Stage 1: Description of the Processes

In this stage, a digital reproduction of the building layout is carried out, which includes the dimension of the areas of operation, the distances between them, the dimensions of the worktables, and the operations assigned to each area. For each part of the process, the production capacity and the resource usage to carry out the activity must be registered.

5.2.2 Stage 2: Develop and Analysis of Value Stream Mapping

Once the part of the process to be mapped has been defined, a scheme is made to identify the flow of information and the process. The construction of value stream map (VSM) integrates the timeline elements (with transportation times and process times), and the arrow transportation is a critical element for the analysis and project improvement, operation, and lockers of process indicators, taking as objective to increase the percentage of added value. For the construction of the VSM, the model proposed by (Thomas et al. 2018; Mei et al. 2018), which is considered, where each stage defines the activities to be carried out, is indicated in Fig. 5.2.

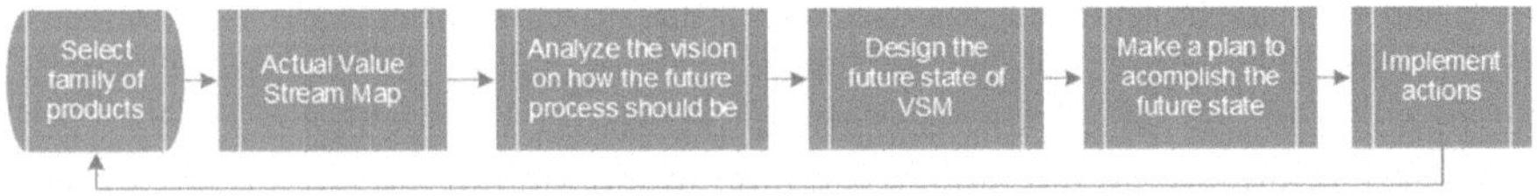

Fig. 5.2 VSM according to Thomas et al. (2018), Mei et al. (2018)

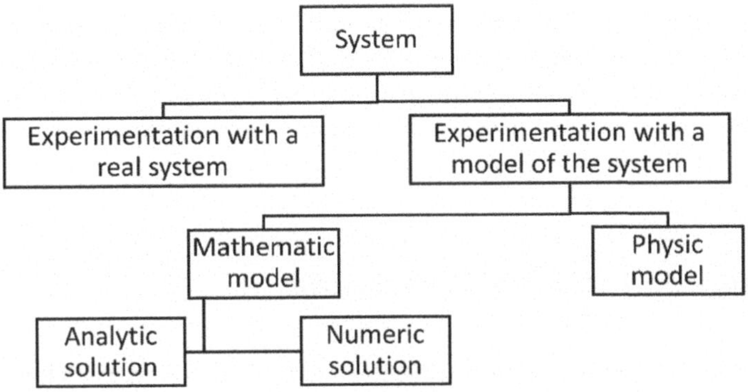

Fig. 5.3 Types of models

5.2.3 *Stage 3: Solution Classification for the System*

Based on the objective to define the structure of the system as a simulation model, Kelton et al. (2017); Cano et al. (2018) methodology was considered; this methodology describes different ways to analyze and create models from a specific system. In addition, Fig. 5.3 illustrates a diagram that summarizes the methodology considered to classify and select the solution system for the seafood packing company. In this case, it is possible to identify a path to build a simulation model and generate a specific solution analytically or numerically, for mathematic models as it is in the current case (Cano et al. 2018; Kelton et al. 2017; Wainer 2017; Khedher and BenOthman 2020; Zhang et al. 2019). Particularly, the model integrates as much as possible elements from a real system, always considering this rule: "If there are more details, there will be more information, and with more information, more complexity." Also, it is important to highlight that these models have analogical characteristics that make their development as if they were real models.

As it is already mentioned, at this stage, alternative solutions will be generated through experimentation of the real system through its reconstruction by digital means following the scheme exposed in Fig. 5.3. In addition, because the economic resource allocated for this project is reduction, it is decided to perform a digital reproduction of the process area as well as the use of Promodel® software for simulation and generation of improvement scenarios.

5.2.4 *Stage 4: Develop of Simulation Model*

Once the information of the process to be modeled is available, a simulation model is developed and a study of operations times is carried out, as well as a data adjustment to determine the probability distributions of these, which allows identify the parameters that should be programmed to perform the simulation. In the same way, the layout is imported with the physical dimensions of the building, which are used for the analysis of routes and for the estimation of the transportation times between workstations. The use of the model facilitates the redistribution of work areas, as well as the analysis of the restrictions associated with the dimensions of the equipment, the quantity of employees, and the material flow.

5.2.5 Stage 5: Analysis of the Scenarios

Once the digital reproduction of the process to be improved is available, variation is made regarding the variable of interest. In this case study, the building cannot be modified, because the relocation of worktables is the areas of operations that are considered. Thus, one scenario considers the relocation of the worktables, another considers the relocation of the worktables and the number of employees per station, while another scenario considers the relocation of the tables and the reconditioning of the supply area. The selection of the best scenario is based on the lower cost of modifying the process, as well as a greater reduction of the distance traveled by the product.

5.2.6 Stage 6: Implementation and Validation of the Improvement

Once the best scenario generated by simulation is available, the modification required to the real process is made, which involve moving the equipment, generating new worktables, modifying existing ones, reallocation workers, modifying or buying containers for the raw material transportation, and reassignment of tasks. Once the modification to the real process has been done, an improvement validation should be done as well. The validation consists in developing a group of pilot tests and then production runs to parameterize the system of operating conditions.

The previous information allows to analyze the improvement behavior and record the main data, which in the first instance and according to VSM, this area is affected by the distance traveled by the product. In this case, a validation is associated with the time consumed physically by the transportation of the product during the process. There is another way to validate the improvement through the reduction of the contaminated product. Nevertheless, this indicator is not essential since there are other factor generators of waiting time during the process.

5.3 Results

The results are presented according to the activities perform in each of the stages that are described above.

5.3.1 Stage 1: Description of the Process

In the description of the process, the variables of interest were identified based on the structure of the building, because walls cannot be demolished or built, neither electrical installation, air, or gas pipelines can be altered. Figure 5.4 illustrates the original distribution that is used for both lines. The packing process is carried out between the slaughterhouse, spoon, sort, and container area. Figure 5.5 shows the route taken by the operators to carry out the process, where it is observed that there are two refrigeration rooms, which are accessed only through a door. In this section, there is no restriction to place the product on one table or another since this operation is performed by the type of sea urchin.

In addition, Fig. 5.5 uses the blue bidirectional arrows to represent the movements that employees take to transport the sea urchin, from the refrigerators to the slaughterhouse area, and from the slaughterhouse area to the spooning area. Also, from the

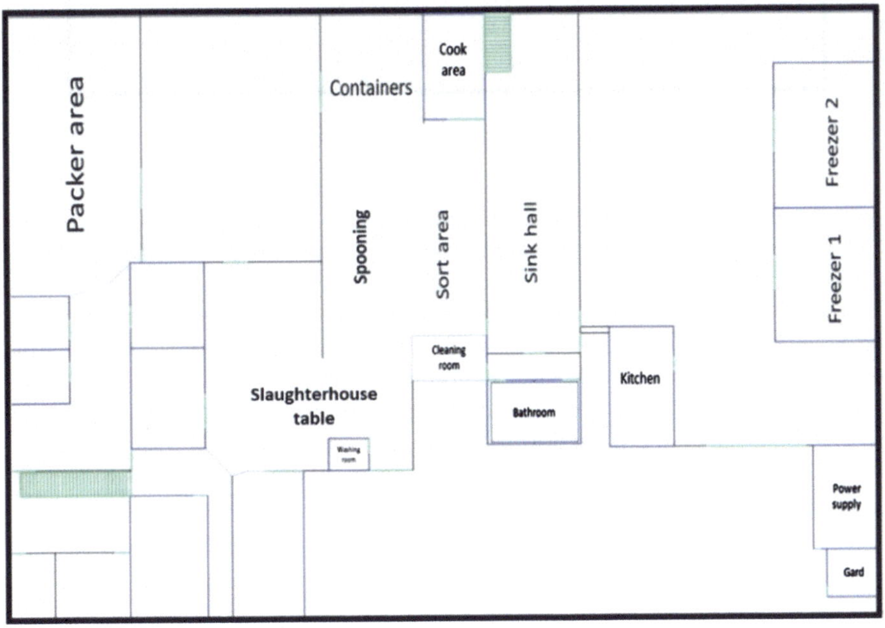

Fig. 5.4 Physical layout of the Seafood Packer Company, Ensenada

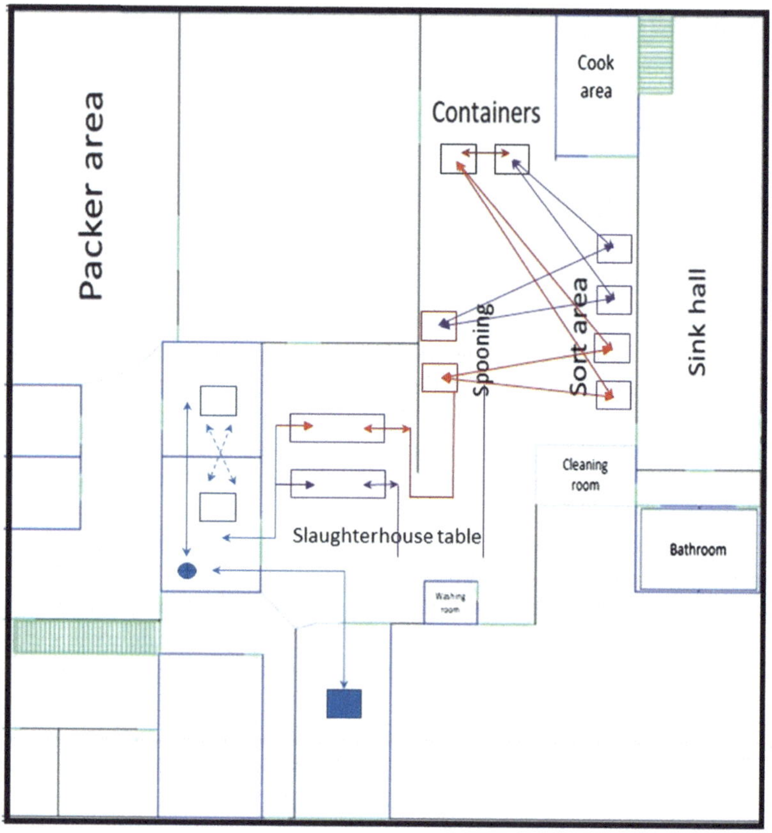

Fig. 5.5 Original physical layout of the Seafood Packer Company

spooning area and cleaning to the sort area, the blue lines are used to describe the flow of the red sea urchin, whereas black lines describe the flow of the purple sea urchin.

Finally, the red arrows represent the transportation of the red sea urchin from the sort area to the containers, where the purple lines represent the flow of the purple sea urchin to the containers. Because the variable of interest depends on the distance traveled by the workers to comply with the process, Table 5.1 is used to illustrate the average distances between the production areas. In addition, it was necessary to create Table 5.2 with information about the capacity of the work area, as well as the percentage of utilization during the packing process.

Table 5.1 Distance cover by employees during the packing activities of red and purple sea urchin

Origin	Destiny	Distance (m)	Employees	
		Red	Purple	
Warehouse	Slaughterhouse	38	36	1
Slaughterhouse	Spooning	22	27	2
Spooning	Cleaning	12	12	2
Cleaning	Sort	20	22	1
Sort	Containers	38	38	

Table 5.2 Capacity versus % of workstation utilization red sea urchin and purple sea urchin

Name	Capacity	% utilization	Name	Capacity	% utilization
Warehouse EZR	500	1.45	Warehouse EZP	650	6.29
Slaughterhouse 1	100	4.54	Slaughterhouse 2	100	70.80
Spooning	100	4.75	Spooning2	20	18.97
Cleaning	40	7.46	Cleaning2	20	14.04
Cleaning3	20	15.46	Celaning4	20	16.09
Sort	10	16.67	Sort2	10	18.72
Supplier	1600	49.12			

5.3.2 *Stage 2: Develop and Analysis of Value Stream Mapping*

Figure 5.6 exposes the VSM for red sea urchin (EZR) and illustrates a waste time between the operations of "Slaughterhouse" and "Spooning" that is 64 min, this time represents the highest time wasting of the process. Also, as a first impression according to its magnitude, this part of the process may be the best alternative to improve them. However, the building characteristics and architecture had to be considered, since one restriction was not meant to modify the facility, and for this reason, it was necessary to analyze the current layout that is portrayed in Fig. 5.4.

Furthermore, regarding the analysis of EZP through VSM, in Fig. 5.8, it is possible to propose by observation of a Kaizen event for the activity between "Slaughterhouse" and "Spooning," which is the same problem for EZR; the magnitude of the waste time in this operation is in average 95.9 min, which is restricted by the building characteristics.

As it has been mentioned, it was impossible to re-design the building, because its structure is old, and it is integrated by different rooms. The first task was to reduce the internal logistic, and to carry out this analysis, it was fundamental to replicate the layout with the paths that employees have tracked when performing production activities. Additionally, Fig. 5.7 presents the original layout with the original routes, which was used during the manufacturing process.

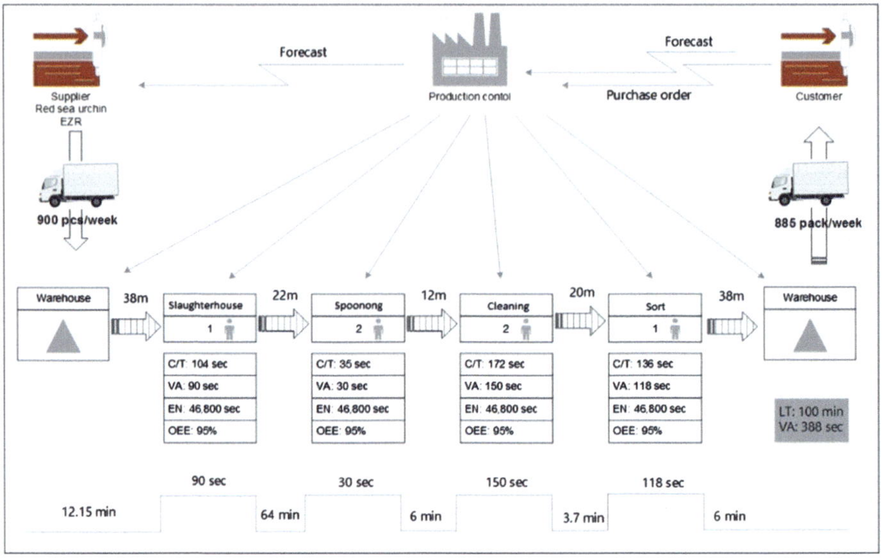

Fig. 5.6 VSM of EZR

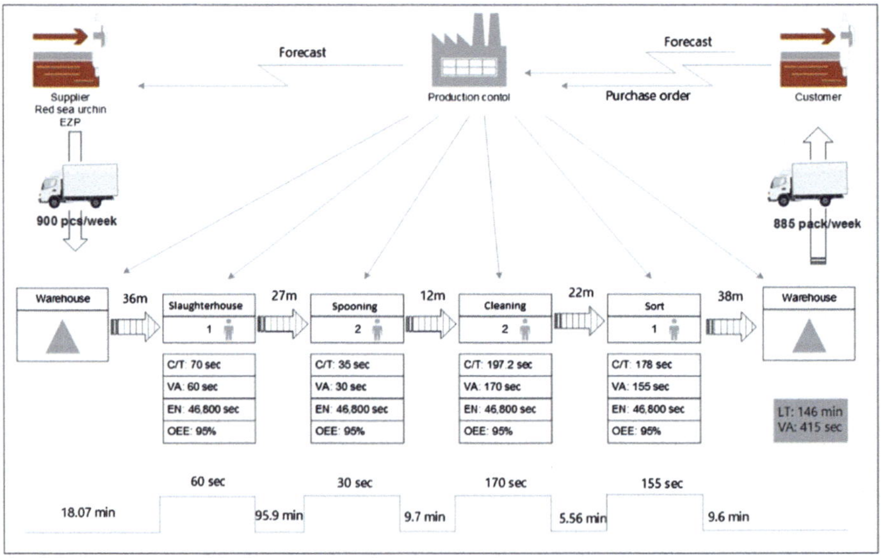

Fig. 5.7 VSM of EZP

Table 5.3 Productivity indicators for EZR and EZP

Operation	Lead time LT (sec)	Value-added time (VA) (sec)		
	EZR	EZP	EZR	EZP
Slaughterhouse	6240	4200	90	60
Spooning	2100	2100	30	30
Cleaning	10,320	11,832	150	170
Sort	8160	10,680	118	155

Table 5.4 Evaluation criteria for system experimentation selection

Criteria	Experimentation in the real system	Experimentation in the system modeled
Required time	160 h	70 h
Allow modifications	No	Yes
Generate scenarios for solution evaluation	No	Yes
Cost of personnel associated with the project	$117,300	$45,000

Hence, the objective is to reduce the process time and increase the time taken to add value to the product. Tables 5.3 and 5.4 were developed summarizing the information portrayed in the VSM.

5.3.3 *Stage 3: Solution Classification for the System*

Table 5.4 shows the evaluation criteria for the selection of the type of experimentation from the system that would be used in the improvement process.

Considering the scheme presented in Table 5.4, it is decided to build the digital model of the system, because it is cheaper and faster to determine the best solution to optimize the system. It is also necessary to declare that the construction of the system using a digital model allows to evaluate several scenarios, such as the distribution of furniture, the quantity of human resources that operate the system, the flow of materials, and the alternatives associated with the flexibility of modification from the system.

5.3.4 *Stage 4: Develop of the Simulation Model*

For the development of the simulation model, the following characteristics presented in the process are considered.

- Setup time, load or unload time, and processing time are average and constant for all processes.
- Scheduling to produce all products is random and planned to meet lead time.
- There are two shifts: one from 6:00 to 13:00 including one break from 10:00 to 10:30, while the second shift is from 14:00 to 21:00 with one break from 18:00 to 18:30, both from Monday through Friday.
- Performance measures:
- Resources utilization: The use of each employee can be analyzed with the maximum utilization of 90%, the company policy predetermines this percentage.

WIP's: The work in process may determine the constrained works areas and guidelines for the required distance on WIP.

As a matter of fact, after defining the VSM and the layout, it was required to define the net used by employees during the process. In addition, the following data was the principal input to design the original layout and the simulation model. Finally, this information is relevant due to the distance covered by employees, which is presented in Table 5.5.

Table 5.5 Distances between nodes, for the EZR network segments

From warehouse to slaughterhouse	From spooning to slaughterhouse				
Origin	Destination	Distance (m)	Origin	Destination	Distance (m)
N1	N2	24.39	N1	N2	11.73
N2	N3	18.05	N2	N3	23.57
N3	N4	23.11	N3	N4	9.65
N4	N2	7.78	N4	N1	27.74
			N1	N3	21.95
			N2	N4	32.10
From spooning to cleaning	From cleaning to sort				
N1	N2	14.32	N1	N2	7.71
N2	N3	7.86	N2	N3	28.35
N3	N4	13.53	N3	N1	34.08
N4	N1	6.81	N1	N4	27.61
N2	N5	7.65	N4	N5	12.44
N5	N6	7.56	N5	N2	5.61
N6	N3	22.72	N2	N4	17.17
N2	N6	14.86	N5	N1	13.29
N5	N3	15.43	N3	N6	11.73
			N6	N1	29.57
			N6	N1	28.84

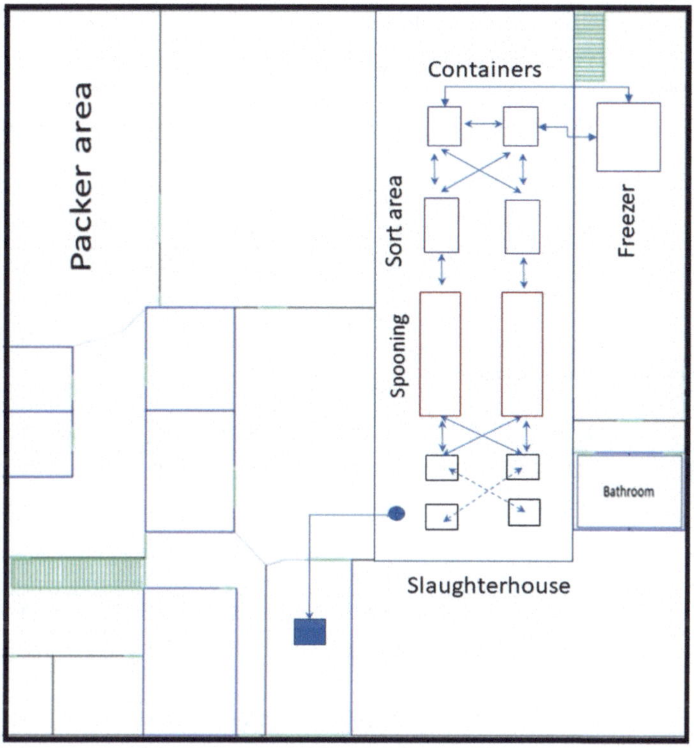

Fig. 5.8 New physical layout of the Seafood Packer Company, Ensenada

After identifying the spaces and equipment availability, the alternative layout was proposed to the CEO; this alternative is shown in Fig. 5.8.

Certainly, along with these alternatives, new networks were estimated; the layout information is described below, where Table 5.6 exposes the distances for the new network.

5.3.5 *Stage 5: Simulation Model*

The model sets the alternative routes according to the physical layout and reproduces the flow process for each product as it is illustrated in the flowchart from Fig. 5.9.

In addition, it is relevant to mention that this model required modules to supply all the input data that was needed to perform the simulation. In addition, the capacity provided in capacity inputs indicates the number of work areas in each process, as well as the schedule cycles for workers to operate. Also, these processed help in determining the amount of capacity that is received; specific data were required for processing each product, such as setup, load-unload time, production rates,

Table 5.6 Distances between nodes, for the new network

From warehouse to slaughterhouse	From spooning to slaughterhouse				
Origin	Destination	Distance (m)	Origin	Destination	Distance (m)
N1	N2	53.55	N1	N2	7.28
N1	N3	14.73	N2	N3	5.61
N3	N4	48.32	N3	N4	8.72
N4	N2	10.75	N4	N1	5.33
			N1	N3	9.83
			N2	N4	9.51
From spooning to cleaning	From cleaning to sort				
N1	N2	7.67	N1	N2	8.85
N2	N3	15.77	N2	N3	25.15
N3	N4	6.73	N3	N1	27.5
N4	N1	12.43	N1	N4	26.88
N2	N5	16.29	N4	N5	12.28
N5	N6	15.14	N5	N2	23.15
N6	N3	18.14	N2	N4	25.81
N2	N6	22.84	N5	N1	24.96
N5	N3	23.29	N3	N6	7.33
			N6	N1	37.67
			N6	N1	14.58

processing batch size, and flow line. Finally, the simulation team could customize the simulation experiment in the model, such as employees, route distance, and other characteristics.

Furthermore, Table 5.7 presents the entity summary before the re-design of the process. It is possible to observe the capacity from one week of production, and under this scene, it is suitable to produce 885 units of EZR and 823 units of EZM, while generating a WIP of 800 EZR and 851 EZM packages. In order to illustrate, Table 5.8 shows that the percentage of time in *move logic* is a critical issue for the lead time.

Table 5.9 represents the location summary before the re-design process. Particularly, the percentage of utilization associated with each location is under 25% in most of them. In addition, it is detected that the location Slaughterhouse 2 is the location with the highest percentage of utilization.

Similarly, Table 5.10 shows the data associated with resources in the original process. It is possible to identify that the employees 4, 5, 6, and 7 have a low percentage of utilization, between 0.5 and 5%. The reason of this low utilization is a consequence of different activities that are performed in separated areas of the process and company, which are carried out by employees.

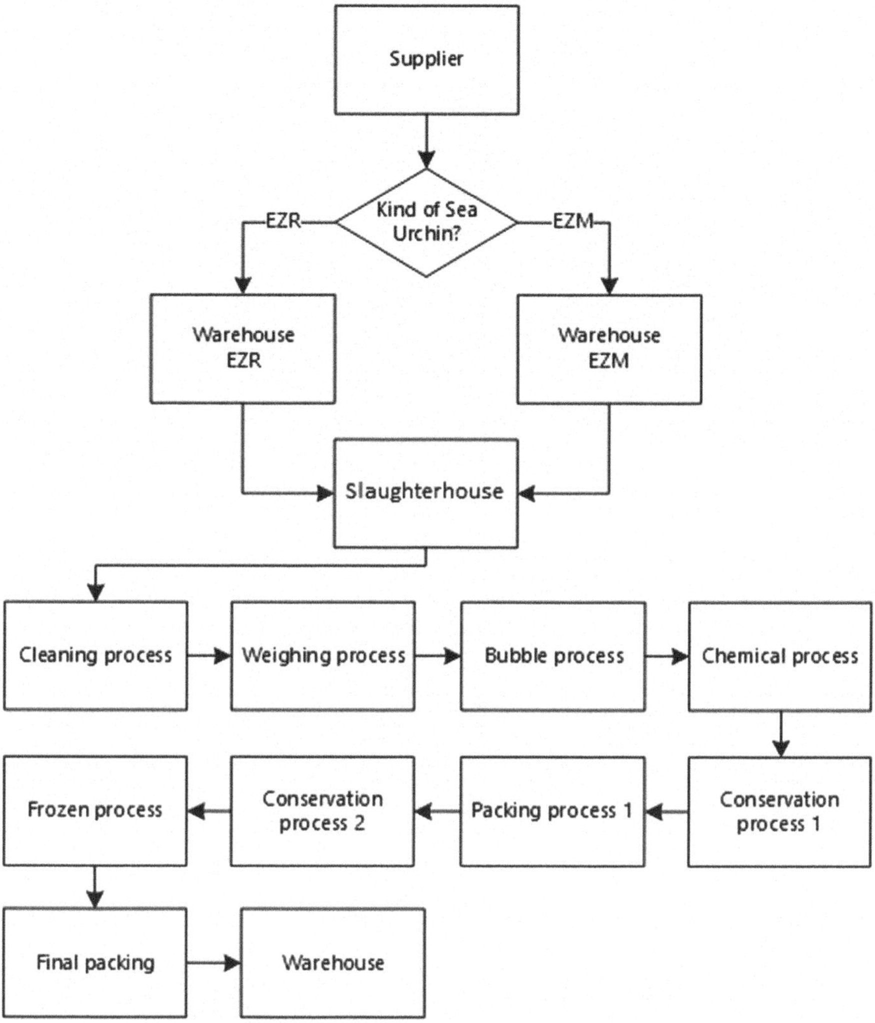

Fig. 5.9 EZR and EZM process flow in the Seafood Packer Company, Ensenada

According to the new design of the logic distribution for the process, the result is favorable to achieve the production objective. In Table 5.11, the number of total exits increases positively for the EZR and EZM finished product, reducing the WIP level. Also, Table 5.12 presents the percentage reduction in move logic and increase the blocked percentage generated by the WIP.

Furthermore, Table 5.13 presents the location summary before the re-design process. The percentage of utilization associated with each location is under 25% in most of them. Also, it is perceived that the Slaughterhouse 2 location got the

Table 5.7 Entity summary

Name	Total exits	Current quantity in system	Average time in system (Min)	Average time in transport (Min)	Average time waiting (Min)	Average time in operation (Min)	Average time blocked (Min)
EZR	885	800	1599.54	1293.35	75.43	6.06	224.69
EZM	823	851	2083.39	1491.46	80.47	6.45	5005.00

Table 5.8 Proportion of times for each entity

Name	% in move logic	% waiting	% in operation	% blocked
EZR	80.85	4.71	0.37	14.04
EZM	71.58	3.86	0.30	24.23

highest percentage of usage, but it is possible to identify a balance between the other facilities.

Moreover, Table 5.14 describes the information associated with the resources in the original process. It is possible to identify that the utilization of the employee 1, 2, and 3 are balanced, but resources 4, 5, 6, and 7 increase their usage in a low percentage including their extra activities. It can be said that this utilization represents an opportunity to reassign their activities in the production flow.

5.3.6 *Stage 6: Analysis of the Scenarios*

In order to reduce the distances traveled by the operators during the process, as well as recovery the time invested in the transportation, it was decided to make three different settings in the operational area. In these settings, the production per family was considered as a first option, production in line without sharing a work mobilizer as a second choice, and online production sharing mobile for both products as a third choice, but packing one product at a time.

In Fig. 5.5, the configuration of the system by family (red sea urchin EZR, purple sea urchin EZP) is exposed. In Fig. 5.10, the configuration of the system in operation online is exposed under the restriction of not sharing furniture. Finally, in Fig. 5.11, the configuration of the system in operation by production in line sharing furniture for the two products is presented. Table 5.15 presents the indicators considered for the selection of the model.

5.3.7 *Stage 7: Implementation and Validation of the Improvement*

During the implementation and validation process, redistribution modification was applied to the area of interest. To identify the impact on the improvement, 14 lead time measurements were obtained for the EZR product and for the EZP product. Figure 5.12 presents the comparison of the simulated lead times against real lead times for both products. This graph shows that the data obtained has an average of 31.88 min with a deviation of 1.77 min, which ensures that the variation associated with the data generated by the simulation model is 6%.

Table 5.9 Location summary

Name	Scheduled time (Min)	Capacity	Total entries	Average time per entry (Min)	Average contents	Maximum contents	Actual contents	% utilization
Warehouse EZR	7560	500	760	72.25	7.26	45	30	1.45
Warehouse EZM	7560	650	720	429.74	40.92	92	37	6.29
Supplier	7560	1600	3040	1954.76	786.04	1574	1560	49.12
Slaughterhouse2	7410	1	683	7.68	0.70	1	1	70.80
Slaughterhouse1	6750	100	730	41.99	4.54	44	1	4.54
Spooning	7470	100	729	48.76	4.75	35	4	4.75
Cleaning	7410	40	72	307.40	2.98	8	2	7.46
Lot	6180	10	20	515.17	1.66	5	0	16.67
Lot2	5760	10	16	674.03	1.87	5	1	18.72
Cleaning2	7440	20	74	282.51	2.80	7	4	14.04
Cleaning3	7470	20	73	316.57	3.09	7	3	15.46
Cleaning4	6960	20	62	361.25	3.21	7	6	16.09
Spooning2	7500	20	682	41.73	3.79	12	2	18.97

Table 5.10 Resources summary

Name	Units	Work time (Min)	Number times used	Number times used (Min)	Average time per usage (Min)	% utilization
Employee1	1	1243.64	578	1.0599	1.0917	29.39
Employee2	1	1681.32	835	0.9919	1.0217	39.73
Employee3	1	349.56	1411	0.1089	0.1388	8.26
Employee4	1	196.20	271	0.3213	0.4027	4.63
Employee5	1	45.74	36	0.6450	0.6255	1.08
Employee6	1	0.00	0	0.0000	0.0000	0.00
Employee7	1	8.05	10	0.3839	0.4218	0.19

Regarding the magnitudes of added value for both products, the graphic in Fig. 5.13 was created. This figure presents a comparison by observation of the data generated through the simulation model and the real data from the real system. In fact, with the information obtained, the data considered for the validation process is observed, with an average of 375.85 s with a deviation of 7.6 s, which allows to ensure that the variation associated with the data generated by the simulation model is only 4%.

Finally, validation and verification evidence was gathered from the simulation results of the simulation tasks, since this was a closed queuing network, there were only two entities that were registered in the system, excepting the indicated entities that are delivered out of the plant as finished products. In addition, the simulation output was verified by the production department, and the proposed model was implemented in the real process. Also, the real data was 99% equal to the simulation model.

5.4 Discussion

The development of this project has faced the paradigm of change associated with the use of lean tools and simulation of the process in a company of small magnitude (according to its dimensions, capabilities, and utilities). Due to this effect, the change of argument is based on the implementation of lean tool and the simulation of the process.

On one hand, the lean implementation has been embraced and thrived in large companies and some medium companies. The impact that these companies have had has been reflected in the growth of their profits and their positioning in the global market according to (Altinkemer et al. 2011; Group 2019; Garretson et al. 2016; Padilla-Pérez and Hernández 2010). However, (Mei et al. 2018; Jiang and

Table 5.11 Entity summary

Name	Total exits	Current quantity in system	Average time in system (Min)	Average time in transport (Min)	Average time waiting (Min)	Average time in operation (Min)	Average time blocked (Min)
EZR	1488	110	556.96	229.21	52.23	6.06	269.46
EZM	675	771	1014.25	191.88	116.25	6.56	699.55

Table 5.12 Proportion of times for each entity

Name	% in move logic	% waiting	% in operation	% blocked
EZR	41.15	9.37	1.08	48.38
EZM	18.91	11.46	0.64	68.97

Qu 2020) mention that the micro, small industry, and the other part of the medium-sized companies encounter the limitations of these tools, which suffer the high risks generated by equipment and workforce failures, as well as the lack of flexibility or margin of error in meeting delivery dates. This translates into losses of process efficiency and customer confidence.

On the other hand, large companies justify the cost of simulation implementation in their use, in order to generate analysis scenarios for decision making (De Steur et al. 2016). Something that micro, small, and medium organizations do not consider because their resources are used for reaction activities, specifically for the customers' needs (Ibarra Cisneros et al. 2017). Assigning human and financial resources to cloning projects of the processes is an activity that does not fit the employers' needs, although they are aware of the benefits of this tool (Ye et al. 2020; Tanenbaum and Holstein 2019).

Due to these limitations, the adoption of these tools should be based on the conviction of the benefits that they generate. Although Ibarra Cisneros et al. (2017); Srivastava (2015); (Thomas et al. 2018; Deshkar et al. 2018; Nagi et al. 2017) show the disadvantages that lean and the simulation of the process represent during its adoption, the benefit obtained in this project differs from these opinions, since there was an increase in production capacity (68%) as well as inventory reduction (5%). In fact, that is translated into economic terms that represent an increase in the utility of the products in 13%, considering that the project was developed, implemented and validated within a small company in only 60 days. Finally, with the results achieved, it is necessary to replicate this methodology, for what it may take for further research, as well as the impact that the present proposal may have on micro and small companies.

5.5 Conclusions

The difference between the two strategies is the distance reduction on the work area and operations, increasing the total exits generated from the new distribution. In addition, there are other effects associated with these modifications, such as resource utilization, work areas, and WIP. Also, the simulation of the process provides its projection considering the layout design and restrictions.

Table 5.13 Location summary

Name	Scheduled time (Min)	Capacity	Total entries	Average time per entry (Min)	Average contents	Maximum contents	Actual contents	% utilization
Warehouse EZR	6810	1000	1252	106.73	19.6231	111	4	1.96
Warehouse EZM	6870	1000	1120	1197.15	195.1697	565	562	19.51
Supplier	6810	INF	2640	467.19	181.11	339	268	0.01
Slaughterhouse2	6750	1	558	10.13	0.83	1	1	83.77
Slaughterhouse1	6690	100	1248	89.15	16.63	84	26	16.63
Spooning	7320	100	1222	25.01	4.17	29	2	4.17
Cleaning	7350	20	126	192.99	3.30	7	0	16.54
Lot	7260	10	34	487.55	2.28	5	4	22.83
Lot2	6690	10	15	836.92	1.87	5	0	18.76
Cleaning2	7350	20	52	435.10	3.07	7	3	15.39
Cleaning3	6870	20	118	188.15	3.23	8	6	16.15
Cleaning4	7440	20	59	538.97	4.27	8	3	21.37
Spooning2	7410	20	557	56.24	4.22	20	2	21.13

Table 5.14 Resources summary

Name	Units	Schedule time (Min)	Work time (Min)	Number times used	Number times used (Min)	Average time per usage (Min)	% utilization
Employee1	1	4141	852.26	869	0.4863	0.4939	20.5796
Employee2	1	4141	904.32	937	0.4728	0.4918	21.8386
Employee3	1	4140	1891.31	1779	0.5280	0.5348	45.6840
Employee4	1	4140	281.55	345	0.3959	0.4203	6.8009
Employee5	1	4140	63.99	48	0.6688	0.6643	1.5457
Employee6	1	4140	0.79	1	0.6790	0.1120	0.0191
Employee7	1	4140	8.88	10	0.4586	0.4300	0.2146

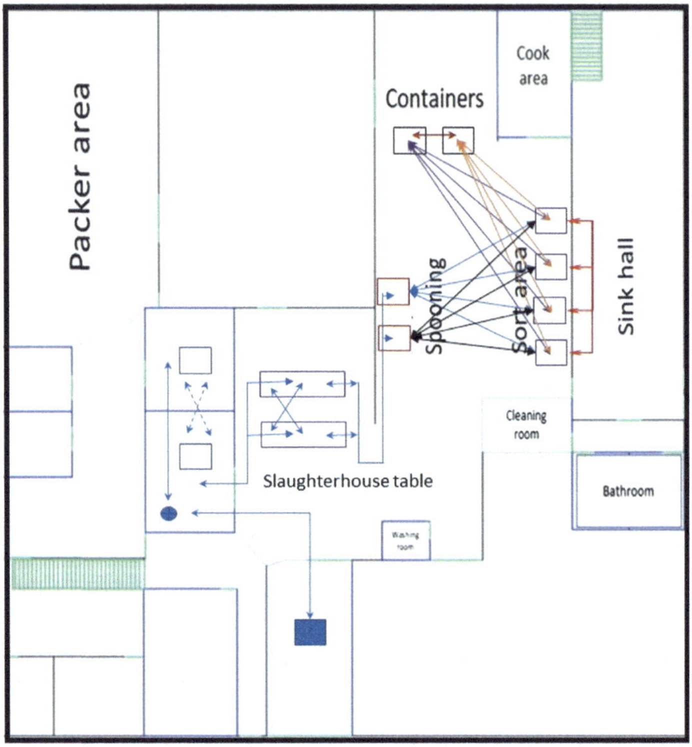

Fig. 5.10 Configuration system for online operation restricted by no mobile shared

In fact, along with the planning process capacity, the simulation model was able to validate the production sequencing and distance distribution between the work areas, considering the product demand. Therefore, developing the simulation model is an advantage for the planners who are also able to use the model to improve their system. It can be stated that these changes expose other alternative scenarios, in other words, the development of the simulation model is meant to provide a planning tool, which provides not only the ability to determine the planning process capacity, but it validates also the capacity to project the simulations and constraints that affect the expansion demand; in order to identify possible issues that may cause some strategic decision-making problems, as well as evaluate the impact of continuous improvement efforts.

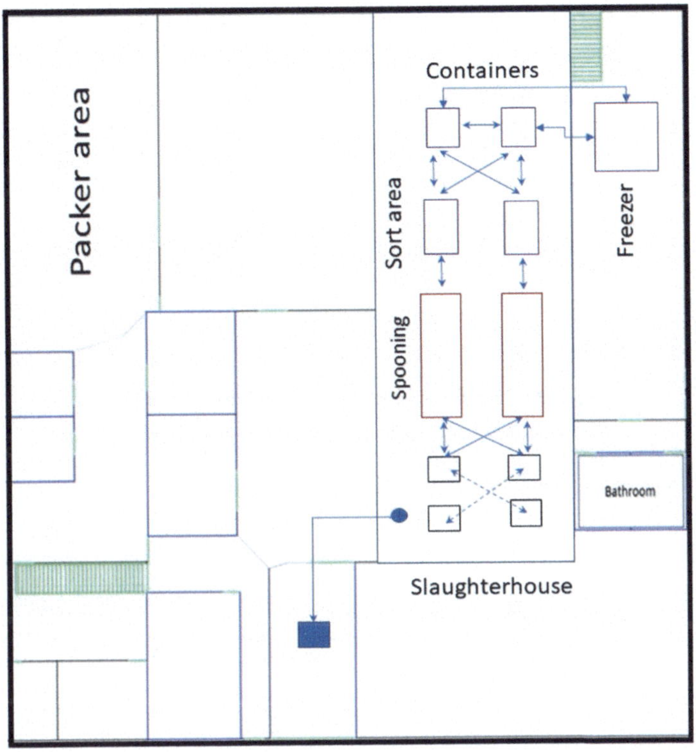

Fig. 5.11 Configuration system for online operation restricted by mobile shared and packing one product at a time

Table 5.15 Lead time and value added, for the scenarios considered

Configuration	Lead time (LT)/ min	Value added (VA)/sec		
	EZR	EZP	EZR	EZP
Original distribution	100	146	388	451
Production by family	90.5	130	383	450
Production online without share furniture	97	128	396	470
Production online, sharing furniture, but packing one product at time	32	47	378	439

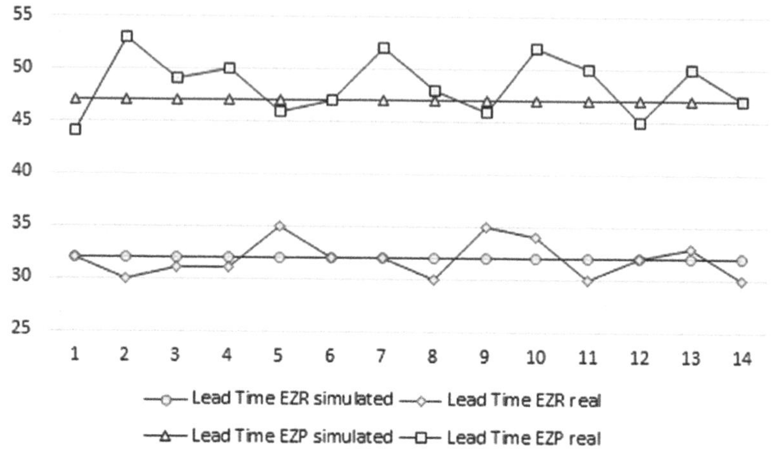

Fig. 5.12 Comparison of simulated lead times versus real lead times

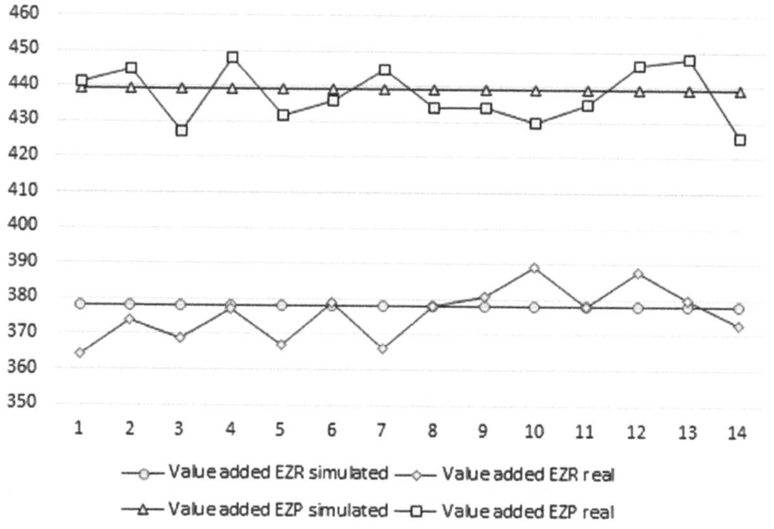

Fig. 5.13 Comparison of simulated value added versus real value added

References

Abdul FW (2018) Lean manufacturing implementation in inventory control as a repair process. J Logistik Indonesia 2(1):31–36. https://doi.org/10.31334/jli.v2i1.216

Altinkemer K, Ozcelik Y, Ozdemir ZD (2011) Productivity and performance effects of business process reengineering: a firm-level analysis. J Manag Inf Syst 27(4):129–162. https://doi.org/10.2753/MIS0742-1222270405

Badurdeen F, Jawahir IS (2017) Strategies for value creation through sustainable manufacturing. Procedia Manufac 8:20–27. https://doi.org/10.1016/j.promfg.2017.02.002

Cano JA, Campo EA, Gómez RA (2018) Discrete event simulation in production planning for modular packaging systems [Ins Spanish: simulación de eventos discretos en la planificación de producción para sistemas de confección modular]. Rev Tecnica 41(1):50–59
Chen JC, Li Y, Shady BD (2010) From value stream mapping toward a lean/sigma continuous improvement process: an industrial case study. Int J Prod Res 48(4):1069–1086. https://doi.org/10.1080/00207540802484911
De Steur H, Wesana J, Dora MK, Pearce D, Gellynck X (2016) Applying value stream mapping to reduce food losses and wastes in supply chains: a systematic review. Waste Manage 58:359–368. https://doi.org/10.1016/j.wasman.2016.08.025
Deshkar A, Kamle S, Giri J, Korde V (2018) Design and evaluation of a Lean Manufacturing framework using Value Stream Mapping (VSM) for a plastic bag manufacturing unit. Mater Today Proc 5(2):7668–7677. https://doi.org/10.1016/j.matpr.2017.11.442
Esmaeilian B, Behdad S, Wang B (2016) The evolution and future of manufacturing: a review. J Manuf Syst 39:79–100. https://doi.org/10.1016/j.jmsy.2016.03.001
Feltrinelli E, Gabriele R, Trento S (2017) The impact of middle manager training on productivity: a test on Italian companies. Ind Relat: a J Econ Soc 56(2):293–318. https://doi.org/10.1111/irel.12174
Ford S, Despeisse M (2016) Additive manufacturing and sustainability: an exploratory study of the advantages and challenges. J Clean Prod 137:1573–1587. https://doi.org/10.1016/j.jclepro.2016.04.150
Garretson IC, Mani M, Leong S, Lyons KW, Haapala KR (2016) Terminology to support manufacturing process characterization and assessment for sustainable production. J Clean Prod 139:986–1000. https://doi.org/10.1016/j.jclepro.2016.08.103
Group WB (2019) The world bank indicators [In spanish: Indicadores del Banco Mundial]. https://data.worldbank.org/topic/private-sector?view=chart. Accessed 02 17 2020 2018
Helleno AL, de Moraes AJI, Simon AT (2017) Integrating sustainability indicators and lean manufacturing to assess manufacturing processes: application case studies in Brazilian industry. J Clean Prod 153:405–416. https://doi.org/10.1016/j.jclepro.2016.12.072
Hong C (2017) Implementing lean six sigma to achieve inventory control in supply chain management. In: AIP conference proceedings, 2017. vol 1 AIP Publishing LLC, p 020037. https://doi.org/10.1063/1.5010654
Ibarra Cisneros MA, González Torres LA, Demuner Flores MdR (2017) Business competitiveness of small and medium manufacturing companies in Baja California [Ins Spanish: Competitividad empresarial de las pequeñas y medianas empresas manufactureras de Baja California]. Estudios fronterizos 18(35):107–130. http://dx.doi.org/https://doi.org/10.21670/ref.2017.35.a06
Jansson H, Söderman S (2013) How large Chinese companies establish international competitiveness in other BRICS: the case of Brazil. Asian Bus Manag 12(5):539–563. https://doi.org/10.1057/abm.2013.17
Jiang J, Qu L (2020) Changes in global trade patterns in manufacturing, 2001–2018. Am J Ind Bus Manag 10(5):876–899. https://doi.org/10.4236/ajibm.2020.105059
Jiménez-García JA, Téllez-Vázquez S, Medina-Flores JM, Rodríguez-Santoyo HH, Cuevas-Ortuño J (2014) Materials supply system analysis under simulation scenarios in a lean manufacturing environment. J Appl Res Technol 12(5):829–838. https://doi.org/10.1016/s1665-6423(14)70589-9
Kavanagh MJ, Johnson RD (2017) Human resource information systems: Basics, applications, and future directions. Human Resources, 4 edn. Sage Publications, Thousand Oaks, CA, USA
Kelton D, Sadowski R, Zupick N (2017) Systems simulation with arena software [In Spanish: Simulación de Sistemas con el software Arena]. McGraw Hill Education, New York, USA
Khedher A, BenOthman K (2020) Modeling, simulation, estimation and boundedness analysis of discrete event systems. Soft Comput 24(7):4775–4789. https://doi.org/10.1007/s00500-019-04231-9

Liu C-S, Lin L-Y, Chen M-C, Horng H-C (2017) A new performance indicator of material flow for production systems. Procedia Manufac 11:1774–1781. https://doi.org/10.1016/j.promfg.2017.07.311

López DCL (2016) Quality factors that affect the productivity and competitiveness of micro, small and medium-sized companies in the metalworking industrial sector [In Spanish: Factores de calidad que afectan la productividad y competitividad de las micros, pequeñas y medianas empresas del sector industrial metalmecánico]. Entre Ciencia E Ingeniería 10(20):99–107

Mei Y, Ye J, Zeng Z (2018) Two-way scheduling optimization of the supply chain in one-of-a-kind production based on dynamic production capacity restrictions. J Manuf Syst 47:168–178. https://doi.org/10.1016/j.jmsy.2018.05.007

Mourtzis D, Doukas M, Bernidaki D (2014) Simulation in manufacturing: review and challenges. Procedia CIRP 25:213–229. https://doi.org/10.1016/j.procir.2014.10.032

Nagi M, Chen FF, Wan H-D (2017) Throughput rate improvement in a multiproduct assembly line using lean and simulation modeling and analysis. Procedia Manufac 11:593–601. https://doi.org/10.1016/j.promfg.2017.07.153

Negahban A, Smith JS (2014) Simulation for manufacturing system design and operation: literature review and analysis. J Manuf Syst 33(2):241–261. https://doi.org/10.1016/j.jmsy.2013.12.007

Oleghe O, Salonitis K (2019) Hybrid simulation modelling of the human-production process interface in lean manufacturing systems. Inter J Lean Six Sigma 10(2):665–690. https://doi.org/10.1108/ijlss-01-2018-0004

Omogbai O, Salonitis K (2016) Manufacturing system lean improvement design using discrete event simulation. Procedia CIRP 57:195–200. https://doi.org/10.1016/j.procir.2016.11.034

Padilla-Pérez R, Hernández RA (2010) Upgrading and competitiveness within the export manufacturing industry in central America, Mexico, and the Dominican Republic. Lat Am Bus Rev 11(1):19–44. https://doi.org/10.1080/10978521003698489

Nithya RP, Haridass H (2019) Modelling and simulation analysis of a bulk queueing system. Kybernetes ahead-of-print (ahead-of-print). https://doi.org/10.1108/K-07-2018-0414

Srivastava PR (2015) A cooperative approach to optimize the Printed Circuit Boards drill routing process using Intelligent Water Drops. Comput Electr Eng 43:270–277. https://doi.org/10.1016/j.compeleceng.2014.10.005

Srivatsan TS, Sudarchan TS (2016) Additive manufacturing. innovations, advances, and applications, vol 1. 1 edn. Taylor & Francis, Boca Raton, Fl, USA

Stahl F (2016) Worker leadership: America's secret weapon in the battle for industrial competitiveness by Fred Stahl, vol 1. MIT, vol 1. MIT, Cambridge, MA,s USA

Swarnakar V, Singh A, Tiwari AK (2020) Assessment of manufacturing process through lean manufacturing and sustainability indicators: case studies in Indian perspective. In: Emerging trends in mechanical engineering. Springer Singapore, Singapur, pp 253–263. https://doi.org/10.1007/978-981-32-9931-3_25

Tanenbaum M, Holstein W (2019) Mass production. Encyclopædia Britannica, inc. https://www.britannica.com/technology/mass-production. Accessed 04 15 2019

Thomas T, Sherman SR, Sawhney RS (2018) Application of lean manufacturing principles to improve a conceptual 238Pu supply process. J Manuf Syst 46:1–12. https://doi.org/10.1016/j.jmsy.2017.10.007

Wainer GA (2017) Discrete-event modeling and simulation: a practitioner's approach. 1 edn. CRC press, Boca Raton, Fl, USA

Ye C, Ye Q, Shi X, Sun Y (2020) Technology gap, global value chain and carbon intensity: evidence from global manufacturing industries. Energy Policy 137:111094. https://doi.org/10.1016/j.enpol.2019.111094

Zhang Y, Cassandras CG, Li W, Mosterman PJ (2019) A discrete-event and hybrid traffic simulation model based on SimEvents for intelligent transportation system analysis in Mcity. Discrete Event Dynamic Syst 29(3):265–295. https://doi.org/10.1007/s10626-019-00286-w

Chapter 6
Effect of Advanced Manufacturing Technology on Responsive Supply Chain Strategy, Pull System and Responsiveness to Market

José Roberto Díaz-Reza, Adrián Salvador Morales-García, and Jorge Luis García-Alcaraz

Abstract Nowadays, companies are requiring a fast response to market, and supply chain must be adapted. This paper presents a structural equation model (SEM) integrating advanced manufacturing technology, pull system, responsiveness to market as independent latent variables, and responsive supply chain strategy as dependent latent variable. Those variables are related using six hypotheses. The SEM is aimed to measure the effect among latent variables and identify the most important activities that have the greatest effect. The SEM is statistically validated using information from 254 responses to a questionnaire applied in the manufacturing industry, and the partial least squares (PLS) technique is used. Findings indicate that advanced manufacturing technologies indirectly support companies to be able to respond to changes in demand and allow them to offer a rapid response in the changing market through the pull system implementation.

Keywords AMT · Pull system · Supply chain · Responsiveness supply chain

6.1 Introduction

Since 1965, long before the North American free trade agreement (NAFTA), favorable regulations have been established with the United States of America (USA) for the maquiladoras. Since then, the proximity to the US market and relatively cheap

J. R. Díaz-Reza
Department of Electric Engineering and Computation, Autonomous University of Ciudad Juárez, Av. Del Charro 450 Norte. Col. Partido Romero. Ciudad Juárez, Chihuahua, México

A. S. Morales-García · J. L. García-Alcaraz (✉)
Department of Industrial Engineering and Manufacturing, Autonomous University of Ciudad Juárez, Av. Del Charro 450 Norte. Col. Partido Romero. Ciudad Juárez, Chihuahua, México
e-mail: jorge.garcia@uacj.mx

J. L. García-Alcaraz
Division of Research and Postgraduate Studies, Tecnológico Nacional de México/Instituto Tecnológico de Ciudad Juárez. Av. Tecnológico, 1340, Fuentes del Valle, 32500. Ciudad Juárez 32500, Chihuahua, México

J. L. Garcia Alcaraz et al. (eds.), *Trends in Industrial Engineering Applications to Manufacturing Process*, https://doi.org/10.1007/978-3-030-71579-3_6

labor have made Mexico one of the most favorable offshore destinations for foreign companies and especially from USA for a long time (Utar and Ruiz 2013). In addition, NAFTA allowed US corporations and other countries import duty-free raw materials, equipment, machinery, spare parts, and other items necessary for the assembly or manufacture of finished products for subsequent export to Mexico (Najera 2008).

The maquiladora industry initially developed as a means to address the efficiency of the supply of manufactured products in the largest market in the world represented by the USA and as a way to increase employment for Mexican workers (Hadjimarcou et al. 2013). Many US companies attempted to develop maquiladora operations to take advantage, in addition to cheap labor, less restrictive Mexican labor laws and unions and the ability to be close to the newer and emerging markets (Dowlatshahi 2008).

In that sense, in Mexico, there are a total of 5121 manufacturing, Maquiladora, and export services industries (IMMEX), being the north of Mexico the region with more participation at national level with a percentage of 58.87% of the companies. Specifically, Ciudad Juarez in the state of Chihuahua is the second city with the highest presence of IMMEX companies with 326. Of these companies, the ones with the most participation in that city are those of the automotive branch with 32% and those of the electronic branch with 29%, plastics and metals 12%, medical 9%, among others. These companies generate a total of 302,584 jobs in the city, and for this reason, it is important that these companies are at the forefront within their supply chains (SC) applying the tools, methodologies, disciplines, and technologies to remain competitive within the market, but, above all, maintain those employment rates for the city (INEGI 2019).

That maquiladora companies are partners in a complex and globalized production systems, and SC perform essential functions in coordination with various commercial entities and in connecting offer with demand (Hum et al. 2018). In addition, SCs play a crucial role in organizations, as they are responsible for guaranteeing the availability of products and services to the final consumer and, consequently, the financial success of the entities involved (Barbosa-Povoa and Pinto 2020). However, the management of such CS in the maquiladora industry is complex due to the multiplicity of material and information flows, diversified characteristics of contradictory entities, and objectives often present (Barbosa-Povoa and Pinto 2020).

Thus, supply chain management(SCM) plays a vital role in maintaining the maquiladoras in the global market, organizing activities from supplier to end customer effectively (Reddy et al. 2019). SCM refers to manage the business from raw material acquisition to manufacturing process, distribution, customer service, and, finally, reprocessing and disposal of products (Reddy et al. 2019); that is, it refers to an integral management in which the entire production cycle is integrated, and therefore, control and supervision arerequired throughout it.

The final SC goal is to improve its performance to meet customer expectations. Therefore, measures and performance metrics are needed to measure its effectiveness and efficiency (Reddy et al. 2019).

In addition, with the growing proliferation of products and shorter technology life cycles, time to market is critical to avoid obsolete inventories (van Hoek 2006), and therefore, manufacturing companies must continually update their product offerings, while remaining competitive, and in that sense, managers must adapt the SC characteristics (Pero 2010), since with a faster pace of life, customers now value time more than ever, and as a result, it is common for manufacturing companies to quote a promised delivery time to customers (Hum et al. 2018).

To achieve the above and maintain or improve the competitive position for the companies, managers must continue to improve their operations (Gules and Burgess 1996) and resort to the implementation of advanced manufacturing technology (AMT) in their SC, which must be aligned with product development decisions that must be designed at low cost and delivered in specific time and quality (Pero 2010). AMT is the latest generation techniques and manufacturing equipment used in the workshop that are integrated with information technology and other practices implemented in relation to the production process (DeRuntz and Turner 2003). Thus, AMT facilitates the development of a successful long-term strategy of the SC with the client (Narasimhan et al. 2008; Díaz-Reza et al. 2019a).

This strategy regarding SC and AMT should be integrated into the production system, and therefore, companies use pull systems, which support the inventory management and their applications are varied. The pull system drives production based on customer demand. Instead of boosting products as allowed by production capacity, in the pull principle, inventory levels are limited and controlled by order requirement signals (or Kanbans) (Puchkova et al. 2016). Pull systems use local information about the status of production and inventories to authorize the order releases (Selçuk 2013).

Fortunately, there are several researches focused on these variables currently in literature; for example, Koufteros (1999) has linked through a structural equation model (SEM) the pull system with preventive maintenance, where he analyzes the setup improvement as an independent variable and delivery reliability as a dependent variable. Also in Díaz-Reza et al. (2019b), AMTs have been related to the benefits that are obtained from their correct implementation within the production systems, while Díaz-Reza et al. (2019c) are analyzing the performance in a production processes.

That is why, it is important that the companies and partners in a SC establish a good strategy to face the increasingly changing market. The SC strategy underpins how each organization within the chain deals with the other entities in that supply chain in their attempts to achieve their individual organizational strategic objectives (Harland et al. 1999). SC strategy is shaped by both the supply and demand characteristics of the product (Nakandala and Lau 2019). A key objective of SC strategy is to enhance the focal firm's supply chain responsiveness with respect to its customers (Melnyk et al. 2010). SC responsiveness is an important indicator of how well the SC strategy fulfills its objectives since it denotes the SC ability to be adapted to changing customer needs and ultimately lead to elevated performance (Blome et al. 2013).

The main problem is that AMT and pull systems tools have not been analyzed together and, in addition, have not been related to the strategy and speed of response of the supply chain. That is why, this research is aimed to quantify their relationship using a SEM and integrating the effect of AMT on a pull system, on the responsiveness of companies, and their strategy on a supply chain within Mexican maquiladora industry, because they have globalized SCs and requiring a fast response to customers around the world. The rest of the article addresses as follows: Sect. 6.2 describes some definitions for each variable in the model, justifying the relationships as hypotheses, Sect. 6.3 addresses the methodology, Sect. 6.4 the results, and Sect. 6.5 refers to conclusions.

6.2 Literature Review and Hypotheses

This section presents a literature review regarding advanced manufacturing technology, pull system, and responsiveness to market; in addition, each of the hypotheses integrated in the SEM are described.

6.2.1 Advanced Manufacturing Technology (AMT)

Advanced manufacturing is understood as the use of modern technologies to deliver existing products and new products to the market and also focuses on the improvement of design and manufacturing processes in all areas, along with the integration of information technology systems throughout the SC (Krot et al. 2019). Thus, to face the competitive pressure of globalized markets, the manufacturing strategy must have an advanced technology component that allows manufacturers to develop competitive capabilities (Birasnav and Bienstock 2019).

Advanced manufacturing technologies (AMT) play a fundamental role in the growth of the industry. The AMT implementation offers the advantage of mass production in customer demand in a short time and allows to maintain the efficiency of the production system (Nath and Sarkar 2017). AMT systems include, among others, flexible, robotic, digital and computer-integrated manufacturing systems, just-in-time systems, which provide opportunities to improve innovation, competitiveness, profitability, and respect for the environment through capabilities highly efficient manufacturing (Liu 2008).

In manufacturing, technology is incorporated into products, in the physical processes by which they are manufactured and, increasingly, in the management systems that control all operations (Sohal Amrik 1992). The benefits gained from AMT are associated to improve cost, quality, flexibility, and delivery times (Sohal Amrik 1992). AMT has operational, technical superiority, and other intangible benefits, compared to traditional systems (Aravindan and Punniyamoorthy 2002; Kaplan 1986; Siha and Linn 1989), such as increased competitiveness, lower production cost, higher value for money, employment of fewer people, minimum inventories, product quality, flexibility, among others.

In addition, AMTs are able to adapt to changes in the variety of products with a very short delivery time, while maintaining efficiency and profitability (Nath and Sarkar 2017), and that is why, selecting an AMT for a productive system, it is important that manufacturing companies analyze the improvement of their manufacturing performance when making capital investment decisions (Chuu 2009). The implementation of AMT requires a substantial investment in technology and, above all, changes in the culture and structure of the organization and requires careful planning at all levels of the organization to ensure that its implementation achieves the desired objectives (Dawal et al. 2015).

Efstathiades et al. (2002) propose a strategic plan, transfer, and implementation of AMT that includes the planning phase, the selection, transfer and pre-installation phase, and a post-installation phase, with the aim of providing a useful guide to managers in justification and correct implementation of AMT to guarantee technical, manufacturing, and commercial success. In the same way, Efstathiades et al. (2000) in their article address the management processes followed during the transfer of technology to the manufacturing environment and the steps taken, both before and after the implementation and productive operation of the technologies.

In the same way, there are several investigations where they analyze AMT, for example, to know which of the technologies is the most appropriate for its implementation.Chuu (2009) builds a group decision-making model using a diffuse analysis of multiple attributes to assess the suitability of manufacturing technology. In Singhry et al. (2016), a model of structural equations is presented in which the AMT is analyzed in effect on the performance of the SC and the concurrent engineering of the product design;Bechtsis et al. (2017) report the role automated guided vehicles on sustainable supply chain management in the digitalization era; Birasnav and Bienstock (2019) report a case of study for measuring the relationship between SC integration, AMT, and leadership; Koh et al. (2011) report the drivers, barriers, and critical success factors for ERP II implementation as AMT in supply chains, among others.

There is no doubt that AMT and its soft technologies have an important role in the performance of the supply chain, since many of them are directly implemented there to reduce accidents, carry out loading and unloading of material, and support in the storage process. In this investigation, AMT is evaluated by the following items, which are associated with the handling of materials and support the SC.

- AMT01. Automated parts loading/unloading
- AMT02. Automated guided vehicles, AGVs
- AMT03. Automated storage–retrieval systems.

6.2.2 Responsive Supply Chain Strategy (RSCS)

Supply chain´s responsiveness is the ability of the SC to respond rapidly to the change in demand, both in terms of volume and mix of products (Holweg 2005a). Quick response enables SC to meet customer demands for ever-shorter lead times (Azaron et al. 2019). The effective strategy SC refers to the process configuration of a supply network so that operation directly supports corporate strategy (Lyons et al. 2012). However, it is likely that the introduction of new products and services or the entry into new markets will be more successful if it is accompanied by innovative CS designs, innovative practices, and technology enablement (Munksgaard et al. 2014) since, according to van Hoek (2006), the impact of the SC on the development of new products and their introduction is important in areas such as

- Shipping the product to the market fast enough (before the product launch dates).
- Ensure enough inventory in the launch data.
- Ensure a flow of parts and components for the manufacture of new products.

In that sense, it is important that companies are able to innovate to be able to introduce new products to the market, since those that innovate gain a competitive advantage over their competitors (Li et al. 2018); since without innovation, companies may be unable to transform resources into products (Hult et al. 2004). Innovating in products, organizations, and SC is crucial for success and competitive advantage (Kim and Chai 2017).

In this investigation, the supply chain strategy is evaluated with the following items:

- RSCS01. Wider product range
- RSCS02. Offer new products more frequently
- RSCS03. Offer more innovative products.

For companies to offer a wider range of products to their customers, as well as offer innovative products, but above all, more frequently, these companies must use the use of AMT since these technologies allow customers to absorb the options to create a new design and transform it into tangible products (Birasnav and Bienstock 2019). A responsive SC uses AMT to improve marketing time, delivery time, cost, flexibility, and quality performance (Roh et al. 2014). Therefore, the capabilities to achieve flexibility of products and processes contribute greatly to maintaining the competitive advantage among the companies that implement these technologies (McDermott et al. 1997).

Specifically, Bechtsis et al. (2017) has found that AGV systems are essential in current production systems, since they have autonomy and favor fast transport processes for raw materials and finished products; Ekren (2020) reports that AR/AS systems support the warehouse efficiency and can be friendly with environment. In that sense, the following hypothesis is established:

H1: AMT has a direct and positive effect on responsive supply chain strategy in maquladora industry.

6.2.3 *Pull System (PS)*

The JIT philosophy was first introduced by Toyota Motor Company, 25 years ago. JIT is essentially a manufacturing approach that seeks to provide the right amount of material when required, thereby reducing work-in-process (WIP) inventories and with the aim of maximizing productivity (Singh 1992) and currently is an important lean manufacturing tool applied in maquiladora industry (García et al. 2014; García-Alcaraz et al. 2016).

Pull system (PS) is a system used in just in time (JIT) that is in opposition to the traditional approach to strictly push (Push) products, which often leads to high levels of inventory in process (WIP) and a prolonged response to customer demands (Marsh and Conard 2008). On the contrary, in a push system (PS), the products are generated following the planned chronograms (depending on the estimated delivery times) and then moved (i.e., pushed) to the next location programmed for use (Li 2005). In PS, jobs are authorized by visible signals (e.g., cards, spaces, etc.) to move from one process to the next; therefore, jobs are "pulled" through production processes according to demand of final products, instead of "forced" jobs along their routes in the push system, that is, they have a limit on the WIP level (Hopp and Spearman 2004).

Likewise, PS is a production system in which the work is released in the production facility based on its current status and includes information associated with the available inventory, WIP, and the demand made (Gass and Fu 2013). The benefits of a pull system are that it would reduce the response time to process orders through the operation (Sakakibara et al. 1997), although the successful application of a pull system has been limited to repetitive production (Marsh and Conard 2008).

To know if a pull system is implemented, the following items are evaluated:

- PS01. Undertaking actions to implement pull production (e.g., reducing batches, setup times, using Kanban systems, etc.)
- PS02. Planned effort to implement pull production (e.g., reducing batches, setup times, using Kanban systems, etc.).

AMT favors the pull system, since the main motivation to invest in this technology is to improve the organization competitiveness, such as responsiveness, quality, and flexibility (Burgess 1997). One of the benefits that AMT brings with it when implementing certain technologies is the flexibility to respond to changes in schedules and the combination of products (Rahman et al. 2009; Mora-Monge et al. 2008), and Lyu et al. (2020) indicate that AMT as AS/RS facilitates warehousing, makes easy inventory management and pull production. Also, Gonzalez et al. (2019) report that AGV systems are supporting flexible manufacturing systems, push production, and just in time (JIT) in an smart manufacturing context. Considering last research, the following hypothesis is proposed:

H2. AMT has a direct and positive effect on pull system in a maquiladora industry.

PSbegins with the customer's order, that is, it is only manufactured to meet the customer's order. In that sense, there will be no inventory to store, and, therefore, there will be no costs to maintain inventory (Khojasteh 2016).

PS is those where order quantities are controlled by the difference between the supplementary stock point and the state of current inventory (Kimura and Terada 1981), maximize the probability of customer demand satisfaction, while maintaining the smallest WIP and inventory (Zhang 2013). In that sense, the following hypothesis can be raised:

H3. Pull system has a direct and positive effect on responsive SC strategy in a maquiladora industry.

6.2.4 Responsiveness to Market (RtM)

Responsiveness corresponds to "the ability to respond and adapt effectively over time based on the ability to read and understand the real signals of the market" (Catalan and Kotzab 2003). Holweg (2005b) defines the ability to respond as the ability to react to a purpose within a period appropriate to customer demandor changes in the market, to achieve or maintain a competitive advantage.

Supply chain responsiveness is the probability of fulfilling customer orders within a promised delivery time (Hum et al. 2018). It corresponds to the speed with which the tasks are performed in which key metrics are used such as order fulfillment, cycle time, delivery cycle time (Ross 2015), product, volume, mix, and delivery that can be related to different temporary hobbies and can be present as a potential or demonstrated response capacity (Reichhart 2007). A high level of response in economic exchanges between a buyer and their key suppliers offers several advantages (Yang 2019), and, in that sense, purchasing and responsible supply improve the response capacity of the SC, which is reflected in the efficiency of the process, the ability to manage customer knowledge, the performance of dyadic quality, and the capacity of innovation and improvement of the buyer–supplier relationship.

Responsiveness in manufacturing is considered as the main source for building responsiveness in the SC (Sandberg 2018). To achieve responsiveness, certain types of flexibility of the manufacturing system itself, as well as supply and logistics subsystems are required (Holweg 2005b). The types of flexibility needed to achieve such responsiveness in the SC depend on the structure and environment of the system and hence the importance of improving the responsiveness of the SC and that companies implement CS practices along with strategies for SC to build receptive SC. In addition, it is important that there is a strategic partnership with suppliers to improve the relationship between the agile SC strategy and the response capacity (Qrunfleh 2013). In that sense, the greater the responsiveness in an SC, the greater the performance of this.

Within the literature, different works can be found in which the ability to respond throughout the SC is analyzed, for example, in Sinkovics et al. (2011), in their study, demonstrate that the use of information technology integration through B2B information technologyand trust can help multinational companies improve the responsiveness of local suppliers in chain relationships of global supply. Also, you can find jobs where the ability to respond has opportunities for improvement, for example, in Catalan (2003), the variable responsiveness is analyzed using the variables of delivery time, the bullwhip effect, the exchange of information, and postponement strategies, this within a Danish mobile phone industry. In this study, the variable responsiveness is evaluated by the following items:

- RtM01. Time to market
- RtM02. Delivery speed
- RtM03. Delivery dependability
- RtM04. Manufacturing lead time.

Responsiveness is gained from different factors. For example, AMTs play a fundamental role for the growth of industry and organization, since mass production can be achieved in customer demand in a short time (Nath and Sarkar 2017). In addition, AMT are able to adapt to changes in the variety of products with a short delivery time, while maintaining efficiency and profitability (Nath and Sarkar 2017). Since AMT favors production, helps shorten production times, and gives companies flexibility, in that sense, the following hypothesis is proposed:

H4: AMT has a direct and positive effect on responsiveness to market in a maquiladora industry.

Also, pull-based manufacturing strives to synchronize production with real-time consumption, which increases on-time delivery performance, reduces shortages, and reduces costly last-minute changes in orders (Kumar 2013). A pull system reduces WIP, releases cash flow and space requirements that allow for expansion, quality and less cycle time (Aghazadeh 2004). Therefore, the following hypothesis can be raised:

H5: Pull system has a direct and positive effect on responsive to market in a maquiladora industry.

The SC plays a crucial role during the execution of the efficient launch and subsequent product performance (Kou 2015). The impact of the SC on the development of new products and the introduction of products is important in areas such as fast product delivery to the market, ensuring sufficient inventory in the launch data and ensuring a flow of parts and components for the manufacture of new products (van Hoek 2006). The SC can improve the process of developing new products, reduces development costs and engineering changes, and improves product quality and time to market (Ragatz et al. 2002). In that sense, the following hypothesis can be established:

H6: Responsive supply chain has a direct and positive effect on responsiveness to market.

Figure 6.1 illustrates the proposed structural equation model, as well as the hypotheses represented graphically with the four latent variables and the observed variables or items.

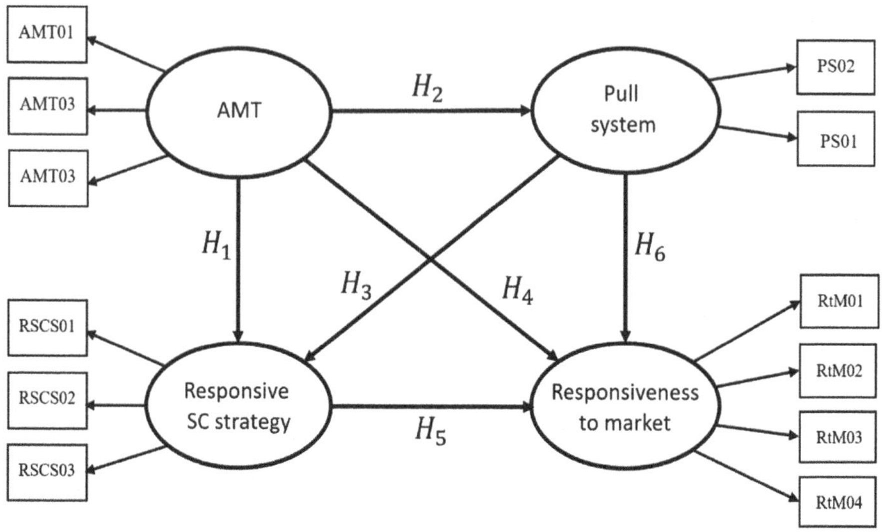

Fig. 6.1 Proposed model

6.3 Methodology

To carry out this research, the following steps are carried out.

6.3.1 Step 1. Survey Development

A literature review was made in databases such as Sciencedirect, Emeraldinsight, Springer, among others with the aim of creating a questionnaire to obtain information from the manufacturing industry. The search was performed using the keywords as supply chain, supply chain strategy, advance manufacturing technology, and supply chain responsiveness. The information collected was classified by latent variables according to their affinity for design a preliminary questionnaire according to a rational validity (Worley and Mitchell 2008). That first questionnaire is reviewed by judges (three academics and four managers in industry) to know adaptability to the maquiladora industry, because every item was translated from other research and adapted to Mexican context.

The final survey was divided into two sections or latent variables. The first section consists of questions related to demographic information corresponding to each of the people who answered the questionnaire. These questions have to do with the years of experience developing the work, the gender of the person, position in the company, and sector to which it belongs. The second section consists of the variables or tools that are used within the supply chain. In this case, the questionnaire consists of eight variables. For this study, only the variables described above were used. Likewise, to answer each question in the final questionnaire, a five-point Likert scale was added where 1 means that the activity is never done and the 5 that the activity is always done.

6.3.2 Step 2. Administration of the Questionnaire

Managers laboring in production and SC departments into Mexican maquiladora industries are contacted via phone calls for agreeing an appointment and responding the questionnaire. Manager must have at least one year in their current job position.

The questionnaire is applied in different sectors of the Mexican maquiladora industry, and this is done through a stratified sample to identify the people involved in the activities related to supply chain. Also, after some interviews, responders some time recommend other colleagues, following a snowball sample technique. Also, if a possible responder cancels three accumulative appointments for interview, then that case is eliminated.

6.3.3 Step 3. Data Screening

The information obtained from questionnaire is collected in a database in the SPSS 25® software, and a debug process is performed to eliminate missing and extreme values (Crambes and Henchiri 2019). If missing values percentage is lower than 10%, then they are substituted by the median because the values are expressed in an ordinal scale, but if the average is higher than 10%, then that case is eliminated (Dray and Josse 2015). For extreme values, every item is standardized, and if that values are higher than four in absolute value, they are replaced by the median.

6.3.4 Step 4. Validation of the Questionnaire

The latent variables in the SEM are validated through the use of some indices such as R^2 and Adjusted R^2 to measure the parametric validity, since they indicate the percentage of variance explained in the latent variable dependent by the independent latent variables (Kock 2017). Values of R^2 coefficients and adjusted R^2 coefficients below 0.02 suggest combined effects of predictors in latent variable blocks that are too weak to be considered relevant from a practical point of view.

Indices such as composite reliability and Cronbach's Alpha are used to measure the internal consistency of the variables, as well as to validate the variables from a parametric point of view for which values equal to or greater than 0.7 are recommended (Kock and Lynn 2012b).

The average variance extracted (AVE) is used to measure discriminant validity in which values equal to or greater than 0.50 are recommended. A measuring instrument has a good discriminant validity if the answers to the questions associated with each latent variable are not confused by the respondents (Kock 2017). Likewise, convergent validity is also measured through cross-loads, and it is indicated that a measuring instrument is suitable if the people who answer each variable's questions understand them in the same way they were designed (Kock 2017). In this case, it is required that the factor loads be equal to or greater than 0.5 and, in addition, that the p-values associated with them be less than 0.05 (Hair et al. 2010).

Collinearity between latent variables is also measured using the VIF index (Kock 2017). VIF is a measure of the degree of "vertical" collinearity (Kock and Lynn 2012a), or redundancy, among the latent variables that are hypothesized to affect another latent variable. Finally, Q^2 is obtained in order to measure the non-parametric predictive validity, whose recommended values are greater than 0.00 and similar to R^2.

6.3.5 Step 5. Structural Equation Model (SEM)

The SEM is used to validate the hypotheses proposed in Fig. 6.1, and the partial least squares technique integrated in WarpPLS 6.0® software is used, extensively recommended to test relationships with ordinal data, without normality and small samples. The model efficiency is evaluated using the average path coefficient (APC) for test, Average R^2 (ARS), and Average adjusted R^2 (AARS), average block VIF (AVIF), Average full collinearity VIF (AFVIF) and Tenenhaus GoF indices (GoF).

Three effects between latent variables are estimated: 1. direct effects that occur between an independent latent variable and a dependent latent variable and that in Fig. 6.1 are represented by arrows, 2. the indirect effects occur between an independent latent variable and a dependent variable but through a mediating variable and 3. total effects, these are the sum of the previous two. The calculations are made with 95% confidence level and the null hypothesis H_0: $\beta = 0$ is tested vs the alternative hypothesis H1: $\beta \neq 0$ to know the relationship between to variables, where β represent a regression coefficient.

6.3.6 Sensitivity Analysis

A sensitivity analysis is performed to obtain estimates based on conditional probabilities of standardized latent variables and to know how the levels of an independent latent variable affect another dependent variable. Specifically, two scenarios are analyzed for each relationship between the variables or hypothesis: high (+) in which the activities in each of the latent variables are carried out on a regular basis, that is, in a favorable scenario and low scenario (−), in which the activities in each of the latent variables are not carried out on a regular basis, that is, an unfavorable scenario. In that sense, if an analysis suggests that two variables are causally linked, producing a trajectory coefficient of 0.25, for example, this essentially means in probabilistic terms that an increase in the predictor variable leads to an increase in the conditional probability(see Eq. 6.1) of which the dependent variable.

$$P(A|B) = \frac{P(A \cap B)}{P(B)} \quad (6.1)$$

This conditional probability will be calculated for each of the hypotheses proposed in the model of Fig. 6.1, and in probabilistic terms, the following scenarios will be obtained:

- The probability that a latent variable is presented independently in a favorable scenario ($Z > 1$). On the contrary, the probability that the activities are not carried out in a favorable manner will be obtained ($Z < -1$).
- The probability of finding the independent latent variable at high level P ($Z_i > 1$) and the probability of having a dependent variable at high level P ($Z_d > 1$) simultaneously.
- The probability of finding the independent latent variable at high level P ($Z_i > 1$) and the probability of having a dependent variable at low level P ($Z_d < -1$) simultaneously.
- The probability of finding the independent latent variable at low level P ($Z_i < -1$) and the probability of having a dependent variable at high level P ($Z_d > 1$) simultaneously.
- The probability of finding the independent latent variable at low level P ($Z_i < -1$) and the probability of having a dependent variable at low level P ($Z_d < -1$) simultaneously.

where Z_i represents the probability in an independent latent variable and Z_d for a dependent.

6.4 Results

The results obtained from the analysis performed by developing the methodology described in the previous section are shown below.

6.4.1 Sasmple Desription

After 3 months applying the questionnaire, 254 valid responses were considered valid and 14 were eliminated. Table 6.1 illustrates that the sector with more participation was the automotive industry with 143, and the most respondents had the position of

Table 6.1 Industrial sector versus position

	Industrial sector							
Position	A	B	C	D	E	F	G	Total
Manager	60	2	7	5	1	3	5	83
Engineer	66	3	12	7	4	0	16	108
Supervisor	17	0	8	2	1	0	6	34
Total	143	5	27	14	6	3	27	225

A Automotive; *B* Aeronautics; *C* electric; *D* Electronics *E* logistics; *F* Machining; *G* Medical

technician with 78 participants, followed by engineers with 77. Note that of the 254 questionnaires analyzed, only 225 people gave demographic information. It can also be noted that of the total of 225 people who answered these questions, 143 are from the automotive sector, that is, 63.5% of the sample, in a way it sounds logical since in the city most of the maquiladora companies are of this sector.

6.4.2 *Latent Variables Validation*

Table6.2 shows the indexes values to validate the latent variables, and it can be concluded that the questionnaire has enough predictive validity from a parametric point of view since it has R^2 and adj R^2 values are higher than 0.2. Also, the latent variables have content validity because the Cronbach composite reliability and alpha index are greater than 0.7. Also, the AVE is greater than 0.5, and it is concluded that there is enough convergent validity.

Also, Fig. 6.2 shows the cross-loads for each of the latent variables. For example, in the AMT variable, there are crossed loads of 0.712, 0.791, and 0.920, so it is concluded that there is convergent validity of the items in the latent variables. In the same way, cross-loads for the other variables in the model in Fig. 6.2 have values greater than the minimum value suggested. Therefore, the questions measure what they intend to measure when the questionnaire is designed. In addition, there are no multicollinearity problems since the VIF values are lower than 3.3, and finally,

Table 6.2 Latent variables validation

	AMT	RSCS	Pull System	RtM
R^2		0.560	0.363	0.718
Adj. R^2		0.556	0.361	0.715
Comp. Rel	0.908	0.933	0.953	0.942
Conbrach alpha	0.846	0.893	0.901	0.918
AVE	0.768	0.823	0.910	0.802
Full collin VIF	1.730	2.353	3.271	3.224
Q-squared		0.547	0.361	0.705

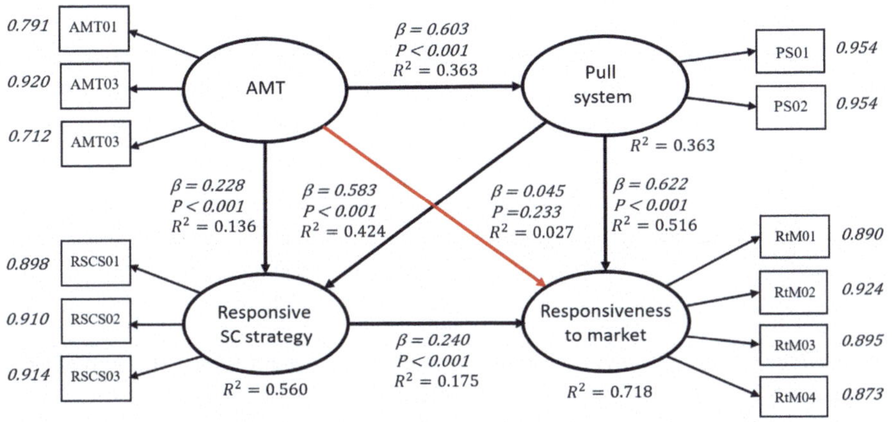

Fig. 6.2 Evaluated model

Q-squared is similar to R^2, and it is concluded that there is non-parametric predictive validity.

6.4.3 Structural Equation Modeling

Table 6.3 shows the quality and efficiency indices of the model presented in Fig. 6.2. It is observed that all the indices have acceptable values. According to the p-values associated with the APC, ARS, and AARS indices, it is concluded that the model has enough predictive validity, there are no collinearity problems, and it fits the data, so it is analyzed.

The model evaluated is illustrated in Fig. 6.2, illustrating the direct effects among latent variables. Based on the p-value associated with β, it is concluded that five hypotheses are statistically significant. For example, for H_1, it is concluded that AMTs have a direct and positive effect on responsive SC strategy; since when the first variable increases its standard deviation by one unit, the second goes up by 0.228 units. H1, H2, H3, H5, and H6 are statistically significant and are accepted since the

Table 6.3 Model fit and quality indexes

Index and criteria	Value	*P*-Value
Average path coefficient (APC, $P < 0.05$)	0.387	$P < 0.001$
Average R^2 (ARS, $p < 0.05$)	0.547	$P < 0.001$
Average adjusted R^2 (AARS, $p < 0.05$)	0.544	$P < 0.001$
Average block VIF (AVIF, acceptable if ≤ 5)	2.032	
AFVIF, acceptable if ≤ 5	2.645	
Tenenhaus GoF, ≥ 0.36 small ≥ 0.1, medium ≥ 0.25, large ≥ 0.36	0.672	

Table 6.4 Hypotheses test

	Independent variable	Dependent variable	β-value	p-value	Conclusion
H1	AMT	Responsive SC strategy	0.228	<0.001	Accept
H2	AMT	Pull system	0.603	<0.001	Accept
H3	Pull system	Responsive SC strategy	0.583	<0.001	Accept
H4	AMT	Responsiveness to market	0.045	0.233	Reject
H5	Responsive SC strategy	Responsiveness to market	0.240	<0.001	Accept
H6	Pull system	Responsiveness to market	0.363	<0.001	Accept

P-value is less than the level of statistical significance with which the hypotheses were tested.

Table 6.4 illustrates a resume regarding the hypothesis.

Figure 6.2 illustrates a value of R^2 in all dependent variables. For example, 0.718 associated to responsive to market is explained in 0.175 by responsive supply chain strategy, in 0.516 of pull system and in 0.027 by AMT. With these results, it can be concluded that the pull system is the most important variable for explain responsive to market because is has the bigger value.

Table 6.5 shows three indirect effects among latent variables and that all of them are statistically significant. It is important to mention that, in Fig. 6.2, the direct effect that exists between the AMT variable on responsiveness to market is not statistically significant, but, the indirect effect has a value of $\beta = 0.513$ with a $p < 0.001$. This means that advanced manufacturing technology has an effect on the responsiveness to market variable; when they introduce products to the market, it also has an effect on the speed of delivery, and this is done through PS and the responsive SC strategy.

Table 6.6 also illustrates the total effects, and all of them are statistically significant, even though a direct effect was not. It is observed that the greatest effect

Table 6.5 Sum of indirect effects

Independent variable	Dependent variable	Beta value	*P*-value	Effect size
AMT	Responsive SC Strategy	0.351	< 0.001	0.210
AMT	Responsiveness to market	0.513	< 0.001	0.302
Pull system	Responsiveness to market	0.140	< 0.001	0.116

Table 6.6 Total effects

Independent variable	Dependent variable	Beta	*P*-value	Effect size
AMT	RSCS	0.579	<0.001	0.346
AMT	PS	0.603	<0.001	0.363
AMT	RtM	0.559	<0.001	0.329
PS	RSCS	0.583	<0.001	0.424
PS	RtM	0.761	<0.001	0.632

is between pull systems and responsiveness to market, this means that, the reduction of the lots, the reduction of the setup times, and the use of Kanbans are of vital importance to achieve a supply chain capable of responding to changes in the market.

6.4.4 Sensitivity Analysis

Table 6.7 shows the results of the sensitivity analysis for each of the hypotheses as shown in Fig. 6.1. The probabilities that each of the variables are presented in their favorable scenario (+) and their unfavorable scenario (-). In the case of PS, it is more likely ($P = 0.224$) that activities do develop favorably within companies. This is

Table 6.7 Sensitivity analysis

			Independent variables					
			AMT		PS		RSCS	
Dependent variables			+	−	+	−	+	−
		P(Z)	0.15	0.193	0.224	0.161	0.118	0.173
PS	+	0.224	& = 0.079	& = 0.008				
			If = 0.526	If = 0.143				
	−	0.161	& = 0.000	& = 0.110				
			If = 0.000	If = 0.571				
RSCS	+	0.118	& = 0.031	& = 0.008	& = 0.055	& = 0.000		
			If = 0.211	If = 0.041	If = 0.246	If = 0.000		
	−	0.173	& = 0.004	& = 0.106	& = 0.004	& = 0.114		
			If = 0.026	If = 0.551	If = 0.018	If = 0.707		
RtM	+	0.118	& = 0.028	& = 0.012	& = 0.071	& = 0.000	& = 0.020	& = 0.004
			If = 0.184	If = 0.061	If = 0.376	If = 0.000	If = 0.176	If = 0.023
	−	0.154	& = 0.000	& = 0.106	& = 0.000	& = 0.126	& = 0.000	& = 0.114
			If = 0.000	If = 0.551	If = 0.000	If = 0.780	If = 0.000	If = 0.659

important because as mentioned with the direct effects, *PS* is the most important variable, and, therefore, managers and administrators must ensure that they implement a *PS* in the production processes within their company.

On the contrary, there is a probability of 0.161 that activities in *PS* do not develop favorably within companies. As mentioned earlier, the direct effect of AMT on RtoM is not statistically significant; however, the indirect effect is presented through the *PS* and *RSCS* variables.

The importance of AMT in pull systems can be verified in the sensitivity analysis; since, if a correct implementation and use of the AMT is made, there is a probability of 0.526 that the pull systems are implemented correctly and by, consequently, that you have a quick response within the market. But, on the contrary, if these AMTs are not implemented correctly, the probability will be 0.571.

6.5 Conclusions and Industrial Implications

According to the values obtained in the evaluated model, as well as the variance values of each of the dependent latent variables, the following conclusions can be obtained:

- Regarding the direct effects between variables, it can be concluded that AMT has a direct and positive effect on the pull system and on the responsive SC strategy, not so on responsiveness to market, but indirectly through the responsive SC strategy and through the *PS*. These results demonstrate what was said by Roh et al. (2014), in which AMTs are used to improve time to market, delivery and cost, and as indicated by McDermott et al. (1997), AMTs help companies maintain the competitive advantage among companies.
- The results show that having a pull system within a production system and using tools such as the Kanban system, reducing batches and setup times help companies cope with sudden changes in the market, also help reduce manufacturing times and therefore, delivery times, but above all, that there is reliability in delivery. In that sense, managers must ensure that the *PS* runs correctly. This agrees with Sakakibara et al. (1997), who mentions that a pull system will reduce the response time to process orders.
- The results also show that AMTs indirectly help companies to be able to respond effectively to offer a rapid response in the changing market through the implementation of a pull system. The above is supported by Gass and Fu (2013), since AMT favors responsiveness, quality, and flexibility. AMT also helps to respond to changes in programming and product combination (Sohal Amrik 1992).

According to the sensitivity analysis performed in the section

- With respect to H1, that is, the effect of AMT on RSCS, there is a probability of 0.150 for AMT to appear in its favorable scenario and 0.118 for RSCS to do so in that same scenario. There is a probability of 0.079 that both variables are presented together in their favorable scenarios. There is a probability of 0.526 that the RSCS variable is present in its favorable scenario if AMT does so in the same scenario. Finally, there is a probability of 0.571 that RSCS appears in its unfavorable scenario if AMT does so in the same scenario. In that sense, it is important that managers ensure that the implementation, use, and maintenance of AMTs are correct so that, in this way, these technologies provide the flexibility that processes need and, consequently, can offer a broader range of products that new products are offered frequently and, above all, that are innovative. Otherwise, if managers do not ensure this and AMTs are not implemented and used correctly, the likelihood that companies are unable to create a strategy in the supply chain is higher.

- With respect to H2, the scenarios are as follows: there is a probability of 0.224 of PS appearing in its favorable scenario and a probability of 0.161 of doing so in its unfavorable scenario. Likewise, there is a probability of 0.526 that PS appears in its favorable scenario if *AMT* does it in the same scenario and a probability of 0.0.571 that it appears in an unfavorable scenario if AMT also does it in that same scenario. There is a probability of 0.079 that they do it jointly in their favorable scenario and 0.110 that they do it jointly in their unfavorable scenario. Similarly, if managers ensure that *AMT* is implemented correctly, companies will have a certain degree of implementation of a pull system; otherwise, they will not be able to work with a *PS*.
- As for H3, if a *PS* is ensured within companies, there is a probability of 0.246 of having a receptive SC strategy, that is, by having a *PS* implemented correctly, companies will be able to offer a range more wide range of products, as well as offering new and more innovative products. On the contrary, there is a probability of 0.707 that there is no strategy of a receptive SC if a PS is not implemented properly.
- As mentioned earlier, AMT does not have a statistically significant direct effect on RtM, however, the direct effect and the total effect is. In that sense, if AMTs are implemented correctly, there is a probability of 0.184 of having speed in delivery, short manufacturing times, reliability in delivery, and fast marketing times. Otherwise, if AMTs are not implemented correctly, there is a probability of 0.551 of not having a responsiveness in the market.
- As for H5, the probability that the companies' capacity to respond is fast is 0.176 if there is a responsive strategy in the supply chain; on the contrary, there is very high probability (0.659) that the capacity of response is slow if the strategy of the responsive SC is bad.
- Finally, the key to the rapid response to the market will depend to a greater extent on the pull system, supported by advanced manufacturing technologies. If you have a well-implemented pull system, there is a probability of 0376 of having a quick response on the market. On the contrary, there is a probability of 0.780 that if the pull system is not implemented, the market response is not quick.

The sensitivity analysis shows the change in the results depending on the degree to which the activities are executed in each of the latent variables. Logically, if the activities are executed correctly within the companies, the results will be positive and that is why the generics must ensure that these activities are carried out in each of the stages of the production process throughout the entire supply chain.

Acknowledgements Special thanks to the National Council for Science and Technology (CONACYT) for the scholarship awarded to carry out this chapter.

References

Aghazadeh SM (2004) Does manufacturing need to make JIT delivery work? Manag Res News 27(1/2):27–42. https://doi.org/10.1108/01409170410784338

Aravindan P, Punniyamoorthy M (2002) Justification of Advanced Manufacturing Technologies (AMT). Int J Adv Manufac Technol 19(2):151–156. https://doi.org/10.1007/s001700200008

Azaron A, Venkatadri U, Doost AF (2019) Designing profitable and responsive supply chains under uncertainty. IFAC-PapersOnLine 52(13):2816–2820. https://doi.org/10.1016/j.ifacol.2019.11.635

Barbosa-Povoa AP, Pinto JM (2020) Process supply chains: perspectives from academia and industry. Comput Chem Eng 132:106606. https://doi.org/10.1016/j.compchemeng.2019.106606

Bechtsis D, Tsolakis N, Vlachos D, Iakovou E (2017) Sustainable supply chain management in the digitalisation era: the impact of Automated Guided Vehicles. J Clean Prod 142:3970–3984. https://doi.org/10.1016/j.jclepro.2016.10.057

Birasnav M, Bienstock J (2019) Supply chain integration, advanced manufacturing technology, and strategic leadership: an empirical study. Comput Ind Eng 130:142–157. https://doi.org/10.1016/j.cie.2019.01.021

Blome C, Schoenherr T, Rexhausen D (2013) Antecedents and enablers of supply chain agility and its effect on performance: a dynamic capabilities perspective. Int J Prod Res 51(4):1295–1318

Burgess TF (1997) Supply-chain collaboration and success in technology implementation. Integr Manuf Syst 8(5):323–332. https://doi.org/10.1108/09576069710179779

Catalan M, Kotzab H (2003) Assessing the responsiveness in the Danish mobile phone supply chain. Int J Phys Distrib Logist Manag 33(8):668–685

Chuu S-J (2009) Group decision-making model using fuzzy multiple attributes analysis for the evaluation of advanced manufacturing technology. Fuzzy Sets Syst 160(5):586–602. https://doi.org/10.1016/j.fss.2008.07.015

Crambes C, Henchiri Y (2019) Regression imputation in the functional linear model with missing values in the response. J Stat Planning Infer 201:103–119. https://doi.org/10.1016/j.jspi.2018.12.004

Dawal SZM, Tahriri F, Jen YH, Case K, Tho NH, Zuhdi A, Mousavi M, Amindoust A, Sakundarini N (2015) Empirical evidence of AMT practices and sustainable environmental initiatives in malaysian automotive SMEs. Int J Precis Eng Manuf 16(6):1195–1203. https://doi.org/10.1007/s12541-015-0154-6

DeRuntz BD, Turner RM (2003) Organizational considerations for advanced manufacturing technology. J Technol Stud 29(1):4–9

Díaz-Reza JR, García-Alcaraz JL, Gil-López AJ, Blanco-Fernández J, Jimenez-Macias E (2019a) Design, process and commercial benefits gained from AMT. J Manufac Technol Manage https://doi.org/10.1108/JMTM-03-2019-0113

Díaz-Reza JR, García-Alcaraz JL, Gil-López AJ, Blanco-Fernández J, Jiménez-Macías E (2019b) Design, process and commercial benefits gained from AMT. J Manufac Technol Manage vol ahead-of-print (No. ahead-of-print.)

Díaz-Reza JR, Mendoza-Fong JR, Blanco-Fernández J, Marmolejo-Saucedo JA, García-Alcaraz JL (2019c) The role of advanced manufacturing technologies in production process performance: a causal model. Appl Sci 9(18):3741

Dowlatshahi S (2008) The role of industrial maintenance in the maquiladora industry: an empirical analysis. Int J Prod Econ 114(1):298–307. https://doi.org/10.1016/j.ijpe.2008.02.009

Dray S, Josse J (2015) Principal component analysis with missing values: a comparative survey of methods. Plant Ecol 216(5):657–667. https://doi.org/10.1007/s11258-014-0406-z

Efstathiades A, Tassou SA, Oxinos G, Antoniou A (2000) Advanced manufacturing technology transfer and implementation in developing countries: the case of the Cypriot manufacturing industry. Technovation 20(2):93–102. https://doi.org/10.1016/S0166-4972(99)00100-5

Efstathiades A, Tassou S, Antoniou A (2002) Strategic planning, transfer and implementation of advanced manufacturing technologies (AMT). Development of an integrated process plan. Technovation 22(4):201–212. https://doi.org/10.1016/S0166-4972(01)00024-4

Ekren BY (2020) A simulation-based experimental design for SBS/RS warehouse design by considering energy related performance metrics. Simul Model Pract Theory 98:101991. https://doi.org/10.1016/j.simpat.2019.101991

García-Alcaraz JL, Macías AAM, Luevano DJP, Fernández JB, López AdJG, Macías EJ (2016) Main benefits obtained from a successful JIT implementation. Int J Adv Manufac Technol 86(9):2711–2722. https://doi.org/10.1007/s00170-016-8399-5

García JL, Rivera L, Blanco J, Jiménez E, Martínez E (2014) Structural equations modelling for relational analysis of jit performance in maquiladora sector. Int J Prod Res 52(17):4931–4949. https://doi.org/10.1080/00207543.2014.885143

Gass SI, Fu MC (2013) Pull System. In: Gass SI, Fu MC (eds) Encyclopedia of operations research and management science. Springer US, Boston, MA, pp 1192–1192. https://doi.org/10.1007/978-1-4419-1153-7_200664

Gonzalez SR, Zambrano GM, Mondragon IF (2019) Semi-heterarchical architecture to AGV adjustable autonomy within FMSs. IFAC-PapersOnLine 52(10):7–12. https://doi.org/10.1016/j.ifacol.2019.10.003

Gules HK, Burgess TF (1996) Manufacturing technology and the supply chain: Linking buyer-supplier relationships and advanced manufacturing technology. Eur J Purchasing Supply Manage 2(1):31–38. https://doi.org/10.1016/0969-7012(95)00014-3

Hadjimarcou J, Brouthers LE, McNicol JP, Michie DE (2013) Maquiladoras in the 21st century: six strategies for success. Bus Horiz 56(2):207–217. https://doi.org/10.1016/j.bushor.2012.11.005

Hair JF Jr, Black WC, Babin BJ, Anderson RE (2010) Multivariate data analysis prentice hall. Upper Saddle River, NJ, USA

Harland CM, Lamming RC, Cousins PD (1999) Developing the concept of supply strategy. Int J Oper Production Manage

Holweg M (2005) The three dimensions of responsiveness. Int J Oper Prod Manag 25(7):603–622. https://doi.org/10.1108/01443570510605063

Holweg M (2005a) The three dimensions of responsiveness. Int J Oper Production Manage

Hopp WJ, Spearman ML (2004) To pull or not to pull: what is the question? Manuf Serv Oper Manag 6(2):133–148

Hult GTM, Hurley RF, Knight GA (2004) Innovativeness: Its antecedents and impact on business performance. Ind Mark Manage 33(5):429–438

Hum S-H, Parlar M, Zhou Y (2018) Measurement and optimization of responsiveness in supply chain networks with queueing structures. Eur J Oper Res 264(1):106–118. https://doi.org/10.1016/j.ejor.2017.05.009

INEGI (2019) Monthly survey of manufacturing (EMIM) [In Spanish]. INEGI. https://www.inegi.org.mx/sistemas/bie/default.aspx?idserPadre=10400100. Accessed 18 Mar 2020 2016

Kaplan RS (1986) Must CIM be justified by faith alone?

Khojasteh Y (2016) Production systems. In: Khojasteh Y (ed) Production control systems: a guide to enhance performance of pull systems. Springer Japan, Tokyo, pp 7–24. https://doi.org/10.1007/978-4-431-55197-3_2

Kim M, Chai S (2017) The impact of supplier innovativeness, information sharing and strategic sourcing on improving supply chain agility: Global supply chain perspective. Int J Prod Econ 187:42–52. https://doi.org/10.1016/j.ijpe.2017.02.007
Kimura O, Terada H (1981) Desiǵn and analysis of Pull System, a method of multi-staǵe production control. Int J Prod Res 19(3):241–253
Kock N (2017) WarpPLS user manual: Version 6.0. ScriptWarp Systems: Laredo, TX, USA
Kock N, Lynn G (2012a) Lateral collinearity and misleading results in variance-based SEM: an illustration and recommendations
Kock N, Lynn G (2012b) Lateral collinearity and misleading results in variance-based SEM: An illustration and recommendations. J Assoc Inf Syst 13(7)
Koh SCL, Gunasekaran A, Goodman T (2011) Drivers, barriers and critical success factors for ERPII implementation in supply chains: a critical analysis. J Strateg Inf Syst 20(4):385–402. https://doi.org/10.1016/j.jsis.2011.07.001
Kou T-C (2015) The influence of supply chain architecture on new product launch and performance in the high-tech industry. J Bus Ind Mark 30(5):677–687. https://doi.org/10.1108/JBIM-08-2013-0176
Koufteros XA (1999) Testing a model of pull production: a paradigm for manufacturing research using structural equation modeling. J Oper Manag 17(4):467–488. https://doi.org/10.1016/S0272-6963(99)00002-9
Krot K, Mazgajczyk E, Rusińska M, Woźna A (2019) Strategy of improving skills of innovation managers in the area of advanced manufacturing technologies. In: Burduk A, Chlebus E, Nowakowski T, Tubis A (eds) Intelligent systems in production engineering and maintenance, Cham, 2019// 2019. Springer International Publishing, pp 296–305
Kumar S (2013) Achieving customer service excellence using Lean Pull Replenishment. Int J Product Perform Manag 62(1):85–109. https://doi.org/10.1108/17410401311285318
Li JW (2005) Investigating the efficacy of exercising JIT practices to support pull production control in a job shop environment. J Manuf Technol Manag 16(7):765–783. https://doi.org/10.1108/17410380510626187
Liu S-T (2008) A fuzzy DEA/AR approach to the selection of flexible manufacturing systems. Comput Ind Eng 54(1):66–76. https://doi.org/10.1016/j.cie.2007.06.035
Li S, Zhao X, Huo B (2018) Supply chain coordination and innovativeness: a social contagion and learning perspective. Int J Prod Econ 205:47–61. https://doi.org/10.1016/j.ijpe.2018.07.033
Lyons AC, Mondragon AEC, Piller F, Poler R (2012) The development of supply chain strategy. In: Lyons AC, Coronado Mondragon AE, Piller F, Poler R (eds) Customer-driven supply chains: from glass pipelines to open innovation networks. Springer London, London, pp 1–19. https://doi.org/10.1007/978-1-84628-876-0_1
Lyu Z, Lin P, Guo D, Huang GQ (2020) Towards zero-warehousing smart manufacturing from zero-inventory just-in-time production. Robot Comput -Integr Manufac 64:101932. https://doi.org/10.1016/j.rcim.2020.101932
Marsh RF, Conard MA (2008) A pull system for delegating knowledge work. Oper Manag Res 1(1):61. https://doi.org/10.1007/s12063-008-0006-y
McDermott CM, Greis NP, Fischer WA (1997) The diminishing utility of the product/process matrix: a study of the US power tool industry. Int J Oper Prod Manag 17(1):65–84
Melnyk SA, Davis EW, Spekman RE, Sandor J (2010) Outcome-driven supply chains. MIT Sloan Manag Rev 51(2):33
Mora-Monge CA, González ME, Quesada G, Rao SS (2008) A study of AMT in North America. J Manufac Technol Manag
Munksgaard KB, Stentoft J, Paulraj A (2014) Value-based supply chain innovation. Oper Manag Res 7(3):50–62. https://doi.org/10.1007/s12063-014-0092-y
Najera M (2008) Dialogue with Mexican maquiladora workers: why do they leave and why do they stay? Int J Commer Manag 18(3):289–302. https://doi.org/10.1108/10569210810907182

Nakandala D, Lau HCW (2019) Innovative adoption of hybrid supply chain strategies in urban local fresh food supply chain. Supply Chain Manage Int J 24(2):241–255. https://doi.org/10.1108/SCM-09-2017-0287

Narasimhan R, Kim SW, Tan KC (2008) An empirical investigation of supply chain strategy typologies and relationships to performance. Int J Prod Res 46(18):5231–5259

Nath S, Sarkar B (2017) Performance evaluation of advanced manufacturing technologies: a De novo approach. Comput Ind Eng 110:364–378. https://doi.org/10.1016/j.cie.2017.06.018

Pero M (2010) A framework for the alignment of new product development and supply chains. Supply Chain Manage Int J 15(2):115–128. https://doi.org/10.1108/13598541011028723

Puchkova A, Le Romancer J, McFarlane D (2016) Balancing push and pull strategies within the production system. IFAC-PapersOnLine 49(2):66–71. https://doi.org/10.1016/j.ifacol.2016.03.012

Qrunfleh S (2013) Lean and agile supply chain strategies and supply chain responsiveness: the role of strategic supplier partnership and postponement. Supply Chain Manage Int J 18(6):571–582. https://doi.org/10.1108/SCM-01-2013-0015

Ragatz GL, Handfield RB, Petersen KJ (2002) Benefits associated with supplier integration into new product development under conditions of technology uncertainty. J Bus Res 55(5):389–400

Rahman AA, Brookes NJ, Bennett DJ (2009) The precursors and impacts of BSR on AMT acquisition and implementation. IEEE Trans Eng Manage 56(2):285–297

Reddy KJM, Rao ANLK (2019) A review on supply chain performance measurement systems. Procedia Manufac 30:40–47. https://doi.org/https://doi.org/10.1016/j.promfg.2019.02.007

Reichhart A (2007) Creating the customer-responsive supply chain: a reconciliation of concepts. Int J Oper Prod Manag 27(11):1144–1172. https://doi.org/10.1108/01443570710830575

Roh J, Hong P, Min H (2014) Implementation of a responsive supply chain strategy in global complexity: the case of manufacturing firms. Int J Prod Econ 147:198–210

Ross DF (2015) Crafting business and supply chain strategies. In: Ross DF (ed) Distribution planning and control: managing in the era of supply chain management. Springer US, New York, pp 83–140. https://doi.org/10.1007/978-1-4899-7578-2_3

Sakakibara S, Flynn BB, Schroeder RG, Morris WT (1997) The impact of just-in-time manufacturing and its infrastructure on manufacturing performance. Manage Sci 43(9):1246–1257

Sandberg E (2018) Retail supply chain responsiveness. Int J Product Perform Manag 67(9):1977–1993. https://doi.org/10.1108/IJPPM-11-2017-0315

Selçuk B (2013) Adaptive lead time quotation in a pull production system with lead time responsive demand. J Manuf Syst 32(1):138–146. https://doi.org/10.1016/j.jmsy.2012.07.017

Siha S, Linn RJ (1989) A zero-one goal programming decision model for selecting technology alternatives. In: International industrial engineering conference, pp 200–206

Singh N (1992) Modelling and analysis of just-in-time manufacturing systems: a review. Int J Oper Prod Manag 12(2):3–14. https://doi.org/10.1108/01443579210009005

Singhry HB, Abd Rahman A, Imm NS (2016) Effect of advanced manufacturing technology, concurrent engineering of product design, and supply chain performance of manufacturing companies. Int J Adv Manufac Technol 86(1):663–669. https://doi.org/10.1007/s00170-015-8219-3

Sinkovics RR, Jean R-JB, Roath AS, Cavusgil ST (2011) Does IT integration really enhance supplier responsiveness in global supply chains? Manag Int Rev 51(2):193. https://doi.org/10.1007/s11575-011-0069-0

Sohal Amrik S (1992) Implementing advanced manufacturing technology: factors critical to success. Logist Inf Manag 5(1):39–46. https://doi.org/10.1108/09576059210011482

Utar H, Ruiz LBT (2013) International competition and industrial evolution: evidence from the impact of Chinese competition on Mexican maquiladoras. J Dev Econ 105:267–287. https://doi.org/10.1016/j.jdeveco.2013.08.004

van Hoek R (2006) From tinkering around the edge to enhancing revenue growth: supply chain-new product development. Supply Chain Manage Int J 11(5):385–389. https://doi.org/10.1108/13598540610682390

Worley JM, Mitchell RM (2008) Assessing content validity in Kaizen event research surveys. In: Proceedings of the IIE annual conference. Institute of Industrial Engineers-Publisher, p 1095
Yang J (2019) Turning responsible purchasing and supply into supply chain responsiveness. Ind Manag Data Syst 119(9):1988–2005. https://doi.org/10.1108/IMDS-01-2019-0029
Zhang L (2013) Kanban-controlled exponential production lines: analysis and design. J Manuf Technol Manag 24(3):358–383. https://doi.org/10.1108/17410381311318873

Chapter 7
Effect of Variability on the Optimal Flow of Goods in Supply Chains Using the Factory Physics Approach

Yamil Mejía-Hernández, Salvador Hernández-González, Aarón Guerrero-Campanur, and José Alfredo Jiménez-García

Abstract In a supply chain, parameters such as demand, service times and the flow between nodes are commonly assumed to be constant quantities for the time horizon or window, while transformation is also assumed to be immediate. However, this is not the way it is. Demand is random; operations require a processing time that is also random, and this causes variations in the links of a chain that generate queues. Cycle time and the amount of work in process are properties that quantify the performance of the flow within a network subject to the effects of the system's innate variability. This document provides a series of steps for analyzing optimal flow in a supply chain by using analytical methods. First, we use the minimum cost flow model to find the optimal solution for the distribution of the product in the chain. Then, assuming that each link in the chain is a G/G/c queueing system, the cycle time and quantity of work in process are quantified for the optimal solution in three scenarios of service time and demand variability. We used for our example the data corresponding to the three-level supply chain, in the context of the vehicle industry. Taking as a reference a scenario where variability is low ($C^2_s <= 0.5$), the cycle time in the network increases its value to six times its value when there is high variability throughout the system.

Keywords Management · Supply chain · Optimal flow · Cycle time · Variability · Queueing networks

7.1 Introduction

A supply chain is a set of activities (suppliers, production, distribution, and consumers) that are carried out several times over several stages. In each stage, raw materials are transformed into finished products with an added value for the consumer. The sources of the raw materials, the plants, and distribution points and

Y. Mejía-Hernández · S. Hernández-González (✉) · J. A. Jiménez-García
Tecnológico Nacional de México, Instituto Tecnológico de Celaya, Celaya, Mexico
e-mail: salvador.hernandez@itcelaya.edu.mx

A. Guerrero-Campanur
Tecnológico Nacional de México, Instituto Tecnológco de Uruapan, Uruapan, Mexico

J. L. Garcia Alcaraz et al. (eds.), *Trends in Industrial Engineering Applications to Manufacturing Process*, https://doi.org/10.1007/978-3-030-71579-3_7

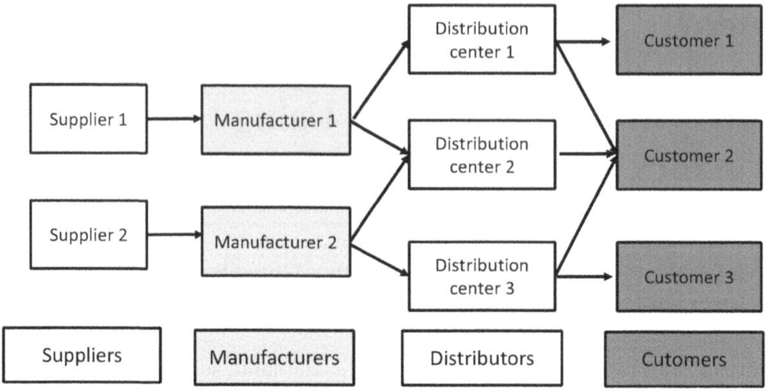

Fig. 7.1 Generic supply chain network

warehouse are normally in geographically separated locations. Both the processes and warehouse locations are integrated into a network known as the supply chain. Supply chains are areas of business that are traditionally assessed in terms of cost, quality of the service or product, response time, and flexibility. Every performance measurement needs the system to have a set of specific capabilities that the company needs to manage in order to have an efficient operation (Hopp and Spearman 2001; Ballou 2003) (Fig. 7.1).

The administration of a supply chain is based on the coordination of activities, from the reception of raw materials up to the point where the product or service is delivered to the customer. The administration of the chain means that it needs to be structured in such a way that the customer benefits, thus maximizing earnings; finding an optimally coordinated activity plan has an important role in the network's operation, as it enables us to lower production and distribution expenses, thus improving the levels of service (Marmolejo et al. 2016). Based on these premises, optimization models adapted for supply chains, such as the minimum cost flow model, are very useful tools in constructing an optimal design for the network's operations where the randomness present in this type of system needs to be taken into account (Hernández-González et al. 2017).

In optimization models, parameters such as demand and capacity for both production and transport are commonly assumed to be constant, unchanging or uniform quantities for the time horizon or window when the minimum cost flow model is adapted, while the transformation is also assumed to be immediate; whereas, in reality, this is not the case, purchase orders, a production process, and, in general, the operations require a processing time. In consequence, a queue of orders that must wait to be served is formed in the links of a chain (in other words, they behave like a queueing system). Supply chains can be treated as a dynamic system where there are events that cause the demand and the times taken in processing, transformation, or attention not to be uniform and act like random variables. This means that they represent a deviation during the operation that is given the generic name of variability.

Variability creates uncertainty and, in consequence, complicates the management and control of supply chains.

Given the complexity of this type of system, the scope of tools such as the minimum cost flow model needs to be broadened, taking into account the randomness of parameters, such as demand and service capacity, to quantify properties like the cycle time and the amount of work in process which are relevant performance measurements when it comes to assessing the efficiency of an operation. These factors as a whole will help to lower the uncertainty associated with the solution's operational efficiency. Does every node in the network have enough capacity to handle the flow of goods? How long does the entity (a production order, a product batch) spend, on average, inside the network? How many units, on average, accumulate in front of each station? How much work in process is there in the entire network? If the variance in the demand and cycle time increases, how is this reflected in the performance of the supply chain?

This paper presents an analysis of the effect that the variability has on the performance of the optimal operating policies of a supply chain. Using the data corresponding to a vehicle industry supply chain as an example, we get the optimal solution for product distribution in the chain by employing the minimum cost flow model. Then, assuming that each node in the chain is a queueing system, the cycle time and quantity of work in process are quantified for the optimal solution in three scenarios of demand variability and service time. These scenarios are used to determine the robustness of the optimal solution for a dynamic scenario. The analytical method shown in this paper can be replicated for other supply chains in different contexts by managers that are responsible for the administration of this type of system. Our research is justified because, although it is well known that increased variability degrades the performance of the system, this is not a valid answer per se, as an administrator has to quantify its effects in order to be able to justify a decision.

7.2 Literature Review and Concepts

The main advantage of analytical models is to provide us with an expression that considers the variables that affect the system. Moreover, the analysis of scenarios takes less computational effort than a simulation model. Furthermore, analytical models are universal and freely available. In particular, when it is necessary to determine the assignment of resources, queueing network models are used to represent manufacturing systems and supply chains, given that they capture, with a degree of accuracy, the effects of the randomness present in these systems (Cruz and van Woensel 2014).

Kerbache and MacGregorSmith (2004) propose a methodology for modeling supply chains such as queueing networks and propose a set of parameters as performance measurements. The methodology is applied to the supply chain of a suitcase manufacturing company. Liu et al. (2004) model an inventory control system for a supply chain. The relevance of their paper lies in the fact that the analysis includes the variability of service time at the nodes.

Gong et al. (2008) establish some properties for the supply chains based on the concept of a closed G/G/c queueing network. Negri-da-Silva and Morabito (2009) construct a model for finding the optimal capabilities of a metallurgical company as well as cost/benefit curves for sensitivity analysis.

Bhaskar and Lallement (2010) carry out an analysis of a supply chain for a textile company using the proposed model to analyze the routes based on the response time. Later, Bhaskar and Lallement (2011) broaden the analysis to a system where the inter-arrival times follow a uniform distribution and incorporate subcontracting into the model.

Vahdani and Mohammadi (2015) develop a bi-objective model for a supply chain by assuming that the network is a closed queueing system with fuzzy parameters, and using expressions from queueing theory, they take the randomness of the system into account while also assuming that the system uses the rule of service disciplines. They had to employ a metaheuristic technique to solve the model because of its size.

Bandaly et al. (2016) analyze the impact of the lead—time variability on supply chains. Pérez-Loaiza et al. (2017) solve a multi-objective model considering the standard deviation of the demand; however, the model does not take into account the fact that the purchase orders form a queue and must wait a certain time to be processed and sent to the customer.

MacGregorSmith and Kerbache (2017) analyze the design of supply chains focusing on M/G/c/K stations with a closed network arrangement, using the system's production as a performance measurement, while the optimization model is integer nonlinear. From their results, the assessment of the additive manufacturing system is of particular interest as is the fact that they confirm the advisability of decentralized designs that favor the production of the system.

Yousefi-Babadi et al. (2017) adapt the expressions from the M/M/c/K system to represent the supply chain of a petrochemical company. The expressions are incorporated into a multi-objective model and determine the optimal operating conditions for minimizing delay, the total cost, and transport costs. Hum et al. (2018) apply Jackson's open networks approach to the optimization of a supply chain where the aim is to maximize the fulfillment of production orders.

The queueing approach has also been applied in contexts other than manufacturing and production: In the field of healthcare, Zahiri et al. (2014) propose a multi-objective optimization model for finding the optimal paths in an organ transplant network. The model requires the implementation of metaheuristic techniques for finding a solution. In the coordination of emergency systems, He and Hu (2014) apply queueing theory models to the coordination of operations in a natural disaster response system, using the network of the city of Shanghai as their case study.

7.2.1 Minimum Cost Flow

The minimum cost flow model is applied to solving problems where it is necessary to schedule the movement of entities (people, goods, information, trains) in a network. It considers a directed network made up by a finite N set of nodes and an A set of arcs connecting said nodes. The arc (i, j) intersects nodes i and j and is directed if the direction of the flow is from node i to node j. Every node is assigned an $f > 0$ value that represents the available amount of an article to be supplied or else a $f < 0$ value that will represent the demand for an item. Finally, when a node only represents a point of transit where there is no demand or supply, it is called a transfer node and $f = 0$.

A certain amount x_{ij} of an item travels through each arc, and there is also a cost c_{ij} associated with the flow of each unit of the item. Let us, for the moment, assume that the sum of the supply and demand is zero: $\sum_i^m f_i = 0$. The problem consists of finding the schedule of dispatches through the network in such a way that the cost is minimized, satisfying the constraints on the available amount of the item and demand. The optimization problem is, in mathematical terms, as shown in Eqs. (7.1–7.3) (Bazaraa et al. 1974; Wagner 1975):

Minimize

$$\sum_{i=1}^{n}\sum_{j=1}^{m} c_{ij}x_{ij} \tag{7.1}$$

Subject to:

$$\sum_{i=1}^{m} x_{ij} - \sum_{k=1}^{m} x_{ki} = f_i, \forall i = 1, 2, \ldots, n \tag{7.2}$$

$$x_{ij} \geq 0 \tag{7.3}$$

where (1) is the cost function and has to be minimized; and the set of constraints (2) corresponds to the network flow conservation equations, where the former is the sum of the outflows or dispatches from the place of origin to destination j and the latter is the sum of the inflows from a node k to node i. Equation (7.3) is the non-negativity constraint for the flow through the arcs. The minimum cost flow model is linear, and the simplex method is normally used for finding the optimal solution. It is also assumed that the transit through a node or station in the model is instant and is not subject to any type of delay (Bazaraa et al. 1974; Wagner 1975).

7.2.2 *Queueing Concepts*

In systems, transformation needs time to convert raw materials into finished product; moreover, people must wait a certain time to complete a transaction. There is a set of entities in a queueing system called customers that arrive at a station to be served at the rate of λ customers per unit of time. If the station is empty, customer enters directly to be served, otherwise they must form a queue to wait their turn. The station has a capacity to serve μ customers per unit of time. A station can have one or more servers in parallel and the service or transformation may require one or more stages. Finally, the service discipline is normally First In, First Out, though it could be random or according to priorities (Gross et al. 2008).

The queue is formed because there is a difference between service rate μ and arrival rate λ, which will be reflected in the formation of a queue in front of the station waiting their turn to be served. The entities waiting are also called work in process (Curry and Feldman 2009). For a system to be stable, the service capacity needs to be higher than the service demand:

$$\varrho = \frac{\lambda}{\mu} = \lambda t_S < 1 \tag{7.4}$$

If demand is higher than the service capacity, then Eq. (7.4) will be equal or greater than 1 and the number of customers in the queue will grow boundlessly. In 1961, Little demonstrated that the amount of work in process depends on demand and the system cycle time (Little 1961):

$$\mathrm{WIP}_S = \lambda \mathrm{CT}_S \tag{7.5}$$

where: CT is the cycle time, the time an entity elapses in the system, WIP is the amount of work in process and λ is the arrival rate. Little's law in Eq. (7.5) applies to individual workstations, multi-step production lines, factories, or even complete supply chains. It is also valid in processes with random demand and random processing time. If we know two of the amounts, we can calculate the third (Jackson 1957). The residence time in Eq. (7.6) in the system is the sum of the cycle time in the queue ($\mathrm{CT}q$) and the expected value of the service time ($\mathrm{E}[t_s]$):

$$\mathrm{CT}_S = \mathrm{CT}_q + \mathrm{E}[t_s] \tag{7.6}$$

7.2.2.1 Queueing Networks

A queueing network is formed when there are at least two stations connected to each other. In a network, the outputs from one process are transformed into the arrivals for another process. Apart from supply chains, other examples of queueing networks are hospitals with their different sections, banks, or fast food outlets.

7.2.3 Flow in an Open Queueing Network

When an entity enters a network, it must go through a set of stages until its service or transformation is completed; there can be several paths in a network, and these paths can be either probabilistic or deterministic (Jackson 1963) assumes that:

1. The system has n stations, each of which has one or more servers.
2. The customers can enter the network through any station at an arrival rate of γ.
3. The service time at each station is t_S.
4. Once transformation in one station is completed, the customer follows their path to another station with probability p_{ij} or leaves the network.
5. There is no limit to the number of customers that can be in the system.

The flow of material is equal to the sum of the flow that arrives at station i and the flow from other stations that feed into station i. This is expressed by Eq. (7.7):

$$\lambda_i = \gamma_i + \sum_{j=1}^{n} \lambda_i p_{ji}, i = 1, 2, \ldots, n \tag{7.7}$$

System (7.7) is set out in matrix form as follows:

$$\lambda = \left(I - P^T\right)^{-1} \gamma \tag{7.8}$$

where I is an identity matrix; P is the flow matrix or process path in the network and gamma the set of flows from outside to each station in the network. When Eq. (7.8) is solved, we get the total number circulating within the network (Whitt 1983).

7.2.4 Probability and Variability

In probability theory, the outcome of an experiment is assumed to be unknown; the entire set of possible outcomes that an experiment can have is called the sample space. The probability space is a set consisting of the sample space, the collection of sample space events, and a number assigned to each sample space event: $(\Omega, \mathcal{F}, Pr)$. The probability space defines an experiment (Curry and Feldman 2009).

A random variable is a function that assigns a real number to each result or realization of an experiment and may be discrete or continuous. A function F is a cumulative distribution of a random variable X if:

$$F(a) = Pr\{X \leq a\} \tag{7.9}$$

Equation (7.9) describes the behavior of the random variables using mass functions in the case of discrete ones, or else with density functions in the case of continuous ones, as shown in Eq. (7.10):

$$\int_{a}^{b} g(u)\mathrm{d}u = Pr\{a < Y < b\} \tag{7.10}$$

For *a, b* in the *Y* range, *g(u)* is the density function as the normal distribution or exponential distribution.

The mean of a random variable is equal to the arithmetic average of an infinitely large number of observations or realizations of said variable. The variance is an indicator for the scatter of observations around the mean; in other words, it is an indicator for the variability of said observations. Both the mean and the variance are defined in terms of the expected value, and in the case of the function of a continuous random variable, it is (Hopp and Spearman 2001):

$$\mathrm{E}[h(X)] = \int_{-\infty}^{\infty} h(s) f(s)\mathrm{d}s \tag{7.11}$$

The mean or $\mathrm{E}[X]$ and the variance or $\mathrm{V}[X]$ of a random variable X are obtained from Eq. (7.11) and are defined as:

$$\text{Mean} = \overline{X} = \mathrm{E}[X] \tag{7.12}$$

$$\text{Variance} = \sigma^2 = \mathrm{E}\left[\left(X - \overline{X}\right)^2\right] \tag{7.13}$$

The expected value or mean in Eq. (7.12) indicates the value that the observations are concentrated around, while variance in Eq. (7.13) is the square of the expected value of the deviation from the mean. The mean is known as the first central moment, and the variance is known as the second central moment. One measurement that connects both parameters is the coefficient of variation expressed in Eq. (7.14) (Curry and Feldman 2009):

$$C = \frac{\sigma}{\overline{X}} \tag{7.14}$$

where $\overline{X}$ is the mean and σ the standard deviation. As the value of σ increases, the value of C also increases when $\overline{X}$ stays constant; alternatively, as shown in Eq. (7.15):

$$C^2 = \frac{\sigma^2}{\left(\overline{X}\right)^2} \tag{7.15}$$

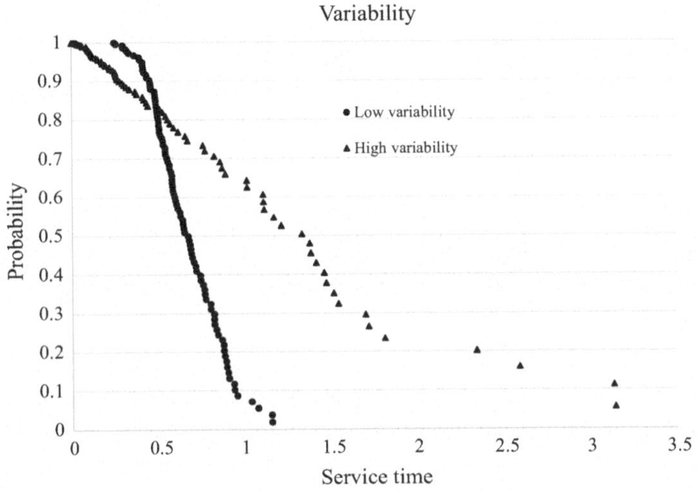

Fig. 7.2 Example of variability. Service time = 0.583 and different variability

Let us consider Fig. 7.2, where we can see two probability distributions obtained from a sample of the service time in two pieces of equipment performing the same operation. In both cases, the mean $t = 0.5833$, however, in one of the pieces of equipment, the variance is 0.04, and in another, the variance is 0.6.

In the case of the service time with greater variance, we observe that there is a higher scatter of service time realizations; in other words, service times farther from the mean are more frequently observed, resulting in a probability distribution with a heavier tail. Variability is a characteristic that is inherent to any system that is subject to randomness. In queues, it is found in both service times and the inter-arrival times of customers in the system. The presence of variability can be seen in the residence time within the system (cycle time) and the amount of work in process (WIP). Quantifying its effect provides the decision maker with the necessary information to improve the performance of a supply chain.

There is a set of factors that break the continuity of the flow in a system; in other words, they generate variability. Several sources of variability are recognized in a system (Hopp and Spearman 2001):

Natural variability: This is typical, for example, of the transformation process, owing to the different needs of consecutive customers.

Variability arising from failures: This is generated by interruptions of the process caused by periods when a piece of equipment is out of service for an unplanned repair. Variability owing to decisions generated by the management policies and rules under which the system operates.

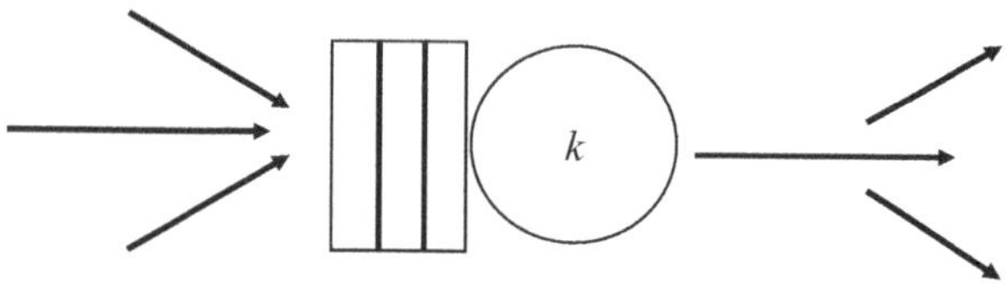

Fig. 7.3 Merge, queue, and split flow in a Jackson network

7.2.5 Approximation of Flow Variability in a Queueing Network

In a queueing network, we consider that the customers enter a station together, the transformation takes place where the queueing phenomenon happens and then the customers leave and take their respective paths with probability p_{ij}. The three moments are taken into account in Eq. (7.16) that calculates the variability of the flow in an open network (Whitt 1983) (Fig. 7.3):

$$C_{a,i}^2 = \frac{\gamma_i}{\lambda_i} C_{a,0i}^2 + \sum_{k=1}^{n} \frac{\lambda_i p_{ki}}{\lambda_i} \times \left[p_{ki}\left(1 - \rho_k^2\right) C_{a,k}^2 + p_{kj} \rho_k^2 \left\{ \frac{C_{S,k}^2 + \sqrt{c_k} - 1}{\sqrt{c_k}} \right\} + 1 - p_{kj} \right]; \quad i = 1, 2, \ldots, n \tag{7.16}$$

7.2.6 Approximation of the Cycle Time

In 1983, Whitt used the Allen–Cunneen approximation for the cycle time:

$$\mathrm{CT}_i = \left(\frac{C_{ai}^2 + C_{Si}^2}{2} \right) \left(\frac{\rho_i^{\sqrt{2(c_i+1)}-1}}{(1 - \rho_i) c_i} \right) t_{Si} + t_{Si} \tag{7.17}$$

Equation (7.17) is exact for the Markov case, i.e., when the inter-arrival time variability and the service time variability are 1.

7.3 Case Study: Analysis of the Supply Chain

Figure 7.4 shows the supply chain to be modeled. As can be observed, this chain has, in first instance, transport costs, capacities, and demands. The demand units for

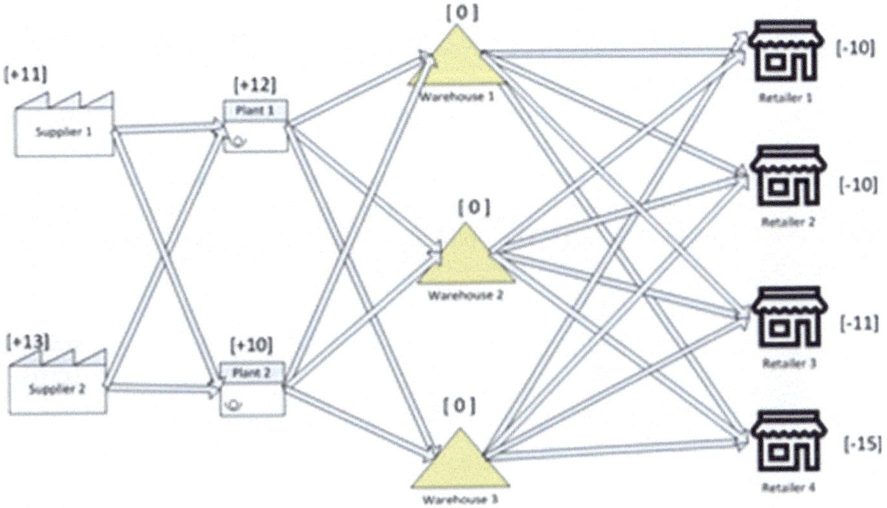

Fig. 7.4 Supply chain case

the four retailers were taken from the research from Guerrero-Campanur (2013) for which one assumption considered for the construction of the model was the capacity units of the supply nodes, which are Supplier 1, Supplier 2, Plant 1, and Plant 2, based on common practice in the vehicle industry (Pérez-Loaiza et al. 2017). The time units considered in the research article are units per week. However, in the results from this article, the magnitude of cycle time is in days, as the pertinent time conversions are considered.

The demand units for the four providers of raw materials are reported in Guerrero-Campanur (2013) and are in pieces per day. The optimization model was constructed for this supply chain using a minimum cost flow approach. The resulting model is as follows:

Min

$$\sum_{i=1}^{n}\sum_{j=1}^{m} c_{ij}x_{ij} \tag{7.18}$$

Subject to:

$$\sum_{j=1}^{m} x_{ij} = O_i,\ i = 1, 2, \ldots, n \tag{7.19}$$

$$\sum_{i=1}^{n} x_{ij} = D_j,\ j = 1, 2, \ldots, m \tag{7.20}$$

$$x_{ij} \geq 0, i = 1, 2, \ldots, n; j = 1, 2, \ldots, m \tag{7.21}$$

where Eq. (7.18) represents the sum of the transport costs and it must be minimized. Equation (7.19) is the set of constraints for the supply, and Eq. (7.20) represents the set of constraints for the demand. The non-negativity constraints correspond to Eq. (7.21). The previous model was put into a spreadsheet, then the solution was found using Lingo 14 (Fig. 7.5). The optimal flow within the supply chain is given in Table 7.1.

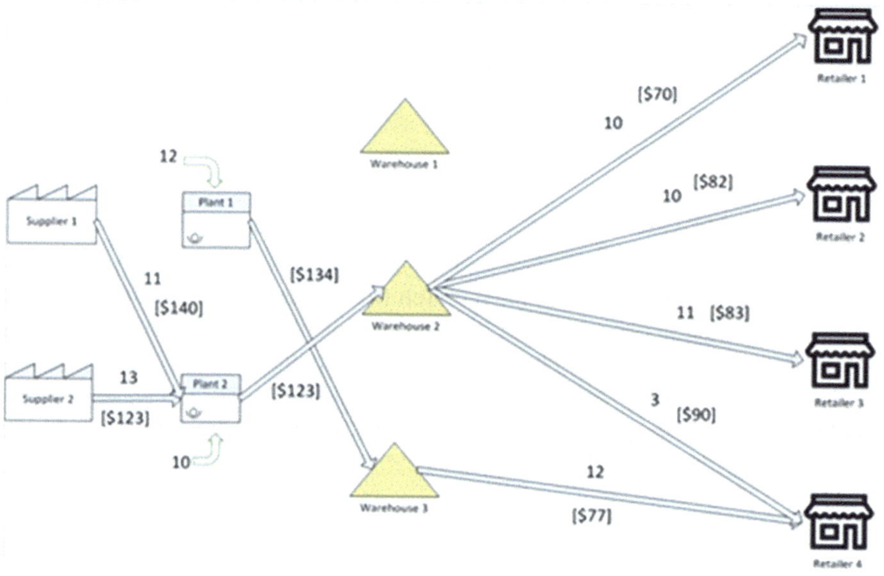

Fig. 7.5 Deterministic solution

Table 7.1 Optimal flow

Source	Target	Flow	%
Supplier 1	Plant 2	11	100
Supplier 2	Plant 2	13	100
Plant 1	Warehouse 3	12	100
Plant 2	Warehouse 2	34	100
Warehouse 2	Customer 1	10	29.41
Warehouse 2	Customer 2	10	29.41
Warehouse 2	Customer 3	11	32.35
Warehouse 2	Customer 4	3	8.82
Warehouse 3	Customer 4	12	100

The cost of the flow of goods in the network is $12,596. We can see that the use of warehouse 1 is rejected in the solution. We also observe that Plants 1 and 2 manufacture extra units (12 and 10, respectively) to meet the scheduled demand. The solution shows that the total stock of Warehouse 2 is distributed among the four customers: 29.41% for Customers 1 and 2, respectively, 32.35% is sent to Customer 3, and 8.82% is sent to Customer 4.

7.3.1 Modeling of the Solution as a Queueing System

Arrivals at a workstation and service time are generally assumed to be deterministic, which is not always fulfilled; both the inter-arrival time and service time are random variables that follow a probability distribution with a t_a and t_s mean and sigma 1 and sigma 2 variance, respectively. Decision makers often must estimate the average time a customer stays within the network, known as the cycle time, as well as the amount of work in process in front of each station and the work accumulated overall.

With the optimal flow obtained using the minimum cost flow model, the cycle time and the quantity of work in process in the chain were then determined by considering that the stages of the supply chain are queueing systems while taking into account the effect of variability on service time and inter-arrival times at the stations. The analysis was done by generating distinct scenarios of variability. The parameter used for this purpose is the squared coefficient of variation; the cycle time and work in process were obtained by analytical means, using the queuec.sce subroutine (Hernández-González and Hernández-Ripalda 2018). Said program applies the decomposition method to estimate the work in process and cycle time in an open Jackson network (Whitt 1983). The subroutine was run in the Scilab program (Scilab 2012).

The variability scenarios studied are:

1. High variability throughout the system.
2. Moderate variability throughout the system.
3. Low variability throughout the system.

It is a known fact that increased variability will deteriorate both the cycle time and the work in process; however, this is an incomplete answer. The people responsible for managing this type of system should numerically measure its impact as this lays the foundations for quantitatively estimating the costs.

To validate the analytical results, a simulation model is constructed using the Arena package; ten simulation runs were carried out to ensure a 95% level of reliability. The inter-arrival times and service times of the simulation model were modeled using a gamma probability distribution with scale parameter α and shape parameter β. When the coefficient of variation is fixed, the value of the variance is fixed and these data are used to obtain the corresponding scale and shape parameters for the distribution.

From the point of view of a production planner of manager, how important is it to take the variability present in a supply chain into account? What do these changes in

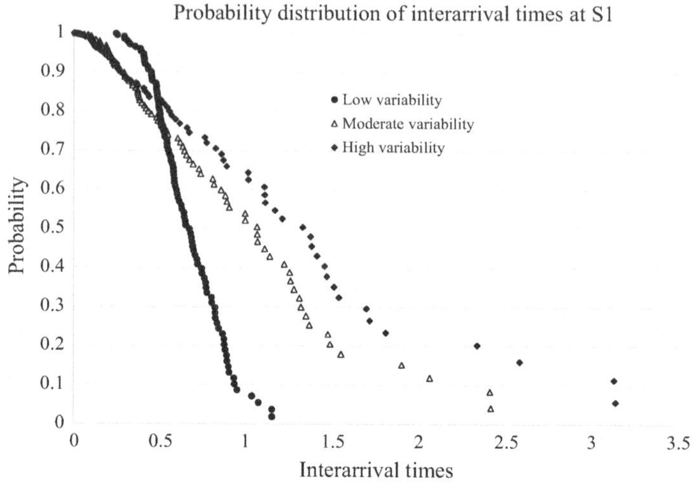

Fig. 7.6 Probability distributions of service time in the three scenarios of variability

the service times and inter-arrival times imply? From the scale and shape parameters for the gamma distribution, the probability distributions of the inter-arrival times at S1 are generated. Figure 7.6 shows the probability distribution graph for the inter-arrival times at S1 for the three scenarios. It can be appreciated that, in the case of demand, the rise in variability shows an increase in the standard deviation for inter-arrival times. This causes there to be a higher probability of times that are increasingly far away from the mean; in other words, the demand becomes increasingly irregular and its cause is, for example, due to changes in batch size, changes in the frequency of orders, or else the need to improve coordination and communication between the supplier and the customer.

In the case of the service times for the stations, Fig. 7.7 illustrates the specific case of W2. The probability distribution was constructed in the same way, using the respective scale shape parameters of the Gamma distribution. We observe in the result that low variability means service times with low scatter and the probability of having service times that are very far from the mean is not great. Figure 7.7 shows that the distribution queue rapidly "drops," whereas when variability increases, the scatter of service times also increases, and this means that there is a higher probability of there being service times farther from the mean. The probability distribution queue in each scenario (moderate variability and high variability) is increasing "heavier."

These are the scenarios that were analyzed, and their effects are quantified below.

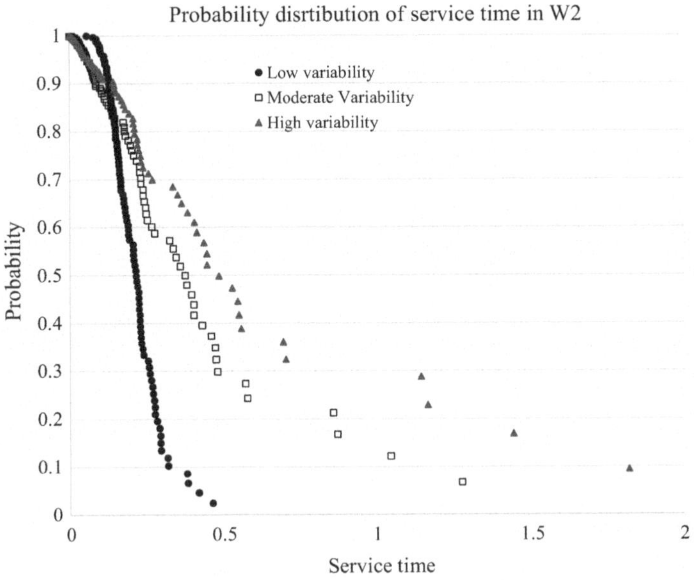

Fig. 7.7 Probability distributions of time between arrivals in the three scenarios of variability

7.3.2 Analysis of the Flow in the Supply Chain Under the Effect of Variability

In the first scenario, it is assumed that the coefficient of variation of the inter-arrival times and service times has a value within the low variability range (Hopp and Spearman 2001; Hopp 2011). Tables 7.2 and 7.3 present the inter-arrival times data for the analytical model and for constructing the simulation model. Demand enters through nodes S1, S2, P1, and P2.

Once the characteristics of the inter-arrival times and service time had been defined, the cycle time and work in process for the optimal solution were calculated analytically. Table 7.4, for example, shows that under these conditions the optimal solution corresponds to a system where the five stages of the chain operate under congestion conditions that are equal to or more than 90%; three stations operate with a load in the 80–90% range, and two stations have a workload in the 70–75% range.

Table 7.2 Mean time between arrivals, coefficient of variation and parameters of gamma distribution, low variability

Time between arrivals (days)	Coefficient of variation	Shape parameter	Scale parameter
0.636	0.35	8.1633	0.07795
0.538	0.35	8.1633	0.06596
0.583	0.35	8.1633	0.07146
0.700	0.35	8.1633	0.08575

Table 7.3 Mean service time, coefficient of variation, and parameters of gamma distribution, low variability

Node	Mean service time (days)	Coefficient of variation	Shape parameter	Scale parameter
S1	0.583	0.35	8.1633	0.07146
S2	0.500	0.35	8.1633	0.06125
P1	0.538	0.35	8.1633	0.06596
P2	0.175	0.35	8.1633	0.02144
W2	0.200	0.35	8.1633	0.02450
W3	0.560	0.35	8.1633	0.06860
R1	0.525	0.35	8.1633	0.06431
R2	0.583	0.35	8.1633	0.07146
R3	0.525	0.35	8.1633	0.06431
R4	0.350	0.35	8.1633	0.04288

Table 7.4 Performance measures of the supply chain as a network of queues

Node	Node flow	Cycle time in the node	Total work in process	Node utilization	
				Analytical (%)	Simulated (%)
S1	11	0.196	2.152	91.67	92.46
S2	13	0.185	2.407	92.86	92.32
P1	12	0.190	2.280	92.31	92.21
P2	34	0.042	1.440	85.00	84.84
W2	34	0.148	5.017	97.14	97.01
W3	12	0.315	3.782	96.00	95.50
R1	10	0.172	1.722	75.00	75.21
R2	10	0.263	2.634	83.33	84.01
R3	11	0.223	2.456	82.50	82.15
R4	15	0.080	1.206	75.00	74.68

The highest quantity of work in process is at station W2 where there would be about five units on average, followed by W3 with 3.78 units.

The cycle time in the network is 3.819 days, and the accumulated work in process is 25.09 units (Table 7.5). It is worth mentioning that cycle time indicates the average

Table 7.5 Global measures of the supply chain

Property	Analytical	Simulated		
		Lower bound	Mean	Upper bound
Average cycle time in the system	3.81	3.33	3.62	4.08
Total work in process	25.09	21.79	23.79	26.99

Table 7.6 Mean time between arrivals, coefficient of variation, and parameters of gamma distribution, moderate variability

Time between arrivals (days)	Coefficient of variation	Shape parameter	Scale parameter
0.636	1	1	0.636
0.538	1	1	0.538
0.583	1	1	0.583
0.700	1	1	0.700

Table 7.7 Mean service time, coefficient of variation, and parameters of gamma distribution, moderate variability

Node	Mean service time (days)	Coefficient of variation	Form parameter	Scale parameter
S1	0.583	1	1	0.583
S2	0.500	1	1	0.500
P1	0.538	1	1	0.538
P2	0.175	1	1	0.175
W2	0.200	1	1	0.200
W3	0.560	1	1	0.560
R1	0.525	1	1	0.525
R2	0.583	1	1	0.583
R3	0.525	1	1	0.525
R4	0.350	1	1	0.350

time that the entity takes within the system from its entry to the first station until its transformation is completed.

The following scenarios that we have analyzed are the equivalent to a loss in the efficiency of the operation, which shows an increase in the system's variability: the average service time at a node may be the same, but not so the standard deviation, which may be higher. Tables 7.6 and 7.7 give the data corresponding to the inter-arrival times and service times in the stages of the chain used for the scenarios with moderate variability in both cases. It must be remembered that changes in variability imply a change in the scatter of data around the available service time.

The low variability results will be taken as the baseline scenario. When the variability in the system increases to a state with moderate variability, the cycle time increases. For example, the cycle time in station 1 increases to 1 day, representing a difference of 410%. For example, observing the work in process is stations W1 and W2, we can now see that, on average, they accumulate 34 and 24 units, respectively, in comparison with 5 and 3.78 units of the baseline scenario (Table 7.8).

As regards the global performance measurements, we can see that the average cycle time is now 17.56 days and that the work in process accumulates 115.381 units, in comparison with 3.81 days (4.6 times more) and 25 units in the baseline scenario (Table 7.9).

Table 7.8 Performance measures of the supply chain as a network of queues, moderate variability

Node	Cycle time in the node	Work in process	Node utilization	
			Analytical (%)	Simulated (%)
S1	1	11	91.67	91.25
S2	1	13	92.86	92.14
P1	1	12	92.31	92.09
P2	0.167	5.66	85.00	85.74
W2	1	34	97.14	96.35
W3	2	24	96.00	94.74
R1	0.3	3	75.00	75.43
R2	0.5	5	83.33	81.98
R3	0.429	4.71	82.50	83.11
R4	0.2	3	75.00	73.03

Table 7.9 Global measures of the supply chain, moderate variability

Property	Analytical	Simulated		
		Lower bound	Mean	Upper bound
Average cycle time in the system	17.558	13.7366	17.1021	22.513
Total work in process	115.381	90.5252	113.51	151.710

The characteristics of the flow within the supply chain are given in Table 7.8. There is no change in the congestion when compared with the previous scenario. This is because the available service time is still the same, so this means that the capacity used in each station is the same; however, the flow within the chain is affected, as we can see an increased amount of work in process piled up front of each station, which in turn increases the cycle time at each link of the chain, seeing as how the customers have to wait longer in the queue for service.

In the following scenario, there is an increased variability in the inter-arrival times and service time (Tables 7.10 and 7.11).

Table 7.10 Mean time between arrivals, coefficient of variation, and parameters of gamma distribution, high variability

Time between arrivals (days)	Coefficient of variation	Shape parameter	Form parameter	Scale parameter
S1	0.636	1.33	0.5653	1.1257
S2	0.538	1.33	0.5653	0.9525
P1	0.583	1.33	0.5653	1.0319
P2	0.7	1.33	0.5653	1.2382

Table 7.11 Mean service time, coefficient of variation, and parameters of gamma distribution, high variability

Node	Mean service time (days)	Coefficient of variation	Form parameter	Scale parameter
S1	0.583	1.33	0.5653	1.0319
S2	0.5	1.33	0.5653	0.8845
P1	0.538	1.33	0.5653	0.9525
P2	0.175	1.33	0.5653	0.3096
W2	0.2	1.33	0.5653	0.3538
W3	0.56	1.33	0.5653	0.9906
R1	0.525	1.33	0.5653	0.9287
R2	0.583	1.33	0.5653	1.0319
R3	0.525	1.33	0.5653	0.9287
R4	0.35	1.33	0.5653	0.6191

Tables 7.12 and 7.13 show that in a scenario of high variability, the performance measurements at each of the network's node show a deterioration in respect of the baseline scenario. In the case of the global cycle time the increase is 677.9% in respect of the baseline scenario as the cycle time is now 29.71 days and the accumulated work in process is 195.23 units (Table 7.13).

Table 7.12 Performance measures of the supply chain as a network of queues, high variability

Node	Cycle time in the node	Work in process	Analytical (%)	Simulated (%)
S1	1.72	18.92	91.67	91.39
S2	1.73	22.48	92.86	90.95
P1	1.72	20.7	92.31	90.37
P2	0.28	9.4	85.00	84.91
W2	1.75	59.49	97.14	95.96
W3	3.48	41.79	96.00	94.96
R1	0.41	4.12	75.00	73.11
R2	0.71	7.08	83.33	80.29
R3	0.61	6.7	82.50	83.49
R4	0.3	4.57	75.00	72.84

Table 7.13 Global measures of the supply chain, high variability

Property	Analytical	Simulated		
		Lower bound	Mean	Upper bound
Average cycle time in the system	29.71	14.082	23.24	37.043
Total work in process	195.238	90.432	154.02	249.29

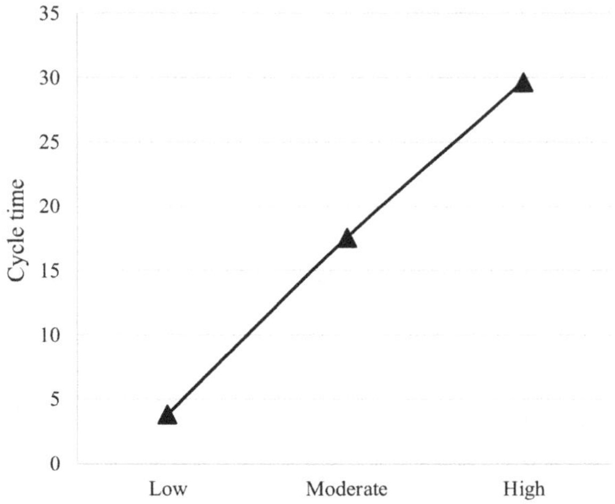

Fig. 7.8 Cycle time trend

Although the solution is optimal when we only take the costs of transport between the components of the supply network into consideration, by estimating the performance of the network as a queueing system that is subject to changes in the variability of service time and flow in the network, we see that residence time in the system as well as the quantity of accumulated work get worse as the characteristics of the flow get less and less like those of a regular process. These changes are indicative of a process whose behavior is increasingly hard to predict. It must be pointed out that in these scenarios, the entire supply chain is considered, which indicates that variability at each link of the chain accumulates as it advances through the process.

Figure 7.8 shows the change of the average cycle time in each variability scenario where we can observe that in the scenario of low variability, the cycle time is 3.819 units; however, in the scenario of high variability, the cycle time increases up to six times its value. The variability is a consequence of the randomness of the system itself; in the case of the design of optimal operating policies, it is advisable to incorporate this phenomenon into the analysis of the robustness of a solution.

7.4 Conclusions

When the operations of a supply chain are optimized, parameters such as demand or service times are normally assumed to be constant; in other words, the effect of variability on the system is ignored; however, it is not always a good idea to assume such behavior, as that could corrupt the solution you get.

In this paper, we used the example of a real supply chain to quantify the effect of variability on the optimal solution. Assuming that the chain is a G/G/c open queueing network, the behavior of the optimal solution was analyzed by altering variability in the flow to the nodes and service time.

The analysis of the results shows that there are no changes in congestion levels in any of the three scenarios; in other words, the same capacity is used as we did not alter the mean service time; however, the cycle time and quantity of work in process rose as the variability of the flow and service time increased. Taking as a reference the scenario where variability is low ($C^2_S <= 0.5$), the cycle time in the network increases to up to six times its value.

Increasing flow variability is equivalent to increasing the scatter of inter-arrival times at every node in the network, while altering the variability of service times is equivalent to increasing the scatter of service times at every node in the network. The changes made can be seen in expressions 16 (variability in a station's output) and 17 (approximation of the cycle time).

From the point of view of control, although the solution is optimal, this is not enough to ensure that the performance of the network is adequate. Incremental changes in variability are typical of systems subject to randomness and arise from faults and breakdowns in the equipment as well as from irregular transfer times between two nodes or reprocessing. Analyzing the capacity of a supply chain is not enough to ensure an adequate operation. This case study shows that we also must take into account the flow of the entity that circulates through the network, particularly if we wish to consider that a supply chain is a system subject to randomness.

This document shows the method to be followed, using analytical means, where the variables are the squared coefficient of variation of the inter-arrival times and the squared coefficient of variation of the service time.

The analytical approach that we used in the case study is easily available to decision makers and includes the variables that affect the system. The same study can be performed using a simulation model; however, this would require more of the administrator's time, and besides, they might not always have access to the necessary computational package.

Although analytical expressions provide the relationship between variables, when the system is very complex, its solution can be complicated; in this case, it is convenient to build a simulation model. It should also be taken into account that analytical models assume that the system is in a stable state, and sometimes, it is necessary to make decisions during the period of system instability.

Supply chains work today by limiting the amount of work in process; a future line of work is to take into account capacity constraints for the queue size.

References

Ballou RH (2003) Supply chain management. 5th edn. Prentice—Hall, New Jersey.

Bandaly D, Satir A, Shanker L (2016) Impact of lead time variability in supply chain risk management. Int J Prod Econ 180:88–100. https://doi.org/10.1016/j.ijpe.2016.07.014

Bazaraa MS, Jarvis JJ, Sherali HD (1974) Linear programming and network flows. Wiley, New Jersey

Bhaskar V, Lallement P (2010) Modeling a supply chain using a network of queues. Appl Math Model 34(2010):2074–2088. https://doi.org/10.1016/j.apm.2009.10.019

Bhaskar V, Lallement P (2011) Queuing network model of uniformly distributed arrivals in a distributed supply chain using subcontracting. Decis Support Syst 51(2011):65–76. https://doi.org/10.1016/j.dss.2010.11.029
Curry GL, Feldman RM (2009) Manufacturing systems modeling and analysis. Springer, Berlin
Cruz FRB, van Woensel T (2014) Finite queueing modeling and optimization: a selected review. J Appl Math. https://doi.org/10.1155/2014/374962
Gong Q, Lai L, Wang S (2008) Supply chain networks: closed Jackson network models and properties. Int J Product Econo 113(2008):567–574. https://doi.org/10.1016/j.ijpe.2007.10.013
Gross D, Shortle JF, Thompson JM (2008) Fundamentals of queueing theory. Wiley, New Jersey
Guerrero-Campanur A (2013) Supplier selection in an inventory location problem for a three-level supply chain. Doctoral thesis, UPAEP, Puebla, Mexico
He X, Hu W (2014) Modelling relief demands in an emergency supply chain system under large-scale disasters based on a queueing network. Sci World J 2014:1–12. https://doi.org/10.1155/2014/195053
Hernández-González S, Hernández Ripalda D (2018) Managerial approaches toward queueing systems and simulations. IGI Global, Pensilvania
Hernández-González S, Flores de la Mota I, Jiménez-García JA, Hernández-Ripalda MD (2017) Numerical analysis of minimum cost network flow with queuing stations: the M/M/1 case. Nova Sci 9(18):257–289. https://doi.org/10.21640/ns.v9i18.840
Hopp WJ (2011) Supply chain science. Waveland Press, Long Grove
Hopp WJ, Spearman ML (2001) Factory physics. Waveland Press, Long Grove
Hum S, Parlar M, Zhou Y (2018) Measurement and optimization of responsiveness in supply chain networks with queueing structures. Eur J Oper Res 264(1):106–118. https://doi.org/10.1016/j.ejor.2017.05.009
Jackson JR (1957) Networks of waiting lines. Oper Res 5(4):518–521. https://doi.org/10.1287/opre.5.4.518
Jackson JR (1963) Job-shop like queueing systems. Manage Sci 10(1):131–142. https://doi.org/10.1287/mnsc.10.1.131
Kerbache L, MacGregor Smith J (2004) Queueing networks and the topological design of supply chain systems. Int J Product Econ 91(2004):251–272. https://doi.org/10.1016/j.ijpe.2003.09.002
Liu L, Liu X & Yao D (2004) Analysis and optimization of a multistage inventory-queue system. Manage Sci 50(3):365–380. https://doi.org/10.1287/mnsc.1030.0196
Little JDC (1961) A proof for the queuing formula: $L = \lambda W$. Oper Res 9(3):383–387. https://doi.org/10.1287/opre.9.3.383
MacGregorSmith J, Kerbache L (2017) Topological network design of closed finite capacity supply chain network. J Manuf Syst 45(2017):70–81. https://doi.org/10.1016/j.jmsy.2017.08.001
Marmolejo JA, Rodríguez R, Cruz-Mejía O, Saucedo J (2016) Design of a distribution network using primal-dual decomposition. Math Probl Eng.https://doi.org/10.1155/2016/7851625
Negri da Silva CR, Morabito R (2009) Performance evaluation and capacity planning in a metallurgical job-shop system using open queueing network models. Int J Prod Res 47:6589–6609. https://doi.org/10.1080/00207540802350732
Pérez-Loaiza RE, Olivares-Benitez E, Miranda-González PA, Guerrero-Campanur A, Martínez-Flores JL (2017) Supply chain network design with efficiency, location and inventory policy using a multiobjetive evolutionary algorithm. Int Trans Oper Res 24:251–275. https://doi.org/10.1111/itor.12287
Scilab Enterprises (2012) Scilab: free and open source software for numerical computation (OS, Version 5.3.3). Available from: https://www.scilab.org
Vahdani B, Mohammadi M (2015) A bi-objective interval-stochastic robust optimization model for designing closed loop supply chain network with multi-priority queueing system. Int J Prod Econ 170:67–87. https://doi.org/10.1016/j.ijpe.2015.08.020
Wagner HM (1975) Principles of operations research, with applications to managerial decisions. Prentice Hall, New Jersey

Whitt W (1983) The queueing network analyzer. Bell Syst Tech J 62(9):2779–2815. https://doi.org/10.1002/j.1538-7305.1983.tb03204.x
Yousefi-Babadi A, Tavakkoli-Moghaddam R, Bozorgi-Amiri A, Seifi S (2017) Designing a reliable multi-objective queuing model of a petrochemical supply chain network under uncertainty: a case study. Comput Chem Eng 100:177–197.https://doi.org/10.1016/j.compchemeng.2016.12.012
Zahiri B, Tavakkoli-Moghaddam R, Mohammadi M, Jula P (2014) Multi-objective design of an organ transplant network under uncertainty. Transp Res Part E 72:101–124.https://doi.org/10.1016/j.tre.2014.09.007

Chapter 8
Mathematical Approach for Street Markets Location in Food Desert Regions: A Case Study in Valparaíso Chile Region

Julián Alberto Espejo-Díaz, Daniela Granados-Rivera, and Gonzalo Mejía

Abstract In this work, we consider facility location decisions for street markets in food desert zones. Street markets operate on certain days of the week in public and open spaces such as local parks or city squares. The implementation of street markets is a decision that involves local governments with limited budget. Therefore, it is crucial to optimize the budget usage while satisfying as much demand as possible and maximizing coverage. In this work, we propose a mathematical model that determines the number, locations, and schedules of street markets considering customers' georeferentiation and its probability of choosing street markets over competitors such as local small retailers. Our model, which is a variant of the facility location problem, integrates different tools such as clustering methods and discrete choice. We conducted computational experiments using information from Valparaíso Chile region. Results show the effectivity of our proposal when locating street markets in food desert zones. The model produces (1) optimal locations of street markets, (2) street markets schedule, (3) quantities to allocate in each street market, and (4) the required budget to be allocated in street market policies.

Keywords Street markets · Facility location problem · Mixed-integer linear programming

8.1 Introduction

A varied and healthy diet, which includes fruits and vegetables (F&Vs), may improve people's health and prevents noncommunicable diseases (World Health Organization 2016). On average, people of vulnerable areas in emerging countries do not consume the minimum daily amount of F&Vs suggested by the World Health Organization, and the risk of disease, undernourishment, and overweight increases (World Health Organization 2017). One of the main reasons people, in particularly in low-income

J. A. Espejo-Díaz (✉) · D. Granados-Rivera · G. Mejía
Faculty of Engineering, Universidad de La Sabana, Autopista Norte de Bogotá, Campus Universitario Puente del Común. Km 7, Chía, Colombia
e-mail: julianesdi@unisabana.edu.co

J. L. Garcia Alcaraz et al. (eds.), *Trends in Industrial Engineering Applications to Manufacturing Process*, https://doi.org/10.1007/978-3-030-71579-3_8

areas, do not have varied and healthy diets, is the lack of access to fresh products such as F&V. Such areas are known in the literature as "food desert" and are mainly characterized by high costs and low availability of healthy foods (Dulin et al. 2018).

The planning and operation of fresh food supply chains in food deserts is a complex task due to a high interdependency between a vast number of suppliers (farmers and other sellers), service providers, and customers (Srivastava et al. 2015). In addition, poor infrastructure, reduced budget of public authorities, and shortage of technology complicate the distribution of fresh foods. Despite these facts, this problem represents an opportunity to produce relevant and impactful research to address food security problems in food desert regions. It is worth noting that in the world, approximately 2 billion people or 26.4% of the world population have moderate or severe levels of food insecurity (FAO et al. 2019).

One strategy that has proven successful to increase the healthy food intake in food deserts is the implementation of street markets (Risica et al. 2018; Widener et al. 2012). Street markets consist of temporal stalls that are set-up and disassembled on the same day. There, local farmers sell their products directly to their customers at lower prices in comparison with small retailers and supermarkets. The location and size of such markets have been traditionally defined by public authorities without quantitative analyses and methods. To the best of our knowledge, most of these implementations do not consider aspects such as geographical location, customer concentration, capacity, and road access among others.

Locating street markets in urban areas represents a challenge. First, street markets must offer a wide range of healthy products at reasonable and accessible prices. Second, street markets must be located near to their customers to avoid long trips. Street markets must also consider the competition from other sellers such as small retailers and central market that change the purchase patterns of households. Third, the population served by these street markets must be educated in the eating habits in order to increase the consumption of healthy foods. Last, street markets must deal with a number of constraints that include traffic restrictions, budget, opposition of the populations of the surrounded areas perhaps affected by noise, traffic, and garbage produced by the street market operation.

The previously problem can be seen as an extension of the "facility location problem" where the optimal location of facilities is determined with the goal of minimizing cost or maximizing coverage and at the same time meeting the demand of their customers (Owen and Daskin 1998). The facility location decisions have several elements (Eiselt and Laporte 1995): (i) the area of the facility to be opened, (ii) the facilities' attributes, (iii) customer demand, and (iv) the customer interactions with the facilities: These interactions depend on the customer assignment to one or multiple facility locations. In this context, the candidate facilities for locating street markets are public open areas such as squares, city parks, and others. The purpose of this work is to develop a computational model that allows establishing the number of street markets to be opened, and its location maximizing the populations' coverage quality. This coverage quality considers different coverage levels that depend on the customer distance to the street markets. The main difference with other facility location models is that the proposed model considers temporal restrictions and probabilities of shopping. As such, street markets have a restriction on the number of days that it can operate in a week and the number of consecutive days that it can operate. Moreover, this model considers the probability of customers choosing the street market over other retailers, calculation based on a discrete choice model that involves street market distance to customers and price levels.

This chapter is organized as follows. Section 2 provides the background of this work by reviewing the main related concepts. Section 3 presents the proposal by formally defining the problem and the mathematical formulation. Section 4 describes the Valparaíso region case study and its implementation. Section 5 shows the main results of the case-study implementation. Finally, in Sect. 6, we discuss the conclusions of this work and we provide some useful suggestions for future work.

8.2 Background

In this section, we give a brief review on the concepts associated with the development of this work.

8.2.1 Street Markets

Street markets, also known as mobile markets or farmers' markets, are shopping facilities that open on one or more days of the week. They are popular because they offer a variety of products and are also meeting places (vom Lehn 2014). Food street markets provide direct contact between farmers and the consumers. These markets are part of what it is known in the literature as a short supply chain (SCC). A SCC aims at eliminating intermediaries to generate higher returns to farmers and providing fresher products at lower prices to their customers (Givens and Dunning 2019). Street markets arise in part from the need of some communities to access specific essential products that are not within their reach, either because these communities are located far from their sources of food or to avoid high prices resulting from several intermediaries interceding in the delivery of the product. The introduction of street markets in certain communities has increased the access and availability of specific products such as F&V, stimulating healthy eating habits in the population (Kahin et al. 2017). For instance, the authors of (Dunn et al. 2012) analyzed the relationship between the consumption of F&V and their costs and availability. They used a multivariate probit regression to study the relationship. Also, in (Ghosh-Dastidar et al. 2014), authors used multivariate logistic regression to study the relation among distance to store, food price, and obesity.

A successful example of street market implementation in a "food dessert" community is studied in (Gary-Webb et al. 2018), where the authors evaluated a pilot test of implementation of green street markets (which exclusively offer F&V) in four communities of the USA. The results of the strategy implementation showed an important improvement in the residents' diet, evidenced by an increase in the average consumption of F&V. Another example is presented in (Hsiao et al. 2018), where the authors assessed the influence of street markets on F&V access in eight neighborhoods, also. The results showed that street markets might influence on F&V access in low-income areas specially helping older adults and people living alone.

In other studies such as (Zepeda et al. 2014), they investigated about the potential of street markets to change attitude and food choices in four areas of New York. They found that street markets had to offer a great variety of items, focus on value and services, and be reliable to become in a potential market. Also, the authors of (Robinson et al. 2016) studied the operational advantages and constraints of street markets to improve food access in low-income areas. They found that these markets can overcome geographical, economic, and social barriers to enhance fresh food access in low-income residents. However, their effects are hampered by operational, political, and economical constraints. Other case is provided by Ylitalo et al. (2019), in which authors described costumers features and barriers to healthy eating. The results indicated that street markets have lower cost and greater flexibility, but they did not serve all consumers. In general, all the reviewed research works mentioned above are related to features to improve healthy eating through street markets.

Other perspective of the problem is a mathematical approach. The authors of (Widener et al. 2012) investigated the location of street markets. The research was made in low-income communities to increase healthy food access considering addressing spatial constraints. The results located street markets according to the geographical distribution from the residents to the selected area. Also, in (Tong et al. 2012), they considered spatial constraints too and incorporated temporal constraints to create a provision planning of street markets. Both used an optimization model.

8.2.2 Facility Location Problem

The facility location problem (FLP) has been applied to many fields over the years. Real applications can be seen in health care systems, solid waste management, production–distribution, education systems, telecommunication, area planning, and on on (Farahani et al. 2014). Its potential high impact is one of the main reasons the problem has been widely studied over the last decades. Another reason for its relevance in the literature is because many private and the public companies are looking for better decision-making tools which allow them to locate their distribution centers in the right place (Farahani et al. 2012). This decision is costly and practically irreversible: opening a distribution center in the wrong place increases the operational costs and reduces undoubtedly the profit of the companies or the profit of operations.

Within the logistics context, the facility location strategies are one of the three main decisions that companies or governments made regarding its supply chain along with transport and inventory decisions (Ballou 2004). The previous configuration of the facility location problem regarding its elements is usually modeled mathematically as the set covering problem as presented in (Farahani et al. 2012), where the objective of the set covering problem corresponds to the minimization of the overall costs of opening facilities subject to customer demand observation.

Finally, in Widener et al. (2012), the authors propose a linear programming model that locates street markets based on the geographical distribution of customers. The programming model is tested with information from Buffalo—New York. Results show that the model increases people's access to healthy food at a relatively low cost. The difference between this proposal and our work lies in the utilization of a clustering methodology for estimating the populations demand within the mathematical model. Additionally, in our work, we include the probability of a customer choosing a street market over a small local retailer. These considerations enhance the classical facility location model previously studied in this context.

8.3 Proposal

In this section, first, we formally define the problem and we present the mixed-integer linear programming (MILP) model for street market location.

8.3.1 Problem Description

It is considered a set of customers (home clusters) $C = \{1, \ldots, j\}$ geographically distributed over a food desert region and must be served in multiple $T = \{1, \ldots, t\}$ periods. Each customer $j \in C$ has a daily demand d_{jt} which have to be fulfilled in the period $t \in T$ by street markets which can be situated in a set of potential locations $M = \{1, 2, \ldots, i\}$. As previously stated, these potential locations are comprised by open areas such as parks or city squares. Each street market location $i \in M$ has a maximum capacity q_i and has a coverage area represented by a_{ij} indicating whether the customer $j \in C$ can be served by the street market location $i \in M$ based on its maximum coverage distance. Each street market location $i \in M$ has a fixed opening cost f_i in any period of the planning horizon and has a coverage level l_{ij} regarding its closeness to customer $j \in C$; the closer the street market is to the customer, the greater l_{ij} value is. We represent this using a decreasing step function. Considering that the candidate places for street market locations may be located near local retailers, each customer $j \in C$ has a probability p_j for choosing a street market over other retailers. Section 8.4.1 shows the calculation of these probabilities. Finally, we consider a budget limit B for investing in street markets in the planning horizon.

8.3.2 Street Markets Facility Location MILP Model

Next, we introduce the facility location model for locating street markets in food desert areas. The model is formulated with the following elements:

Sets

M: Set of street markets potential locations $M = \{1, \ldots, i\}$.
C: Set of customers $C = \{1, \ldots, j\}$.
T: Set of periods.

Parameters

l_{ij}: Value of coverage level for customer j being served by street market i.
d_{jt}: Expected demand of customer j in period t.
a_{ij}: 1 if customer j is in the coverage area of street market location i and 0 otherwise.
f_i: Fixed cost for establishing street market i.
q_i: Capacity of street market i.
p_j: Probability of customer j choosing to purchase in street markets.
B: Available budget in the planning horizon.
GG: Size of set T.
$BIGM$: Big value for logical constraints.

Decision variables

y_{ijt}: Demand percentage of customer j covered by street market i in period t.
x_{it}: 1 if street market i is selected in period t, 0 otherwise.

The objective function in Eq. 8.1 maximizes the customer coverage quality level.

$$\text{Max}\left(\sum_{i \in M} \sum_{j \in C} \sum_{t \in T} l_{ij} d_{jt} p_j y_{ijt}\right) / S \tag{8.1}$$

Subject to:

$$\sum_{j \in C} d_j y_{ijt} \leq q_i x_{it} \quad \forall i \in M, t \in T \tag{8.2}$$

$$\sum_{i \in M} y_{ijt} a_{ij} \leq 1 \ \ \forall j \in C, t \in T \tag{8.3}$$

$$y_{ijt} \leq a_{ij} \quad \forall j \in C, t \in T \tag{8.4}$$

$$x_{it} + x_{it+1} \leq 1 \quad \forall i \in M, t \in T | t < G \tag{8.5}$$

$$\sum_{i \in M} \sum_{t \in T} f_i x_{it} \leq B \tag{8.6}$$

$$y_{ijt}, s_{ijt}, w \geq 0 \quad \forall i \in M, j \in C, t \in T \tag{8.7}$$

$$x_{it} \in \{0, 1\} \quad \forall i \in M, t \in T \tag{8.8}$$

Constraint set (8.2) observes the capacity of street markets' potential locations. Constraint set (8.3) establishes that each customer does not receive more than its demand. Constraint set (8.4) establishes that a customer can only be served by a street market if the market is within the maximum range of coverage. Constraint set (8.5) avoids opening a street market in consecutive periods. Constraint set (8.6) observes the available budget for the planning horizon. Finally, constraint sets (8.7) and (8.8) correspond to the nonnegativity and binary constraints for the decision variables.

8.4 Case Study

This chapter uses data collected from the municipality of Valparaíso, Chile. Valparaíso is a seaport city which has approximately 300.000 inhabitants and is located in the central zone of Chile (Instituto Nacional de Estadística 2017). Despite being a seaport city with a leading role in the country's trade, most of its population belong to the lowest socioeconomic classes of Chile. The city consists of about 45 steep hills that create difficulties in road access to a large part of the population (Municipalidad de Valparaíso 2020). As such, there is a lack of access to essential products such as F&V especially in these remote areas (Saavedra Manríquez 2016). Currently, the average intake of F&V in the Valparaíso region averages 200 g of F&V per day (Ministerio de Salud de Chile 2010). This figure is well below, the 400 g recommended by the WHO.

The greater part of F&V comes from central valley of Chile to the wholesalers. In Valparaíso, the most relevant wholesaler is Mercado Cardonal (Saavedra Manríquez 2016). It is the principal supply source for nanostores, public institutions, and individual buyers, among others. However, going to the Cardonal requires transportation. Currently, there are several operating street markets as seen in Table 8.1. These operate with the frequencies shown also in Table 8.1. Additionally, Fig. 8.1 shows its location in the Valparaiso map.

Initial studies (FAO Regional Office for Latin America and the Caribbean 2013) by the Ministry of Agriculture of Chile and the Food and Agriculture Organization characterized these markets. Despite their popularity, the current street markets can

Table 8.1 Valparaíso city street markets current configuration, source (Saavedra Manríquez 2016)

Street Market	Day of the week
Los Copihues	Thursday and Sunday
Playa Ancha	Thursday and Sunday
Avenida Argentina	Wednesday and Saturday
La Conquista Placeres	Thursday and Sunday
Placilla	Thursday and Sunday

Fig. 8.1 Current street market locations

only supply a small percentage of the population (around 30%) based on our estimates of area, number of stalls, and frequency of operation (FAO Regional Office for Latin America and the Caribbean 2013).

8.4.1 Data Collection and Preparation

The potential locations of the street markets were established by scanning Google Maps™ and considering the knowledge of the authors of this book chapter. We are aware that further refining is required with the municipality authorities, but for the sake of having a first approximation, we believe this is acceptable. We determined 55 potential locations in public parks and open spaces throughout the city. Each street market location has a capacity determined by the available area. In the case of the existing street markets, we used data retrieved from (Observatorio Feria Libre 2013). The opening cost for each street market was estimated according to their number of stalls (Modo Beneficios 2019).

The demand points were randomly generated in residential areas of the hills of Valparaíso. We generated 75,000 demand points that correspond roughly to the number of households in the city, assuming four persons per household. In order to have a runnable mathematical model, we reduced the number of demand points using the classical k-means clustering strategy. Such a k-means algorithm has been widely used for its simplicity and quickly processing great volumes of data (Yunoh et al. 2017). This algorithm has been applied to classify information in different sciences as decision sciences (Gbadoubissa et al. 2018), statistical methods (Asgharinia and Petroselli 2020), among others. It consists of the divided dataset in k predefined sets of clusters (Ramalingam and Thangarajan 2020). In this case, we use this method implemented in the software R-project to aggregate the households' demand in 500 clusters. The clustering was done according to these geographical locations defined by the latitude and longitude. The average daily demand per cluster was generated randomly taking as input the average daily demand from (Ministerio de Salud 2010). In this clustering strategy, the Valparaíso population homes were grouped in 500 geospatial points with their respective aggregated demand.

As mentioned above, customers may do their shopping of F&V in facilities different from the street markets. Previous studies have evidenced that consumers' purchase behavior of F&V is determined according to several parameters such as price, assortment, diet, and distance, among other preferences.

In this study, we established the choice of market probability derived from multinomial logit model with price and distance as parameters which are the most important for food shopping in food deserts (Dunn et al. 2012). Multinomial logit model consists in select an alternative h of H alternatives with a specific probability, which depends on K attributes that either has a weight $\beta_k \beta_k$. The choice of the consumers jj is given by $W_{jh} W_{jh}$ according to $Z_{jhk} Z_{jhk}$, a parameter known (Cramer 2003). The probability of selecting an alternative uu is given by Eq. (8.9).

$$\Pr\left(W_{ju}\right) = \frac{\exp\left(\beta^T Z_{ju}\right)}{\sum_{h=1}^{H} \exp\left(\beta^T Z_{jh}\right)} \tag{8.9}$$

This value (Pr) is employed as a fixed parameter that determines the probability that a household cluster select a street market instead of other sources. The probability seeks to give greater weight to locate street markets in food desert areas due to the available purchase places are limited, giving a higher probability of choosing street markets. In this case, we determined the β_k according to previous studies (Ministerio de Agricultura de Chile 2009; Saavedra Manríquez 2016).

Fig. 8.2 F&V local small retailers' location in Valparaíso region

In this study, we considered 140 small retailers (corner shops) as competitors of the street markets and one central market. The georeferentiation of the corner shops was obtained from (Aránguiz et al. 2018) and is shown in Fig. 8.2. The choice probability is different for each household cluster as it cluster has J alternatives to buy (see Eq. 8.9). J is established according to the number of corner shops within a range of 2 km around each household. The prices used in Eq. 8.5 for corner shops were retrieved from (Centro de Estudios del Retail (CERET) 2013) and for street markets from (ODEPA Ministerio de Agricultura 2019). The location of street markets, corner shops, and households was established using Google Maps™ application. The distances between these were calculated using the Google Maps™ application.

In the tests, the planning horizon corresponds to one week (seven days). We set a maximum street market coverage area of 2 km. The coverage levels were established using a decreasing step function as presented in Table 8.2. For instance, for a customer located 500 m away from the street market, their coverage level corresponds to 80%.

Table 8.2 Street markets' coverage level

Range	Coverage level l_{ij} (%)
0–400 m	100
400–800 m	80
800 m–1.2 km	60
1.2–1.6 km	40
1.6–2 km	20
>2 km	0

8.5 Results

The mathematical model was implemented in the software IBM ILOG CPLEX Optimization Studio Version: 12.8.0.0 in a laptop with 30 GB RAM and 2 GHz processor. The experimentation tests were configured in the following way: We tested several policies varying the available budget B. Figure 8.3 shows the policies' objective function (coverage quality) and covered demand with respect to the total customers' demand.

Next, Table 8.3 presents the number of street markets opened in the planning horizon. Clearly as the budget increases, so does the number of street markets, the coverage quality, and the covered demand. However, the increase is not proportional, as seen in Fig. 8.3.

For instance, with a weekly budget of US $32,000.00 (solution 7), the coverage level reaches 76% with the objective function in 41.56%. Figure 8.4 depicts the georeferentiation of the proposed and current street market with this budget and coverage level.

Now, we analyze the scenario of solution 7. Figure 8.5 shows that in that solution, the street markets would be mainly located in food desert zones. For instance, street markets #7–9, #14, #18, and #20 would be located in the Valparaíso hills (southwest part), where there are few local retailers and they cannot satisfy the nearby

Fig. 8.3 Percentage of coverage quality and covered demand varying the budget

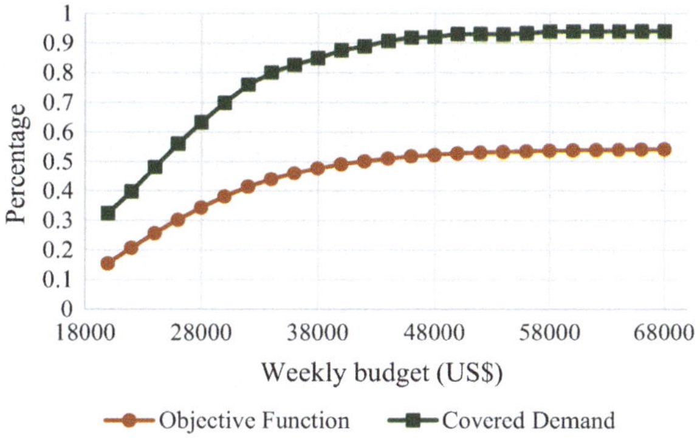

Table 8.3 Number of street markets in the planning horizon

Solution	Budget	Objective function (%)	Covered demand (%)	Number of street markets opened in a week
1	$20,000.00	15.62	32.51	17
2	$22,000.00	20.84	39.82	27
3	$24,000.00	25.79	48.06	31
4	$26,000.00	30.31	56.08	38
5	$28,000.00	34.49	63.31	45
6	$30,000.00	38.19	69.92	50
7	$32,000.00	41.56	76.00	58
8	$34,000.00	44.11	80.10	65
9	$36,000.00	46.14	82.72	74
10	$38,000.00	47.81	85.00	86
11	$40,000.00	49.23	87.60	93
12	$42,000.00	50.30	88.99	97
13	$44,000.00	51.22	90.78	98
14	$46,000.00	51.95	91.92	106
15	$48,000.00	52.42	92.15	112

customers demand. The same situation can be seen with street markets #1 (downtown), #3 (west), #13 (west), #17 (northwest), and #21 (northwest). The remaining street markets supply F&V to customers located in zones with either a small number of local retailers or in zones where the current street markets do not have capacity to cover the customers' demand.

Next, we analyze the scenario of solution 13. This solution achieves a covered demand level of 90.78% with a coverage quality (objective function) of 51.22%. However, this solution requires additional funding of 37.5% and increases the covered demand in 15% with respect to solution #7.

Figure 8.6 depicts the location of the additional opened street markets with the current ones and the small local retailers. In this solution, new Valparaiso zones are served. For instance, street market #2 and #18 serves the population which is located in the extreme orient in the city. Besides, with this solution, mobile market #6 and #22 supply the uncovered demand of Placilla street market in the south part of the city. It is up to the decision-maker to decide the number of streets markets to be opened based on the available budget and the desired coverage level.

Fig. 8.4 Proposed street markets location with 76% of customers demand coverage

8.6 Conclusions and Future Work

The low consumption of F&V as part of a healthy diet is a worrying issue in developing countries. Nonetheless, the implementation of street markets represents a plausible strategy to address food insecurity problems in food desert zones. They represented an alternative to offer F&V cheap in low-income communities that usual are in food deserts. By assigning a specific budget, governments can assist local farmers in setting-up street markets. In this manner, they can offer their products directly to customers without the need of intermediaries at lower costs. Frequently, the assigned budget is limited, and therefore, there is a need to optimize this scarce financial resource. Street markets function on specific days of the week and can be located in available open spaces in the region. The most frequent candidate open spaces used in street market location comprises city parks, city squares, or unused spaces. The candidate spaces apart from its location or capacity may have competence near such as local retailers.

In this research, we developed a facility location model that determines the number and locations of street markets and their weekly schedule to increase the coverage of F&V households' demand in the city of Valparaíso, Chile. Our computational experiments indicate that the model successfully locates street markets in food desert areas

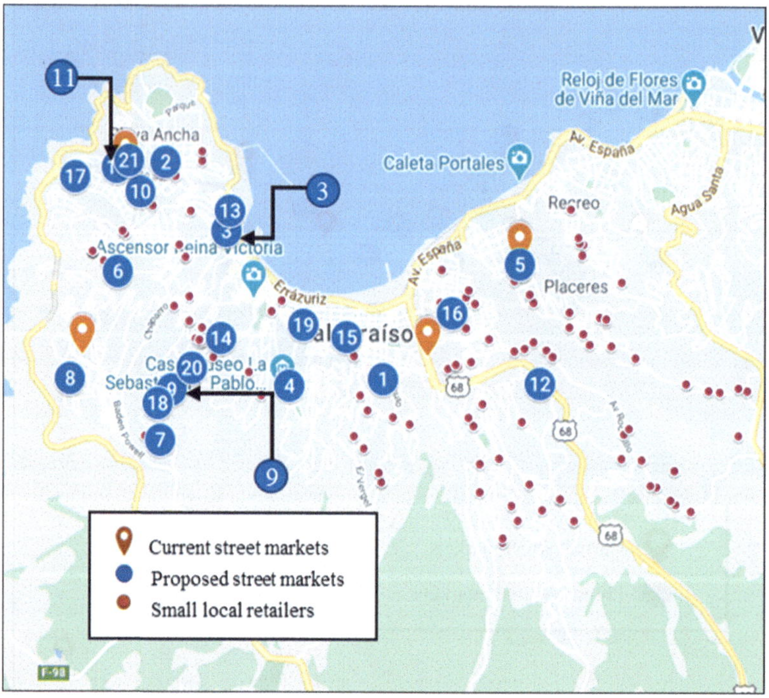

Fig. 8.5 Proposed street market locations in Valparaíso food desert zones

considering customers' demand, capacity, and the currently enabled street market. It is worth noting that the model aims to maximize the coverage quality. For that, we observe the interesting effect of the probability on the solution, which allowed increasing the number of street markets located in food deserts. It is showed in Figs. 8.5 and 8.6 that the western area of Valparaiso had a few nanostores, and for it, a great number of street markets were located.

This research, however, is subject to the following limitations that could be addressed in future research. First, we did not consider the transportation costs between street markets' potential locations and farmers' locations, which could change the proposed street markets' optimal location. Second, the probability estimation of customers choosing a street market over a local retailer only contemplates street markets and nanostores as buy options. In that way, we are not considering the existence of other stakeholders such as wholesalers and supermarkets, which can alter the customer choice. Last, the probability of a customer choosing a street market was calculated with two factors: distance and price. Other features can also determine the buy choice, such as assortment, eating habits, the service level of sellers, and socioeconomic factors. Considering the above, future works can include transportation costs to street markets candidate locations, and the probability estimation can include more factors such as the mentioned before and other stakeholders.

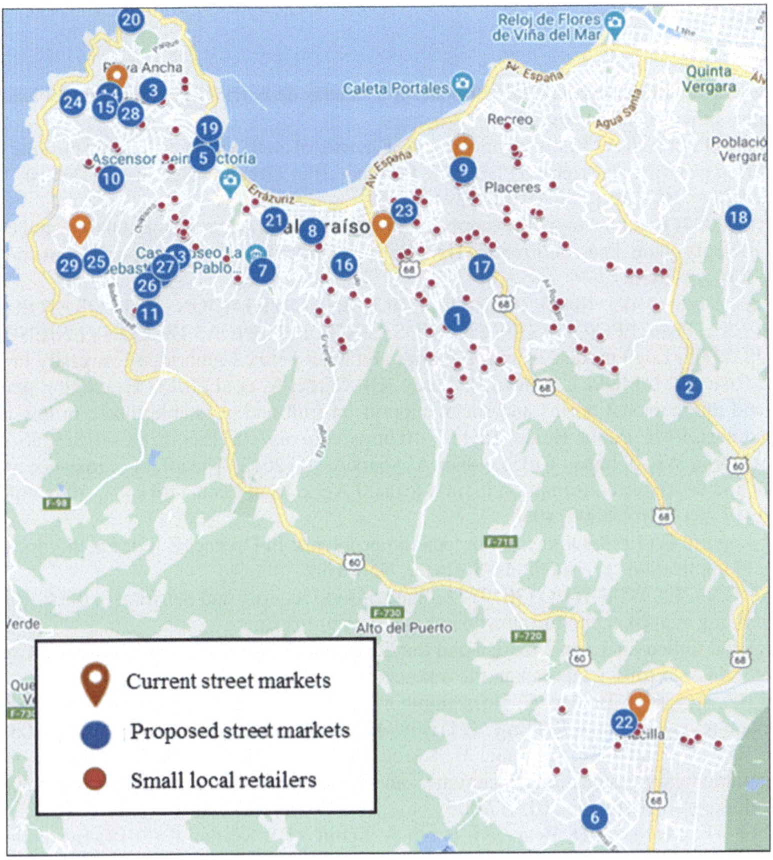

Fig. 8.6 Proposed street markets location with 90.78% of customers demand coverage

Additionally, in this work, we presented a MILP for deterministically locating street markets in food desert areas. The logical next step in future works is to address uncertainty in the proposal. It is suggested to use stochastic or robust methodologies to cope with uncertainty in parameters such as the customers' demand.

Finally, future work should be directed to use more sophisticated methodologies for capturing in more detail customers' decisions and the repercussions in the F&Vs supply chain. In addition, these methodologies can also include other stakeholders such as supermarkets and wholesalers. Agent-based modeling and simulation can consider customers' individual behaviors and their interactions with other stakeholders.

Acknowledgements The authors of this work would like to thank the Faculty of engineering and the research group Sistemas Logisticos at Universidad de La Sabana for the support of this work. Additionally, J.A. Espejo and D. Granados would like to thank Universidad de La Sabana for the Teaching Assistant Scholarship for their master's studies.

References

Aránguiz R, Mejía G, Mejía C (2018) Encuesta a tiendas de barrio y estrategia de Ubicación de Mercados móviles en Valparaíso

Asgharinia S, Petroselli A (2020) A comparison of statistical methods for evaluating missing data of monitoring wells in the Kazeroun Plain, Fars Province, Iran. Groundwater Sustain Dev 10:100294. https://doi.org/10.1016/j.gsd.2019.100294

Ballou RH (2004) Business logistics/supply chain management: planning, organizing, and controlling the supply chain. Pearson/Prentice Hall. Retrieved from https://books.google.com.co/books?id=sgsdQAAACAAJ

Centro de Estudios del Retail (CERET) (2013) INVESTIGACIÓN EXPLORATORIA DE NIVELES Y PERCEPCIONES DE PRECIOS. SECTOR TIENDAS DE CONVENIENCIA

Cramer JS (2003) Logit models from economics and other fields. Cambridge University Press

Dulin A, Risica PM, Mello J, Ahmed R, Carey KB, Cardel M et al (2018) Examining neighborhood and interpersonal norms and social support on fruit and vegetable intake in low-income communities. BMC Public Health 18(1):1–10.https://doi.org/10.1186/s12889-018-5356-2

Dunn RA, Dean WR, Johnson CM, Leidner A, Sharkey JR (2012) The effect of distance and cost on fruit and vegetable consumption in rural texas. J Agric Appl Econ 44(4):491–500. https://doi.org/10.1017/s1074070800024068

Eiselt H, Laporte G (1995) Objectives in location problems. In Drezner Z (ed) Facility location: a survey of applications and methods. Springer, New York

FAO, Ifad, UNICEF, WFP, WHO (2019) The state of food security and nutrition in the world 2019. Safeguarding against economic slowdowns and downturns Rome

FAO Regional Office for Latin America and the Caribbean (2013) FAO: ferias libres de Chile son un motor de desarrollo y un gran aporte a la seguridad alimentaria

Farahani RZ, Asgari N, Heidari N, Hosseininia M, Goh M (2012) Covering problems in facility location: a review. Comput Ind Eng 62(1):368–407. https://doi.org/10.1016/j.cie.2011.08.020

Farahani RZ, Hekmatfar M, Fahimnia B, Kazemzadeh N (2014) Hierarchical facility location problem: models, classifications, techniques, and applications. Comput Ind Eng 68(1):104–117. https://doi.org/10.1016/j.cie.2013.12.005

Gary-Webb TL, Mendez DD, Bear TM, Fabio A, Schiff MD, Keenan E (2018) Evaluation of a mobile farmer's market aimed at increasing fruit and vegetable consumption in food deserts: pilot study to determine evaluation feasibility. Health Equity 2(1):375–383. https://doi.org/10.1089/heq.2018.0003

Gbadoubissa JEZ, Ari AAA, Gueroui AM (2018) Efficient k-means based clustering scheme for mobile networks cell sites management. J King Saud Univ Comput Inf Sci. https://doi.org/10.1016/j.jksuci.2018.10.015

Ghosh-Dastidar B, Cohen D, Hunter G, Zenk SN, Huang C, Beckman R, Dubowitz T (2014) Distance to store, food prices, and obesity in urban food deserts. Am J Prev Med 47(5):587–595. https://doi.org/10.1016/j.amepre.2014.07.005

Givens G, Dunning R (2019) Distributor intermediation in the farm to food service value chain. Renew Agric Food Syst 34(3):268–270. https://doi.org/10.1017/S1742170517000746

Hsiao BS, Sibeko L, Wicks K, Troy LM (2018) Mobile produce market influences access to fruits and vegetables in an urban environment. Public Health Nutr 21(7):1332–1344. https://doi.org/10.1017/S1368980017003755

Instituto Nacional de Estadística (2017) Censos de Población y Vivienda. Retrieved from https://reportescomunales.bcn.cl/2017/index.php/Valparaíso#Poblaci.C3.B3n_total_Censo_2002_y_Censo_2017

Kahin SA, Wright DS, Pejavara A, Kim SA (2017) State-level farmers market asctivities. J Public Health Manag Pract 23(2):96–103. https://doi.org/10.1097/PHH.0000000000000412

Ministerio de Agricultura de Chile (2009) Ferias libres de la Región de Valparaíso, comuna de Valparaíso

Ministerio de Salud (2010) Encuesta Nacional de Salud ENS Chile 2009–2010, 1064

Ministerio de Salud de Chile (2010) Encuesta Nacional de Salud ENS Chile 2009–2010. Encuesta Nacional de Salud (vol 1). https://doi.org/10.1007/s11883-013-0322-z
Modo Beneficios (2019) Fondo de Desarrollo de Ferias Libres | Bonos, Subsidios, Becas, Fondos Concursables, Beneficios del Gobierno
Municipalidad de Valparaíso (2020) Valparaíso una síntesis de lo que somos
Observatorio Feria Libre (2013) Características económicas y sociales de ferias libres de Chile [Economic and social characteristics of farmer markets in Chile]
ODEPA Ministerio de Agricultura (2019) Boletín diario de precios y volúmenes de frutas y hortalizas en mercados mayoristas (Vol. 2019)
Owen SH, Daskin MS (1998) Strategic facility location: a review. Eur J Oper Res 111(3):423–447. https://doi.org/10.1016/S0377-2217(98)00186-6
Ramalingam M, Thangarajan R (2020) Mutated k-means algorithm for dynamic clustering to perform effective and intelligent broadcasting in medical surveillance using selective reliable broadcast protocol in VANET. Comput Commun 150:563–568. https://doi.org/10.1016/j.comcom.2019.11.023
Risica PM, Gorham G, Dionne L, Nardi W, Ng D, Middler R et al (2018) A multi-level intervention in worksites to increase fruit and vegetable access and intake: rationale, design and methods of the 'Good to Go' cluster randomized trial. Contemp Clin Trials 65:87–98.https://doi.org/10.1016/j.cct.2017.12.002
Robinson JA, Weissman E, Adair S, Potteiger M, Villanueva J (2016) An oasis in the desert? The benefits and constraints of mobile markets operating in Syracuse, New York food deserts. Agric Hum Values 33(4):877–893
Saavedra Manríquez CC (2016) Vegemas: la experiencia del comer saludable. Pontifica Universidad Católica de Valparaiso
Srivastava SK, Chaudhuri A, Srivastava RK (2015) Propagation of risks and their impact on performance in fresh food retail. Int J Logist Manag 26(3):568–602. https://doi.org/10.1108/IJLM-02-2014-0032
Tong D, Ren F, Mack J (2012) Locating farmers' markets with an incorporation of spatio-temporal variation. Socioecon Plann Sci 46(2):149–156
vom Lehn D (2014) Timing is money: managing the floor in sales interaction at street-market stalls. J Mark Manag 30(13–14):1448–1466. https://doi.org/10.1080/0267257X.2014.941378
Widener MJ, Metcalf SS, Bar-Yam Y (2012) Developing a mobile produce distribution system for low-income urban residents in food deserts. J Urban Health 89(5):733–745. https://doi.org/10.1007/s11524-012-9677-7
World Health Organization (2016) Promoting fruit and vegetable consumption around the world. Retrieved from https://www.who.int/dietphysicalactivity/fruit/en/index2.html
World Health Organization (2017) Increasing fruit and vegetable consumption to reduce the risk of noncommunicable diseases. Retrieved from https://www.who.int/elena/titles/fruit_vegetables_ncds/en/
Ylitalo KR, During C, Thomas K, Ezell K, Lillard P, Scott J (2019) The veggie van: customer characteristics, fruit and vegetable consumption, and barriers to healthy eating among shoppers at a mobile farmers market in the United States. Appetite 133:279–285. https://doi.org/10.1016/j.appet.2018.11.025
Yunoh MFM, Abdullah S, Saad MHM, Nopiah ZM, Nuawi MZ (2017) K-means clustering analysis and artificial neural network classification of fatigue strain signals. J Braz Soc Mech Sci Eng 39(3):757–764. https://doi.org/10.1007/s40430-016-0559-x
Zepeda L, Reznickova A, Lohr L (2014) Overcoming challenges to effectiveness of mobile markets in US food deserts. Appetite 79:58–67

Chapter 9
Optimization of Supply Management in a Fishing Sector Company

Carlos Gonzales, Natalí Zavala, Yuri Diaz, Manuel Peña, Ana Luna, and Mario Chong

Abstract The Peruvian anchovy fishery sector is the largest in the world, producing about 30% of the fishmeal and oil, as well as its derivatives. Given this context, the internal economy benefits from the importance of this industry with decisions related to design, plan, and execute their operation plans in the whole value network. This research shows how companies in this sector can seize the opportunities presented by the market through decisions related to internal strengths. The current study is applied in a company dedicated to the extraction, production, and commercialization of marine food and ingredients, which are processed strategically in four production plants located along the Pacific Ocean. The objective of this research is to impact positively on the corporate strategy, through the characterization, evaluation, and improvement of the procurement process.

Keywords Strategy · Food chain · Global fishery · Healthy food

9.1 Introduction

Society faces the challenge of providing food to the population while addressing the impacts of climate change and environmental degradation (Rockström et al. 2009; Sundström et al. 2014; FAO 2018a, b). Food and agriculture are critical elements to achieve sustainable and inclusive development of society (Barsukova 2017). Fishery and aquaculture are important for the food, nutrition, and employment of many people (Naylor et al. 2000; Lacroix et al. 2010; FAO 2018a, b; Nong 2019). In 2016, total fish production reached a record of 171 million tons, 88% was used for human consumption (it increased from 9.0 kg in 1961 to 20.2 kg in 2015), with an average growth of 1.5% per year (FAO 2018a, b). The annual growth of fish product consumption has

C. Gonzales · N. Zavala · Y. Diaz · A. Luna · M. Chong (✉)
Universidad del Pacífico, Jr. Sánchez Cerro N° 2141, Jesús María, Perú
e-mail: m.chong@up.edu.pe

M. Peña
Universidad Nacional Mayor de San Marcos, Calle Germán Amézaga N° 375—Edificio Jorge Basadre, Lima, Perú

J. L. Garcia Alcaraz et al. (eds.), *Trends in Industrial Engineering Applications to Manufacturing Process*, https://doi.org/10.1007/978-3-030-71579-3_9

doubled population growth since 1961, which demonstrates the importance of the fishing sector (Godfray et al. 2010) in reducing hunger and malnutrition (Sundström et al. 2014; FAO 2018a, b). Likewise, the sector's contribution to economic growth and the fight against poverty is incrementing (Rockström et al. 2009; Gephart et al. 2017). Strengthened demand and higher prices increased the value of world fish exports in 2017 to 152 billion US$, which represents the participation of developing countries by 54% (FAO 2018a, b).

In Peru, the 2017 GDP related to fishing reached 550 million US$, representing an increase of 4.7% over the previous year. Similarly, 1611 thousand tons were exported and generated foreign currency for the country of 927 million US$, representing 6.5% of the total value of Peruvian exports (Ministerio de la Producción 2018). In this way, the fishing export sector recovered with a 31.4% increase in the value of exports, compared to 2016, after falls since 2015 (Majluf et al. 2017; Ministerio de la Producción 2018). The internal sales in the sector were for a total of 728.4 thousand tons, 6.2% less than the volume recorded in 2016 as a result of the decrease in sales of fishery products for indirect human consumption (−32.7%) (Ministerio de la Producción 2017); in contrast, the sale of products for direct human consumption grew by 1.6%, evidencing a significant contribution of 86.8% of the total internal sales of fishery products (Ministerio de la Producción 2018).

Regarding the anchovy fishing industry, 3209 tons of anchovy were landed in 2017, increasing by 16% compared to 2016 (Ministerio de la Producción 2017, 2018). Only in the first anchovy fishing season of 2017 (between April 22 and July 10) in Peru, a total of 2.28 million tons were captured, 81.6% of the production quota established. This represents a significant advance of the activity. The first anchovy fishing season of 2019 in Peru began with 2.1 million tons, 36% less than 2018, but this is due to the influence of the *El Niño* phenomenon. The Peruvian Ministry of Production aims to protect the sustainability of the resources and juvenile anchovy specimens, among others (Majluf et al. 2017). Peru exported a total of 57,800 tons of fish oil around the world in the first quarter of 2019, with Denmark being its largest consumer importing 15,500 tons of Peru's fish oil. During this period, Peru's total fish oil export to the world was 30,700 tons higher than in 2018.

In this context, it would be important to take advantage of the fishing sector's growth through the improvement in the operations of the companies that comprise it. This research aims to present a framework to characterize the procurement in a company and thus show the current state of the process within the organization, as well as contribute to the improvement of supply chain management. The object of study is an organization dedicated to the extraction, production, and marketing of marine food and ingredients, which are processed strategically in production plants located along the Peruvian coast.

The sections of the research are organized as follows: A theoretical framework is presented where the most representative concepts of the research are reviewed; then, the employed methodology is described; afterward, the results obtained are shown and discussed, and finally, the conclusions are presented.

9.2 Literature Review

A supply chain is made up of all the parties involved in the fulfillment of a customer's order (Lourenço and Ravetti 2018), and it also includes all the functions engaged in the reception and satisfaction of the customer (Laari et al. 2016). Its objective is to maximize the total value generated, that is, the difference between what a customer pays for the final product and the costs (Chopra and Meindl 2013) incurred by the chain to satisfy the order (Chase and Jacobs 2014). The competitive strategy and the supply chain strategy must have aligned goals to achieve a strategic adjustment (Qi et al. 2011), to gain consistency between the customer's priorities, that the competitive strategy wishes to meet, and the supply chain capabilities that the strategy of the organization tries to build (Seuring and Müller 2008; Perez-Franco et al. 2016). To reach such consistency, it is necessary to: achieve the adjustment between the competitive and functional strategies to form a coordinated strategy (Perez-Franco et al. 2016), properly structure the processes and resources of an organization's functions to execute the strategies successfully (Phadnis et al. 2016), and align the design of the entire supply chain and the role of each stage to support the chain's strategy (Chopra and Meindl 2013; Perez-Franco et al. 2016; Phadnis et al. 2016; Perez-Franco 2018).

The balance between the cycle and the push/pull views is achieved from an adjustment in the supply chain to support the organization's competitive strategy (Chopra and Meindl 2013; Qrunfleh and Tarafdar 2014). The logistic drivers of a supply chain are facilities, inventory, transport, information, sourcing, and pricing (Chopra and Meindl 2013); these allow the improvement of the supply chain's performance and the interaction to determine the same, taking into account both the capacity response (Zhang and Riemann 2014) and the financial effect. Provisioning is understood as the set of business processes necessary for the acquisition of products and services (Chang et al. 2013). Managers must decide if each task will be carried out by a source at the manufacturer/supplier interface (Pereira et al. 2014) that includes the inputs necessary to ensure the production according to the schedule (Chopra and Meindl 2013). The procurement function is to buy the materials necessary for the organization and store them while the production or marketing process begins (Seuring and Müller 2008). Therefore, the acquisition has two very important functions. The first is the cost or expense, and it has a direct effect on the final result of a company. The purchase or acquisition cost must be reduced to have a direct effect on the final result of your company's earnings. The second is the effect the acquisition has on the operational side of the business. The acquisition must meet three (three) characteristics and in a certain order to be successful, (a) quality, (b) delivery, and (c) cost. If the acquisition does not offer quality and on-time delivery, operations will affect customer satisfaction, sales, and the bottom line of the business. Thus, acquisitions affect both the company's costs and the efficiency of the production of services, which will ultimately affect the company's results. In the study organization, the purchasing department performs the procurement function, and its objective is to supply the production department for manufacturing and the sales department for product marketing.

Competency in procurement and supply management focuses on effective innovation from internal and external customers (Pereira et al. 2014). It plays a pivotal role in increasing overall organizational competitiveness. In the case of internal supply management, the aim is to optimize skills and capacities related to product development and innovation in transforming supply inputs into more valuable products (Chang et al. 2013; Perez-Franco and Phadnis 2018). Developments such as sourcing innovation (Luzzini et al. 2015; Schiele 2010, 2012), handling potential supplier disruptions (Wieland et al. 2016), ensuring sustainability in the supplier network (Montabon et al. 2016; Schneider and Wallenburg 2012; Wilding et al. 2012), and the changes due to increased digitization are challenging about what competencies the professional will require. On the other hand, innovation in external supply includes competencies associated with contracting, through mergers, acquisitions, and joint venture work (Rossi et al. 2013), as well as the ability to make sound supplier-related decisions, and then develop them effectively (Perez-Franco et al. 2016; Perez-Franco and Phadnis 2018).

This research seeks to improve the procurement management of an organization of the fishing sector in Peru that has been in the market for 30 years. It is a leading company and its main products are fishmeal and fish oil. Currently, the organization has delays in acquisitions and many urgent orders that jeopardize the availability of ships or production plants at that time. A day without production could mean up to $161,000 of lost production and sales. Purchases are made reactively for each request from internal customers, in the plant and fleet maintenance areas.

9.3 Methodology

The stages that make up the framework which was carried out to improve the process in the organization are developed in this section. After conducting a set of interviews, collecting information from different sources, and reviewing the documents, we diagnosed the problem, analyzed possible solutions, and presented the final proposal (Perez-Franco et al. 2016; Perez-Franco 2018).

9.3.1 Step 1. Organizational Assessment

The organization's value proposition focuses on providing customers with highly digestible fishmeal of excellent nutritional qualities (Seafish 2018). That is why the organization periodically measures customer satisfaction (Radziwill 2015). Regarding the organization, over 1200 collaborators are under the leadership of four (4) managers. The fleet management is responsible for the vessel coordination, and its main function is to transport the largest amount of raw material; the operations management is responsible for the production plants, and its main function is to process the extracted and purchased raw material, ensuring quality and managing business relationships with customers; the human resources management is responsible for human talent and hiring; and the administration and finance management is in charge of the areas of finance, treasury, purchasing, warehouses, and transportation.

The organization's primary activities consist of the inbound logistics, which includes the processes of reception and storage of the supplies (Demir and Lenhart 2019), spare parts and materials in the central warehouse in Callao until their distribution to the production plants; the organization also performs fishing for their raw material (Kabu and Tira 2015). The operations include the production of fishmeal and fish oil activities (Fréon et al. 2017). The outbound logistics include the distribution of finished products, transferred in bags to be exported. The marketing and sales are activities necessary for the launching of products, brand promotions, and customer service (Rosales et al. 2017). The support activities are related to the organization infrastructure (four production plants, a central warehouse, and administrative offices); human resources management, composed of highly trained personnel who watch over the organization's work climate and carry out constant training; the technology development consisting of activities such as the implementation of the enterprise resource planning (ERP) and other information technologies; and the procurement management to integrate the alliance between the suppliers and the organizations' systems (Dyer and Singh 1998).

The organization needs to perform various types of processes, from administrative, operational, and logistics, in order to distribute the bags of fishmeal to the world. The administrative processes are composed of purchasing management, which includes negotiation with suppliers and category management. These purchases are stored in the transit warehouse, which is located in Callao, and then distributed to the production plants according to the user's requirement. Concerning the regulations, the beginning of the fishing season establishes the fishing area and quota per authorized vessel. With this information, the fishing planning area generates the production plan and the recommended extraction. The production plan constraints are the plants' capacities, raw materials, consumables, equipment, and others, which are based on the composition of the raw material and the quota. The purchasing area, with this information, generates requests for goods and services. When the raw material arrives at the plants, the production plan activities are executed, bearing in mind the global standards, which are recognized in the international market. Then, the final product is bagged and placed in containers until they arrive at their final destination. Figure 9.1 shows the supply chain of the organization.

9.3.2 Step 2. Critical Processes Identification

The organization has four main macrologistic processes, inbound logistics, which is related to purchases; internal logistics, consisting of reception, storage, inventories, and internal transport; process logistics, which is related to the plant operation and

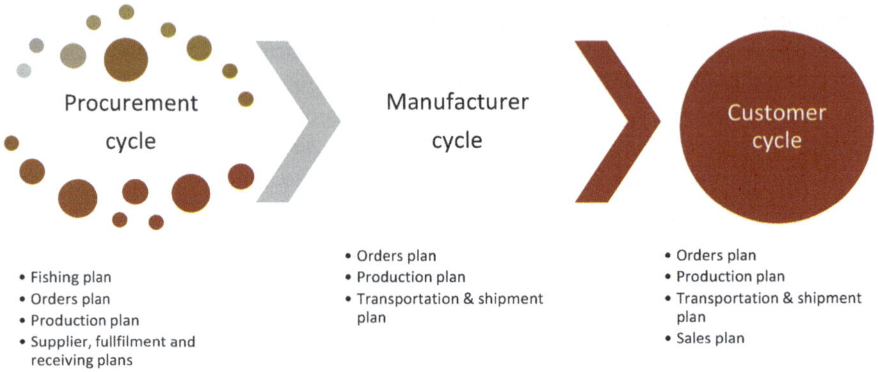

Fig. 9.1 Supply chain. *Source* Own elaboration (2019)

fleet; and outbound logistics, consisting of order preparation, dispatch and export. The main macrologistics processes are detailed in Fig. 9.2.

For this case study, the logistic macroprocesses will be analyzed up to the level of threads and thus achieve a specific analysis. The strategic matrix showed the relation between improve the operative efficiency (A2) and the process systematization (A1)

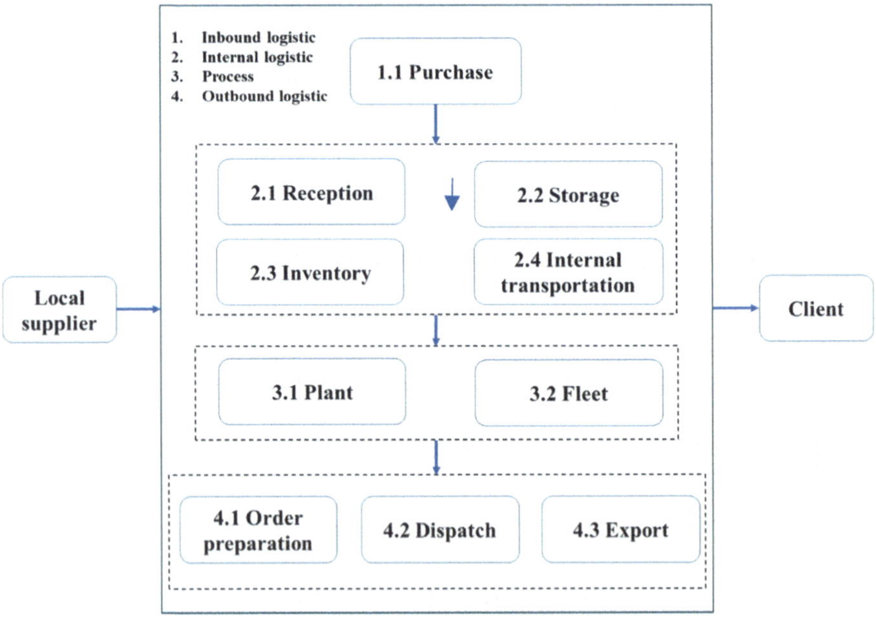

Fig. 9.2 Level process map. *Source* Own elaboration (2019)

(Table 9.1). The process systematization (A1) has a relation with the strategic objectives: purchase process automatization (B2) and collaborative management with the supplier (B4) as shown in Table 9.2.

The purchase process automatization (B2) has a relation with the collaborative supplier management and reduces the ordering costs (B7) and the storage cost (B8) as shown in Table 9.3. The criteria to be evaluated are a virtual platform with real-time stock (C1), digital contact management (C2), reduce the quotation process time (C3), and reduce the manual quotation (C4). A confrontation matrix was performed, to determine the weights of each criterion scores (Perez-Franco et al. 2016; Perez-Franco 2018), a prioritization matrix was developed, as shown in Table 9.4, and it was noted that the most representative threads are purchase order management and inventory control. The analysis focused on the most relevant thread is the one with the highest weighting, in this case, the purchase management.

The purchasing process is responsible for identifying the requirements in the user area and the final delivery. The purchasing policies for amounts less than 1000 US$, one (01) quotation are required; for amounts greater than 1000 US$ and less than 5000 US$, two (02) quotations are required; and for amounts greater than 5000

Table 9.1 Strategic matrix

Matrix 1	A1	A2	A3
A1			
A2	2		
A3	1.2	1.6	

Source Own elaboration (2019)

Table 9.2 Strategic objectives matrix

Matrix 2	B1	B2	B3	B4	B5	B6	B7	B8
A1	0.7	1.9	1.8	1.9	0.3	1.1	0.1	−0.8
A2	0.2	1.9	0	−1.0	0.4	−0.6	−1.0	0.2
A3	−0.6	−0.6	−0.7	1.1	1.7	1.8	1.1	0.5
Matrix 2	B1	B2	B3	B4	B5	B6	B7	B8
A1	10	1	4	1	13	7	16	22
A2	14	1	17	23	12	18	23	14
A3	18	18	21	7	6	4	7	1
Matrix 2	B1	B2	B3	B4	B5	B6	B7	B8
A1		+		+				
A2		+		–			–	
A3								

Source Own elaboration (2019)

Table 9.3 Synergic between objectives matrix

Matrix 3	B1	B2	B3	B4	B5	B6	B7	B8
B1		1.2	0.1	2.0	0.4	-0.5	1.5	0.1
B2	2.3		−0.5	2.4	1.1	2.1	2.4	2.2
B3	0.1	0.9		−0.4	−0.3	2.3	−0.3	1.9
B4	0.4	0.5	−0.2		−0.3	2.1	2.5	−0.6
B5	0.6	1.5	−0.2	2.4		1.7	−0.8	1.6
B6	1.9	0.8	−0.8	2.1	1.8		−0.1	0.3
B7	0.8	0.1	0.8	−0.1	1.3	1.2		2.5
B8	0.7	1.4	1.3	−0.8	0.9	0.1	2.3	

Matrix 3	B1	B2	B3	B4	B5	B6	B7	B8
B1		24	38	13	35	51	19	38
B2	6		51	3	26	10	3	9
B3	38	27		50	47	6	47	14
B4	35	34	45		47	10	1	53
B5	33	19	45	3		17	54	18
B6	14	29	54	10	16		43	37
B7	29	38	29	43	22	24		1
B8	32	21	22	54	27	38	6	

Matrix 3	B1	B2	B3	B4	B5	B6	B7	B8
B1								
B2	+			+		+	+	+
B3						+		
B4						+	+	
B5				+			–	
B6			-	+				
B7								+
B8				-			+	

Source Own elaboration (2019)

US$, three (03) quotations are required. The purchase order management process has twelve (12) activities that depend on the organization's policies. According to the maturity model (Reck and Long 1988), the organization is in phase 3—intermediate, because they are focused on a collaborative system. In the medium term, one year, the objective is to reach phase 4—Systematized as shown in Fig. 9.3.

The organization purchases around 103 million U$, which are divided into both categories, services, and products. During the last year, the purchase of products represented 77% and the services represented 23%. This amount varies due to the

Table 9.4 Confrontation matrix

	C1	C2	C3	C4	C5	C6	C7	C8	C9	C10	C11	C12	C13
B1	+2	+2	+2										
B2				+2	+1								
B3						+1							
B4							+1	+2					
B5									+2				
B6										+2			
B7											+2	+1	
B8													+1

Matrix 4	C1	C2	C3	C4	Score	%
C1		1	0	1	2	25
C2	1		0	1	2	25
C3	1	1		1	3	38
C4	0	1	0		1	13

Source Own elaboration (2019)

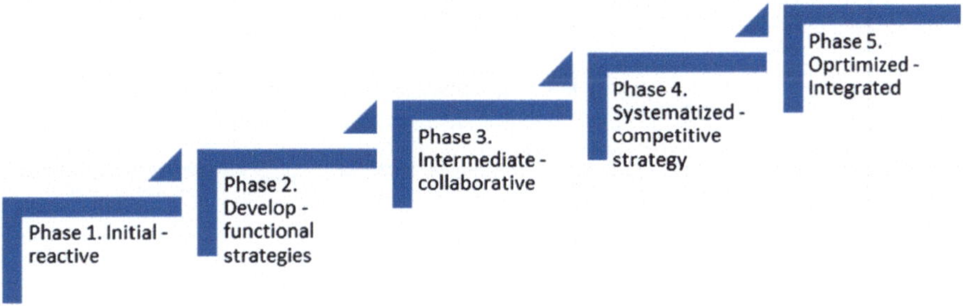

Fig. 9.3 Maturity model. *Source* Own elaboration

availability of raw materials and their compositions. The highest percentage represents the fishing process and its services, followed by a lower percentage of inputs, spare parts, and machinery.

9.3.3 Step 2. Elaborate the Functional Strategic Map

The development of the functional strategy map considers the pillars of the company, the increase in operational efficiency, and the systematization of processes. Besides, it contemplates the specific objectives of the organization, and its current operational objectives, which are shown in Fig. 9.4 (Suwasono and Rosana 2013; Perez-Franco et al. 2016).

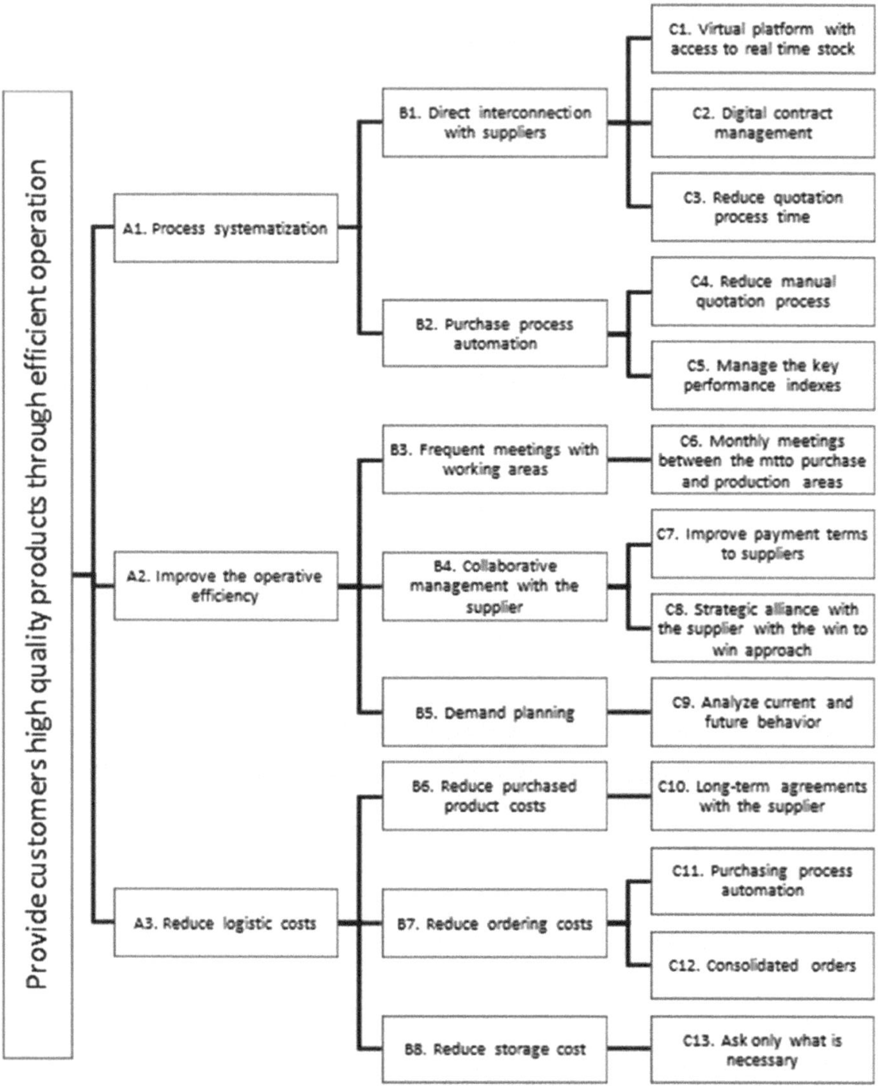

Fig. 9.4 Current functional strategy map. *Source* Own elaboration (2019)

The empathy maps were used for deeper insights and the identification of the problems. Those were identified at each stage and then the solutions were proposed, considering the organization structure, the internal processes, and the information technology. We established the operational inefficiency of procurement management as the main problem. The causes are the high percentage of urgent purchases, order delays in lead time, and policies that are not aligned. It is important to highlight that

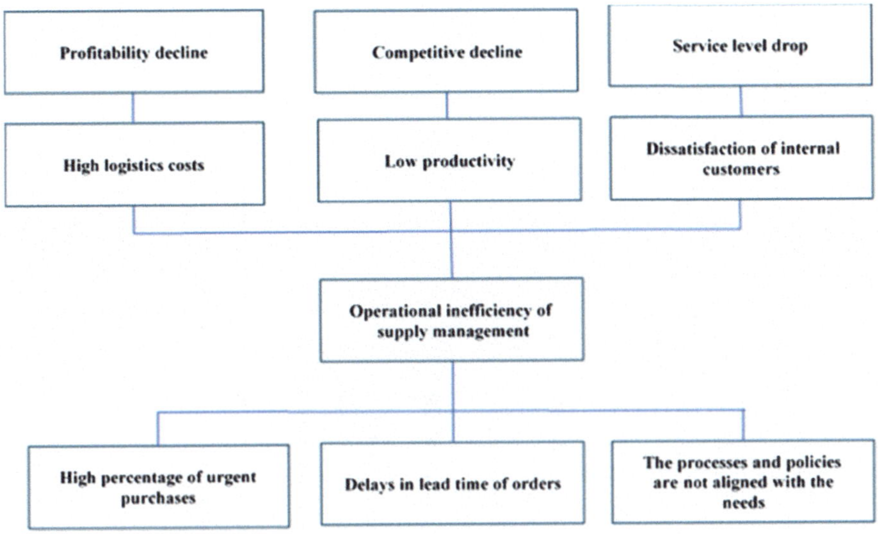

Fig. 9.5 Problem statement. *Source* Own elaboration (2019)

the main problem generates logistics costs, reduces productivity, and increases the dissatisfaction of the internal customers, as shown in Fig. 9.5.

The operationalization of the problem, the general objective, and the two variables studied are shown in Table 9.5. The variables' dimensions are systems, processes, and people. Each one of them is measured through different indicators.

9.4 Results

After identifying the problem, potential improvements were proposed for procurement management. The recognition of those improvements was made using a cause-strategy matrix and a project matrix including policies and procedures for each purchase flow, a supplier system management, and a new strategic position (see Table 9.6).

The project contemplates human resources as a fundamental component aligned with the organization's functional and strategic plans. Due to these facts, the project needs to: generate function manuals, dimension the activities, generate a development plan, evaluate the collaborators, request a new strategic position, and establish an integral training plan.

The process evaluation is another critical aspect that covers this project. The processes should support the activities and correctly focus on maximizing their performance, managing the goods and services acquisition, as well as their policies.

The implementation of a new information system will also be considered as a collaborative system with the suppliers: generate inventory policies, manage and

Table 9.5 Operationalization matrix

Problem	Objective	Variable	Dimension	Indicators	M.U	Goal	Actual
Operational inefficiency of supply chain management	Reduce operational inefficiency, through optimization of provisioning management	Operational inefficiency of AG	Time	Order lead time	Days	15	30
			Quantity	Orders delivered with right quantity	%	100%	95%
			Quality	Orders delivered with right quality	%	100%	95%
			Save	Savings generated in negotiations	%	1%	0%
			Reject	Orders rejected by the approver	%	5%	0%
		Optimization	System	Contribution of the system in AG processes	%	90%	20%
			Process	Automated provisioning processes	%	90%	40%
			People	Productivity per buyer	oc/buyer (month)	1200	0
				Staff training	Trainings/year	4	0

Source Own elaboration (2019)

Table 9.6 Projects proposal

Project	Percentage
Planning area strategic	08
Supplier system management	22
New strategic position	17
Management program	03
Purchase policies and procedures	50

Source Own elaboration (2019)

Table 9.7 Amount of the project

Amount 120 000 US$	
1. Personnel (relocation, unlinking and new entry)	$33,000
2. SAP Cloud platform subscription	$55,000
3. iProvider system integration	$17,000
4. Other costs	$7000
5. Contingency reserve	$5000
6. Management reserve	$3000

Source Own elaboration (2019)

optimize the flow processes, dimension the activities, and consider the families of products, as well as the purchasing and operational policies. The SAP Ariba, Full-step, and CSTI iProvider platforms were proposed and evaluated based on public information, organization inquiries, and industry references.

This project is aligned with the supply chain area objectives: improve the area efficiency and effectiveness, implement a e-commerce solution and maintain the annual budget, considering the Table 9.7 breaks down the costs considered for the project.

9.5 Discussion

The fishing sector is growing globally. Within a fish derivatives manufacturing organization, the operations are critical to establishing a balance between responsiveness and efficiency in the supply chain. The benefits of the purchasing process are supported by achieving operational efficiency and improving the organization, considering people, processes, and technology. However, the implementation of the proposal involves negative risks as a possible stoppage of the project, due to the priorities during a fishing stage, the variation in scope, the delay in the valuations' approval, the poor performance of the technology, and the resistance to change. Likewise, the project represents opportunities that must be exploited to contribute to operational efficiency such as reducing subjectivity in the processes.

Carrying out these improvements implies an initial investment of US $65,000 with annual costs of US $18,333. The expected savings are US $198,000, generating a positive annual income of US $169,667, and a positive NPV of US $383,583 with an IRR of 24%, considering two years of return on investment.

9.6 Conclusions and Recommendations

The ability to respond to climatic and state conditions is crucial in the fishing sector because it has a dependent relationship with the quality of the raw materials. For this reason, the supply area is fundamental, considering the operational activities of the organization. This document proposes a framework in order to identify the organization's problem which is focused on characterizing the procurement, as the area to generate value from the operational functional plan that is supported and aligned to the corporate strategies. Using the data from the organization, the methodology has several steps, organizational assessment, value chain and value network evaluation, identify and analyze the critical process, elaborate the functional strategic map, and finally identify the problem statement and improvement.

This chapter contributes to the literature with the review of the procurement cycle (manufacturer, supplier, distribution) in which dynamic properties are specific address in a fishing company transition from an intermediate interaction with the suppliers to the systematized to a more customer focus organization. Finally, this chapter presents a method for addressing a real organizational problem and several practical implications for different industries as an example of how to approach similar problems.

One of the recommendations to take into account is to approach, in the short term, the acquisition of a technological environment for the detection and capture of the company's main product that benefits the fishing sector in time and costs.

References

Barsukova S (2017) Food and agriculture. In: Russia: strategy, policy and administration. Palgrave Macmillan, pp 241–255. https://doi.org/10.1057/978-1-137-56671-3_22

Chang H, Tsai Y, Hsu C (2013) E-procurement and supply chain performance. Surg Endosc Other Interv Tech 18(1):34–51. https://doi.org/10.1108/13598541311293168

Chase RB, Jacobs F (2014) Operations and supply chain management, 14th edn. Global Edition, pp 347–362

Chopra S, Meindl P (2013) Administración de la cadena de suministro. Quinta edición. Pearson educación, México

Demir M, Lenhart S (2019) Optimal sustainable fishery management of the Black Sea anchovy with food chain modeling framework. Nat Resour Model. https://doi.org/10.1111/nrm.12253

Dyer J, Singh H (1998) The relational view: cooperative strategy and sources of interorganizational competitive advantage. Acad Manag Rev 23(4):660–679. https://doi.org/10.5465/AMR.1998.1255632

Food and Agriculture Organization of the United Nations—FAO (2018a) Fishery and aquaculture statistics 2016. FAO yearbook, p 104. Food and Agriculture Organization of the United Nations. https://doi.org/10.5860/CHOICE.50-5350
Food and Agriculture Organization of the United Nations—FAO (2018b) The state of world fisheries and aquaculture 2018—meeting the sustainable development goals FAO. Rome
Fréon P, Durand H, Avadí A, Huaranca S, Orozco R (2017) Life cycle assessment of three Peruvian fishmeal plants: toward a cleaner production. J Clean Prod 145:50–63. https://doi.org/10.1016/j.jclepro.2017.01.036
Gephart J, Deutsch L, Pace M, Troell M, Seekell D (2017) Shocks to fish production: identification, trends, and consequences. Glob Environ Chang 42:24–32. https://doi.org/10.1016/j.gloenvcha.2016.11.003
Godfray H, Beddington J, Crute I, Haddad L, Lawrence D, Muir J et al (2010) Food security: the challenge of feeding 9 billion people. Science. https://doi.org/10.1126/science.1185383
Laari S, Töyli J, Solakivi T, Ojala L (2016) Firm performance and customer-driven green supply chain management. J Clean Prod 112:1960–1970. https://doi.org/10.1016/j.jclepro.2015.06.150
Lacroix D, Hubert B, Treyer S (2010) Aquaculture and prospective: a simple past and its future past. AgroParisTech, Paris (France)
Lourenço H, Ravetti M (2018) Supply chain management. In: Handbook of heuristics (2–2), pp 1241–1258. Springer International Publishing. https://doi.org/10.1007/978-3-319-07124-4_54
Luzzini D et al. (2015) The path of innovation: purchasing and supplier involvement into new product development. Indus Market Manage 47:109–120. https://doi.org/10.1016/j.pursup.2019.100572
Kabu E, Tira D (2015) Value chain analysis towards sustainability: a case study of fishery business in Kota Kupang, Indonesia. Int J Econ Financ Issues 5:150–156
Majluf P, De la Puente S, Christensen V (2017) The little fish that can feed the world. Fish Fish 18(4):772–777. https://doi.org/10.1111/faf.12206
Ministerio de la Producción (2017) Anuario estadístico pesquero y acuícola 2016. Ministerio de la producción, Lima, Perú
Ministerio de la Producción (2018) Anuario estadístico pesquero y acuícola 2017. Ministerio de la producción, Lima, Perú
Montabon F, Pagell M, Wu Z (2016) Making sustainability sustainable. J Supply Chain Manag 52(2):11–27. https://doi.org/10.1111/jscm.12103
Naylor R, Goldburg R, Primavera J, Kautsky N, Beveridge M, Clay J et al. (2000) Effect of aquaculture on world fish supplies. Nature. https://doi.org/10.1038/35016500
Nong D (2019) Potential economic impacts of global wild catch fishery decline in Southeast Asia and South America. Econ Anal Policy 6:213–226. https://doi.org/10.1016/j.eap.2019.04.004
Pereira C, Christopher M, Da Silva A (2014) Achieving supply chain resilience: the role of procurement. Surg Endosc Other Interv Tech 19:626–642. https://doi.org/10.1108/SCM-09-2013-0346
Perez-Franco R (2018) Proposed set of criteria for supply chain strategy evaluation: a multi-country perspective.https://doi.org/10.1108/978-1-78756-803-720181002
Perez-Franco R, Phadnis S (2018) Eliciting and representing the supply chain strategy of a business unit. Int J Logist Manag. https://doi.org/10.1108/IJLM-05-2016-0128
Perez-Franco R, Phadnis S, Caplice C, Sheffi Y (2016) Rethinking supply chain strategy as a conceptual system. Int J Prod Econ 182:384–396. https://doi.org/10.1016/j.ijpe.2016.09.012
Phadnis S, Caplice C, Sheffi Y (2016) How scenario planning influences strategic decisions. MIT Sloan Manag Rev 57(4):24–27
Qi Y, Zhao X, Sheu C (2011) The impact of competitive strategy and supply chain strategy on business performance: the role of environmental uncertainty. Decis Sci 42(2):371–389. https://doi.org/10.1111/j.1540-5915.2011.00315.x
Qrunfleh S, Tarafdar M (2014) Supply chain information systems strategy: Impacts on supply chain performance and firm performance. Int J Product Econ 147 (PART B):340–350. https://doi.org/10.1016/j.ijpe.2012.09.018

Radziwill N (2015) Value proposition design. Qual Manag J 22(1):61–61. https://doi.org/10.1080/10686967.2015.11918419
Reck R, Long B (1988) Purchasing: a competitive weapon. J Purchas Mater Manag 24(3):2–8
Rockström J, Steffen W, Noone K, Persson Å, Chapin F, Lambin et al (2009) A safe operating space for humanity. Naturehttps://doi.org/10.1038/461472a
Rossi S, Colicchia C, Cozzolino A, Christopher M (2013) The logistics service providers in eco-efficiency innovation: an empirical study. Surg Endosc Other Interv Tech. https://doi.org/10.1108/SCM-02-2012-0053
Rosales R, Pomeroy R, Calabio I, Batong M, Cedo K, Escara N Sobrevega M (2017) Value chain analysis and small-scale fisheries management. Marine Policy 83:11–21.https://doi.org/10.1016/j.marpol.2017.05.023
Schiele H (2010) Early supplier integration: the dual role of purchasing in new product development. R&d Manag 40(2):138–153. https://doi.org/10.1111/j.1467-9310.2010.00602.x
Schiele H (2012) Accessing supplier innovation by being their preferred customer. Res Technol Manag 55(1):44–50. https://doi.org/10.5437/08956308X5501012
Schneider L, Wallenburg CM (2012) Implementing sustainable sourcing—does purchasing need to change? J Purch Supply Manag 18(4):243–257. https://doi.org/10.1016/j.pursup.2012.03.002
Seafish (2018) Fishmeal and fish oil facts and figures. Seafish (October) 30
Seuring S, Müller (2008) From a literature review to a conceptual framework for sustainable supply chain management. J Clean Prod 16(15):1699–1710.https://doi.org/10.1016/j.jclepro.2008.04.020
Sundström J, Albihn A, Boqvist S, Ljungvall K, Marstorp H, Martiin C et al (2014) Future threats to agricultural food production posed by environmental degradation, climate change, and animal and plant diseases—a risk analysis in three economic and climate settings. Food Secur 6(2):201–215.https://doi.org/10.1007/s12571-014-0331-y
Suwasono B, Rosana N (2013) The application of GIS-AHP to develop a strategic planning for an urban farming: fishery and aquaculture. RSW publications. https://doi.org/10.13033/isahp.y2013.084
Wieland A, Handfield RB, Durach CF (2016) Mapping the landscape of future research themes in supply chain management. J Bus Logist 37(3):205–212. https://doi.org/10.1111/jbl.12131
Wilding R et al (2012) Sustainable purchasing and supply management: a structured literature review of definitions and measures at the dyad, chain and network levels. Supply Chain Manag Int J 17(5):478–496
Zhang W, Reimann M (2014) Towards a multi-objective performance assessment and optimization model of a two-echelon supply chain using SCOR metrics. CEJOR 22(4):591–622. https://doi.org/10.1007/s10100-013-0294-7

Chapter 10
Decision Support Model for Solid Waste Management in a Closed-Loop Supply Chain

Vivian Lorena Chud-Pantoja, Claudia Cecilia Peña-Montoya, and Juan Carlos Osorio-Gómez

Abstract The importance of solid waste management is a relevant issue in supply chain management because of the accelerated increase in waste generation, pollution and the need to properly manage waste in companies in a particular way and throughout the whole supply chain in a general way. We present a model that supports decision-making regarding solid waste management (particularly glass packaging), for a manufacturer, considering the perspective of the closed-loop supply chain. The model uses the AHP methodology to define the multiple criteria to be considered in the decision and for the selection of the best alternative available according to those criteria. This model was applied at a glass packaging manufacturer in Colombia. The result of the implementation of the model delivered the waste management strategy (used packaging) that best meets the objectives of the echelon under consideration.

Keywords AHP · Closed-loop supply chain · Decision support · Solid waste management

10.1 Introduction

As the world evolves, societies change their structures, as well as their production and consumption patternssss. Technological development and present consumer patterns have led to an increase in waste volumes generated on all continents for some decades (MAVDT 2007). Hence, waste management is a matter of concern on

V. L. Chud-Pantoja
Escuela de Ingeniería Industrial, Programa de Ingeniería Industrial, Universidad del Valle, Sede Zarzal, Zarzal, Valle del Cauca, Colombia

C. C. Peña-Montoya
Departamento de Operaciones y Sistemas, Facultad de Ingeniería, Universidad Autónoma de Occidente, Cali, Valle del Cauca, Colombia

J. C. Osorio-Gómez (✉)
Escuela de Ingeniería Industrial, Universidad del Valle, Cl. 13 #100-00, Cali, Valle del Cauca, Colombia
e-mail: juan.osorio@correounivalle.edu.co

J. L. Garcia Alcaraz et al. (eds.), *Trends in Industrial Engineering Applications to Manufacturing Process*, https://doi.org/10.1007/978-3-030-71579-3_10

the planet, becoming an environmental problem that is largely the responsibility of waste-generating organizations.

According to Vego et al. (2008), the rapid increase in the volume of solid waste, site-specific problems and economic, ecological, operational and management needs in solid waste management require the commitment of decision-makers to find new methods and tools which improve the effectiveness of the decision-making process.

The Sustainable Development Goals (SDGs), promulgated by the United Nations Development Program (2015) also aim at reducing solid waste and better using the resources. For example, SDG 12 for responsible production and consumption aims at reducing waste generation through prevention, reduction, recycling and reuse activities by 2030.

In this way, it is a priority for adequate solid waste to manage and change the paradigm that consider final disposal as a single option (PGIRS 2019). Therefore, it has to consider the valuation of solid waste which is "the set of recovery and treatment processes that allow waste to be put into technical and economic conditions to continue in the productive cycle" (MAVDT 2010). Currently, in Colombia, the recovery and recycling of post-consumer waste are carried out under conditions of informality, to the detriment of the recyclers' welfare, especially in social, economic and health aspects.

Once a product has lost the functional value for which it was processed and has been discarded by the consumer, it can be considered as waste, which must be treated to minimize the environmental impact it may cause. When waste rejoins the production cycle, regardless of the type of waste, it is known as closed-loop supply chain (CLSC) (Soto 2007).

A CLSC considers reverse flow activities, i.e., those where the product is returned (for different reasons) to the producer after being used; these reverse flow activities are called reverse logistics in which the purpose is to design, develop and control systems for the efficient management of the flow of return of products out of use, from the consumer to the producer, in order to use the value that these products still incorporate. This process of recovery of economic value incorporated by out-of-use products is called economic recovery and is carried out through the reuse, remanufacturing and/or recycling (Rubio et al. 2007). Reverse logistics activities can optimize all urban solid waste management and provide reuse, remanufacturing, recycling and proper disposal (LOURENÇO et al. 2016). In general, producers are looking for efficient ways to integrate reverse logistics into their supply chains, mainly to recover the economic value of their returned products and to reduce the disposal costs of non-recoverable waste (Realff et al. 2000; Corum 2020). However, it is a challenge for decision-makers to decide how best to recover the value of waste considering environmental, social, economic and technical aspects.

Regarding the residue of glass containers, Ko et al. (2012) state that glass is one of the most important materials that needs to be controlled. In addition, according to Hekkert et al. (2000a, b), improving the management of such materials can impact the environment, reducing energy consumption and CO_2 emissions to the environment. Likewise, it is important to recognize that glass is 100% recyclable and usable, and in Colombia it has not been used 100% (Departamento Nacional de Planeación 2016a);

therefore, an adequate management of this waste, returning it to its production cycle, is mandatory. Glass is considered as bulky material, which when it is not adequately managed, then it ends up in landfills and therefore decreases the life span of them.

10.2 Literature Review

10.2.1 Solid Waste Management

According to Das et al. (2019), solid waste generation varies between countries and the composition according to the economic status of countries; sustainable solid waste management requires the participation of social, economic, institutional, legal, technical and environmental aspects. Due to the increase and complexity of solid waste, programs involving zero waste and the diversion of waste from landfills are gaining popularity in response to increase urban densification and the growth of space value in the world's largest cities (Paes et al. 2020).

Solid waste management has been linked to issues such as sustainability and circular economy. For example in sustainability, da Silva et al. (2019) proposed sustainability indicators for the management of urban solid waste in medium and small cities in Brazil. Difficulties were found in the availability of information and weaknesses for its evaluation, mainly in infrastructure, technical and human resources investment. However, these indicators are a strategic tool for implementing global policies such as the Sustainable Development Goals.

Similarly, Phonphoton and Pharino (2019) proposed a decision support system based on sustainability principles to mitigate the impact and setup of municipal solid waste management during the Bangkok floods. Using the AHP, the authors found that adapting waste trucks was the most important alternative. This research contributes to the compliance of the SDG No 11 of sustainable cities and communities.

Regarding circular economy, CONPES 3874 2016 (Departamento Nacional de Planeación 2016b) issued for the integral solid waste management in Colombia states that waste should be kept in the production cycle for as long as possible generating other management options. Waste is thus considered as a resource that can be used repeatedly before considering its final disposal.

Other research involves frameworks for the assessment of solid waste management. According to Tsai et al. (2020a), technical integration and social acceptability are the most important aspects for the reference framework for effective municipal solid waste management in two cities in Vietnam. In addition, the capacity for innovation in treatment, safety and health, economic benefits, functionality and technological convenience deserves special attention.

Similarly, Tsai et al. (2020b) evaluated solid waste management in Vietnam using the sustainable balanced scorecard. This tool has been promoted to help decision-makers by providing a holistic vision of the organization to improve the efficiency of solid waste management. Decisive aspects in decision-making were financial investment, the stakeholder's participation and the capacity for innovation.

Other aspects that contribute to improve the operational performance of solid waste management, according to Tsai et al. (2020b), are costs efficiency, collaboration between stakeholders, flexibility/adaptability for environmental changes, availability of local technical skills, knowledge acquisition and information technologies.

10.2.2 Reverse Supply Chain

A reverse supply chain consists of a series of activities required to collect a product used from a consumer and reprocess it (the product used) to recover its market value or to be properly disposed. The implementation of a reverse supply chain requires at least three parts: collection centers where consumers return used products, reprocessing recovery facilities (remanufacturing or recycling) and demand centers where customers buy reprocessed products (Pochampally et al. 2009b).

(García and Bermeo 2017) define reverse logistics as the process of planning, developing and efficient control of the flow of materials, reverse products and information from the place of consumption to the place of origin, so that the needs of the consumer are met, recovering the waste obtained and managing it in such a way that it is possible to reintroduce it into the supply chain, obtaining an added value and/or achieving an adequate elimination of it.

Figure 10.1 presents a generic reverse supply chain, which shows the reverse flow from consumers to demand centers. Likewise, Monroy and Ahumada (2006) state that when the process of "reverse logistics" is in supply chains, it becomes a closed system; in this way, there are different types of chains that are handled in a reverse logistics process (depending on who retrieves the product and what it is recovered for), which can be seen in Fig. 10.2. The first type is given when the product is recovered by the same production company; in the second, the production company recovers its own product and that of the competition; in the third, when the company recovering the entire product is different from the producer, the waste is used for the same product; and finally, the case in which the company recovering the product is different from the producer, this is used for a production process completely different from the original.

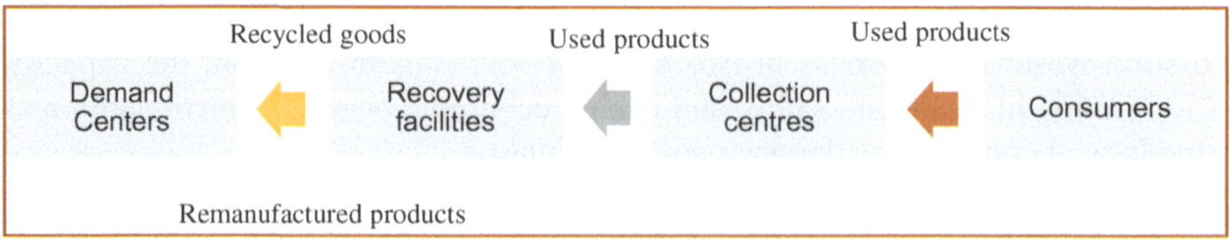

Fig. 10.1 Generic reverse supply chain (Pochampally et al. 2009b)

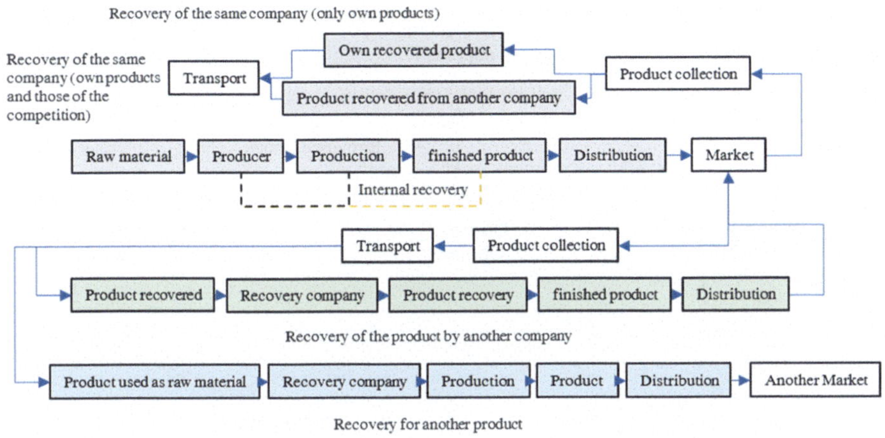

Fig. 10.2 Types of supply chains in a reverse logistics process (Monroy and Ahumada 2006)

10.2.3 Closed-Loop Supply Chains

A closed-loop supply chain (CLSC) is a combination of a traditional direct flow supply chain with a reverse supply chain (Pochampally et al. 2009b). In general, a CLSC can be seen as the traditional forward flow supply chain complemented by reverse operations for recovery products that are reprocessed and eventually reintroduced into the supply chain (Senthil et al. 2012).

Figure 10.3 presents the generic structure of a closed-loop supply chain; if analyzed from consumers, they consume new remanufactured products or products (obtained from demand centers), thus generating used products, which are collected and taken to collection centers, from where they are shipped to the different facilities or recovery plants, here new products, recycled goods or remanufactured products are produced, which are transported to the demand centers, which sends the new products and the remanufactured products to consumers and the recycled goods send them to suppliers of raw materials, and this cycle continues to be repeated in this closed cycle.

According to the literature review conducted by Stindt and Sahamie (2014), the implementation of CLSC strategies allows companies to reduce the negative environmental impacts of their businesses, associated with protecting primary resources, saving space in landfills and confining the effects of toxic substances. Authors explain that manufacturing goods from secondary materials generally requires less energy and emits less CO_2, also economic viability is often realized through cost savings, a competitive advantage and intangible benefits such as reputation.

As mentioned in the previous section, there are different types of waste management options, which represent options in a CLSC. Authors like (De Brito and Dekker 2004; Fleischmann et al. 1997; Thierry et al. 1995), mentioned by Stindt and

Fig. 10.3 Structure of a generic supply chain (Pochampally et al. 2009b)

Sahamie (2014), agree on the following management options: reuse, repair, restoration, recovery and recycling. In accordance with the types of reverse logistics chains suggested by Monroy and Ahumada (2006), the sources of the goods or waste to be recovered may be customer returns, distribution and manufacturing, as presented in Fig. 10.4. A characteristic can be observed that the flow of the products to be recovered or the waste meet a cycle, returning to the main producer and returning from an external flow or an internal flow.

In general, there are significant uncertainties that present a serious challenge to the planning and actual operation of reverse logistics activities compared to the traditional supply chain, which is why it is a subject of high-impact research (Stindt and Sahamie 2014). Below is a review of some of the most important works developed on waste management from the application of mathematical techniques, mainly multi-criteria decision models.

10.2.4 Multi-criteria Techniques in Waste Management

The literature reports the development of models for solid waste management since the 1970s (Quintero et al. 2007), among which they have used optimization, simulation and multi-criteria tools. However, as Pochampally et al. (2009a) point out, the

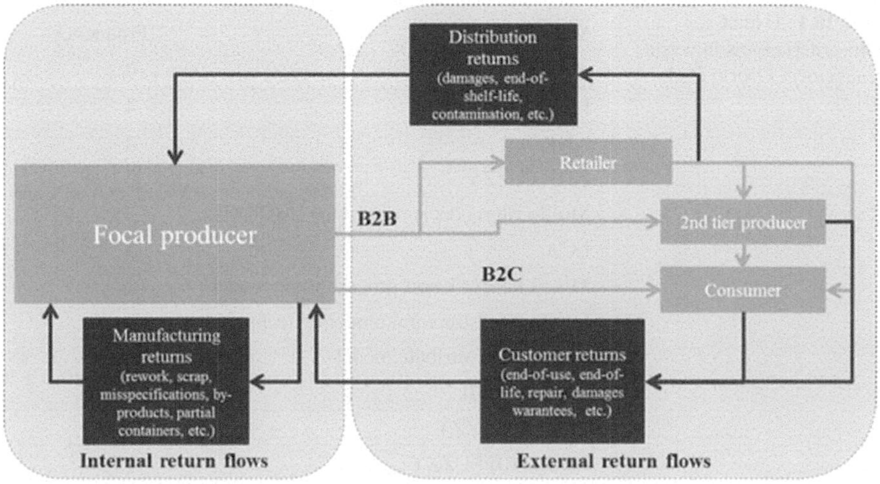

Fig. 10.4 Return sources in a closed-cycle supply chain (Stindt and Sahamie 2014)

problem of waste management is complex and has different levels of decision, and it requires making decisions considering multiple actors involved, different criteria and alternatives. In this sense, using multi-criteria decision tools is important, as it allows to consider the criteria of the supply chain, the alternatives of reverse logistics, as well as the inclusion in the decision process to the different actors involved.

Multi-criteria tools that have been used in waste management issues are analytic hierarchy processAHP), analytic network process (ANP), ranking organization method for enrichment evaluation (PROMETHEE), ELimination and Choice Expressing Reality (ELECTRE), decision-making of fuzzy multiple attributes and multi-matrix decision analysis tools. In this sense, Quintero et al. (2007) propose a conceptual model that supports decision-making regarding the selection of sustainable technologies (SELTEC RS) during the planning phase of a comprehensive solid waste management project. Vego et al. (2008) make use of multi-criteria decision-making through the PROMETHEE and GAIA in the strategic management of municipal solid waste in Dalmatia, Croatia. Similarly, Barker and Zabinsky (2011) present a multi-criteria decision-making model for reverse logistics using AHP, which evaluates a hierarchy of criteria and subcriteria including costs and the relationship of companies to network design regarding critical decisions. Similarly, Phonphoton and Pharino (2019) used the AHP and found that the alternative of adapting trucks is the most important for municipal solid waste management during the Bangkok floods.

In relation to solid waste management, different authors have used AHP to develop their research, in some cases they use hybrids, with the purpose of improving the performance in decision-making associated for example to the different alternatives of waste management, the design of reverse logistics networks, the contracting of suppliers for waste management, among others.

Table 10.1 Use of multi-criteria tools in waste management Osorio et al. (2013)

Methodology	Frequency
AHP	7
FUZZY	4
ANP	4
AHP FUZZY	3
Multi-criteria decision-making (MCDM)	3
LCA	1
Mixed integer linear programming model	1
Interpretive structural modeling technique	1
Fuzzy multi attribute model	1
ELECTRE III	1
TOPSIS FUZZY	1
ELECTRE FUZZY	1
Euler	1
Total	29

Considering the literature review by Achillas et al. (2013), it can be inferred that the application of AHP in decision-making related to waste management has a greater focus on municipal solid waste management. This research will address a strategic analysis, as strategic decisions related to reverse logistics will be analyzed in a CLSC.

Osorio et al. (2013) conducted a literature review about the application of multi-criteria decision models in solid waste management (Table 10.1). This defines that the most commonly used tool is AHP and some extensions (hybrids). For this reason, the AHP tool was used in this work as a multi-criteria model to evaluate the criteria and alternatives of reverse logistics for waste management in the CLSC of glass.

Table 10.2 presents some applications of multi-criteria tools according to the literature review, showing that more than half of the revised papers focus on waste management. However, it is also important to note that there is an interest on both reverse logistics management and closed-loop supply chains.

On the other hand, according to the literature review carried out by Achillas et al. (2013), the application of AHP in decision-making related to waste management has a greater focus on municipal solid waste management, that is, 13 articles out of 22, as can be seen in Table 10.3. Also, the most discussed topics are localization of landfill and strategies of waste management, with five and three articles, respectively. This concludes two important aspects: The first one is the importance of the use of AHP to make decisions related to solid waste management, and the second one is that decisions have focused on the location of landfills and there are not many researches related to strategic issues such as those related to reverse logistics in closed-loop supply chains which is the central subject of this chapter.

Table 10.2 Multi-criteria tool application in waste management

No.	Classification topic	No. papers	Authors
1	Supplier selection	9	Liu et al. (2019), Jayant et al. (2014), Büyüközkan and Çifçi (2012), Chen et al. (2010), Divahar and Sudhahar (2012), Liao and Chiu (2011), Ho et al. (2010), Kannan et al. (2009), Efendigil et al. (2008)
2	Closed-loop supply chains	4	Govindan et al. (2020), Ahmadi and Amin (2019), Amin and Zhang (2012a), Olugu and Wong (2012)
3	Reverse logistics	5	Akdoğan and Coşkun (2012), Barker and Zabinsky (2011), Chiou et al. (2012), Ravi et al. (2005), Senthil et al. (2012)
4	Waste management	32	Bollinger and Pictet (2008), Chen et al. (2010) Chang et al. (2008), Contreras et al. (2008), Dhouib and Elloumi (2011), Ekmekçioğlu et al. (2010), Garfì et al. (2009), Gomes et al. (2008), Hanandeh and El-Zein (2009), Hatami-Marbini et al. (2013), Herva and Roca (2013), Hung et al. (2007), Iakovou et al. (2009), Karmperis et al. (2012), Kaya (2012), Khadivi and Ghomi (2012), Khan and Faisal (2008), Liamsanguan and Gheewala (2008), Liao and Chiu (2011), Nas et al. (2010), Önüt and Soner (2008), Pires et al. (2011), Roussat et al. (2009), Shiue and Lin (2012), Su et al. (2007), Subramoniam et al. (2013), Tsai et al. (2020b), Tseng and Lin (2009), Vego et al. (2008), Yeh and Xu (2013), Coban et al. (2018), Phonphoton and Pharino (2019)
5	Green supply chain management	6	Kainuma and Tawara (2006), Lin (2013), Lu et al. (2007), Tseng and Chiu (2013), Tseng et al. (2012), Xi et al. (2010)
6	Comparison of the application of multi-criteria methodologies to environmental problems	1	Salminen et al. (1998)
7	Managing a supply chain	7	Agarwal et al. (2006), Ayağ and Özdemir (2009), Büyüközkan and Berkol (2011), Cruz (2009), Erol et al. (2011), Nagurney and Toyasaki (2003), Nasiri and Huang (2008)
	Total papers	63	

Table 10.3 Multi-criteria methods related to solid waste management problems

Waste classification	Year	Authors	Method	Subject
Municipal solid waste (13 papers)	2019	Phonphoton and Pharino (2019)	AHP	Solid waste management during flooding
	2018	Coban et al. (2018)	TOPSIS PROMETHEE I and II	Municipal solid waste management
	2011	Tavares et al. (2011)	AHP, GIS	Location of the MSW facility
	2010	De Feo and De Gisi (2010)	AHP	
	2010	Aragonés-Beltrán et al. (2010)	AHP, ANP	Location of the solid waste management plant
	2009	Garfì et al. (2009)	AHP	Waste management strategy
	2009	Wang et al. (2020)	AHP	Location of the landfill site
	2008	Contreras et al. (2008)	AHP	Waste management strategy
	2008	Önüt and Soner (2008)	AHP, TOPSIS	Location of the transshipment site
	2007	Gemitzi et al. (2007)	AHP	Location of the landfill site
	2007	Su et al. (2007)	AHP	Waste management strategy
	1997	Charnpratheep et al. (1997)	AHP	Location of the landfill site
	1996	Siddiqui et al. (1996)	AHP	Location of the landfill site
Wastewater	2011	Bottero et al. (2011)	AHP, ANP	Waste management strategy
Nuclear/radioactive (3 papers)	2005	Taji et al. (2005)	AHP	Location of the nuclear waste disposal site
	1995	Bowen (1995)	AHP	Location of the nuclear waste disposal site
	1982	Saaty and Gholamnezhad (1982)	AHP	Nuclear waste management strategy

(continued)

Table 10.3 (continued)

Waste classification	Year	Authors	Method	Subject
Industrial (2 papers)	1996	Alidi (1996)	AHP	Waste management strategy in the petrochemical industry
	2006	Cram et al. (2006)	AHP	Location of the hazardous waste disposal site
Hospital (3 papers)	2010	Karagiannidis et al. (2010)	AHP	Hospital waste management strategy
	2007	Brent et al. (2007)	AHP	Hospital waste management strategy
	2007	Karamouz et al. (2007)	AHP	Hospital waste management strategy

Adapted from Achillas et al. (2013)

10.3 Methodology

The analysis initially focused on identifying the actors involved in the decision-making of solid waste management in the closed-cycle glass supply chain in Colombia. Subsequently, the definition of the criteria and decisions to be considered takes as a reference of those involved in the chain, among them the company that manufactures glass containers from virgin or recycled material glass; the bottler, a company that uses the container to pack its drinks; representatives of the government (mayor), environmental regulators such as the Regional Autonomous Corporation of Valle del Cauca, representatives of universities and the recycler sector, all of them located in Valle del Cauca, Colombia. The city object of this study was developing a comprehensive solid waste management plan, in which the whole community was involved to improve the decision-making process regarding waste management.

Figure 10.5 presents the methodology that was followed to obtain the proposal for the multi-criteria model. Initially, a literature review was conducted to identify criteria, subcriteria and alternatives, and next the application of AHP was carried out to prioritize the criteria and subcriteria through a paired comparison of the members of the supply chain such as the manufacturer, the bottler, higher education institutions and recycler associations participated on the comparison.

Then, the reverse logistics alternatives proposed for the manufacturer were prioritized. A total of 31 paired comparison matrices were obtained, and comparison data was processed using Expert Choice software, which support the results for criteria priority and alternative preference. Finally, a sensitivity analysis was performed using the same software, including the variation of each criterion, increasing each weight

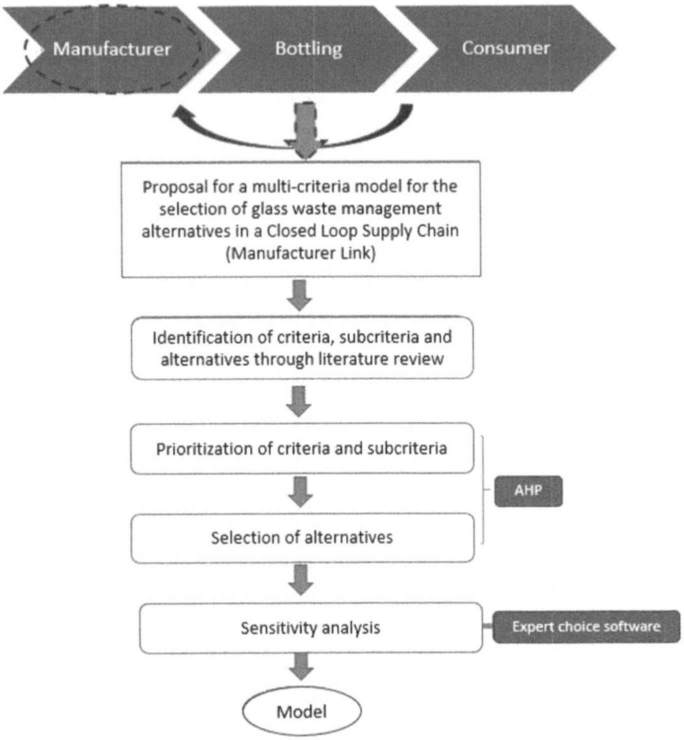

Fig. 10.5 Proposed methodology (Chud et al. 2018)

by 50%, starting with the economic criterion, then the environmental, social and technical, in order to identify some type of variation in the preference results of the alternatives and, in this way, validate the choice of the best management alternative. It should be noted that the consistency for each of the comparison matrices was analyzed using the same software. As a result, the multi-criteria model was obtained to choose of reverse logistics alternatives for waste management considering the manufacturer link.

10.4 Results

The criteria and management alternatives are presented as results throughout the CLSC of glass container, as well as the detailed analysis of decisions for glass container manufacturer link, since they are fundamental in the returning process of waste to the production process.

Figure 10.6 shows the actors involved in this chain and the flow of material between them, which can be containers or cullet (glass residue when the container

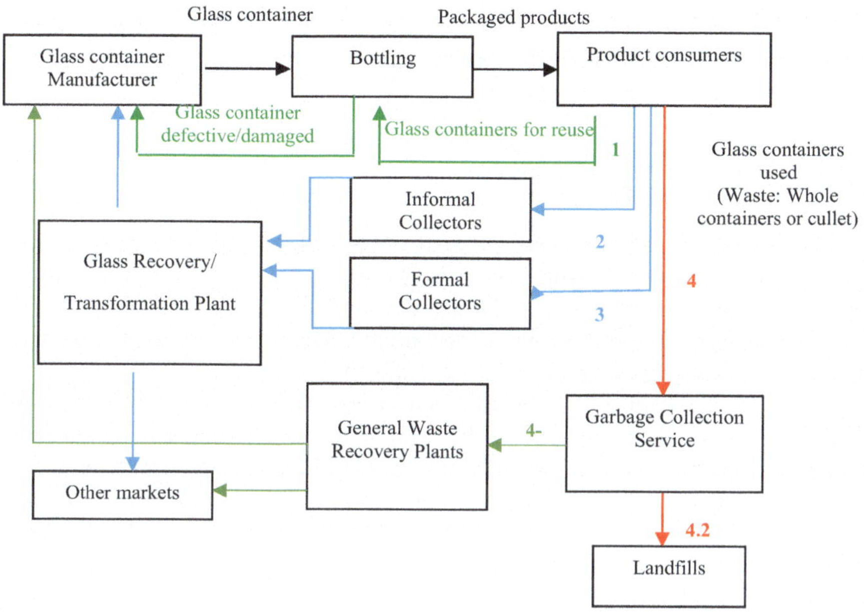

Fig. 10.6 CLSC for glass container adapted from (Chud et al. 2018)

is broken). Initially, a direct flow begins when the manufacturer makes the different glass containers, which are used for different sectors, for example, for fizzy drinks or juices. In this sector, the products are bottled that will be sent from the bottlers to the consumers, and it can also be the case that the products are damaged in the packaging process and sent to the manufacturer so that they can be used as raw material and thus make new glass container.

After the packaged products are consumed, the packaging is generated as waste, the entire container or the cullet can be considered (when the container breaks), and at that time different options can be generated:

(1) That the container be returned to the bottler of the product to be reused.
(2) Let the container be collected by an informal collector
(3) It is collected by a formal collector (by the company, outsourcing of the company or other companies dedicated to this business), in these last two cases, waste is taken to a recovery or treatment plant to turn them into raw material for the manufacturer or for other markets.
(4) The packaging is collected by the garbage collection service company:

    (4.1) Containers are sent to a general waste recovery plant and treated to convert them back into raw material and sent them to glass container manufacturers or other markets, or
    (4.2) Containers can be taken to the landfill.

It should be noted that Fig. 10.6 presents the possible options, considering that some of them can generate a greater environmental, social or economic impact.

10.4.1 Criteria and Subcriteria

According to the literature review and analysis of criteria, different categories or groups of criteria were classified as it is shown in Table 10.4 as are environmental, economic, social, technical, human development, legislation and corporate social responsibility. Environmental and economic categories are commented more frequently in the articles with 11 and 9, respectively.

According to the results of Table 10.4, five criteria were considered for the AHP application model (C1: Economic, C2: Environmental, C3: Social, C4: Technicians, C5: Legal). Each criterion included subcriteria, in total 25 subcriteria were included, which are presented with their respective notation in Table 10.5. This result was obtained from a literature review and validated by those involved in the decision-making process.

Table 10.4 Categories of criteria involved in waste management

Categories	Authors
Environmental	Amin and Zhang (2012b), (2013), Büyüközkan and Berkol (2011), Chiou et al. (2012), Cruz (2009), Erol et al. (2011), Garfi et al. (2009), Hanandeh and El-Zein (2009), Kaya (2012), Khadivi and Ghomi (2012), Pires et al. (2011)
Economic	Akdoğan and Coşkun (2012), Büyüközkan and Berkol (2011), Chiou et al. (2012), Erol et al. (2011), Garfi et al. (2009), Khadivi and Ghomi (2012), Liao and Chiu (2011), Pires et al. (2011), Senthil et al. (2012)
Social	Büyüközkan and Berkol (2011), Chiou et al. (2012), Erol et al. (2011), Garfi et al. (2009), Khan and Faisal (2008), Liao and Chiu (2011), Pires et al. (2011), Su et al. (2007)
Technical	Garfi et al. (2009), Kannan et al. (2009), Khadivi and Ghomi (2012), Khan and Faisal (2008), Pires et al. (2011), Su et al. (2007)
Human development	Garfi et al. (2009), Liao and Chiu (2011)
Legislation	Akdoğan and Coşkun (2012)
Corporate social responsibility	Akdoğan and Coşkun (2012)

Adapted from Chud (2015)

Table 10.5 Subcriteria identification

Subcriteria		Criteria
Identification	Description	
SC1.1	Cost of removing waste from glass	Economic
SC1.2	Reduced investment in virgin glass, including recovered and recycled glass	
SC1.3	Glass recovery and recycling costs	
SC1.4	Inventory costs (used glass, cullet or recycled glass)	
SC1.5	Glass transport costs	
SC1.6	Investment in facilities and labor for recovery and/or recycling activities	
SC1.7	Tax benefits or incentives	
SC2.1	Energy consumption in glass waste management	Environmental
SC2.2	Reducing the extraction of natural resources as virgin raw material for glass	
SC2.3	Increased life span of final disposal sites or landfills	
SC2.4	Water consumption in glass waste management	
SC3.1	Human resource management related to glass waste management activities	Social
SC3.2	Corporate social responsibility	
SC3.3	Participation of all those who have an impact on generation and management of glass waste	
SC3.4	External community	
SC3.5	Social aspects	
SC3.6	Advertising and image of the organization by glass waste management activities	
SC4.1	Returning recycled glass to the production cycle	Technical
SC4.2	Quality of glass products reused or made from recycled glass	
SC4.3	Technical standards related to the management of glass waste for production and/or administrative processes	
SC4.4	Adequate facilities and technology	
SC4.5	Transmission of technical information from the glass waste treatment process	
SC5.1	Existence of plans for the management of glass waste	Legal
SC5.2	Decrees and regulations for the return of glass products	
SC5.3	Laws and resolutions to apply specialized techniques for the management of glass waste	

10.4.2 Alternatives for Waste Management in a CLSC Glass Containers

There are several alternatives for the final disposal of the packaging in the CLSC of glass containers, some alternatives are related to each link in the chain, and some others are related to direct responsible actors for the environmental impact generated by their product (i.e., glass packaging manufacturer).

Alternatives for the manufacturer

Recycling. It involves the manufacturer collecting (recovering) containers discarded or converted into a cullet (both considered glass waste) and recycling them into raw materials to use in the production of their own products. The packaging manufacturer may have different options for recycling, as listed below:

1. To collect glass containers through formal collectors (the same company), transform them into cullet; recycle the glass and turn it into raw material for their product.
2. Collect glass containers through informal collectors (other waste management companies), transform the containers into cullet, and recycle the glass and turn it into raw material for their product.
3. Obtain the cullet (collect it or buy it), recycle the glass, and convert it into raw material for their product.

General alternatives

General alternatives are considered as those that apply to all links in the chain. As mentioned above, some options that any company may consider are to allow that the waste management companies perform the final disposal in landfills or to allow that the competition or any other companies carry out the collection and recycling:

Do not perform any waste management

1. Final disposal of containers (whole or cullet) in landfills
2. Allow the competition to carry out collection and recycling.

Three alternatives, related to solid waste management, were considered which are in charge of the manufacturer of the glass containers; they are AF1: Recycling with direct collection (reprocessing); AF2: Recycling with indirect collection (reprocessing); AF3: Do nothing, i.e., final disposal in the filling (or that other companies recycle their products).

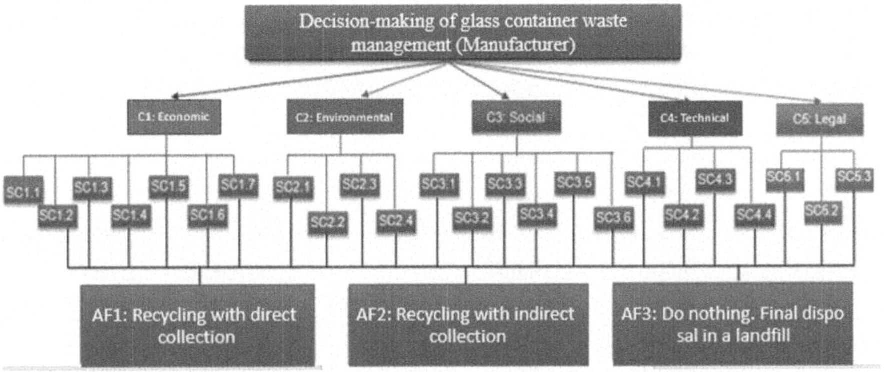

Fig. 10.7 Hierarchical structure for glass container manufacturer decisions (Chud et al. 2018)

10.4.3 Hierarchical Structure for Decision in the Manufacturer Link

The AHP hierarchical structure was established for the packaging manufacturer link decisions presented in Fig. 10.7 showing the five criteria, 25 subcriteria and the three alternatives that were considered.

10.4.4 Prioritization of Criteria and Subcriteria

The results of prioritization of criteria and subcriteria were obtained after applying the AHP, and these are presented in Fig. 10.8. The criterion with the highest priority is legal with 51.2%, followed by social with 23.1%. According to research conducted by Barker and Zabinsky (2011), the decision about how to collect return products is mainly determined by the supplier's classification for business relationships; in the case of this investigation, the decision on the management of the waste of glass container is mainly determined by the legal aspects in which the manufacturer is involved.

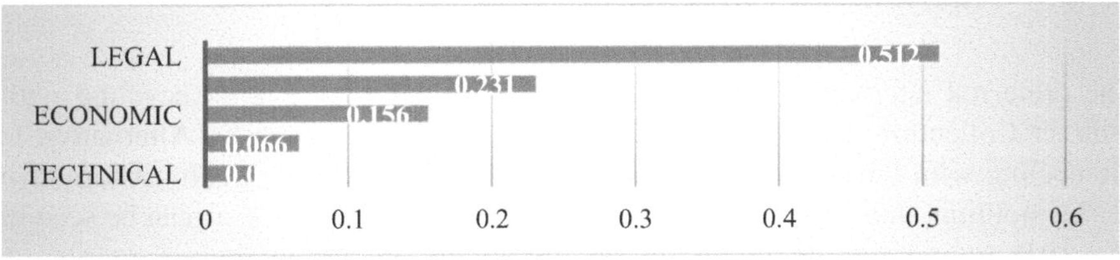

Fig. 10.8 Result of multi-criteria priorities for the manufacturer

Table 10.6 Priorities of subcriteria in the AHP multi-criteria model for the manufacturer

Criteria	Subcriteria	Priority	Criteria	Subcriteria	Priority
Economic	SC1.7	0.364	Social	SC3.1	0.341
	SC1.1	0.208		SC3.6	0.224
	SC1.3	0.143		SC3.2	0.177
	SC1.5	0.126		SC3.3	0.108
	SC1.2	0.072		SC3.5	0.085
	SC1.6	0.063		SC3.4	0.064
	SC1.4	0.024	Technical	SC4.3	0.401
Environmental	SC2.4	0.450		SC4.1	0.327
	SC2.1	0.415		SC4.2	0.173
	SC2.2	0.070		SC4.4	0.063
	SC2.3	0.064		SC4.5	0.035
			Legal	SC5.2	0.490
				SC5.3	0.451
				SC5.1	0.059

On the other hand, Table 10.6 shows the priority that each subcriterion has within the category to which it belongs. If the economic criteria are analyzed, those subcriteria with the greatest weight are SC1.7 and SC1.1 with 36.4% and 20.8%, respectively; for environmental criteria, the highest priority is to SC2.4 and SC2.1 with 45% and 41.5%, respectively; for social criteria the prioritized subcriteria were SC3.1 and SC3.6 whose priorities were 34.1% and 22.4%, respectively; technical criteria were SC4.3 and SC4.1 with 40.1% and 32.7%, respectively; for its part in legal criteria those subcriteria that represent the highest priority were SC5.2 with 49% and SC5.3 with 45.1%.

In this sense, Barker and Zabinsky (2011) reported greater preference for subcriteria savings in recycling costs in two of the three cases studied. Conversely, in this study, the main criterion is legal and the subcriteria decrees and regulations to return glass products (SC5.2) presented the highest priority among all the subcriteria.

10.4.5 Alternative Preferences

The preferred alternative for the manufacturer is Alternative 2: "Recycling with Indirect Collection" with a final valuation of 43.7%, followed by Alternative 1: "Recycling with Direct Collection" with a rating of 39.6% and finally Alternative 3: "Do nothing, final disposal in a landfill" with a score of 16.7%, as can be seen in Fig. 10.9.

The manufacturer has a glass container production plant where he/she collects the resulting waste of the glass. Also, there are collection centers located in different

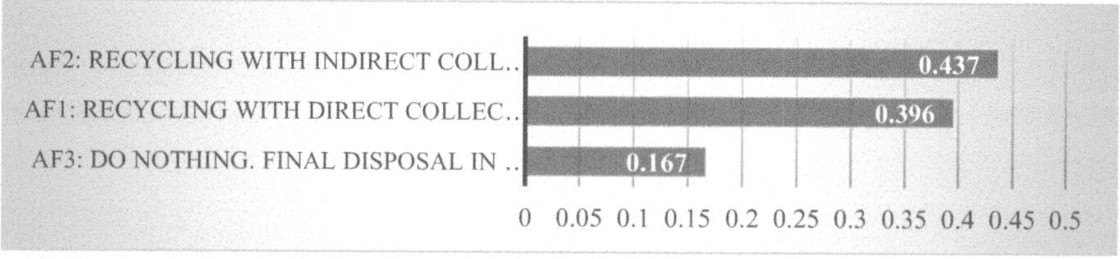

Fig. 10.9 Preferences of alternatives in manufacturer's multi-criteria model with expert choice

parts of the country, and this collection is done by other entities or natural people who carry the glass containers to their collection centers which are then purchased by the manufacturer. Similarly, manufacturer's customers, bottling companies, also collect the product that meets its life cycle and becomes cullet to sell to the packaging supplier.

It can be indicated that the model of decision-making of waste management applied to the CLSC of glass especially for the packaging manufacturer is relevant and provides meaningful results to choose alternatives according to the company objectives.

10.4.6 Sensitivity Analysis of the Decision-Making Model for the Manufacturer Link

A sensitivity analysis proposal was made to review the impact of varying the weight of each criterion when choosing the respective alternatives. Figure 10.10 shows on the left-hand side that the legal criterion is the one with the greatest weight (51.2%) and that Alternative 2 is the most preferred alternative in each comparison considering all the criteria, corresponding to the red line above the other two lines. On the right-hand side, the figure is showing the specific priority percentages of each criterion and the resulting preference for alternatives.

Table 10.7 shows the results of the sensitivity analysis for the manufacturer's model, presents the initial state and each of the priority variations of the criteria

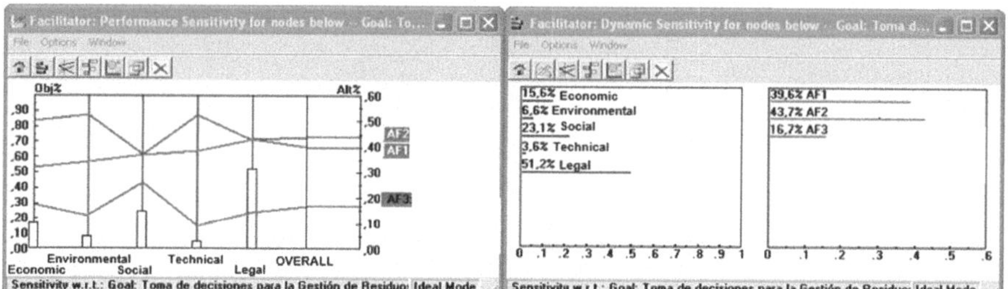

Fig. 10.10 Initial status of manufacturer's alternative preferences considering the priority of criteria

Table 10.7 Summary of sensitivity analysis scenarios for the manufacturer's multi-criteria model (%)

Stage	C. Economic	C. Environmental	C. Social	C. Technical	C. Legal	Alt preference first	Alt preference second	Alt. preference third
Initial status	15.6	6.6	23.1	3.6	51.2	A2 = 43.7	A1 = 39.6	A3 = 16.7
C. Economic 50%	50	26	15	7.8	1,2	A2 = 48.9	A1 = 34	A3 = 17
C. Environmental 50%	4	50	28.8	14.9	2.3	A2 = 47.7	A1 = 35.8	A3 = 16.5
C. Social 50%	6.9	13.1	50	25.9	4.1	A2 = 44.2	A1 = 36.9	A3 = 19
C. Technical 50%	13.44	25.3	3.5	50	7.8	A2 = 50.9	A1 = 36.8	A3 = 12.2

with the preference results of the alternatives. As mentioned above, in any of the scenarios, the preference for alternatives maintains the same order, first alternative 2, second alternative 1 and in third and last place, alternative 3. The preference for alternative 2 in all cases was higher than 43%, for alternative 1 higher than 34% and for alternative 3 higher than 12%. In addition, it can be observed that the highest preference for alternative 2 is achieved when the priority of the technical criterion is 50%.

10.5 Conclusions

- The options that actors within supply chains have to contribute to sustainability are different; nevertheless, making the best decision will not always correspond to an easy task because although objectives might be clear, there are so many factors or criteria involved in the decision that disparity can arise between the trials issued by some of them. In this sense, multi-criteria tool such as the AHP is important to make a decision between different alternatives in a multi-criteria environment.
- In the multi-criteria model of the glass container manufacturer, the most important criteria are legal followed by the social and economic with 51.2%, 23.1% and 15.6% priority, respectively. Being the best alternative recycling with indirect collection, which is consistent, since it complies with the aspects determined as the use of the waste, returns to the production cycle, and it is linked in the social aspect by the existence of companies and associations that can dedicate themselves to the collection and/or recovery of the waste to produce an income and selling to the manufacturer of packaging. Even if the priority of each criterion is changed by up to 50% (in different scenarios), recycling with indirect collection remains as the alternative selected by using AHP.
- In Colombia, there are policies that promote the integral management of solid waste such as glass, and this material turns out to be a 100% usable material; however, recycling rate of glass waste still is low. Currently, companies and municipalities are generating independent contributions represented in actions to mitigate the negative social and environmental impacts, aimed at the sustainability of cities.
- The closed-loop supply chain concept is fundamental and contributes directly to two organizational purposes aimed at minimizing environmental impact. On the one hand, the reuse of waste, in this particular case glass containers, helps to reduce the load of landfill and emissions and different impacts to the environment; on the other hand, the incorporation of these elements into the production process minimizes the use of raw materials that in this particular case correspond to non-renewable materials and therefore prevent the environmental impact associated with their exploitation and the possible depletion of such materials in the environment.

- Finally, it is important to analyze each link in the chain independently of the other, because in each one it is possible to make independent and non-exclusive decisions, in addition there may be different interests and conditions and that would impact the waste management.

References

Achillas C, Moussiopoulos N, Karagiannidis A et al (2013) The use of multi-criteria decision analysis to tackle waste management problems: a literature review. Waste Manag Res 31:115–129. https://doi.org/10.1177/0734242X12470203

Agarwal A, Shankar R, Tiwari MK (2006) Modeling the metrics of lean, agile and leagile supply chain: an ANP-based approach. Eur J Oper Res 173:211–225. https://doi.org/10.1016/j.ejor.2004.12.005

Ahmadi S, Amin SH (2019) An integrated chance-constrained stochastic model for a mobile phone closed-loop supply chain network with supplier selection. J Clean Prod 226:988–1003. https://doi.org/10.1016/j.jclepro.2019.04.132

Akdoğan MŞ, Coşkun A (2012) Drivers of reverse logistics activities: an empirical investigation. Proc Soc Behav Sci 58:1640–1649. https://doi.org/10.1016/j.sbspro.2012.09.1130

Alidi AS (1996) A multiobjective optimization model for the waste management of the petrochemical industry. Appl Math Model 20:925–933https://doi.org/10.1016/S0307-904X(96)00106-0

Amin SH, Zhang G (2012a) An integrated model for closed-loop supply chain configuration and supplier selection: multi-objective approach. Expert Syst Appl 39:6782–6791.https://doi.org/10.1016/j.eswa.2011.12.056

Amin SH, Zhang G (2012b) A proposed mathematical model for closed-loop network configuration based on product life cycle. Int J Adv Manuf Technol 58:791–801.https://doi.org/10.1007/s00170-011-3407-2

Amin SH, Zhang G (2013) A multi-objective facility location model for closed-loop supply chain network under uncertain demand and return. Appl Math Model 37:4165–4176. https://doi.org/10.1016/j.apm.2012.09.039

Aragonés-Beltrán P, Pastor-Ferrando JP, García-García F, Pascual-Agulló A (2010) An analytic network process approach for siting a municipal solid waste plant in the Metropolitan Area of Valencia (Spain). J Environ Manag 91:1071–1086. https://doi.org/10.1016/j.jenvman.2009.12.007

Ayağ Z, Özdemir RG (2009) A hybrid approach to concept selection through fuzzy analytic network process. Comput Ind Eng 56:368–379. https://doi.org/10.1016/j.cie.2008.06.011

Barker TJ, Zabinsky ZB (2011) A multicriteria decision making model for reverse logistics using analytical hierarchy process. Omega 39:558–573. https://doi.org/10.1016/j.omega.2010.12.002

Bollinger D, Pictet J (2008) Multiple criteria decision analysis of treatment and land-filling technologies for waste incineration residues. Omega 36:418–428. https://doi.org/10.1016/j.omega.2006.07.008

Bottero M, Comino E, Riggio V (2011) Application of the analytic hierarchy process and the analytic network process for the assessment of different wastewater treatment systems. Environ Model Softw 26:1211–1224. https://doi.org/10.1016/j.envsoft.2011.04.002

Bowen WM (1995) A Thurstonian comparison of the analytic hierarchy process and probabilistic multidimensional scaling through application to the nuclear waste site selection decision. Socioecon Plann Sci 29:151–163. https://doi.org/10.1016/0038-0121(95)00003-5

Brent AC, Rogers DEC, Ramabitsa-Siimane TSM, Rohwer MB (2007) Application of the analytical hierarchy process to establish health care waste management systems that minimise infection risks in developing countries. Eur J Oper Res 181:403–424. https://doi.org/10.1016/j.ejor.2006.06.015

Büyüközkan G, Berkol Ç (2011) Designing a sustainable supply chain using an integrated analytic network process and goal programming approach in quality function deployment. Expert Syst Appl 38:13731–13748. https://doi.org/10.1016/j.eswa.2011.04.171

Büyüközkan G, Çifçi G (2012) A novel hybrid MCDM approach based on fuzzy DEMATEL, fuzzy ANP and fuzzy TOPSIS to evaluate green suppliers. Expert Syst Appl 39:3000–3011. https://doi.org/10.1016/j.eswa.2011.08.162

Chang N-B, Parvathinathan G, Breeden JB (2008) Combining GIS with fuzzy multicriteria decision-making for landfill siting in a fast-growing urban region. J Environ Manag 87:139–153

Charnpratheep K, Zhou Q, Garner B (1997) Preliminary landfill site screening using fuzzy geographical information systems. Waste Manag Res 15:197–215. https://doi.org/10.1006/wmre.1996.0076

Chen H-W, Chang N-B, Chen J-C, Tsai S-J (2010) Environmental performance evaluation of large-scale municipal solid waste incinerators using data envelopment analysis. Waste Manag 30:1371–1381. https://doi.org/10.1016/j.wasman.2010.02.002

Chiou CY, Chen HC, Yu CT, Yeh CY (2012) Consideration factors of reverse logistics implementation—a case study of Taiwan's electronics industry. Proc Soc Behav Sci 40:375–381https://doi.org/10.1016/j.sbspro.2012.03.203

Chud VL (2015) Modelo multicriterio soporte a la tomma de decisiones para la gestión de residuos sólidos en una cadena de suministro de ciclo cerrado. Universidad del Valle

Chud VL, Osorio JC, Peña CC (2018) Modelo multicriterio como soporte la gestión de residuos de vidrio en una cadena de suministro de ciclo cerrado. Rev Espac 39:20–35

Coban A, Ertis IF, Cavdaroglu NA (2018) Municipal solid waste management via multi-criteria decision making methods: a case study in Istanbul, Turkey. J Clean Prod 180:159–167. https://doi.org/10.1016/j.jclepro.2018.01.130

Contreras F, Hanaki K, Aramaki T, Connors S (2008) Application of analytical hierarchy process to analyze stakeholders preferences for municipal solid waste management plans, Boston, USA. Resour Conserv Recycl 52:979–991. https://doi.org/10.1016/j.resconrec.2008.03.003

Corum A (2020) Remanufacturing, an added value product recovery strategy: In: Waste management: concepts, methodologies, tools, and applications. IGI Global, pp 1432–1452

Cram S, Sommer I, Morales L-M, et al (2006) Suitability of the vegetation types in Mexico's Tamaulipas state for the siting of hazardous waste treatment plants. J Environ Manag 80:13–24. https://doi.org/10.1016/j.jenvman.2005.08.013

Cruz JM (2009) The impact of corporate social responsibility in supply chain management: Multicriteria decision-making approach. Decis Support Syst 48:224–236

da Silva L, Marques Prietto PD, Pavan Korf E (2019) Sustainability indicators for urban solid waste management in large and medium-sized worldwide cities. J Clean Prod 237:117802. https://doi.org/10.1016/j.jclepro.2019.117802

Das S, Lee SH, Kumar P et al (2019) Solid waste management: scope and the challenge of sustainability. J Clean Prod 228:658–678. https://doi.org/10.1016/j.jclepro.2019.04.323

De Brito M, Dekker R (2004) A framework for reverse logistics. Reverse Logist 3–27

De Feo G, De Gisi S (2010) Using an innovative criteria weighting tool for stakeholders involvement to rank MSW facility sites with the AHP. Waste Manag 30:2370–2382. https://doi.org/10.1016/j.wasman.2010.04.010

Departamento Nacional de Planeación (2016a) Informe nacional de aprovechamiento

Departamento Nacional de Planeación D (2016b) Documento CONPES Social 3874. Política Nacional Para La Gestión Integral de Residuos Sólidos. Bogotá D.C

Dhouib D, Elloumi S (2011) A new multi-criteria approach dealing with dependent and heterogeneous criteria for end-of-life product strategy. Appl Math Comput 218:1668–1681. https://doi.org/10.1016/j.amc.2011.06.046

Divahar SR, Sudhahar C (2012) Selection of reverse logistics provider using AHP. Proc Eng 38:2005–2008. https://doi.org/10.1016/j.proeng.2012.06.242
Efendigil T, Önüt S, Kongar E (2008) A holistic approach for selecting a third-party reverse logistics provider in the presence of vagueness. Comput Ind Eng 54:269–287. https://doi.org/10.1016/j.cie.2007.07.009
Ekmekçioğlu M, Kaya T, Kahraman C (2010) Fuzzy multicriteria disposal method and site selection for municipal solid waste. Waste Manag 30:1729–1736. https://doi.org/10.1016/j.wasman.2010.02.031
Erol I, Sencer S, Sari R (2011) A new fuzzy multi-criteria framework for measuring sustainability performance of a supply chain. Ecol Econ 70:1088–1100. https://doi.org/10.1016/j.ecolecon.2011.01.001
Fleischmann M, Bloemhof-Ruwaard JM, Dekker R et al (1997) Quantitative models for reverse logistics: a review. Eur J Oper Res 103:1–17
García JO, Bermeo JA (2017) Logística emprearial. Primer, Machala-Ecuador
Garfì M, Tondelli S, Bonoli A (2009) Multi-criteria decision analysis for waste management in Saharawi refugee camps. Waste Manag 29:2729–2739. https://doi.org/10.1016/j.wasman.2009.05.019
Gemitzi A, Tsihrintzis VA, Voudrias E et al (2007) Combining geographic information system, multicriteria evaluation techniques and fuzzy logic in siting MSW landfills. Environ Geol 51:797–811. https://doi.org/10.1007/s00254-006-0359-1
Gomes CFS, Nunes KRA, Xavier LH et al (2008) Multicriteria decision making applied to waste recycling in Brazil. Omega 36:395–404
Govindan K, Mina H, Esmaeili A, Gholami-Zanjani SM (2020) An integrated hybrid approach for circular supplier selection and closed loop supply chain network design under ncertainty. J Clean Prod 242:118317. https://doi.org/10.1016/j.jclepro.2019.118317
Hanandeh A, El-Zein A (2009) Strategies for the municipal waste management system to take advantage of carbon trading under competing policies: the role of energy from waste in Sydney. Waste Manag 29:2188–2194. https://doi.org/10.1016/j.wasman.2009.03.002
Hatami-Marbini A, Tavana M, Moradi M, Kangi F (2013) A fuzzy group electre method for safety and health assessment in hazardous waste recycling facilities. Saf Sci 51:414–426. https://doi.org/10.1016/j.ssci.2012.08.015
Hekkert MP, Joosten LA., Worrell E, Turkenburg WC (2000a) Reduction of CO_2 emissions by improved management of material and product use: the case of primary packaging. Resour Conserv Recycl 29:33–64. https://doi.org/10.1016/S0921-3449(99)00056-7
Hekkert MP, Joosten LAJ, Worrell E (2000b) Reduction of CO_2 emissions by improved management of material and product use: The case of transport packaging. Resour Conserv Recycl 30:1–27.https://doi.org/10.1016/S0921-3449(00)00046-X
Herva M, Roca E (2013) Ranking municipal solid waste treatment alternatives based on ecological footprint and multi-criteria analysis. Ecol Indic 25:77–84. https://doi.org/10.1016/j.ecolind.2012.09.005
Ho W, Xu X, Dey PK (2010) Multi-criteria decision making approaches for supplier evaluation and selection: a literature review. Eur J Oper Res 202:16–24. https://doi.org/10.1016/j.ejor.2009.05.009
Hung ML, Ma HW, Yang WF (2007) A novel sustainable decision making model for municipal solid waste management. Waste Manag 27:209–219. https://doi.org/10.1016/j.wasman.2006.01.008
Iakovou E, Moussiopoulos N, Xanthopoulos A et al (2009) A methodological framework for end-of-life management of electronic products. Resour Conserv Recycl 53:329–339
Jayant A, Gupta O, Garg S, Khan M (2014) TOPSIS-AHP based approach for selection of reverse logistics service provider: a case study of mobile phone industry. Proc Eng 97:2147–2156
Kainuma Y, Tawara N (2006) A multiple attribute utility theory approach to lean and green supply chain management. Int J Prod Econ 101:99–108. https://doi.org/10.1016/j.ijpe.2005.05.010

Kannan G, Pokharel S, Kumar PS (2009) A hybrid approach using ISM and fuzzy TOPSIS for the selection of reverse logistics provider. Resour Conserv Recycl 54:28–36. https://doi.org/10.1016/j.resconrec.2009.06.004

Karagiannidis A, Papageorgiou A, Perkoulidis G, et al (2010) A multi-criteria assessment of scenarios on thermal processing of infectious hospital wastes: a case study for Central Macedonia. Waste Manag 30:251–262. https://doi.org/10.1016/j.wasman.2009.08.015

Karamouz M, Zahraie B, Kerachian R, et al (2007) Developing a master plan for hospital solid waste management: a case study. Waste Manag 27:626–638. https://doi.org/10.1016/j.wasman.2006.03.018

Karmperis AC, Sotirchos A, Aravossis K, Tatsiopoulos IP (2012) Waste management project's alternatives: a risk-based multi-criteria assessment (RBMCA) approach. Waste Manag 32:194–212. https://doi.org/10.1016/j.wasman.2011.09.001

Kaya I (2012) Evaluation of outsourcing alternatives under fuzzy environment for waste management. Resour Conserv Recycl 60:107–118. https://doi.org/10.1016/j.resconrec.2011.12.006

Khadivi MR, Ghomi SF (2012) Solid waste facilities location using of analytical network process and data envelopment analysis approaches. Waste Manag 32:1258–1265. https://doi.org/10.1016/j.wasman.2012.02.002

Khan S, Faisal MN (2008) An analytic network process model for municipal solid waste disposal options. Waste Manag 28:1500–1508. https://doi.org/10.1016/j.wasman.2007.06.015

Ko YD, Noh I, Hwang H (2012) Cost benefits from standardization of the packaging glass bottles. Comput Ind Eng 62:693–702. https://doi.org/10.1016/j.cie.2011.11.026

Liamsanguan C, Gheewala SH (2008) LCA: a decision support tool for environmental assessment of MSW management systems. J Environ Manage 87:132–138. https://doi.org/10.1016/j.jenvman.2007.01.003

Liao C-H, Chiu ASF (2011) Evaluate municipal solid waste management problems using hierarchical framework. Proc Soc Behav Sci 25:353–362. https://doi.org/10.1016/j.sbspro.2011.10.554

Lin R-J (2013) Using fuzzy DEMATEL to evaluate the green supply chain management practices. J Clean Prod 40:32–39. https://doi.org/10.1016/j.jclepro.2011.06.010

Liu A, Xiao Y, Lu H, et al (2019) A fuzzy three-stage multi-attribute decision-making approach based on customer needs for sustainable supplier selection. J Clean Prod 239:118043. https://doi.org/10.1016/j.jclepro.2019.118043

Lourenço J, Barbosa M, Cirne L (2016) A logística reversa aplicada a gestão de resíduos sólidos urbanos no Município de Campina Grande-PB. Espacios 37:13

Lu LYYY, Wu CHH, Kuo T-CC (2007) Environmental principles applicable to green supplier evaluation by using multi-objective decision analysis. Int J Prod Res 45:4317–4331. https://doi.org/10.1080/00207540701472694

MAVDT (2007) Gestión Integral de residuos o desechos peligrosos. Bases Conceptuales, República de Colombia

MAVDT (2010) Politica Nacional de Producción y Consumo Sostenible

Monroy N, Ahumada M (2006) Logística reversa: Retos para la Ingeniería Industrial. Rev Ing 23. Doi: 10.16924%2Friua.v0i23.349

Nagurney A, Toyasaki F (2003) Supply chain supernetworks and environmental criteria. Transp Res Part D Transp Environ 8:185–213. https://doi.org/10.1016/S1361-9209(02)00049-4

Nas B, Cay T, Iscan F, Berktay A (2010) Selection of MSW landfill site for Konya, Turkey using GIS and multi-criteria evaluation. Environ Monit Assess 160:491

Nasiri F, Huang G (2008) A fuzzy decision aid model for environmental performance assessment in waste recycling. Environ Model Softw 23:677–689. https://doi.org/10.1016/j.envsoft.2007.04.009

Olugu EU, Wong KY (2012) An expert fuzzy rule-based system for closed-loop supply chain performance assessment in the automotive industry. Expert Syst Appl 39:375–384. https://doi.org/10.1016/j.eswa.2011.07.026

Önüt S, Soner S (2008) Transshipment site selection using the AHP and TOPSIS approaches under fuzzy environment. Waste Manag 28:1552–1559. https://doi.org/10.1016/j.wasman.2007.05.019
Osorio JO, Chud V., Peña CC, et al (2013) Una aproximación multicriterio a la gestión integral de residuos sólidos en las cadenas de suministro. In: X Optima, VI Red - M. Universidad de Concepción. Chile
Paes MX, de Medeiros GA, Mancini SD et al (2020) Municipal solid waste management: integrated analysis of environmental and economic indicators based on life cycle assessment. J Clean Prod 254:119848. https://doi.org/10.1016/j.jclepro.2019.119848
Phonphoton N, Pharino C (2019) Multi-criteria decision analysis to mitigate the impact of municipal solid waste management services during floods. Resour Conserv Recycl 146:106–113. https://doi.org/10.1016/j.resconrec.2019.03.044
Pires A, Chang N-B, Martinho G (2011) An AHP-based fuzzy interval TOPSIS assessment for sustainable expansion of the solid waste management system in Setúbal Peninsula, Portugal. Resour Conserv Recycl 56:7–21. https://doi.org/10.1016/j.resconrec.2011.08.004
PGIRS (2019) Plan de Gestión Integral de Residuos Sólidos de Santiago de Cali, Colombia. https://www.cali.gov.co/planeacion/publicaciones/32970/plan_de_gestin_integral_de_residuos_sli dos_pgirs/
Pochampally K, Gupta S, Govindan K (2009a) Metrics for performance measurement of a reverse/closed-loop supply chain. Int J Bus Perform Supply Chain Model 1:8–32.https://doi.org/10.1504/IJBPSCM.2009.026263
Pochampally K, Nukala S, Gupta S (2009b) Strategic planning models for reverse and closed loop supply chains. Londres
Quintero P, Galvis A, Marmolejo LF, Collazos H (2007) Modelo conceptual de la selección de tecnologías para el manejo integral de residuos sólidos en localidades colombianas con menos de 50000 habitantes. In: Conferencia Latinoamericana de Saneamiento, Seminario Gestión de Residuos Sólidos. Cali
Ravi V, Shankar R, Tiwari MK (2005) Analyzing alternatives in reverse logistics for end-of-life computers: ANP and balanced scorecard approach. Comput Ind Eng 48:327–356. https://doi.org/10.1016/j.cie.2005.01.017
Realff MJ, Ammons JC, Newton D (2000) Strategic design of reverse production systems. Comput Chem Eng 24:991–996. https://doi.org/10.1016/S0098-1354(00)00418-X
Roussat N, Dujet C, Méhu J (2009) Choosing a sustainable demolition waste management strategy using multicriteria decision analysis. Waste Manag 29:12–20. https://doi.org/10.1016/j.wasman.2008.04.010
Rubio S, Miranda FJ, Chamorro A, Valero V (2007) A reverse logistics system at grupo industrial alfonso gallardo. Universia Bus Rev 88–99
Saaty TL, Gholamnezhad H (1982) High-level nuclear waste management: analysis of options. Environ Plan B Plan Des 9:181–196
Salminen P, Hokkanen J, Lahdelma R (1998) Comparing multicriteria methods in the context of environmental problems. Eur J Oper Res 104:485–496
Senthil S, Srirangacharyulu B, Ramesh A (2012) A decision making methodology for the selection of reverse logistics operating channels. Prorc Eng 38:418–428. https://doi.org/10.1016/j.proeng.2012.06.052
Shiue Y-C, Lin C-Y (2012) Applying analytic network process to evaluate the optimal recycling strategy in upstream of solar energy industry. Energy Build 54:266–277. https://doi.org/10.1016/j.enbuild.2012.07.032
Siddiqui MZ, Everett JW, Vieux BE (1996) Landfill siting using geographic information systems: a demonstration. J Environ Eng 122:515–523
Soto P (2007) Cadenas de suministro de ciclo cerrado. La Trib Del GREL 57:18–19
Stindt D, Sahamie R (2014) Review of research on closed loop supply chain management in the process industry. Flex Serv Manuf J 26:268–293. https://doi.org/10.1007/s10696-012-9137-4

Su J-P, Chiueh P-T, Hung M-L, Ma H-W (2007) Analyzing policy impact potential for municipal solid waste management decision-making: a case study of Taiwan. Resour Conserv Recycl 51:418–434.https://doi.org/10.1016/j.resconrec.2006.10.007

Subramoniam R, Huisingh D, Chinnam RB, Subramoniam S (2013) Remanufacturing Decision-Making Framework (RDMF): research validation using the analytical hierarchical process. J Clean Prod 40:212–220. https://doi.org/10.1016/j.jclepro.2011.09.004

Taji K, Levy JK, Hartmann J et al (2005) Identifying potential repositories for radioactive waste: multiple criteria decision analysis and critical infrastructure systems. Int J Crit Infrastructures 1:404–422

Tavares G, Zsigraiová Z, Semiao V (2011) Multi-criteria GIS-based siting of an incineration plant for municipal solid waste. Waste Manag 31:1960–1972. https://doi.org/10.1016/j.wasman.2011.04.013

Thierry M, Salomon M, Van Nunen J, Van Wassenhove L (1995) Strategic issues in product recovery management. Calif Manag Rev 37:114–136

Tsai FM, Bui TD, Tseng ML, et al (2020a) A performance assessment approach for integrated solid waste management using a sustainable balanced scorecard approach. J Clean Prod 251:119740.https://doi.org/10.1016/j.jclepro.2019.119740

Tsai FM, Bui TD, Tseng ML, Wu KJ (2020b) A causal municipal solid waste management model for sustainable cities in Vietnam under uncertainty: a comparison. Resour Conserv Recycl 154:104599.https://doi.org/10.1016/j.resconrec.2019.104599

Tseng M-L, Chiu ASF (2013) Evaluating firm's green supply chain management in linguistic preferences. J Clean Prod 40:22–31

Tseng M-L, Lin YH (2009) Application of fuzzy DEMATEL to develop a cause and effect model of municipal solid waste management in Metro Manila. Environ Monit Assess 158:519

Tseng M-L, Tan K-H, Lina R-J, Gengb Y (2012) WITHDRAWN: multicriteria analysis of green supply chain management using interval-valued fuzzy TODIM

United Nations Development Program (2015) United nations program for the sustainable development goals. https://www.undp.org/content/undp/es/home/sustainable-development-goals.html

Vego G, Kučar-Dragičević S, Koprivanac N (2008) Application of multi-criteria decision-making on strategic municipal solid waste management in Dalmatia, Croatia. Waste Manag 28:2192–2201. https://doi.org/10.1016/j.wasman.2007.10.002

Wang D, Tang YT, Long G et al (2020) Future improvements on performance of an EU landfill directive driven municipal solid waste management for a city in England. Waste Manag 102:452–463. https://doi.org/10.1016/j.wasman.2019.11.009

Xi BD, Su J, Huang GH et al (2010) An integrated optimization approach and multi-criteria decision analysis for supporting the waste-management system of the City of Beijing, China. Eng Appl Artif Intell 23:620–631. https://doi.org/10.1016/j.engappai.2010.01.002

Yeh C-H, Xu Y (2013) Sustainable planning of e-waste recycling activities using fuzzy multicriteria decision making. J Clean Prod 52:194–204. https://doi.org/10.1016/j.jclepro.2013.03.003

Part II
Manufacturing and Quality

Chapter 11
Manufacturing Execution System State-Of-The-Art: Its Evolution and Dynamism Focused on Industry 4.0

Saúl Manuel Favela-Camacho, Javier Molina-Salazar, and Lázaro Rico-Pérez

Abstract The utilization of applied knowledge, specifically related with information and communication technologies (ICT) for supporting production processes, from top management to operational levels, is a trend that has been increased throughout the time. Its use seeks support in functions like organization planning, management and controlling devices related with manufacturing processes. Likewise, the manufacturing execution system (MES) has become the bridge that connects enterprise resource planning systems with operations within organizations, becoming a *Digital Twin*, helping to know virtually what is happening on productive processes in real time. Like any practical concept, an updated state-of-the-art is needed to understand its evolution, functions, importance level, and dynamism, particularly nowadays when we are immersed in Industry 4.0 paradigm. This chapter will expose an outline of the state-of-the-art about MES, its implications in the manufacturing processes, and its transformation as a dynamic productive system, because sometimes it is used as a tool for quality management. Based on the global trends, standards, and industrial behavior, we can have a baseline for the design of new and related technologies that will support the core processes within businesses.

Keywords Manufacturing execution systems · State-of-the-art · Industry 4.0 · Quality management system

11.1 Introduction

It is well known that manufacturing industry currently is going through significant changes related to its adaptation to Industry 4 (Schwab 2016). This new industrial revolution not only poses a change in the way production processes are defined, but also influences the way we live, work, and interact. For example, people are continuously connected due to the technology of mobile devices such as cell phones; robots can perform surgical interventions using artificial intelligence (Strickland

S. M. Favela-Camacho (✉) · J. Molina-Salazar · L. Rico-Pérez
Department of Industrial Engineering and Manufacturing, Universidad Autónoma de Ciudad Juárez. Av. Del Charro 450 Norte. Col. Partido Romero. Juárez, Chihuahua, México
e-mail: saul.favela@uacj.mx

J. L. Garcia Alcaraz et al. (eds.), *Trends in Industrial Engineering Applications to Manufacturing Process*, https://doi.org/10.1007/978-3-030-71579-3_11

2016); autonomous vehicles are increasingly used (Anderson et al. 2016); energy storage technologies have had great advances (Oberhofer and Meisen 2012); and there are solutions for the factories of the future (Rodriguez et al. 2019; Witzel et al. 2019).

The above are based on new concepts, such as *Internet of Things* (IOT), which refers to the connectivity of everyday devices with the Internet for decision making (Velez-Martinez 2017); *Cyber-physical* systems, where computer systems are used in networks to monitor and control physical processes (Mabkhot et al. 2018); or the so-called *Smart Factory,* where flexible cyber and physical systems solve problems in productive environments and optimize processes (Radziwon et al. 2014).

Despite the evolution of these technologies, some of these are in developing stages. However, they are expected to be implemented in everyday life, searching the integration and emergence of new technologies in both cyber and physical world.

The integration of these technologies changes the dynamics of companies and industries, since they stimulate the emergence of new business models, redefining the production, consumption, transport, and delivery systems, at the same time offering the potential for the support of product manufacturing systems.

Industry 4.0 has led companies to assimilate new and advanced technologies to turn them into competitive companies, where automation, interconnectivity and access to information are integrated into their production processes, thus turning them into smart factories, a fundamental concept for Industry 4.0 (Mantravadi and Møller 2019).

Although there have already been some information and interconnectivity systems that have been applied to production systems (such as in the case of *Enterprise Information Systems*) for almost 50 years with the emergence of the computer, many of them have been transformed as there are new advances in science, research, and technology. One of the systems that are important because it connects the planning systems to the resources available for operations is the *Manufacturing Execution System* or *MES* (McClellan 1997).

11.2 Methodology

This conceptual review is based on an exploration of the literature. The primary source of information was scientific articles, indexed journals, books, and academic texts whose theme included aspects about the history, evolution, regulations, classification, and trends of Industry 4.0 and MES. Various databases were used, including *Google Scholar, Scopus,* and *Elsevier*; the existence of texts in the library of the Institute of Engineering and Technology of the Ciudad Juarez Autonomous University was also reviewed. Most of the texts were written in English, but texts in Spanish and German were also found. Information published from 1997 to 2019 was considered in order to understand the evolution of the concept. The literature review conducted in this exploration was descriptive, where the reader is provided with an update on the useful concepts in constantly evolving areas, which is useful for teaching and for the interest of the parties involved (Guirao Goris 2015). The literature review is divided into the background of manufacturing execution systems, temporal evolution, regulatory standards and guidelines, taxonomy, its use in quality operations, and their dynamics in the context of Industry 4.0.

11.3 Literature Review

11.3.1 Background of Manufacturing Execution Systems

The beginnings of the manufacturing execution system concept can be traced from the starting of the use of computers in organizations and with the emergence of *Enterprise Information Systems* (EIS), which are information systems to support functional activities of the company, such as planning, manufacturing, sales, finance, human resources, project management, and others (Rashid et al. 2002). Its widespread use was caused by the integration and extension of business processes between the own organizational functions and those of other companies, including the dynamics of the globalized economy (Xu 2011).

As the technology evolved in terms of sophistication, the EIS were actively used in the key processes of the organizations, including in the processes related to material planning, inventory systems and forecasts. In the seventies of the last century, the first *Information Management System* (IMS) appeared, while more EIS were developed related to *Material Requirements Planning* (MRP) and *Manufacturing Resource Planning* (MRPII) in the following years. In contrast, the first commercial software packages for *Enterprise Resource Planning* (ERP) emerged in the 1990s, evolving at the beginning of the new century into advanced systems where there was a collaboration between all the functions of the organization (Møller 2004).

Meanwhile, ERP systems have included various integrated modules for the functions of logistics, purchasing, sales, marketing, human resources, and finance, where their use has been extended at the online level, where the application of entities such as the Internet, social networks, and Big Data, in order to have a broader picture of corporate performance (Romero and Vernadat 2016). However, ERP systems such as MRP and MRPII were focused on the planning and control of materials and production and were not able to have a real-time management of what happened in the manufacturing processes and production stations. That is where a new EIS entity emerged: the manufacturing execution system or MES (Kletti 2007).

11.3.2 Manufacturing Execution Systems Evolution

To assist in the execution of production, the MES connects the planning systems with the control systems and uses the information generated in the manufacturing processes to support the same processes. Like any information system, the MES has evolved over time by integrating several extensions to carry out various manufacturing activities using the advances in computer technology (Mcclellan 2001).

The MES conceptually emerged in 1992 in Boston by AMR Research Inc. as the level of execution of manufacturing activities that connects business functions and direct process control systems, providing visibility and functional control (Salazar 2009).

For example, previously some companies adapted information systems tailored to the production areas, so that they could compile information in databases, which made consolidation and maintenance difficult. As it was implemented as a dynamic system, the MES was developed with the purpose of integrating multiple points of the system and companies that developed software were able to gather production execution functions in the form of an MES software (Saenz de Ugarte et al. 2009).

However, manufacturing systems, with the aim of constantly being in continuous development, have implemented automated tools and systems that work in real time to avoid the expense and use of material resources. As progress is made over time and with technologies, extensive use of smart or intelligent factories is expected, where wireless technology and mobile information and communication technologies are the key to the future industry. Also, intelligent factories will rely on information and communication systems that allow them to have a real-time management of productive systems, where the MES would have a great role in that scheme (Sauer 2014).

Romero and Vernadat (2016) have classified the MES also based on this temporal evolution. The first generation MES (MES/I) was developed to support data collection, centralization and processing to offer support in planning, programming, traceability, quality assurance, and generation of reports in manufacturing processes by providing visualization in real time for management and operators. The second generation MES (MES/II) became a dynamic system of integral information control to control the execution of manufacturing operations using real-time data to control operations from when orders are generated until delivery of the product. The third generation MES (MES/III) is currently inserted in the context of Industry 4.0 and in the paradigms of mass customization, where its functionality is directly related in detailed manufacturing processes to have a flexible capacity, turning them in smart production systems, so that there is an integration between data and applications along with materials, tools, and machines.

11.3.3 Manufacturing Execution Systems in the Normative Setting

To contextualize the concept of what the manufacturing execution system is from a normative and standardized perspective, several models are taken into consideration. First, the vision proposed by the International Automation Society (ISA), which as an international organization, oversees establishing the guidelines and standards for the instrumentation of industrial processes. The model is exposed in two of its standards: ISA-S88 and ISA-S95.

ISA-S88. The first regulation, which was also included as a national standard in the USA at the American National Standards Institute (ANSI), defines models for batch control in industrial processes, where process relationships are specified. The core of everything is to identify what is called a "recipe" of the mode of operation, which allows the division between the structure of the plant and the processes involved. A batch creates a defined quantity of a product, which is composed of one or more components and is manufactured in one or more devices in a defined order. The production instructions are indicated in that "recipe," which contains the set of operations, which are also made up of a set of functions or phases. Each phase has its own parameters depending on the requirement in each function. The recipe management is related with production planning and scheduling and production information management, which depend from the output and input of the process management systems, where MES is one of them. Some of the functions under process management control are how control recipes must be generated, how batches must be initiated and supervised, what unit activities require coordination, and which logs and reports must be generated. All the systems included in the model proposed by ISA-S88 are related with control activities. The proposed model shows the main relationships achieved via information flow between systems. Figure 11.1 depicts the control activity model relationship of recipe management in the ISA-S88 regulation (ISA 2006; Meyer et al. 2009).

ISA-S95. In contrast, the ISA-S95 standard specifies both the terminology and the models that are used to integrate the ERP systems at the business level with the automated systems at the production level. In this regulation, there is a model of levels, which are counted from zero to four, where at the highest level (4), the systems responsible for operations and tasks related to the administration of raw materials, components, are defined. Other things that can be managed are energy resources, production plans, supply management, and optimization. Some commercial ERP systems are used in this level (SAP®, Microsoft Dynamics®, Odoo®, Oracle ERP®, Baan®, JDEdwards®, GPAO®, APS®, GPI®, etc.). On the next level (3), there are the manufacturing management systems, which are responsible for evaluating the relevant data generated in the production, inventory, personnel, raw material, replacement of parts, and energy processes. The same regulation includes functions or activities of managing information about the schedules, use, capability, definition, history, and status of the previous processes. Manufacturing execution systems are included here. In the last levels (2, 1, 0) the monitoring, supervision systems (such

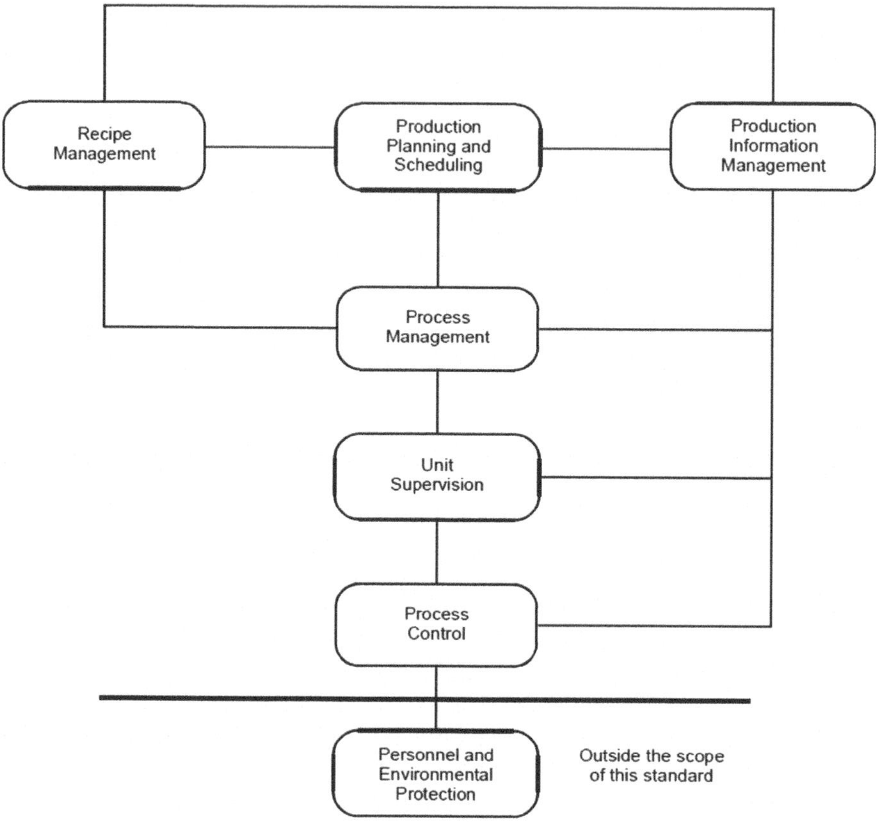

Fig. 11.1 Control recipe model based on the ISA-S88 (ISA 2006)

as process data historians, HMI/SCADA, OPC servers, relational databases such as Oracle, SQL Server, and others), control of the process, sensors, process manipulation, and current production processes are included (ISA 2005). Figure 11.2 depicts the levels of the systems according with ISA-S95.

Manufacturing operations management in level 3 is subdivided into four categories: production operations management, maintenance operations management, quality operations management, and inventory operations management. The ISA-S95 shows a manufacturing operations management model and relationships between their categories that are depicted in Fig. 11.3.

MESA. On the other hand, the guidelines proposed by the *Manufacturing Enterprise Solutions Association* (MESA) are considered to define all the operations that are included in the whole manufacturing processes that MES can manage: genealogy and traceability of the product, status and allocation of resources, performance analysis, process management, data acquisition and collection, quality management, labor management, shipments of production units, logistics, and controls (MESA 2019). The same association defines the MES as "a dynamic information system

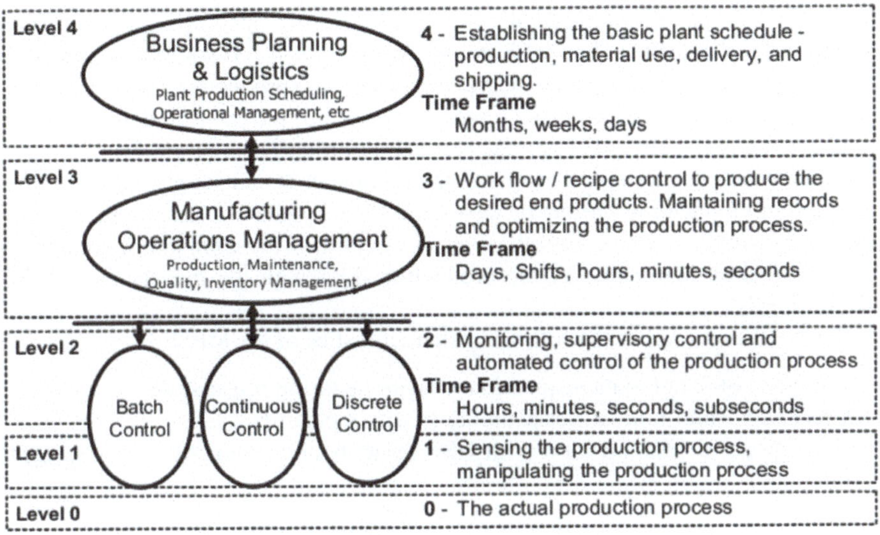

Fig. 11.2 Multi-level functional hierarchy of activities based on the ISA-S95 (ISA 2005)

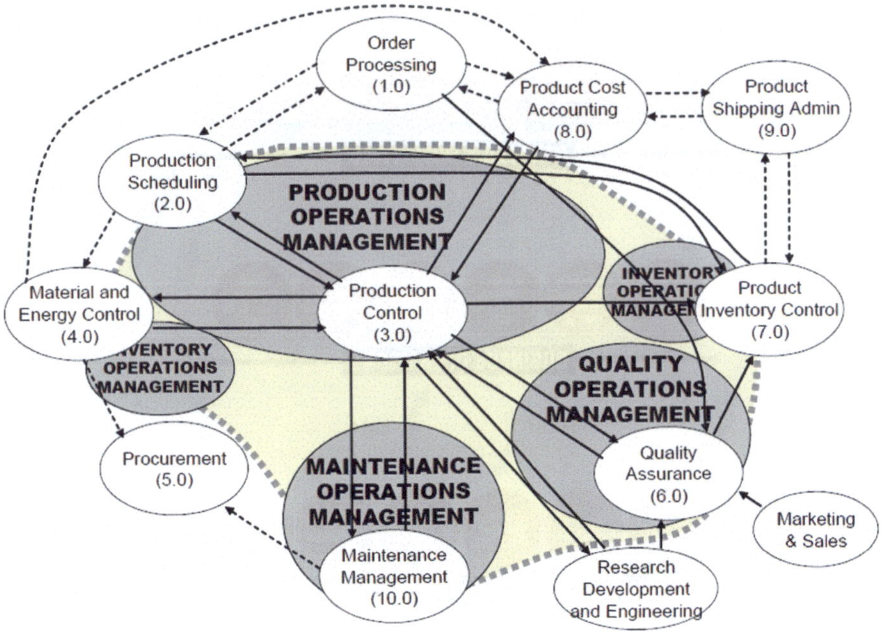

Fig. 11.3 Manufacturing operations management model based on ISA-S95 (ISA 2005)

application that handles the execution of manufacturing operations, and by using current and accurate data, the MES guides, triggers, and reports factory activities as events. The set of functions of the MES manages the production operations from the point where the order to manufacture is released to the point of delivery of the product as finished. The MES provides critical information about production activities to other related systems throughout the organization and supply chain through two-way communication. In general terms, the MES is defined as a stratum that integrates the business systems with the control systems of the companies, commonly referred to as an integration of the production floor with the management floor" (MESA 1997).

This definition implies the following characteristics of the MES:

- High level of detail (data acquisition of manufacturing processes);
- A relatively short horizon for planning (reactive planning);
- Bidirectional communication with ERP systems and production floor systems (interface).

MESA has been developed a model that shows the relationship between the company's strategic initiatives from business operations to plant operations. This model is exposed in Fig. 11.4.

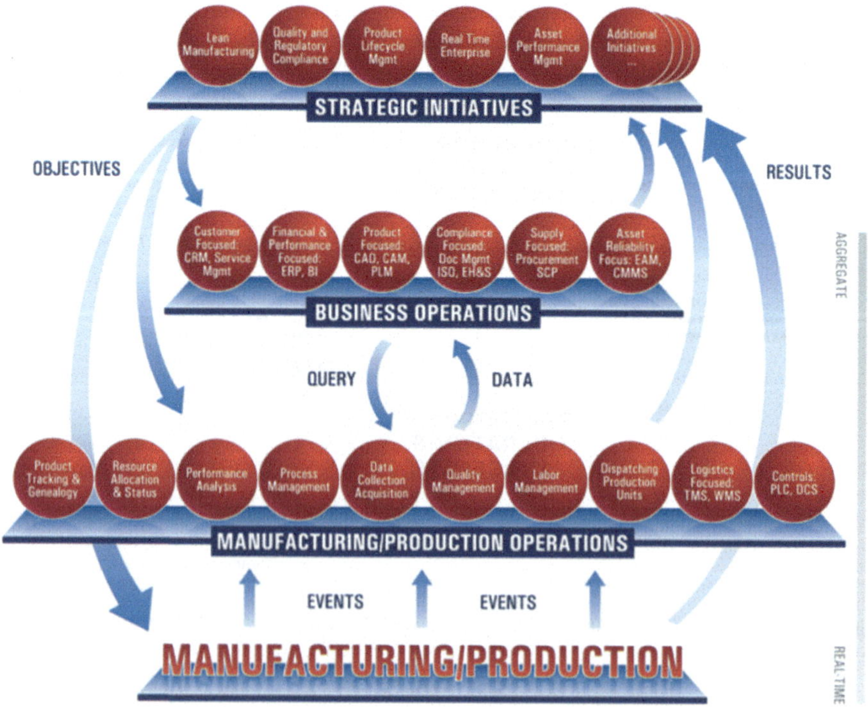

Fig. 11.4 MESA model based on business strategic initiatives (MESA 2019)

VDI. Another normative guideline is that proposed by the Association of German Engineers or *Verein Deutscher Ingenieure* (VDI), which is published in its regulation number 5600, from which information on the functions of the MES and its integration into the company operations. Based on the standards and regulations together with the findings and the development of the market, VDI developed such regulations to give the MES a fixed meaning preventing the indiscriminate use by some companies who have used any new buzzword for purposes of marketing. The goal is to maintain a perception oriented to the eyes of industrial manufacturing leaders. This guideline seeks to attract attention to distinguish the requirements and functions of the MES between different types of manufactures (Kletti 2007). The guidelines highlight the importance of the dependence and coupling that the MES has with the machines and devices that are used in the production lines. Without this union, specific MES tasks cannot be executed correctly. For this reason and due to the fact that there is heterogeneity in each of the organizations together with the connectivity between the elements of their processes, the regulations provide users with the possibility of standardizing the contents of the data that need to be exchanged between the machines and the MES (VDI 2016). In that sense, the MES needs to provide the following functions:

- Data extraction and synchronization with planning data to have the ability to respond to changing conditions in real time.
- Have a connection with the level of automation to achieve vertical integration.
- Have an exchange of data with other applications at the MES level, such as logistics applications, which could be considered as a horizontal integration.
- Carry out a comprehensive evaluation of the current data in the MES by means of data mining to auto-optimize the production system, as, for example, by allowing the MES to identify connections between quality data and process parameters and adjust the latter when necessary, and;
- Identify interrelated data from multiple volumes of either the MES or operational processes and combine them for meaningful information (Sauer 2014).

VDMA. The *Industrial Association of Mechanical Engineering* or *Verband Deutscher Maschinen und Anlagenbau* (VDMA, for its acronym in German) presented a regulation under number 66412, where the MES guidelines are expressed under the context of Industry 4.0 (VDMA 2009) which also they are associated with the standards of the *Architecture References Model for Industry 4.0* or *Referenzarchitekturmodell Industrie 4.0* (RAMI) (ZVEI 2015). These reference guidelines consist of the description of the crucial aspects of Industry 4.0, describing the levels of hierarchies (such as the one presented in the ISA-S95 standard) for information technology systems and control systems. They also provide the required elements including an orientation to trace what is necessary in each of the production sectors at national and international level to be able to define and develop Industry 4.0.

IEC. The International Electrotechnical Commission (IEC) has an international standard under number 62264 Enterprise Control System Integration, which describes the domain (level 3 according to ISA-S95) of manufacturing operations management and its activities, including the interfaces and associated transactions between the same level 3 and between level 3 and level 4. This description allows the integration between manufacturing operations and the control domain (levels 3, 2, and 1) and the domain business or business (level 4). Its objective is to increase the uniformity and consistency with the terminology of the interfaces and reduce the risk, costs, and errors associated with the implementation of those interfaces. For the application of the MES system, this regulation is for its effective application (IEC 2013). The standard IEC 62264 is considered the de facto MES standard (Karadgi 2014) and covers production, maintenance, quality, and inventory domains (Arab-Mansour et al. 2017).

NAMUR. The *Association of Users of Automated Technology in Industrial Processes* (NAMUR by its acronym in German) is a group of users who are involved in the chemical and pharmaceutical industry. The association presented the regulations under number 94 and gave some recommendations beyond those based on the definitions of ISA-S95 to make specific distinctions in the flow of information and functions. The recommendations made by NAMUR are a practical representation of information to configure and adapt the MES system in an environment oriented to industrial processes (NAMUR 2003).

11.3.4 Manufacturing Execution Systems Taxonomy

Arica and Powell (2018) proposed a taxonomy to characterize MES, which can be used for its selection or design. Taxonomy consists of two main categories of factors:

- Business and manufacturing factors.
- Technological factors.

Next, each of the main categories of factors will be analyzed.

11.3.4.1 Business and Manufacturing Factors

Business and manufacturing factors classify MES systems according to their approach, scope, and functionality. In relation to the classification of approach, there are significant differences in each industry in the way in which MES deals with manufacturing monitoring processes and control tasks, due to the manufacturing of each product consists of a series of products, data, machinery, and unique systems. For this reason, the adjustment that the MES has with a certain industry is an important factor for its implementation. That is why there are some implementations of the MES system in various companies, for example, in a car engine manufacturing company (Huang and Liu 2012); in a steel company (Govindaraju and Putra 2016); in a microcircuit production line (Zhang et al. 2017); including also in software development (Naedele et al. 2015).

In relation to the classification of scope, the MES has two associated factors: business level and operational level. The first refers to the coverage that the MES has within the value chain system that can be applied to a single plant or to several (Helo et al. 2014). The second refers to the activities or functions that the MES can operate, and as already mentioned in relation to the ISA-S95 regulations, it can support production, maintenance, quality, and inventory operations (ISA 2005). However, most MES systems are designed based on contextual requirements, they may cover part, or all of the functions suggested by ISA-S95 (Wang et al. 2010; Cottyn et al. 2011; Köksal and Tekin 2012; Menezes et al. 2018).

For the classification of functionality, there are three factors that characterize MES: functional configuration, functional integration, and structural design. As explained before, some standards had been created in order to have a basis in the functional configuration of the MES. However, some manufacturing companies have customized or configured the systems long before the standards were published. Making a comparison of what is published in the standards and what has been done, it is imperative that organizations select the appropriate system according to their integral and core functions. In relation to the functional structure, it can be categorized into centralized/hierarchical and decentralized/heterarchical structures. But decentralized systems are being widely used due to the reaction to the dynamic external changes of organizations (Trentesaux 2009).

11.3.4.2 Technological Factors

The technological factors, which are predominant when selecting the right MES for the organization, include aspects of data management and communication and support in the logical decision and the user interface. Regarding the aspect of data management and communication, the primary functionality of the MES is data collection and communication in real time. Regardless of the type of industry, companies primarily use MES as a tool to collect data. For data collection to be done correctly from the organization's devices, you need to obtain the desired data and transmit it efficiently and accurately within the MES system. For that reason, schemes have been designed for data collection using radio frequency identification devices or through advanced location algorithms (Lee et al. 2012; Yang et al. 2016). Therefore, the communication and integration of the MES with other systems is also an important factor, due to functional overlaps, information exchange, requirements, and interfaces. In this regard, improvements in storage technology are useful for the capabilities of the MES, as is the case with the use of cloud-based systems (Jiang et al. 2015), or communication protocols for the flow of material in a simulated system (Timo et al. 2016).

Like any decision support system, MES contains various types and techniques. For example, Zhong et al. (2013) proposed the use of decision support through optimization using real-time programming techniques. Grauer et al. (2010) proposed an MES that uses data mining algorithms.

A critical component for the MES is the user interface for its successful implementation and use. Studies on ERP systems suggest that one of the important keys to its implementation is the ease of use (Ratkevičius et al. 2012). This is because the users of the system would generally be the staff of the production floor and its use would be frequent.

One of the uses of the MES can also be described in terms of transparency in manufacturing processes and as a result establishes horizontal and vertical (closed) control circuits (Kletti 2007). These control circuits allow a rapid reaction to incidents in the production floor, as the information feeds the planning systems to trigger respective measures to the subsequent manufacturing stages (horizontal integration) (Schmidt et al. 2010).

11.3.5 Manufacturing Execution Systems as a Quality Tool

According with the models and standards reviewed (like IEC 62264 and MESA), one of the important tasks that MES helps in businesses is in the process of quality management by performing real-time management and analysis of product specifications and requirements in order to identify any deviation or non-conformance established on the company's quality system. Such deviations could influence in a negative way all the processes involved. So, it is important that any MES used in quality management has the characteristic of being a real-time system, which means that speed, time, and responsiveness should help the system to be able to adjust to the internal and external changes, remain alert to any special event, to react within the time constraints or deadlines (Karadgi 2014). Also, it helps the user/operator not only to know the current state of the production process, but also to identify any abnormal or critical state that compromise the quality of the product (Kulcsar et al. 2005).

In the actions of improving the quality of the manufacturing system, MES can provide information for the interested parts, so the problems can be solved in a quick manner to reduce interruptions in the production system. Managers, supervisors, or users can be notified to act or proceed properly (Engelbrecht 2007).

Some of the quality functions that can be performed by MES as part of quality management are the following: (1) statistical process control or SPC: specific data acquisition of measured values, comparing the measured values against standard values, warnings and indication of different values or that are out of tolerance, and tracking trends; (2) tracking non-conformance events: sometimes products that are not complaint with the quality system need to be registered and stored for analysis, and information about their technical aspects, manufacturing conditions, input materials, root-cause events, and traceability can be managed by a MES; (3) quality checks and inspection of incoming and finished goods: MES is capable of getting information about the characteristics and dimensions of the inputs and outputs of manufacturing processes and provides alarms if certain values are not in complaint with the quality system; (4) tool and resource management: a MES can perform inspection, measurement, and test of equipment, machines and tools used in manufacturing processes to ensure that they meet the required specifications and have the correct setups to operate under the desired conditions; (5) process data processing:

MES can be implemented in a way that can gather information from different factors and circumstances of the production process directly, verifying them against standard limits through correlations, and in events of deviations being capable to act with countermeasures (Kletti 2007); and (6) document control: MES is useful to control the records needed in the production process, like work instructions, drawings, standard operation procedures, part programs, batch records, engineering change notices, policies, standards, and regulations (Van Dyk and Van Schoor 2012).

There are some examples of MES that had been applied for quality management processes in different industries, such in a way to improve the quality of oil refinery process (Kuvykin and Petukhov 2019); as a tool of integration of process and quality control using multi-agent technology in a home appliance company production line (Cristalli et al. 2013); and to comply with the regulations of the manufacturing of pharmaceutical products (Blumenthal 2004).

11.3.6 Manufacturing Execution Systems and Industry 4.0

The MES have been essential in the performance, quality, and agility necessary for the challenges that arise in the business world of the globalized manufacturing which will continue to be in the future. However, a new generation of systems is required to meet the new challenges that Industry 4.0 has brought. Below are some of those challenges and opportunities that can be applied in MES.

Internet of Things. This term is used to refer to a global network of interconnection of intelligent objects through the Internet, making use of connectivity technologies like radio frequency identification devices, sensors/actuators, and inter-machine connection devices (Patel et al. 2016). This trend offers the potential to develop MES solutions for the entire supply and value chain instead of simply providing functionality in terms of planning and control. *Digital twins* could also be created, which is a virtual representation of a production system capable of running in different simulation disciplines and is characterized by synchronization between virtual and real systems, thanks to the data generated by the connectivity between sensors and devices, mathematical models, and real-time monitoring (Cimino et al. 2019), that can be used for simulations and scenario analysis (Golfarelli et al. 2006).

Decentralization and vertical integration. What is sought in future systems is that they have a logical decentralization, that the devices can identify themselves and that they connect in a centralized physical system, providing their position and status within that system. For this reason, the new MES trends will require decentralized connectivity with connection possibilities with other entities, for example, intelligent materials. It is also sought that the connectivity of the MES not only be at the horizontal level, but also serves as a bridge between the ERP systems of the business and the productive processes, with an orchestration of that system with those of quality, logistics, engineering, and operations (Almada-Lobo 2016).

Virtual Reality. It is the field of acquisition, analysis, and synthesis of visual information through computers that provide a virtual experience of a real situation

(Posada et al. 2015). MES can be designed to allow the interaction of the real system through the so-called augmented reality through interfaces, which can be used for training or control of production systems (Posada et al. 2018).

Industrial automation. It is the use of control systems, such as computers or robots and information technologies to handle different processes and machines in an industry to replace the human being. It is the second step beyond mechanization in the scope of industrialization. A higher level of industrial automation would result in complex levels of control and management that could be controlled and manipulated through the MES (Filipov and Vasilev 2016).

Autonomous Robots. It is an important part of artificial intelligence and robotics. They are used for the creation of complex intelligent networks, capable of learning, reasoning, and acting based on the information gathered during the industrial process (Dopico et al. 2016). Smart robotics will work as a support mechanism where machines can use real-time information from the MES to reconfigure the production system and the external supply chain.

Cybersecurity. It can be described in terms of requirements of Industry 4.0 technologies to avoid unauthorized access to production systems to prevent economic, environmental, and human damage (Drath and Horch 2014). The new MES systems should have the ability to be secure between the internal and external connectivity of companies.

Big Data. It refers to the set of data that, due to its size, is far from the capacity of databases and software tools to capture, store, manage, and analyze (Yin and Kaynak 2015). MES can be developed to have the ability to perform those functions that cannot be done by typical and common systems (Li et al. 2019).

Mobile technology. Communication technologies have evolved to meet industry demand by providing real-time information on production, predictive maintenance, and automation. The communication technology that is required by the needs of the industry is mobile. Some of them, such as GSM, LTE, or mobile applications, have been the trends used in today's factories (Meredith and Pope 2018). The new MES must have the ability to interact with mobile technology (Meyer 2012).

Cloud Computing. As a model to have access to different computing resources (networks, servers, storage, applications, and services) that can be provided quickly without efforts in their administration and reducing costs, MES systems can be configured to have the ability to use those computing resources. You can even combine cloud storage to analyze a large amount of data (Vitliemov 2016; Atobishi et al. 2019).

Smart MES. There is a proposal for a smart MES that can be used in small and medium-sized manufacturing companies, where features of the Internet of Things are applied for connectivity of manufacturing machines, through the use radio frequency identification tags together with sensors embedded in critical components, tools, and controllers of the machines, which together generate data that can be sent in real time for analysis to the interested parties (Menezes et al. 2018). Also, it can be applied under sustainability trends (Larreina et al. 2013).

System integration. Known also as intelligent or cyber-physical systems, which cover the hardware and software, as well as integrated physical components that interact closely with each other, where physically mechanical devices interact with IT systems, hardware, software, digital, and electronical components (Chukalov 2017). A cyber-physical system architecture MES had been applied in a shop floor (Liu and Jiang 2016).

Additive Manufacturing. Defined as "the process of joining materials to make objects from 3D model data, usually layer upon layer, as opposed to subtractive manufacturing methodologies, such as traditional machining" (ASTM International 2012), there are technologies of the use of design for additive manufacturing, which uses MES architecture, as information framework able to real-time acquire, analyze, and synthesize process and product data (D'Antonio et al. 2017).

11.4 Conclusions and Industrial Implications

This work proposed a review on MES, based on an analysis of the literature. Within the functions of the business and manufacturing processes, MES has played a vital role in companies in achieving its objectives. However, as technology and knowledge evolve, it is also seen that the MES has evolved, even within the Industry 4.0 paradigm. This evolution is not only palpable in the way of the application of technology, but also in its interpretation within the normative and standardized context, as it was shown when presenting some of the regulations that are used for its implementation.

The fundamental opportunity for Industry 4.0 is to create new information value in offering new services to customers and increasing efficiency in internal operations. Through the new technologies of data collection, analysis and communication functions through the value chain, the MES would serve as a platform to implement Industry 4.0 technologies.

References

Almada-Lobo F (2016) The industry 4.0 revolution and the future of manufacturing execution systems (MES). J Innov Manag 3:16–21. https://doi.org/10.24840/2183-0606_003.004_0003

Anderson J, Kalra N, Stanley K, et al (2016) Autonomous vehicle technology: a guide for policymakers. RAND Corporation

Arab-Mansour I, Millet PA, Botta-Genoulaz V (2017) A business repository enrichment process: a case study for manufacturing execution systems. Comput Ind 89:13–22. https://doi.org/10.1016/j.compind.2017.03.006

Arica E, Powell DJ (2018) Status and future of manufacturing execution systems. In: IEEE international conference on industrial engineering and engineering management 2017 Dec, 2000–2004. https://doi.org/10.1109/IEEM.2017.8290242

ASTM International (2012) ASTM: F2792–12a. Standard technology for additive manufacturing technologies, West Conschohocken

Atobishi T, Gábor Szalay Z, Bayraktar S (2019) Cloud computing and big data in the context of industry 4.0 : opportunities and challenges. https://doi.org/10.20472/iac.2018.035.004

Blumenthal R (2004) Manufacturing execution systems to optimize the pharmaceutical supply chain. Pharm Ind 66:1414–1424

Chukalov K (2017) Horizontal and vertical integration, as a requirement for cyber-physical systems in the context of industry 4.0. Int Sci J "Industry 40" 2:155–157

Cimino C, Negri E, Fumagalli L (2019) Review of digital twin applications in manufacturing. Comput Ind 113:103130. https://doi.org/10.1016/j.compind.2019.103130

Cottyn J, Van Landeghem H, Stockman K, Derammelaere S (2011) A method to align a manufacturing execution system with Lean objectives. Int J Prod Res 49:4397–4413. https://doi.org/10.1080/00207543.2010.548409

Cristalli C, Foehr M, Jager T et al (2013) Integration of process and quality control using multi-agent technology. IEEE Int Symp Ind Electron 374–379.https://doi.org/10.1109/ISIE.2013.6563737

D'Antonio G, Bedolla JS, Chiabert P (2017) A novel methodology to integrate manufacturing execution systems with the lean manufacturing approach. Procedia Manuf 11.https://doi.org/10.1016/j.promfg.2017.07.372

Dopico M, Gomez A, De la Fuente D, et al (2016) A vision of industry 4.0 from an artificial intelligence point of view. In: Proc 2016 international conference on artificial intellgence ICAI 2016—WORLDCOMP 2016, pp 407–413

Drath R, Horch A (2014) Industrie 4.0: hit or hype? IEEE Ind Electron Mag 8:56–58. https://doi.org/10.1109/MIE.2014.2312079

Engelbrecht GJ (2007) Successfully implementing a manufacturing execution systems (MES) solution

Filipov V, Vasilev P (2016) Manufacturing operations management—the smart backbone of industry 4 . 0. Int Sci Conf "Inudstry 40" 27/213:8–13

Golfarelli M, Rizzi S, Proli A (2006) Designing what-if analysis: Towards a methodology. Dol Proc ACM Int Work Data Warehous Ol 51–58.https://doi.org/10.1145/1183512.1183523

Govindaraju R, Putra K (2016) A methodology for manufacturing execution systems (MES) implementation. IOP Conf Ser Mater Sci Eng 114.https://doi.org/10.1088/1757-899X/114/1/012094

Grauer M, Karadgi S, Metz D, Schäfer W (2010) An approach for real-time control of enterprise processes in manufacturing using a rule-based system. MKWI 2010—Multikonferenz Wirtschaftsinformatik 2010, pp 301–302

Guirao Goris SJA (2015) Utilidad y tipos de revisión de literatura. Ene 9:. https://doi.org/10.4321/S1988-348X2015000200002

Helo P, Suorsa M, Hao Y, Anussornnitisarn P (2014) Toward a cloud-based manufacturing execution system for distributed manufacturing. Comput Ind 65:646–656. https://doi.org/10.1016/j.compind.2014.01.015

Huang D, Liu M (2012) Design and implementation of manufacturing execution system (MES) for automobile main gear reducer assembly line. Adv Mater Res 468–471:111–114.https://doi.org/10.4028/www.scientific.net/AMR.468-471.111

IEC (2013) 62264 Enterprise-control system integration—part 1: models and terminology. Geneva

ISA (2005) Enterprise control system integration part 3 : activity models of manufacturing operations management. International Society of Automation

ISA (2006) Batch control, part 1: models and terminology, ISA-88.01–1995 (R2006)

Jiang P, Zhang C, Leng J, Zhang J (2015) Implementing a WebAPP-based software framework for manufacturing execution systems. IFAC-PapersOnLine 28:388–393. https://doi.org/10.1016/j.ifacol.2015.06.112

Karadgi S (2014) A reference architecture for real-time performance measurement. Springer, Berlin

Kletti J (2007) Manufacturing execution system—MES. Springer-Verlag, Berlin

Köksal A, Tekin E (2012) Manufacturing execution through e-FACTORY system. Proc CIRP 3:591–596. https://doi.org/10.1016/j.procir.2012.07.101

Kulcsar G, Hornyak O, Erdelyi F (2005) Shop floor control decision supporting and MES functions in customized mass production. JOUR

Kuvykin VI, Petukhov MY (2019) Improving the quality of process models in oil refinery information systems. Int J Qual Res 13:539–552. https://doi.org/10.24874/IJQR13.03-03
Larreina J, Gontarz A, Giannoulis C et al (2013) Smart manufacturing execution system (SMES): the possibilities of evaluating the sustainability of a production process. In: The 11th global conference on sustainable manufacturing, pp 517–522
Lee SW, Nam SJ, Lee JK (2012) Real-time data acquisition system and HMI for MES. J Mech Sci Technol 26:2381–2388. https://doi.org/10.1007/s12206-012-0615-0
Li P, Jiang P, Liu J (2019) Mini-MES: A microservices-based apps system for data interconnecting and production controlling in decentralized manufacturing. Appl Sci 9. https://doi.org/10.3390/app9183675
Liu C, Jiang P (2016) A cyber-physical system architecture in shop floor for intelligent manufacturing. Proc CIRP 56:372–377. https://doi.org/10.1016/j.procir.2016.10.059
Mabkhot M, Al-Ahmari A, Salah B, Alkhalefah H (2018) Requirements of the smart factory system: a survey and perspective. Machines 6:23. https://doi.org/10.3390/machines6020023
Mantravadi S, Møller C (2019) An overview of next-generation manufacturing execution systems: how important is MES for industry 4.0? In: Procedia manufacturing
McClellan M (1997) Applying manufacturing execution systems. Lucie Press, Boca Raton, FL, St
Mcclellan M (2001) Introduction to manufacturing execution systems Baltimore, Maryland. MES Conf Expo 1–12
Menezes S, Creado S, Zhong RY (2018) Smart manufacturing execution systems for small and medium-sized enterprises. In: Procedia CIRP
Meredith JM, Pope M (2018) How mobile IoT is changing the industrial landscape
MESA (1997) MES explained : a high level vision. Pittsburgh
MESA (2019) MESA model. In: Manufacturing enterprise solutions association. https://www.mesa.org/en/modelstrategicinitiatives/MESAModel.asp. Accessed 17 Nov 2019
Meyer H (2012) OS independent mobile solutions for manufacturing execution systems. In: WMSCI 2012—16th world multi-conference systemics, cybernetics and informatics, Proc 1, pp 168–171
Meyer H, Fuchs F, Thiel K (2009) Manufacturing execution systems: optimal design, planning , and deployment. The McGraw-Hill Companies, Inc.
Møller C (2004) ERP II-Next-generation extended enterprise resource planning. In: Organizing for networked information technologies: readings in process integration and transformation
Naedele M, Chen H, Kazman R et al (2015) The journal of systems and software manufacturing execution systems: a vision for managing software development. J Syst Softw 101:59–68. https://doi.org/10.1016/j.jss.2014.11.015
NAMUR (2003) NA 94 functions and examples of operations control level solutions
Oberhofer A, Meisen P (2012) Energy storage technologies. https://www.geni.org/globalenergy/research/energy-storage-technologies/Energy-Storage-Technologies.pdf
Patel KK, Patel SM, Scholar PG (2016) Internet of Things-IOT: definition, characteristics, architecture, enabling technologies, application and future challenges. Int J Eng Sci Comput 6:1–10. https://doi.org/10.4010/2016.1482
Posada J, Toro C, Barandiaran I et al (2015) Visual computing as a key enabling technology for industrie 4.0 and industrial internet. IEEE Comput Graph Appl 35:26–40. https://doi.org/10.1109/MCG.2015.45
Posada J, Zorrilla M, Dominguez A et al (2018) Graphics and media technologies for operators in industry 4.0. IEEE Comput Graph Appl 38:119–132. https://doi.org/10.1109/MCG.2018.053491736
Radziwon A, Bilberg A, Bogers M, Madsen ES (2014) The smart factory: exploring adaptive and flexible manufacturing solutions. Proc Eng 69:1184–1190. https://doi.org/10.1016/j.proeng.2014.03.108
Rashid M, Hossain L, Patrick J (2002) The evolution of ERP systems: a historical perspective. Enterp Resour Plan Solut Manag 1–16.https://doi.org/10.4018/978-1-931777-06-3
Ratkevičius D, Ratkevičius Č, Skyrius R (2012) Erp selection criteria: theoretical and practical views. Ekonomika 91:97–116. https://doi.org/10.15388/ekon.2012.0.893

Rodriguez I, Mogensen RS, Khatib EJ et al (2019) On the design of a wireless MES solution for the factories of the future, pp 1–6. https://doi.org/10.1109/giots.2019.8766419
Romero D, Vernadat F (2016) Enterprise information systems state of the art : past, present and future trends.https://doi.org/10.1016/j.compind.2016.03.001
Saenz de Ugarte B, Artiba A, Pellerin R (2009) Production planning and control: the management of operations manufacturing execution system—a literature review. Prod Plan Control Manag Oper 20:525–539. https://doi.org/10.1080/09537280902938613
Salazar V (2009) Análisis de la Integración de los Sistemas MES—ERP en industrias de manufactura. In: Energy and technology for the Americas: education, innovation, technology and practice. Latin American and Caribbean Conference for Engineering and Technology, San Cristobal, Venezuela, pp 1–6
Sauer O (2014) Information technology for the factory of the future—state of the art and need for action. Proc CIRP 25:293–296. https://doi.org/10.1016/j.procir.2014.10.041
Schmidt A, Otto B, Kussmaul A (2010) MES services in the automotive industry. St, Gallen
Schwab K (2016) The fourth industrial revolution. Geneva
Strickland E (2016) Autonomous robots surgeon bests humans in world first. In: IEEE Spectr. https://spectrum.ieee.org/the-human-os/robotics/medical-robots/autonomous-robot-surgeon-bests-human-surgeons-in-world-first
Timo IF, Mónica RL, Christian B et al (2016) Agent-based communication to map and exchange shop floor data between MES and material flow simulation based on the open standard CMSD. IFAC-PapersOnLine 49:1526–1531. https://doi.org/10.1016/j.ifacol.2016.07.796
Trentesaux D (2009) Distributed control of production systems. Eng Appl Artif Intell 22:971–978. https://doi.org/10.1016/j.engappai.2009.05.001
Van Dyk L, Van Schoor CD (2012) Realising the potential of manufacturing execution systems. South African J Ind Eng 12:101–118. https://doi.org/10.7166/12-1-358
VDI (2016) 5600 part 1: manufacturing execution systems. Beuth-Verlag
VDMA (2009) 66412—Manufacturing execution systems kennzahlen. Berlin
Velez-Martinez C (2017) Internet de las Cosas. Gac Inst Ing 1:20–21
Vitliemov P (2016) From cloud manufacturing to cloud-based manufacturing execution system. In: International scientific conference "Industry 4.0." pp 7–10
Wang M, Dai Q, Zhang X, et al (2010) A RFID-enabled MES for real-time pharmaceutical manufacturing supervision. In: Proceedings of 2010 IEEE international conference on RFID-technology and applications, RFID-TA 2010, pp 49–53. https://doi.org/10.1109/RFID-TA.2010.5529858
Witzel O, Wilm S, Karimanzira D, Baganz D (2019) Controlling and regulation of integrated aquaponic production systems—an approach for a management execution system (MES). Inf Process Agric 6:326–334. https://doi.org/10.1016/j.inpa.2019.03.007
Xu LD (2011) Enterprise systems: State-of-the-art and future trends. IEEE Trans Ind Informatics 7:630–640. https://doi.org/10.1109/TII.2011.2167156
Yang Z, Zhang P, Chen L (2016) RFID-enabled indoor positioning method for a real-time manufacturing execution system using OS-ELM. Neurocomputing 174:121–133. https://doi.org/10.1016/j.neucom.2015.05.120
Yin S, Kaynak O (2015) Big data for modern industry: challenges and trends. Proc IEEE 103:143–146. https://doi.org/10.1109/JPROC.2015.2388958
Zhang J, Jiang Y, Jiang W (2017) Quality management of manufacturing process based on manufacturing execution system, 020025. https://doi.org/10.1063/1.4979757
Zhong RY, Dai QY, Qu T et al (2013) RFID-enabled real-time manufacturing execution system for mass-customization production. Robot Comput Integr Manuf 29:283–292. https://doi.org/10.1016/j.rcim.2012.08.001
ZVEI (2015) Reference architecture model industrie 4.0 (RAMI4.0)

Chapter 12
Enablers and Barriers for a Quality Management System Implementation in Mexico: An Exploratory Analysis

Marcos Alberto Sanchez-Lizarraga, Jorge Limon-Romero, Diego Tlapa, Yolanda Baez-Lopez, Lizbeth Puerta-Sierra, and Marco Maciel-Monteon

Abstract Organizations need to find the best practices in their operations to develop products or services that meet customer specifications; therefore, implementing a quality management system such as the ISO 9001 standard could provide these organizations with a framework to develop the continuous improvement of their work systems to meet or even exceed customer requirements while accomplishing the organization objectives. This chapter presents the ISO 9001 standard and the analysis performed to identify the critical success factors (CSFs) that enable its implementation based on a surveying instrument previously designed taking as reference the seven quality management principles (QMPs) stated by the ISO 9001:2015. To identify these enablers, an exploratory factor analysis (EFA) was carried out based on information collected from organizations in the manufacturing sector in Mexico. Results showed that engagement of people, customer satisfaction and decision making as well as leadership are the main factors that integrate the QMPs to facilitate the adoption of the standard in the manufacturing sector. Furthermore, the main barriers and the benefits related to the implementation of the standard are presented and discussed.

Keywords Critical success factors · ISO 9001 standard · Manufacturing sector quality management system · Survey

---

M. A. Sanchez-Lizarraga · J. Limon-Romero (✉) · D. Tlapa · Y. Baez-Lopez
Facultad de Ingeniería, Arquitectura y Diseño, Universidad Autónoma de Baja California, Carretera Transpeninsular Ensenada-Tijuana #3917, Colonia Playitas, 22860 Ensenada, Baja California, Mexico
e-mail: jorge.limon@uabc.edu.mx

L. Puerta-Sierra
Facultad de Economía y Negocios, Universidad Anáhuac México, Av. Universidad Anáhuac 46, 52786 Colonia Lomas Anáhuac Huixquilucan, Estado de México, Mexico

M. Maciel-Monteon
Unidad Académica San Luis, Universidad Estatal de Sonora, Carretera Sonoyta-San Luis Rio Colorado Km, 6.5, Parque Industrial, 83500 Sonora, Mexico

J. L. Garcia Alcaraz et al. (eds.), *Trends in Industrial Engineering Applications to Manufacturing Process*, https://doi.org/10.1007/978-3-030-71579-3_12

12.1 Introduction

The fulfillment of the characteristics that provide quality to the products or services is one of the main features that organizations pursuit into a level that satisfies the customer's needs and requirements. To accomplish these features, a quality management system (QMS) has been a strategy implemented by many organizations to meet them, and at the same time achieve customer satisfaction. Hallberg et al. (2018) conceptualize the QMS as systems certified to ensure that the company has processes in place for various routines and theorize that the QMS could delimit the selection of qualified suppliers, thereby demonstrating an upstream pressure for such QMS applied. Also, Garza-Reyes et al. (2015) described that the use of a QMS is essential to support the performance of organizations, provides a range of benefits for improvement, and therefore, positively affects the organization. Likewise, Gargasas et al. (2019) describe that QMS involves aspects such as occupational safety and environmental protection, focus on customer and employee health and observance of the needs of society. Finally, Bravi et al. (2019) state that quality is a philosophical perspective; therefore, the implementation of a QMS and its subsequent certification is a voluntary process, supported by the motivations, objectives and policies of the organization.

Many QMS are related to quality methodologies and strategies such the Malcolm Baldrige National Quality Award (MBNQA), European Foundation for Quality Management (EFQM), Six Sigma (SS), Total Quality Management (TQM), Lean Manufacturing (LM) and the ISO 9001 standard among others. In this chapter, the ISO 9001 standard is the main focus to describe it as a QMS. The chapter will be divided into six main topics: (1) history, revisions, definition and structure of the standard; (2) the ISO 9001 standard framework; (3) the critical success factor for its implementation; (4) a development of a measurement instrument to recognize the critical success factors in the implementation in the manufacturing sector in Mexico; (5) discussion; and (6) conclusion.

12.1.1 A Brief History of the International Standardization Organization and ISO 9001 Standard

The International Organization for Standardization (ISO) was created from the union of two organizations—one was the ISA (International Federation of the National Standardizing Associations), established in New York in 1926, and administered from Switzerland, and the other was the United Nations Standards Coordinating Committee (UNSCC), established in 1944, and administered in London. The conference of national standardizing organizations which established the ISO took place in London 1946. The ISO is an independent, non-governmental international organization established in Geneva, Switzerland, with a membership of 164 national standards bodies (ISO 1997). The ISO is a network of national standards bodies

(NSB) which each member represents ISO in its country. ISO standards are developed by groups of experts which form technical committees (TC). Each TC deals with a different subject and is made up of representatives of industry, non-governmental organizations (NGOs), governments and other stakeholders, who are put forward by ISO's members (Hoyle 2018). But ISO is much more; it is the key component in a worldwide standardization system, and it is accepted as such, because it respects the competence and autonomy of all the other elements of the system, while being careful to keep in view the necessary synergies and compatibilities (ISO 1997).

For many years, ISO was participating in discussions dealing with testing a quality control, but not until 1978 the ISO start its own program by "Spike" Spickernell, Director General of the British Standards Institution, and stimulated a lot of discussion before the title and scope for a new technical committee were agreed upon. The result was a Technical Committee 176 (TC 176) for quality management and quality assurance which would eventually produce one of most spectacular standards: the ISO 9000 series (ISO 1997). The ISO 9000 family addresses various aspects of quality management and contains some of ISO's best known standards providing guidance and tools for organizations who want to ensure that their products and services consistently meet customer's requirements, and that quality is consistently improved (ISO 2015).

The ISO Survey of Certifications is an annual survey of the number of valid certificates to ISO management system standards worldwide such as ISO 9001 and ISO 14001. The latest results of the survey are from 2018 which show an estimation of the number of valid certificates. According to the ISO Survey (2018), the popularity of the ISO 9001 standard can be measured with the number of organizations adopting the standard worldwide; the first 10 countries with more certified organizations are listed in Table 12.1, where China, Germany, Japan, Italy and UK and Northern Ireland are the top five countries with the most certified organizations.

Likewise, the ISO Survey (2017) shows the total number of organizations certified worldwide. Figure 12.1 shows noticeable variability in the worldwide growth from year 2007 to 2017 of the ISO 9001 standard, where year 2016 presents the highest number of certifications obtained with 1,105,937 organizations certified; this could have happened due to the transition from the ISO 9001:2008 to 9001:2015 in organizations. After almost 25 years and many versions, the expectation caused by the ISO 9001 standard endorses the standard as a reference for organizations that wish to implement an efficient and effective QMS.

12.2 The ISO 9001 Standard

The management of the standard in Mexico comes directly from the General Direction of Standardization (DGN, for its acronym in Spanish) that is an internal institution of the Economic Ministry responsible to arrange the Mexican Norms (NOM, for its acronym in Spanish) to coordinate the standardization and conformity assessment systems to promote the competitiveness of industry and commerce in the national

Table 12.1 Countries with the highest number of certified organizations with ISO 9001:2015 standard

Rank	Country	Total
1	China	160587
2	Germany	30312
3	Japan	29398
4	Italy	22982
5	UK of Great Britain and Northern Ireland	17433
7	India	15974
8	Spain	12992
9	USA	11047
10	France	7704
11	Australia	6101
12	Czech Republic	5953
13	Brazil	5591
14	Thailand	5192
15	Switzerland	5075
16	Israel	4453
17	Netherlands	4451
18	Taiwan, Province of China	4444
19	Korea (Republic of)	4360
20	Malaysia	4283
21	Poland	3794
22	Argentina	3560
23	Indonesia	3075
24	Hungary	3059
25	Portugal	2974
26	Bulgaria	2832
27	Canada	2641
28	Colombia	2603
29	Mexico	2577
30	Romania	2545

Source ISO survey (2018)

and international scope. The DGN with The Mexican Institute of Standardization and quality assurance standard certification (IMNC, for its acronym in Spanish) created the NOM series NMX-CC-9001-IMNC-2015 which is in accordance with the ISO 9001:2015 standard in every guideline stablished by the ISO. Finally, the Mexican Accreditation Body (EMA, for its acronym in Spanish) is a private institution with governmental permission to be responsible for the accreditation of the Conformity Assessment Bodies to evaluate and certificated the organizations once accomplishing the requirements for the ISO 9001 standard/NMX-CC-9001-2015.

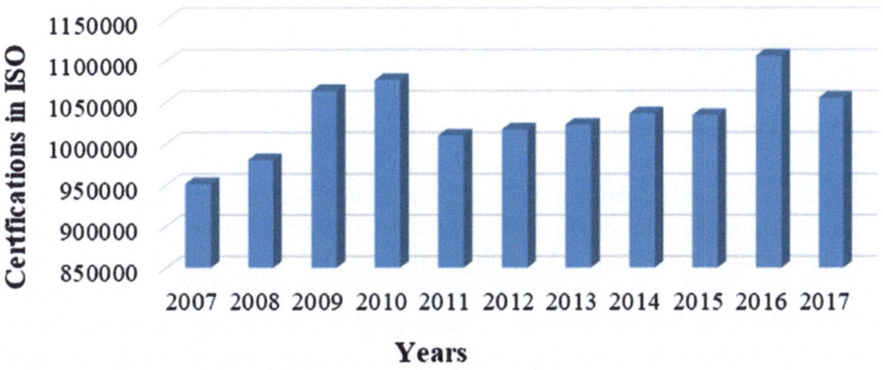

Fig. 12.1 Organizations with ISO 9001 certification worldwide over time. *Source* ISO survey (2017)

12.2.1 History and Follow up of the ISO 9001 Standard

12.2.1.1 ISO 9001:1987

Although the ISO 9000:1987 is not recognized as the first edition of the standard, it was the first to include the guidelines, requirements and terminology of a QMS mainly based on BS-5750 series from the British Ministry of Defense (Harrigan 1993; Lopez 2016; Wilson and Campbell 2016). The emphasis of the application of the ISO 9000 series on this version was between customers and suppliers, taking into account that the BS-5750 was emphasized between governments and manufactures (Lopez 2016). The series embraces the following standard in its structure:

- ISO 8402:1986 Quality—Vocabulary.
- ISO 9000. Quality Management and Quality Assurance Standard—Guidelines for Selection and Use.
- ISO 9000-2. Quality Management and Quality Assurance Standards—Part 2: Generic Guidelines for the Application of ISO 9001, ISO 9002 and ISO 9003.
- ISO 9000-3. Quality Management and Quality Assurance Standards—Part 3: Guidelines for the Application of ISO 9001 to the Development, Supply and Maintenance of Software.
- ISO 9001. Quality Systems—Model for Quality Assurance in Design/Development, Production, Installation and Servicing.
- ISO 9002. Quality Systems—Model for Quality Assurance on Production, Installation and Servicing.
- ISO 9003. Quality System—Model for Quality Assurance in Final Inspection and Test.
- ISO 9004:1987. Quality Management and Quality System Elements—Guidelines.
- ISO 10011-1. Guidelines for Auditing Quality Systems—Part 1: Auditing.

- ISO 10011-1. Guidelines for Auditing Quality Systems—Part 2: Qualification criteria for quality systems auditors.

12.2.1.2 ISO 9001:1994

ISO 9001:1994 standard emphasized quality assurance via preventive actions, instead of just checking the final product and continued to require evidence of compliance with documented procedures. As with the first version, the disadvantage was that companies tended to implement its requirements by creating shelf-loads of procedure manuals, and becoming burdened with an ISO bureaucracy (Wolniak 2018); however, at this time, the success of the standard was unstoppable in many countries. Also, the ISO 9001 standard became known as "reference standard" for other certifiable management standards such as the ISO 14001 (for environmental management) or OHSAS 18001 (for occupational safety and health management) among others (Lopez 2016). In 1997, ISO carried out a survey to more than thousand users of the standard to know their opinions and recollect proposals for the improvement of the ISO 9001:2000 standard. The ISO 9001:1994 contained following standards:

- ISO 8402:1994 Quality management and quality assurance—Vocabulary.
- ISO 9001:1994 Quality systems—Model for quality assurance in design, development, production, installation and servicing.
- ISO 9002:1994. Quality systems—Model for quality assurance in production, installation and servicing.
- ISO 9003:1994. Quality systems—Model for quality assurance in final inspection and test.
- ISO 9004-1:1994 Quality management and quality system elements—Part 1: Guidelines.
- ISO 9004-2:1994. Quality management and quality system elements—Part 2: Guidelines for services.
- ISO 9004-3:1993 Quality management and quality system elements—Part 3: Guidelines for processed materials.
- ISO 9004-4:1993 Quality management and quality system elements—Part 4: Guidelines for quality improvement.

12.2.1.3 ISO 9001:2000

In this version, many of the standards in the ISO 9000 series were discontinued in the publication of the new edition of the basic standards. The original 9002 and 9003 standards were merged with ISO 9001, and ISO 8402 was merged and designated as ISO 9000. The standards ISO 9004-1, 9004-2 and ISO 9004-3 merged and designated only as ISO 9004. ISO 10011 on auditing were merged with auditing standards for environmental management designated as ISO 19011 (Wolniak 2018). Another of the most important novelties of this edition is the adoption of the eight quality management principles (QMPs) that continue as a frame of reference for the requirements of the standard and in the unquestionable pillars of any QMS. Figure 12.2 shows the eight quality principles established by the ISO in the ISO 9000:2000 series (ISO 2000). Other novel change in the standard was the process approach applied to develop structural processes using the PDCA cycle created by Deming relating all areas of the organization, since the identification of processes is a requirement of the standard.

Minor changes in this version were to simplify documentary load, simplify the contents to be more understandable, amplify the standard approach to apply to services companies and in the organization management, include the prevention approach in the problem correction and include the continuous improvement approach (Lopez 2016). With the changes described, it can be seen that the standard experienced a remarkable transformation in the aspect of quality management, not only in fulfilling the client's requirements, but also in integrating audit functions to the

Fig. 12.2 Eight quality management principles in the ISO 9001:2000 standard

organization's management to continuously improve the QMS. The ISO 9001:2000 contained the following standards:

- ISO 9000:2000. Quality Management Systems—Fundamentals and Vocabulary
- ISO 9001:2000. Quality Management Systems—Requirements
- ISO 9004:2000. Quality Management Systems—Guidelines for Performance Improvements
- ISO 19011: Guidelines for Auditing Management Systems

12.2.1.4 ISO 9001:2008

The survey for the improvement of the standard did not provide great changes needed for the 2008 revision. Changes were mainly made in the redaction of some requirements, just some descriptive notes and modifications in the text for a greater clarity. Consequently, transition from the ISO 9001:2000 to the 2008 version was simple to carry out because no structural changes were necessary either in the requirements of the quality management.

12.2.1.5 ISO 9001:2015

The draft of the new version of the standard started with the compilation of the improvement needs given by expert and certification entities. The ISO 9001:2015 was finally recognized as an international standard in September 2015. Some strategic goals on which the TC 176 based its work for the improvement of the standard were:

- **Clarity in manual writing**: Without losing sight of the importance of the description of the requirements, there is a clear effort to simplify and clarify the wording.
- **Applicability of the standard**: The purpose is that the standard is applicable to all type of companies. The content of some requirements has been made more flexible, and also, the concept of exclusion disappears but is replaced by non-applicability.
- **Process approach**: Even though the process approach comes from the ISO 9001:2000 standard, this version is reinforced and distinct in the QMS determining a right management of the processes.
- **Risk based thinking**: This approach reinforces the preventive nature of the QMS that was implicitly in the previous versions of the standard. Now preventive actions are eliminated and replaced by a more effective tool such as risk analysis. The ISO 9001:2015 does not include any requirement about the risk management methodology used by the organization.
- **Flexibility in documentation**: This version of the standard focus in the elimination of the conceptual documentation, procedure, instruction or register and its replacement with the term "documented information." Also, the obligation to

Table 12.2 Major differences in terminology between ISO 9001:2008 and ISO 9001:2015

ISO 9001:2008	ISO 9001:2015
Products	Products and services
Exclusions	Not used
Management representative	Not used (similar responsibilities and authorities are assigned but no requirement for a single management representative)
Documentation, quality manual, documented procedure, records	Documented information
Work environment	Environment for the operation of processes
Monitoring and measuring equipment	Monitoring and measuring resources
Purchased product	Externally provided products and services
Supplier	External provider

Source (ISO 2015)

maintain certain documentation such as the manual of quality or the old documented procedures manual disappears, leaving to the organization the criterion on the type and size of its documentary system.

In the ISO 9001:2015, quality management requirements manually displays a differentiation matrix against ISO 9001:2008 that can be shown in Table 12.2.

12.2.2 Standardization Process of ISO 9001 Standard

All ISO standards are reviewed by their TC every five years in order to analyze if the standard needs any modification or update to maintain its validity and relevance; in this case, the TC 176 is responsible for carrying out this activity (Lopez 2016). The revision for the standardization process of the ISO 9001 standard is divided into six phases according to Lopez (2016) and Hoyle (2018):

1. *Systematic review (NWIP)*: In the New Work Item Proposal (NWIP) phase, the TC performs the evaluation of the standard to determine if it requires a change or modification in any of the requirements. The TC considers the recommendations, observations and proposals from all the participants involved in the development of the standard such as organizations and certification houses.
2. *Design specification (WD)*: With the proposed content accepted, the expert group begins a discussion to prepare a working draft (WD). Each section of the draft usually is commissioned to a specific group or subcommittees (SC) including opinions of experts from around the world of recognized prestige.
3. *Committee draft (CD)*: The WD is shared with the TC and with ISO CS (Central Secretariat), the TC comment and vote about the content of every draft made

until reaching a concession, then a final committee draft (CD) is made and distributed to the TC as an internal draft.

4. *Draft international standard (DIS)*: The final CD is submitted to vote by the TC to be registered as a Draft International Standard (DIS). The DIS is shared with all ISO national members, who have three months to comment. The TC as well the SC should notice all comment from the members of the ISO and review the DIS; when this happens, another review period of a new CD is established.
5. *Final projects of international standard (FDIS)*: In this approval stage, the Final Draft of the International Standard (FDIS) is review again for the ISO national members for two months for analysis and vote. The standard is approved by a two-thirds majority of the participation members which is in favor and not more than one-quarter of the total votes are negative. Only editorial corrections are made to the final text.
6. *International Standard*: Once the FDIS is accepted, its technical content cannot be modified and finally is published as an ISO international standard.

12.2.3 The ISO 9001:2015 Standard

One of the most recognized and implemented QMS worldwide is the ISO 9001 standard to facilitate international and national recognition of quality requirements in organizations (Salagean et al. 2014). This standard helps organizations to satisfy customers, meet regulatory requirements and achieve continuous improvement regardless the size or activity of the organization (ISO 2015); hence, the ISO 9001 standard disposes the guidelines to improve the quality system in the company. The most recent version of the standard is the ISO 9001:2015 that emphasizes the adequate management of activities to accomplish and exceed the customer requirements (Wecjenmann et al. 2015); in others words, organizations use the ISO 9001 standard to demonstrate to customers their capability and performance within its working systems to offer the confidence of satisfy their needs and expectations while committing to continuous improvement.

The ISO 9001:2015 standard uses three principal scopes for its functionality: the process approach which incorporates the PDCA cycle (Plan-Do-Check-Act) enabling an industry to plan its process related with its resources and internals interactions; risk-based thinking that enables an industry to find circumstances that could cause troubles and malfunctions in its QMS placing preventive controls to minimize negative effects and make the maximum usage of opportunities as they occur (Manders et al. 2016); and the seven quality management principles (Fig. 12.3) (Farinha et al. 2016; Manders et al. 2016). Likewise, Heuvel (2007) and Kutnjak et al. (2019) detail how the standard takes the process approach to improve organizational and financial performance with a specific focus on quality management, process control and quality assurance techniques to achieve planned outcomes and prevent unsatisfactory performance or non-conformance. Also, Kutnjak et al. (2019) affirm that quality in business operations models by the ISO 9001 standard is one of the ways of improving

Fig. 12.3 Seven quality management principles in ISO 9001:2015 standard

competitiveness in the entire benchmark or industry sector. None of the guidelines of the standard contain requirements with which a product or service can comply, and there are no product acceptance criteria in the ISO 9000 series; in this wise, the standard cannot inspect a product against the standard terms and guidelines, providing that the main interest of the ISO 9001 standard is only the QMS of the business (Salagean et al. 2014).

As described above, it is recognized that the ISO 9001:2015 standard provides theoretical and practical knowledge for the development of a system that manages the quality of an organizations based on rising awareness of all work areas into the continuous improvement of processes and activities performed to the fulfillment of the requirements and needs of customers. The greatest value obtained of the standard is by using the entire set of standards in an integrated manner; it is highly recommended that an organization using the ISO 9001 standard become acquainted with the fundamental concepts, principles and normative vocabulary of a QMS before adoption of the standard to achieve an effective level of performance (ISO 2016).

12.2.3.1 High-Level Structure in Management System Standards of the ISO

The ISO has published many standards in almost all important topics by the industry or government, and the management system standards of the ISO (MSS) are among the most widely used and recognized documents such as ISO 9001, ISO 14001 and ISO 50001. In an organization, especially the big companies have five or six certifications in quality management, environmental management, logistics, safety, information safety management or energy management to mention a few. Aware of the problem of integrating the different management standards and avoiding as far as possible duplications and inconsistencies between the systems, the Joint Technical Coordination Group (JTCG) of the ISO developed a structure of a "generic" management system to guarantee the compatibility with other systems. The concept of a high-level structure (HLS) is that management standards are structured in the same way, regardless of the domain of application. Users who are familiar with one MSS will immediately feel ease with another, even when using if for the first time, in other words, the standard with a HLS has the same structure and contains many of the same terms and definitions. The basic requirements for a HLS are: (1) equal number of chapters; (2) equal introductory text for all manuals; (3) identical statements for identical requirements and; (4) common terms. The HLS can be found as SL annex in every requirement standard manual to play a key role in the interoperability and user friendliness of standards for countless users of ISO management standards (Lopez 2016).

The HLS has been applied in some quality management systems standards that can be consulted at the ISO webpage: https://www.iso.org/management-system-standards-list.html.

12.2.3.2 High-Level Structure in the ISO 9001:2015

The HLS of the ISO 9001:2015 standard consists of ten closures or chapters. The first three closures describe generalities of the standard: where can be implemented, what normative can be referenced and the main terms and definitions to understand the standard and, from closure 4 to 10 the requirements of the QMS are defined. So, to exemplify what is the HLS in the ISO 9001:2015 is the arrangement of how standard is written and structured in the requirement manual. Table 12.3 shows the closures and its name.

Table 12.3 HLS of the ISO 9001:2015 standard

Closure number	Name
Closure 1	Scope
Closure 2	Normative references
Closure 3	Terms and definitions
Closure 4	Context of the organization
Closure 5	Leadership
Closure 6	Planning
Closure 7	Support
Closure 8	Operation
Closure 9	Performance evaluation
Closure 10	Improvement

Source ISO 9001:2015 quality management system—Requirements

12.2.4 Structural Standards of the ISO 9000 Family

The international quality management ISO 9000 series standards has earned a global status as a basis for establishing effective and efficient QMS. The need for international standards makes more organizations operate in the global economy by selling or buying products and services from sources outside their domestic surroundings. According to the quality management system requirement manual of the ISO 9001:2015, the standard is a set of almost 20 complementary standards that provide the theoretical and practical support to organizations; however, in this section only five of them will be described as main core standards because they describe and define all the basics and methodologic procedures to its adaptability in an organization.

1. *ISO 9000:2015 Quality management system—Fundamentals and vocabulary*: This standard describes the fundamental concepts of quality management which are universally applicable to organizations seeking sustained success through the implementation of a QMS. The ISO 9000 provides concepts and vocabulary used in the entire ISO 9000 series as well the introduction of with the seven QMPs and the use of the process approach to achieve continual improvement.
2. *ISO 9001:2015 Quality management system—Requirements*: This standard specifies the requirements against which QMS can be certified by an external body. The ISO 9001 recognizes that the term "products and services" applies to services, processed material, hardware and software intended for customer. The requirements in all sections of ISO 9001 are applicable. Organizations will need to provide justification for any requirement of this standard that the organization determines is not applicable to the scope of its QMS.
3. *ISO 9004:2018 Management for the sustained success of an organization*: the ISO 9004:2018 gives guidance on a wider range of objectives of a QMS than ISO

9001; in other words, the ISO 9004 manages the long-term success of an organization. The ISO 9004 is designed to serve as a guide for QMS implementation, maintenance and enhancement in any type of organization and is clearly oriented toward directing top management to accept responsibility for QMS performance improvement. Likewise, Jarvis and Palmes (2018), define the standards as an encouragement to go beyond the fundamental QMS requirements to extend the benefits of the ISO 9001 in pursuit of systematic and continual improvement of the performance of an organization; however, the standard is not intended for certification or contractual purposes.

4. *ISO 10001:2018 Quality management—Customer satisfaction—Guidelines for code of conduct for organizations*: This standard provides guidance for planning, designing, developing, implementing, maintaining and improving customer satisfaction codes of conduct. Mohammad et al. (2015) describe how the term "code" is used by the standard as a promises made to customers by an organization concerning its behavior that are aimed at enhanced customer satisfaction, the code of the standard is consistent with the term service guarantee.
5. *ISO 19011:2018—Guide for systems audit administration*: this standard covers the area of auditing the QMS and environmental management systems. Sukoco et al. (2012) describe that the ISO 19011 emphasizes the importance of audits as a management tool for monitoring and verifying the effective implementation quality policy. This standard provides an overview of how an audit should operate. Effective audits ensure the QMS implementation meeting the requirements specified in ISO 9001 standard.

12.2.5 Quality Standards Based on the ISO 9001 Standard

Although the ISO 9001 standard is described as the most recognized certification in the world covering all functional areas by standardizing production processes to meet customer requirements, there are productive sectors where processes need a higher degree of demand to meet quality and production requirement that the ISO 9001 standard does not emphasize. That is why standards have been created based on ISO 9001 to meet requirements to these sectors cover the needs in their products or services with a greater exigency to accomplish the quality requirements. Each standard is listed below, and Table 12.4 shows the ISO Survey (2018) results as well the amount of certified organizations and its variability against years. It should be noted that, although AS 9100 and ISO/TC 16949 are not propriety of ISO, these standards are created based on the ISO 9001 standard. For further information about each standard, the ISO webpage can be consulted.

- ISO 13485:2016 quality management in medical products
- ISO 14001:2015 environmental management system—requirements with guidance for use
- ISO 18091:2019 quality management to apply ISO 9001 in government
- ISO 22000:2018 food quality management

Table 12.4 Organizations certified with standards based on ISO 9001 standard worldwide

Standard	Year		Variation (%)
	2017	2018	
ISO 9001:2015	758,344	739,206	−3
ISO 13485:2003 and 2016	15,840	14,618	−8
ISO 14001:2015	251,343	258,566	3
ISO 22000:2005 and 2018	26,652	27,091	2
ISO IEC 27001:2013	15,848	16,523	4
ISO 50001:2011	13,827	14,549	5

- ISO/IEC 27001:2013 information technology—security techniques—information security management systems—requirements
- ISO/TS 29001:2010 quality management for oil, petrochemical and natural gas companies
- ISO 50001:2018 energy management system—requirements with guidance for use
- AS 9100:2004 quality management system—requirements for aviation, space and defense organizations
- ISO/TS 16949:2009 quality management: requirements to apply ISO 9001:2008 in the automotive sector.

12.3 Critical Success Factors for ISO 9001 Standard Implementation

In the search for an adequate implementation of a QMS in work processes and systems, organizations are continuously searching which factors are important or are determined as critical for a successful implementation of QMS, these factors in literature are conceptualized as critical success factors (CSFs). According to Rockart (1979) cited by Kumar et al. (2009), defines the CSFs as factors for an organization to achieve success, that is, if these factors are not taken into account as essential, completed or fulfilled in the development of a project or activity, the failure of the organization is highly probable. Likewise, Brotherton and Shaw (1996) consider that the CSFs can be defined as the essential components that the organization needs to fulfill in order to obtain the greatest advantages and competencies to be able to achieve the proper implementation of a QMS; in this manner, the CSFs are the characteristics or elements that organizations need to develop and fulfill to a degree that the implementation of the QMS is satisfactory in all the processes and systems of the organization (Garza-Reyes et al. 2015). Even for Banuelas Coronado and Antony (2002); Jeyaraman and Kee Teo (2010) and Lande et al. (2016), the CSFs are the components that the organization targets to identify and develop in its processes to recognize which areas of the organization will produce the greatest competitive advantages. A vast amount of literature was consulted to find the CSFs that enable organizations to implement a QMS based on the ISO 9001 standard (Quazi and Padibjo 1998; Sun 2000; Tarí 2005; Aggelogiannopoulos et al. 2007; Soltani and Lai 2007; Gotzamani et al. 2007; Dowlatshahi and Hooshangi 2010; Psomas et al. 2010; Sıtkı İlkay and Aslan 2012; Sumaedi and Yarmen 2015; Ismyrlis et al. 2015; Wilson and Campbell 2016; Farinha et al. 2016; Militaru and Zanfir 2016; Anttila and Jussila 2017a, b) remark different essential factors that allow the implementation of the ISO 9001 standard in organizations; however, the CSFs considered by them essentially agree with the QMPs presented in the ISO 9001:2015 standard; consequently, it is possible to recognize the QMPs as CSFs define them as the necessary characteristics that must be met by an organization for the adequacy and implementation of the standard. The QMPs and the corresponding definitions of the new version are shown in Table 12.5.

Table 12.5 QMPs in ISO 9001:2015 standard

Quality management principle	Definition
Leadership (LD)	Leaders at all levels establish unity of purpose and direction and create the conditions in which people engage the business's objectives
Customer focus (CF)	Primary focus of quality management is the satisfaction of customer requirements and the effort to exceed their expectations
Engagement of people (EP)	Essential for the business that people are competent and empowered at all levels to enhance its capability to create and deliver value
Process approach (PA)	Consistent and predictable results are achieved more effectively and efficiently when activities are understood and managed as interrelated processes that function as a coherent system
Improvement (IMP)	Successful businesses have an ongoing focus on improvement
Evidence-based decision making (EDM)	Decision based on the analysis and evaluation of data is more likely to produce desired results
Relationship management (RM)	For sustained success, a business manages its relationships with interested parties such suppliers

Source ISO 9001:2015 Quality Management—Requirements and Farinha et al. (2016)

12.4 The ISO 9001 Standard Implementation in Mexican Companies

12.4.1 Methodology

The target population for the study involved all the companies within the manufacturing sector in Mexico (MSM) that have or have had the ISO 9001 standard certified preferably in the 2008 and 2015 versions within the SMEs and large company classification. A survey was developed for the needed data based on literature with 55 items in a five-point Likert scale (1: never, 2: rarely, 3: regularly, 4: almost always, 5: always) related to the level of use of each QMP. This survey was divided into four main sections: the first section with ten items for demographic data; the second correspond to the items for the QMPs (seven items for LD, four items for CF, six items for EP and PA and five items for IMP, EDM and RM); the third section with seven items related to benefits; and the last section with seven items associated with barriers. The survey could be answered by any personnel related to the quality system of the organization.

Over a period of two years, a total of 531 responses were obtained; however, only 183 responses were complete and usable for this study representing the 34.46% of the sample obtained. The software SPSS was used to perform the exploratory factor analysis (EFA). Factor Analysis (FA) is an interdependence technique whose main objective is to define the underlying structure between the variables in the analysis (Hair et al. 2009). In this research, the EFA is used to identify the structure of the measurement instrument and identify the potential factors that could emerge grouping items that are correlated and poses certain information in common. To identifying if FA is applicable to the database, the Bartlett's test of sphericity and the Kaiser–Mayer–Olkin (KMO) will be performed. The Bartlett's test of sphericity tests the null hypothesis that the correlation matrix of the variables is equal to the identity matrix and the Kaiser–Mayer–Olkin test is used to verify the correlation among the variables, KMO values greater than 0.7 are considered regular, values above 0.8 are great, and values beyond 0.9 are very good (Kaiser and Rice 1974), that is, if the KMO test results in a small number or very close to zero, the sample cannot be used for a FA (Hair et al. 2009). Finally, results corresponding to barriers and benefits related to the adoption of the standard are described in the discussion.

12.4.2 Results

12.4.2.1 Benefits of the ISO 9001 Standard Implementation

Many benefits can be acquired implementing the ISO 9001 standard in all types of business or industry; the benefits are related to the effective implementation of the standard, for example, Gonzalez et al. (2001) reported a decrease in manufacturing times reducing waste which means, an increase improvement of the quality system. Some benefits by the ISO 9001 standard are related to the improvement of the manufacturing systems which leads to produce quality products, as the productivity and profit of the organization increase (Turner et al. 2000; Aggelogiannopoulos et al. 2007; Boiral 2012). Also, Psomas et al. (2013) defined how food enterprises can achieve some benefits with the standard such as continuous improvement, prevention of non-conformities and customer satisfaction focus with its implementation. Likewise, Anttila and Jussila (2017b) and Wilcock and Boys (2017) described that, when the industries increase productivity, the customer complaints tend to minimize; as a result, more customer satisfaction is achieved. According to the results obtained in the exploratory analysis, among the most reported benefits is the increase to customer satisfaction, the reductions of errors or defects, the integration of the continuous improvement culture and the decrease of waste and activities that do not add value to products.

12.4.2.2 Barriers for the ISO 9001 Standard Implementation

Sometimes the implementation of the ISO 9001 standard is not always as simple as it could be in the organizations, certain difficulties or barriers make the implementation impossible, in other words, barriers can be associated with various aspects such as the lack of focus on the CSFs for an adequate implementation (Santos et al. 2015). There are two common classifications for barriers: the ones related with leadership and the others related to the engagement of people. Barriers related to engagement of people creates problems such as: lack of people cooperation and involvement, lack of discipline following work methods, resistance to change, insufficient training and lack of motivation to improve (Turner et al. 2000; Ab Wahid 2012).

Destitute leadership carries problems such as misunderstanding ISO requirements, lack of communication, financial issues, lack of strategic thinking, poor audit systems, stationary issues and lack on continuous improvement focus (Taylor 1995; Maza and Ramírez 2005; Sampaio et al. 2009). Also, the large time of implementation to develop this kind of projects or limited resources for a short amount of time can be conceptualized as a barrier (Capmany et al. 2000; Turner et al. 2000). In relation to the results of the main barriers in the MSM, the most mentioned were the lack of the demand for the standard by the customers, followed by the high investment it requires. Also, organizations reported that its current QMS is better than the standard; hence, the standard becomes unnecessary. Finally, the bureaucratic paper work that the standard demands is another barrier reported.

12.4.2.3 Exploratory Factor Analysis and Enablers Definition

The Bartlett's test of sphericity was tested obtaining a chi-square approximation of 2259.23 with 153 degrees of freedom and a P-value = 0.00; therefore, the null hypothesis cannot be accepted indicating that the correlation matrix is different from the identity matrix. Likewise, from the Kaiser–Mayer–Olkin test resulted a value of 0.949 which denotes that the sample is suitable for a FA. Finally, once that the evaluation of the database was suitable to factorize, the EFA was performed using the SPSS software and the principal components extraction method and Varimax rotation which are broadly reported in the literature (Psomas and Pantouvakis 2015; Xiong et al. 2016; Young et al. 2017; Macias-Velasquez et al. 2019). Table 12.6 shows the results obtained through the EFA.

Results of the EFA identified 18 variables related to the QMPs with a significant loading in three different factors, emerging this way the enablers. These loadings are greater than .4 considered for determining the influence of the item on the factor relative to the sample size (Hair et al. 2009). The grouping of the variables is notably structured by their own approach. For example, the items belonging to EP and IMP load in the same factor; therefore, it should be considered as one factor named *Engagement of People and Improvement (EPI).* In the same sense, the items from PA, CF, RM and EDM load together; hence, by the approach that this QMPs have in the standard, this factor could be named *Customer Satisfaction and Decision Making*

Table 12.6 Results of the EFA

Items	Factor			Eigenvalues	Cronbach's alpha
	EPI	CSDM	LID		
EP3	0.775			10.272	0.887
EP4	0.764				
EP1	0.753				
EP6	0.734				
EP2	0.725				
IMP4	0.684				
PA6		0.776		1.193	0.921
CF3		0.682			
RM2		0.659			
PA1		0.649			
EDM2		0.638			
PA4		0.623			
EDM4		0.620			
RM1		0.617			
LD2			0.839	1.010	0.902
LD4			0.694		
LD7			0.685		
LD3			0.657		
% Variance	24.536	18.582	26.192		
% Total variance	24.53	43.11	69.31		

(CSDM). Finally, the items related to LD loaded together on the same factor and because of this, and the original name of *Leadership (LD)* remained. Besides, from this table the corresponding Cronbach's alpha values for each one factor identified are shown. Values in the Alpha of Cronbach between .80 and 1.0 demonstrate an acceptable reliability and consistency (Cronbach 1951) which, in this case, the alpha values are within the values recommended. Finally, taking into account the three constructs obtained, the total variance explained is 69.31% which means that more than a half of the total phenomenon can be explain by the selected factors.

12.5 Discussion

The results of the EFA reveal that three main factors that integrate the seven QMPs that enable the implementation of the ISO 9001 standard in the MSM. Regarding EPI, CSDM and LD, similar factors have been reported in the literature. For example, Poksinska et al. (2006) describe how the top management leadership and the engagement of people provide the implementation of the standard. Also, Psomas et al. (2013) Manders et al. (2016); Almeida et al. (2018) and Stainslaus et al. (2018) describe the leadership as one of the main characteristics to arrange a QMS into the work procedures developing the organization goals and objectives. The customer focus, human resources management and the strategic planning of the quality should have the highest consideration when implementing the standard in an organization (Texeira and Fernandez 2013).

Finally, Xiong et al. (2016) and Dowlatshahi and Hooshangi (2010) reveal that the organizations certified in the ISO 9001 standard have an excellent relationship management, and with engaged personnel, a significant approach is accomplishing to fulfill the standard guidelines. The loading structure of the QMPs took place in the EFA made possible its classification into three factors; thus, it can be concluded that the EPI, CSDM and LID can be considered as the essential elements that need to be develop and accomplish for an adequate integration of the ISO 9001 standard in the manufacturing sector in Mexico.

According to the benefits reported in the EFA, these are similar to the benefits reported by Turner et al. (2000), Ahmad et al. (2006), Boiral (2012) and Psomas et al. (2013) describing how the implementation of the standard increases customer satisfaction and productivity. Likewise, Gonzalez et al. (2001) and Magd and Curry (2003) report that the implementation of the ISO 9001 integrates a continuous improvement approach increasing the productivity into the work systems, and as consequence, errors and waste are reduced increasing profit which resembles with the benefits reported in the MSM. In the matter of barriers, Prodromos et al. (2015) mention how the lack of commitment by the top management and financial or resources limitation hinders the assessment of the standard. Moreover, Rogala (2016) and Bravi and Murmura (2017) mentioned that the bureaucracy of the standard into de process, the lack of interest of the management, the high investment and the long time required for its implementation are obstacles and limitations. According to the EFA, the high investment and the bureaucracy agree with the results obtained; however, the main barrier reported in the MSM is the lack of demand of the standard by the customers unlike the literature consulted where this is not reported as a barrier.

12.6 Conclusions

Currently, organizations seek to find the best practices to manage the activities that provide the best productive conditions in the manufacture of a product or delivering a service. The quality management system (QMS) that the ISO 9001 standard develops in an organization offers the necessary bases to plan, organize and control such activities considerate critical to meet the requirements of customers, with the support of other standards that cover the essential features of growth and organizational strengthening. This growth can be measured thanks to the benefits that the standard can contribute; however, there are difficulties or barriers that inhibit its implementation due to limitations that organizations may suffer.

Likewise, the ISO 9000 series has served as a basis format to create standard to manage the quality systems in a specific industrial sector that need more rigorous and demanding control. This shows the flexibility and adaptation of the standard to cover all types of industrial activities. On the other hand, comparing the results of the ISO Survey of 2017 and 2018, it can be observed a decrease of certified companies worldwide, and this may be possible due to the evolution that the same standard asks to develop in the certified QMS; this means that organizations have applied tools and controls that improve even more their quality systems obtaining more or better benefits than with the ISO 9001 standard. Also, it would be interesting to find the causality of this phenomenon and recognize which others QMS the organization are applying. Even as an extra contribution, it would be advantageous for the ISO to know which standards are most often implemented in conjunction with the ISO 9001 standard to describe in the new version better and novel requirements.

According to the implementation of the standard in the manufacturing sector in Mexico, the EFA demonstrates how the measuring instrument is suitable for measuring the implementation of the standard in this industrial sector explaining the 69.31% of the phenomenon. With this results, organizations can recognize which factors are essential to successfully implement the ISO 9001 standard and begin obtaining the benefits it provides; however, the results do not show a model that represents the relationship between the factors, so it would be necessary to apply structural equation modeling (SEM) to obtain a structural model that indicates the significant impact between factors associated with the benefits of the standard to recognize exactly the relationship of the constructs that make possible the implementation of the standard ISO 9001 in the manufacturing sector or any business activity.

References

Ab Wahid R (2012) Beyond certification: a proposed framework for ISO 9000 maintenance in service. TQM J 24:556–568. https://doi.org/10.1108/17542731211270115

Aggelogiannopoulos D, Drosinos EH, Athanasopoulos P (2007) Implementation of a quality management system (QMS) according to the ISO 9000 family in a Greek small-sized winery: a case study. Food Control 18:1077–1085. https://doi.org/10.1016/j.foodcont.2006.07.010

Ahmad T, Singh R, Rai A (2006) Comparison of bootstrap methods for missing survey data: A simulation study. Model Assist Stat Appl 1:43–49

Almeida D, Pradhan N, Muniz JJ (2018) Assessment of ISO 9001:2015 implementation factors based on AHP: case study in Brazilian automotive sector. Int J Qual Reliab Manag 34:231–250

Anttila J, Jussila K (2017a) ISO 9001:2015—a questionable reform. What should the implementing organizations understand and do? Total Qual Manag Bus Excell 28:1090–1105. https://doi.org/10.1080/14783363.2017.1309119

Anttila J, Jussila K (2017b) Understanding quality—conceptualization of the fundamental concepts of quality. Int J Qual Serv Sci 9:251–268. https://doi.org/10.1108/IJQSS-03-2017-0020

Banuelas Coronado R, Antony J (2002) Critical success factors for the successful implementation of six sigma projects in organizations. TQM Mag 14:92–99

Boiral O (2021) ISO 9000 and organizational effectiveness: a systematic review. Qual Manag J 19:16–37

Bravi L, Murmura F, Santos G (2019) The ISO 9001:2015 quality management system standard: companies' drivers, benefits and barriers to its implementation. Qual Innov Prosper 23:64–82. https://doi.org/10.12776/QIP.V23I2.1277

Brotherton B, Shaw J (1996) Towards an identification and classification of critical success factors in UK hotels Plc. Int J Hosp Manag 15:113–135. https://doi.org/10.1016/0278-4319(96)00014-X

Capmany C, Hooker NH, Ozuna T, Van Tilburg A (2000) ISO 9000—a marketing tool for U.S. agribusiness. Int Food Agribus Manag Rev 3:41–53

Cronbach LJ (1951) Coefficient alpha and the internal structure of tests. Psychometrika 16:297–334

Dowlatshahi S, Hooshangi S (2010) Enabling quality management systems in the maquiladoras: an empirical analysis. Int J Qual Reliab Manag 27:981–1001

Farinha L, Lourenço J, Caroço C (2016) Guidelines for the implementation of a quality management system in industrial companies. Rom Rev Precis Mech Opt Mechatronics 1:195–201

Gargasas A, Samuolaitis M, Mugiene I (2019) Quality management system in logistics. Manag Theory Stud Rural Bus Infrastruct Dev 41:290–304. https://doi.org/10.5772/intechopen.71431

Garza-Reyes JA, Rocha-Lona L, Kumar V (2015) A conceptual framework for the implementation of quality management systems. Total Qual Manag Bus Excell 26:1298–1310. https://doi.org/10.1080/14783363.2014.929254

Gonzalez Torre P, Adenso-Diaz B, Gonzalez BA (2001) Empirical evidence about managerial issues of ISO certification. TQM Mag 13:355–360. https://doi.org/10.1108/EUM0000000005861

Gotzamani KD, Tsiotras GD, Nicolaou M, Nicolaides A, Hadjiadamou V (2007) The contribution to excellence of ISO 9001: the case of certified organizations in Cyprus. TQM Mag 19:388–402. https://doi.org/10.1108/09544780710817838

Hair JFJ, Black WC, Babin BJ, Anderson RE (2009) Multivariate data analysis, 7th edn. Pearson Prentice Hall

Hallberg P, Hasche N, Kask J, Öberg C (2018) Quality management systems as indicators for stability and change in customer-supplier relationships. IMP J 12:483–497. https://doi.org/10.1108/imp-01-2018-0006

Harrigan WF (1993) The ISO 9000 series and its implications for HACCP. Food Control 4:105–111

Hoyle D (2018) ISO 9000 quality systems handbook: increasing the quality of an organization's outputs. Routledge, Londond and New York, Seventh

International Organization for Standardization (1997) Friendship among equals: recollections from ISO's first fifty years. Switzerland

International Organization for Standardization (2000) ISO 9001:2000 quality management systems—requirements
International Organization for Standardization (2015a) ISO 9000:2015 Quality management systems—fundamentals and vocabulary
International Organization for Standardization (2015b) ISO 9001:2015 Quality management systems—requirements
International Organization for Standardization (2015c). Quality management principles https://doi.org/ISBN978-92-67-10650-2
International Organization for Standardization (2016) Selection and use of the ISO 9000 family of standards
International Organization for Standardization The ISO Survey (2017) https://www.iso.org/the-iso-survey.html. Accessed 8 May 2019
International Organization for Standardization The ISO Survey (2018) https://www.iso.org/the-iso-survey.html. Accessed 8 May 2019)
Ismyrlis V, Moschidis O, Tsiotras G (2015) Critical success factors examined in ISO 9001:2008-certified Greek companies using multidimensional statistics. Int J Qual Reliab Manag 32:114–131. https://doi.org/10.1108/IJQRM-07-2013-0117
Jarvis A, Palmes PC (2018) Business sustainability: going beyond ISO 9004:2018. ASQ Quality Press
Jeyaraman K, Kee Teo L (2010) A conceptual framework for critical success factors of lean six sigma: implementation on the performance of electronic manufacturing service industry. Int J Lean Six Sigma 1:191–215. https://doi.org/10.1108/20401461011075008
Kaiser HF, Rice J (1974) Little jiffy, mark IV. Educ Psychol Measure 34:11–117. https://doi.org/10.1177/001316447403400115
Kumar M, Antony J, Douglas A (2009) Does size matter for Six Sigma implementation?: findings from the survey in UK SMEs. TQM J 21:623–635. https://doi.org/10.1108/17542730910995882
Kutnjak G, Miljenović D, Mirković A (2019) Improving competitiveness of small and medium-sized enterprises with the application of quality management system. Sci J Marit Res 33:11–21. doi:https://doi.org/10.31217/p.33.1.2
Lande M, Shrivastava RL, Seth D (2016) Critical success factors for lean six sigma in smes (small & medium enterprises). TQM J 28:1–32. https://doi.org/10.1108/TQM-12-2014-0107
Lopez LP (2016) Novedades ISO 9001:2015. Editorial Fundación Confemetal Madrid
Macias-Velasquez S, Baez-Lopez Y, Maldonado-Macías AA, Limon-Romero J, Tlapa D (2019) Burnout syndrome in middle and senior management in the industrial manufacturing sector of Mexico. Int J Environ Res Public Health 16:1467. https://doi.org/10.3390/ijerph16081467
Magd H, Curry A (2003) ISO 9000 and TQM: are they complementary or contradictory to each other? TQM Mag 15:244–256. https://doi.org/10.1108/09544780310486155
Manders B, De Vries HJ, Blind K (2016) ISO 9001 and product innovation: a literature review and research framework. Technovation 48–49:41–55. https://doi.org/10.1016/j.technovation.2015.11.004
Maza Rubio MT, Ramírez Arias V (2005) Study of main motivations and discouraging factors for the implementation of ISO 9000 standards in Spanish agribusiness sector. J Int Food Agribus Mark 17:229–243. https://doi.org/10.1300/J047v17n02
Militaru C, Zanfir A (2016) The vision of new ISO 9000:2015 standards. Knowl Horizons Econ 8:131–135
Mohammad A, Rahman K, Stanislav K (2015) Establishing an ISO 10001-based promise in inpatients care. Int J Health Care Qual Assur 28:100–114
Murmura F, Bravi L (2017) Empirical evidence about ISO 9001 and ISO 9004 in Italian companies. TQM J 29:650–665. https://doi.org/10.1108/TQM-11-2016-0097
Poksinska B, Eklund JAE, Jörn Dahlgaard J (2006) ISO 9001:2000 in small organisations: lost opportunities, benefits and influencing factors. Int J Qual Reliab Manag 23:490–512. https://doi.org/10.1108/02656710610664578

Prodromos C, Chatzoudes D, Kipraios N (2015) The impact of ISO 9000 certification on firms' financial performance. Int J Oper Prod Manag 35:145–174. https://doi.org/10.1108/IJOPM-07-2012-0387

Psomas E, Pantouvakis A (2015) ISO 9001 overall performance dimensions: an exploratory study. TQM J 27:519–531. https://doi.org/10.1108/TQM-04-2014-0037

Psomas EL, Fotopoulos CV, Kafetzopoulos DP (2010) Critical factors for effective implementation of ISO 9001 in SME service companies. Manag Serv Qual An Int J 20:440–457

Psomas EL, Kafetzopoulos DP, Fotopoulos CV (2013) Developing and validating a measurement instrument of ISO 9001 effectiveness in food manufacturing SMEs. J Manuf Technol Manag 24:52–77. https://doi.org/10.1108/17410381311287481

Quazi HA, Padibjo SR (1998) A journey toward total quality management through ISO 9000 certification—a study on small- and medium-sized enterprises in Singapore. Int J Qual Reliab Manag 155:489–508. https://doi.org/10.1108/02656719810196225

Rogala P (2016) Identification of barriers to improving quality management systems. The management representatives' perspective. TQM J 28:79–88. https://doi.org/10.1108/TQM-05-2014-0047

Salagean HC, Gârbacea RD, Emmanouilidis E, Marian O (2014) From ISO standards To TQM philosophy. Manag Challenges Contemp Soc 7:93–98

Sampaio P, Saraiva P, Guimarães Rodrigues A (2009) ISO 9001 certification research: questions, answers and approaches. Int J Qual Reliab Manag 26:38–58

Santos G, Rebelo M, Lopes N, Rui RA, Silva R (2015) Implementing and certifying ISO 14001 in Portugal: motives, difficulties and benefits after ISO 9001 certification. Total Qual Manag Bus Excell 27:1211–1223. https://doi.org/10.1080/14783363.2015.1065176

Sıtkı İlkay M, Aslan E (2012) The effect of the ISO 9001 quality management system on the performance of SMEs. Int J Qual Reliab Manag 29:753–778. https://doi.org/10.1108/026567112 11258517

Soltani E, Lai P-C (2007) Approaches to quality management in the UK: survey evidence and implications. Benchmarking An Int J 14:429–454

Stainslaus RL, Premaratne S, Tritos L (2018) Quality management capabilities of manufacturing industries in the Western Sydney region: comparative analysis for quality improvement. Int J Qual Reliab Manag 35:1232–1252

Sukoco A, Marzuki, Cucus A (2012) Concept of quality measurement system software based on standard ISO 9126 and ISO 19011. In: Proceeding of 2012 international conference on uncertainty reasoning and knowledge engineering, URKE 2012. IEEE, pp 105–108

Sumaedi S, Yarmen M (2015) The effectiveness of ISO 9001 implementation in food manufacturing companies: a proposed measurement instrument. Procedia Food Sci 3:436–444. https://doi.org/ 10.1016/j.profoo.2015.01.048

Sun H (2000) Total quality management, ISO 9000 certification and performance improvement. Int J Qual Reliab Manag 17:168–179. https://doi.org/10.1108/02656710010304573

Tarí JJ (2005) Components of successful total quality management. TQM Mag 17:182–194. https:// doi.org/10.1108/09544780510583245

Taylor WA (1995) Organizational differences in ISO 9000 implementation practices. Int J Qual Reliab Manag 12:10–27. https://doi.org/10.1108/02656719510093529

Texeira Quirós J, Ferndandes Justino MDR (2013) A comparative analysis between certified and non-certified companies through the quality management system. Int J Qual Reliab Manag 30:958–969. https://doi.org/10.1108/IJQRM-04-2011-0059

Turner CR, Ortmann GF, Lyne MC (2000) Adoption of ISO 9000 quality assurance standards by South African agribusiness firms. Agribusiness 16:295–307. https://doi.org/10.1002/1520-629 7(200022)16:3%3c295:AID-AGR3%3e3.0.CO;2-P

Van Den Heuvel J (2007) The effectiveness of ISO 9001 and six sigma in healthcare. Erasmus University Rotterdam, Alphen aan den Rijn

Wecjenmann A, Akkasoglu G, Werner T (2015) Quality managment—history and trends. TQM J 27:281–293. https://doi.org/10.1089/can.2016.0016

Wilcock AE, Boys KA (2017) Improving quality management: ISO 9001 benefits for agrifood firms. J Agribus Dev Emerg Econ 7:1–37. https://doi.org/10.1108/JADEE-10-2013-0040
Wilson JP, Campbell L (2016) Developing a knowledge management policy for ISO 9001: 2015. J Knowl Manag 20:829–844. https://doi.org/10.1108/JKM-11-2015-0472
Wolniak R (2018) The history of ISO 9001 series up to ISO 9001:2000. Zesz Nauk Politech ŚLĄSKIEJ 119:331–338. doi:https://doi.org/10.29119/1641-3466.2018.119.24
Xiong J, He Z, Ke B, Zhang M (2016) Development and validation of a measurement instrument for assessing quality management practices in hospitals: an exploratory study. Total Qual Manag Bus Excell 27:465–478. https://doi.org/10.1080/14783363.2015.1012059
Young DKW, Ng PYN, Pan J, Fung T, Cheng D (2017) Validity and reliability of recovery assessment scale for cantonese speaking Chinese consumers with mental illness. Int J Ment Health Addict 15:198–208. https://doi.org/10.1007/s11469-016-9657-3

Chapter 13
Gestation of the Genetics of a Company

Jorge Vera Jiménez

Abstract The definition of the genetic aspects of the organizations contributes to their success, the personnel that provide their services in them, will generate effective results using resources efficiently. Applying the principles of coherence and cohesion, useful formats were generated in the design of the basic aspects required to develop the genetic information that organizations will generate, through the design of systems with a focus on processes. It includes the way to achieve the introjection of genetic information in personnel, key to the permanent survival of companies over time.

Keywords Purpose · Mission · Functions · Vision · System · Culture · Policy

13.1 Introduction

Maintaining an organization with a chain of successes in its performance depends on the documented ideological expression of the fundamental aspects that give rise to its creation.

The different types of information taken as a reference in the structuring of an organization are commonly used, both to leading organizations and those that live solving problems, that information serves as a guide to shape its characteristics and their actions; however, some have high levels of performance and some have low levels of performance; the difference lies in the way of conceiving their kind of information, in the application or not of the principles of coherence and cohesion, in the design of the system based on processes and, its assimilation by all the personnel that provide their services in the organization.

Public, private, and social companies should review the basic precepts that gave origin to its creation, they need to consider a model that guides the focus of the gestation of their genetic information that, when considered, will adapt their concepts to their changing reality, allowing them to stay over time.

J. V. Jiménez (✉)
Department of Industrial Engineering, TecNM Campus Oaxaca, Oaxaca, Mexico

J. L. Garcia Alcaraz et al. (eds.), *Trends in Industrial Engineering Applications to Manufacturing Process*, https://doi.org/10.1007/978-3-030-71579-3_13

Every organization has a purpose for which it was created, could have been technically designed or, be the product of a series of logical occurrences based on experience or well-intentioned assumptions about its best structure and its ideal production factors; whatever the situation, with the formats developed here, you can redesign the information that supports your existence, in order to make it efficient and effective. The continuous fulfillment of the organizational purpose is ensured with the dissemination to personnel, they need to understand it, this will regulate their performance.

Organizational purpose and policy are genetic aspects of the organization and essential information for its gestation; this work describes them in detail; formats included will support those interested in its development; the conceptualization of the genetic aspects is based on the design of the useful formats so that each organization can manage its genetic information with coherence, cohesion and capable of being assimilated by all the persons that provide their services in some activity in or for the organization; the guarantee of generating good results is supported by the design of the organization consisting of a system with a focus on processes.

13.1.1 Genetic Information of an Organization

The characteristics of an organization of both its infrastructure and its mode of production will be generated by the information defined by senior management, and this information that encourages the start of the implementation of actions for the creation of a company was given the qualification of "Genetics" is the basis for the emergence of an organization, will determine its size, competencies, functions, will contribute to its gestation.

The first information to be specified in the creation of an organization is that corresponding to its purpose, the reason or reasons for which a physical or moral legal entity will exist; subsequently, it is necessary to determine what this person's mission will be, in addition to identifying the means that will contribute to fulfilling the purpose or purposes.

The fulfillment of the mission is achieved by operating the organization, and it is necessary to diligently choose its functional areas so that it can operate efficiently and generate effective results.

The identification of the functions feeds the intelligence of those in charge of the design of a company and induces them to use the technical knowledge to anticipate how it will be organically structured, and the functions are necessary and indispensable for the development of the vision.

The performance when developing the functions, the structural organs of the companies, is associated with the cultural environment product of the behavior of its personnel; their cultural values set the pattern for the way they act.

The aspects required to begin the birth of an organization must be documented. Maintaining the unification of the personnel's criteria in making decisions about what is appropriate requires collective intelligence; when there are documents, they

Table 13.1 Genetic information of an organization

Genetic information	Have the intention
Purpose	Unification
Mission Functions Vision	Characterization
Culture	Performance
Policies	Orientation

regulate the judgment of the human factor, what to do is known; but in the presence of confusing cases, of situations not determined in the documented information, decisions will be made based on the judgment of the personnel; organizational policy is the unifying reference for the criteria for personnel performance when there are no pre-established precepts.

Development of employee activities is confined between the provisions of the purpose and organizational policy; leads the actions of those responsible for functional areas; induces to determine the skills to be trained in personnel, and these ideological elements are part of the genetic information of organizations (Table 13.1).

The knowledge of the genetic elements of the organizations, besides gestating them, keeps them operating under controlled conditions, the purpose and the policy are like the chains of business DNA between which they are kept in the consciousness of the service providers of the organizations. Mission, functions, vision, and determination to maintain an organizational culture are conducive to the fulfillment of the purpose. There are other organizational DNA formats such as the one proposed by Lozano et al. (2020), referring to the gestation of organizational strategies, they do not refer to the information that gave rise to them.

The texts of the genetic information need to contain the assignment of responsibility for the achievement of the purpose by personnel, as well as other aspects must be clear, concise, and understandable by all persons; must identify their responsibility within the aspects of genetic information, they must contribute to make them aware of the relevance of their work in achieving the objectives, for this reason, all genetic aspects must include the text "all the persons that provide their services in or for the organization", the precepts will have a focus on people; another approach to genetic information is that of processes; in the vision of the organization, the functional areas identified, give rise to business bodies, should operate by implementing interrelated processes (Table 13.2).

Table 13.2 Approaches to genetic aspects

Focus	Description
To the people	All persons who provide their services in or for the organization are responsible for the fulfillment of the purpose
To processes	The organization operates with an interrelated process structure, each process emerges from each identified functional area

13.2 The Purpose

The main idea of generating a personal benefit for investors, workers, or any other person is the seed of the emergence of the need for the existence of an organization; this idea of obtaining benefits can be propitiated by the existence of some opportunity to take advantage of situations observed in the environment or some other relevant reasons.

The generation of benefits is the purpose of any organization, it is the reason for its creation, its existence or the decision to have a company, Muñoz (2020), indicates that it has the core values of the organization, it is conceptualized as:

Purpose	Reason that originates the idea of creating an organization

The reason that generates the intention to create a company is to obtain benefits; the purpose can also be conceived as a formulated objective of obtaining or providing benefits; any other objective that is not related to the idea of having a direct or indirect benefit as a human being cannot be considered as an organizational purpose.

The theoretical concept of purpose is the basis for the development of the specific purpose of each of the organizations, it focuses on the benefits; its construction begins with the identification of the people to benefit. Strichow (2013) considers that the purpose of a company is to create and maintain customers, however it indicates that the ultimate purpose of a product's lifecycle is to create profits.

In commercial organizations, the client is the main beneficiary, will generate the income to the investor, who will benefit from the profits; obviously the service providers in the company will also benefit from receiving a salary; in the Mexican Republic, there is a legal benefit called Participation of the Worker in the Utilities; the government will benefit from receiving taxes; in Mexico, the client pays the value added tax (IVA) for his purchases, the investor and the workers pay the income tax (ISR); the producers of goods and services of the environment will benefit from the purchases of their products and/or services to meet their needs that the workers and investors of the company will make when they have income, the suppliers of the organization will pay the ISR, and the organization when you buy them you will pay the (IVA), they will benefit the government with the payment of their taxes (Table 13.3).

All beneficiaries have in common the existence of an underlying need, the purpose establishes the willingness to satisfy it. There are two types of beneficiaries, the internal ones and the external ones, to some it satisfies it directly, it generates a satisfactory one, to the clients it delivers the product to them or it provides the expected service to them; service providers in (or for) the organization will pay their salary, providers will pay their products and/or services; the organization is the source of the monetary income of the direct beneficiaries, with the exception of the client who receives the product and/or service, from it comes the money to pay to the other beneficiaries (Table 13.3).

Table 13.3 Beneficiaries of operating a commercial organization

Beneficiary	Type of beneficiary	
	Interns	External
Customers		Direct benefit
Investors	Direct beneficiary	
Service providers in the organization	Direct beneficiary	
Suppliers	Direct beneficiary	
Government		Indirect beneficiary
Society		Indirect beneficiary

Indirect beneficiaries also receive money through direct beneficiaries; there is a phenomenon that multiplies the benefits, those who buy, those who sell pay taxes, those who have their own income, make purchases benefiting the productive sector of the region.

It is clear that a single type of beneficiary is the one who provides the goods for the others, the client; when paying for the product and/or service it receives, it provides the monetary resource to be distributed to the remaining beneficiaries; in exchange for your money, he receives the satisfaction of his needs. The exchange provides each beneficiary with the product that will satisfy his needs in a "win-win" situation.

Every organization has or not defined its purpose, satisfies the human being, whether previously defined or not, it is required to document its conception. Within the conceptualization of the purpose, the personnel that provide their services in (or for) the organization must be included, it must identify its intervention in its fulfillment, it is the one who will generate the value that will satisfy the needs of the clients.

A complete statement of purpose would be: "To satisfy the needs of the clients over time by the services provided by all the people who work in the organization"; has a structure, contains elements ordered this declaration in its wording (Table 13.4).

The reason or cause of the "existence" of an organization is the "existence" of needs, both coexists, the rationale of the companies comes from their surroundings; the satisfaction of the needs is a cyclical duty of each of the people hired by the

Table 13.4 Structure of the organizational purpose

Order of elements of purpose	Elements
I	Leave satisfied
II	The needs (indicate them)
III	Of the customers (generic name)
IV	Through time
V	For the service provided by the entire personnel
VI	In the organization (Name)

organization. An example of purpose for a shoe manufacturing company based on Table 13.4 would be declared as follows: "Satisfy the needs of having a comfortable protection when walking of men and women from one year onwards in the Central Valley region of the State of Oaxaca over time for the services provided by all the personnel of the manufacturer La Favorita SA de CV".

This genetic information of the company presents the basis for defining the initial market area, the characteristics of the satisfiers, for example, it is the type of footwear, it awakens the awareness of the personnel to make their purpose their own (Baldoni 2011); it makes them recognize their responsibility direct to satisfy the needs of customers; by indicating that it is through time, it makes it clear that the provision of its services is continuing to satisfy the needs.

13.3 Mission

A mission can only be accomplished by an intelligent being who can make autonomous decisions in the face of changes that arise; since the intelligence of an organization lies in its personnel, the mission must be assigned to all service providers regardless of the hierarchical level they have in the organizational structure, everyone must be explicitly assigned the responsibility of generating benefits. The beneficiaries receive the benefits, tangible and/or intangible assets valued by them, expected by them for their satisfaction. The mission will influence and give meaning to the actions of the personnel every day (Taiwo et al. 2016).

This genetic information implies the knowledge of the benefits expected by the direct and indirect beneficiaries (Table 13.3), customers need to have the products and/or services at their disposal, work to be carried out by the performance of some work team associated with the sales function.

The mission has to do with the valuable elements to be generated directly or indirectly by the action of the personnel that provide their services in the organization; these elements are of value to different types of people, they constitute the satisfying factors of their needs, through them they are provided a benefit; Barraza (2019) includes within the mission the indication of the temporality of its scope.

The concept of mission implies the fact of generating benefits, being able to express its general theoretical concept as

Mission	Responsibility assigned to the personnel serving in an organization to generate profits for different types of people

From the theoretical concept, it is possible to structurally detach a concept of organizational mission; his statement would be: "The people who provide their services to the company (register the name), have the responsibility of generating products and services (indicate the name), to satisfy the needs (write them) of their clients (register the generic name), making them available to them, satisfying all their requirements, complying with legal provisions, taking care of sustainability over time" (Table 13.5).

Table 13.5 Mission structure

Order of elements of purpose	Elements
I	The personnel who provide their services
II	Legal person's name (company name)
III	He has the responsibility
IV	To generate (name of products and/or services)
V	To satisfy the needs (indicate them)
VI	Of your customers (generic name)
VII	Putting them at your disposal
VIII	Satisfy all your requirements
IX	With complying with the legal provisions
X	With a sustainable approach
XI	Through time

The mission of the organization contains the benefits to be provided to each different type of beneficiary; for a commercial enterprise, "profitability" must be included using the internal rate of return, companies whose primary purpose is not profit will omit this term and will include the specific benefits to be generated, in the case of public, social organizations and civil associations must identify their beneficiaries of each of them and will determine the benefits that have the mission of generating the personnel of their organizations. The structure of Table 13.5 is likely to be used to write the mission of organizations.

An example of writing the mission taking as reference Table 13.5 for a manufacturing company producing footwear, it would be: "The personnel who provide their services in the shoe manufacturer La Favorita SA de CV, have the responsibility of producing walking shoes to satisfy the needs of comfort and protection of the feet, of men and women from one year onwards in the Central Valley region of the State of Oaxaca, making them available in shopping centers, fully complying with their variety requirements, having all the measures, offering accessible prices, different forms of payment and, good service, generating profitability, complying with the legal provisions of safety for workers, taxes and, ecological, with sustainability over time".

The mission when described makes explicit the requirements of the clients, the needs to be satisfied and the legal precepts; the information contained in the mission does not come from genius of thought or from inspirations or occurrences; it is based on real information; research provides certainty, objectivity, meaningful and representative knowledge of reality.

Consistency with the policy is noticeable, fundamentally in this the beneficiaries are identified, in the mission it takes them as a reference to indicate the benefits to be provided to each of them.

13.4 Functions

In order for the personnel that provide their services in an organization to fulfill the mission of the organization, it is necessary to identify the indispensable functions for the generation of the products and/or services destined to the clients, they are the generators of the cascading benefits, that is to say, from the payment for the satisfaction of their needs, the clients will give back to the organization the monetary value that will be generated by the deployment of benefits to investors, workers, government, suppliers, and society; there are basic functions in accordance with the classical theory of administration (Mendoza 2013); to particularize the additional functions required of an organization, the company can be broken down into modules (Wu et al. 2006).

From the structure of the mission, the functional areas to be included in the organization emerge, the phrase "the personnel that provide their services" gives rise to an administrative function of attention to human resources, also, within this area the tasks will be taken of compliance with legal precepts and the collection of resources for all other functional areas of the organization.

The generation of products and/or services implies the inclusion of a production function, and this function must be supported by the others functions, the result of its development will be obtained tangible and/or intangible goods designed to satisfy the needs of customers; it is the reason for the generation of its expected value, it is the means to exchange products and/or services for its monetary value at market prices.

The part of the mission text of "satisfy needs", "the generic name of customers", "satisfying the requirements of customers", gives rise to the need for a sales function.

The assignment of responsibilities, sustainability, and representation of the legal entity requires an organizational leadership function.

Guaranteeing customers, the satisfaction of their needs is important for the survival of the organization, from the phrase "over time", they leave the functional area of verification of the organization's performance; in case things are not being done as planned, it will be required to have a functional area to correct, that area will be called "improvement"

The phrase "over time" also involves considering changes in likings, needs, emergence of new products and/or services of the competition, more efficient and effective technology, innovative materials, etc. This leads to the integration of a functional planning area responsible for maintaining adequate genetic information as future changes occur.

The functional areas to develop (Table 13.6) obtained from the mission are indispensable, contribute to the operational efficiency of the organization, are the means

Table 13.6 Functions to develop

Functional area	Link to mission text
Administrative	"Compliance with legal precepts" "Collection of resources for the operation of the entire organization" "Cost-effectiveness"
Production	"Generate the products and/or services"
Sales	"Make them available to the customer" "Satisfaction of needs" "Customer identification" "Compliance with customer requirements"
Leadership	"Assignment of responsibilities" "Sustainability" "Representation of the legal entity"
Check	"Guarantee the satisfaction of customer needs" "Survival of the organization through time"
Improvement	"Correct deficiencies" "Innovation"
Planning	"Changes of likings, variation in needs, new products and/or services of the competition, new technology, innovative materials"

to generate the products and/or services effectively. The names of the functions may vary; however, the functions need to exist in all organizations, if some are missing, their absence will be manifested in the expected results, there will be a performance not in accordance with the planned.

All functions are important, the tasks performed in each of them must be oriented toward the fulfillment of the mission, the main function, which requires more attention is that of production, generates the satisfaction of the needs of the client with the sales function you will be able to be giving positive results over time, these two functions are basic.

The administration's function directly supports those of production and sales, it also provides resources to all functional areas of the organization for its operation.

The verification and improvement functions contribute to the creation of loyal customers, ensure that all their requirements are satisfied, in addition to the personnel that provide their services in them, will strive to provide products and/or services with qualities and quantities that exceed the expectations of your satisfaction.

All functional areas must be at the service of production and sales; the planning function will be in charge of the existence of a harmonic coordination of actions between the functional areas, its task is that there be synergy, it will be necessary to generate the documents that regulate the action.

In every organization, there must be a command unit, the functional area where the decisions of the convenience of the structure of the organization will be taken, the changes, its expansion, its development, the innovations to be introduced, changes in the design of the products and/or services, will be the functional area of leadership.

13.5 Policy

The identification of the functions is necessary in the logistics of the gestation of the organizations, this information is not enough, the guidelines that will guide them toward the fulfillment of the mission and the unity of the scope of the purpose, the ideology of a dictated policy is required by senior management will lead to the aforementioned purposes.

The organizational policy must be developed by the highest competent authority responsible for the operation of the organization, if the investors do not have the competence, they will delegate the task of drafting it to the person to whom they will assign in the direction of the organization, who will help them in their conception; if there is insecurity in the description of the policy, it is advisable to hire a professional advisor; will provide the technical guide for the preparation of the policy, he will not develop it, he will supervise that the operational intentions of both the investors and the person who will be the authority in the organization are properly documented.

The essence of the policy lies in the induction of all personnel who provide their services in the organization, become aware of their commitment to contribute to the achievement of the purpose, and must identify how they participate with their services to fulfill the mission, must keep in mind the importance of satisfying the client's needs, seeking to exceed their expectations, having in mind to improve the organization over time; with these basic ideas, the concept of politics can be written as follows

Policy	The intentions of the investors, structured by the director of the organization, of having personnel committed to the fulfillment of the purpose taking into account the approaches on the form of operation predefined by them

The theoretical concept of politics is taken as a reference to structure the statement of the politics of an organization; it would look like this: "The people who provide their services to the company (name of the organization), is committed to fully meeting the needs (indicate them) of their customers (register the generic name), striving to exceed their expectations (indicate the requirements) and, improve the organization's system, maintaining sustainability in its operation, complying with legal requirements and making the company profitable over time" (Table 13.7).

The highest authority of the organization, director, manager, superintendent, president, as it is called, must expose the policy to all personnel who provide services to the organization, must be understood; each worker must deduce how he contributes directly or indirectly to it; it will be aware of the participation of its services for its scope and also of the consequences of the inefficiency of its work in its scope.

In the description of the policy of an organization, Table 13.7 can be used, for example, for a shoe manufacturing company. The text of its policy would be: "The personnel that provides their services for the manufacturing company La Favorita SA de CV, undertakes to satisfy the needs of protection and comfort of the feet when walking of men and women from one year onwards in the Central Valley region of Oaxaca, striving to exceed your expectations of variety, availability, price, all sizes,

Table 13.7 Policy

Order of elements of purpose	Elements
I	The personnel who provide their services
II	For company name (name of organization)
III	Commit their self
IV	To fully meet the needs (indicate each one)
V	From your customers (register the generic name)
VI	Striving to exceed their expectations (indicate customer requirements)
VII	And to improve the organizational system
VIII	Maintaining sustainability in your operation
IX	Complying with legal requirements
X	And making it profitable
XI	Through time

different models and good service; to improve the organization's system, maintaining sustainability in the operation, complying with legal requirements and, to make the company profitable over time".

The policy must live in the mind of all personnel all the time, each person will make it their own, express it in their own words, identify their work in its implementation, should know how to apply it in their work; it will help you make decisions without consulting your superior authority (Corvo 2019), it will make you more efficient in the development of your functions. Politics is the means of making workers understand the intentions of senior management (Ziegler 1966).

13.6 Vision

It is necessary to clarify the conceptual sense defined in this treatise of the vision and to differentiate it from the general concept managed in the literature of the works referring to the vision of the organizations; (Economic Encyclopedia 2020) coincides with them only in the expression of the future time. The leader of the company is committed to generating the vision, and Snayder et al. (1994) indicate that they must also design the process for his scope, under this idea, it is necessary to have a vision of the set of processes that the organization will develop.

In the logic of the gestation of organizations using their genetic information, it is necessary to have knowledge of the organic structure of the company; it is required to have a holistic view of its elements and functions. The concept of vision will lead to the provision of the most appropriate organizational system to fulfill the purpose; the personnel that provide their services in (and for) the organization must identify in which organ of the company it is located, where it is assigned the responsibility to carry out its activities for the development of the functions defined for the implementation of policy; under the above, the vision will be understood as:

Vision	Representation, as a whole, of the structure of the elements of the organization capable of contributing to the fulfillment of its purpose

Making the comparison with the traditional concepts of vision, the existing difference lies in the design of the organization through its schematization in the form of a system, identifying its elements and interrelating them: when operating the system, the expected results will be generated successfully, they will fully satisfy customers' needs, all your requirements will be satisfied.

The traditional generalized concept consists in the prose description of the organization; the scheme of its system is not made. The vision that the workers of a company have contributes to the attitude that they adopt in the company (Darbi 2012). The essential aspect of the vision is the determination of the system, and the application of the principle of coherence of genetic information implies the identification of its dependence on the nuclei of genetic knowledge; the vision depends on the functions defined after the mission and the policy.

The policy establishes the disposition to have an organization operating under the modality of a system, each of the functions requires organs that develop them, are elements to be incorporated into the system. Under the premise that efficient and effective systems are designed with the process approach, each of the organs of the system will be processes (Table 13.8).

Each of the processes in Table 13.8, correspond to the elements of the organization system, this is called "process approach"; the principle of cohesion induces the interrelation of processes, it is the result of the implementation of a process, and the product of one process will be input to another process. The relations in the processes have an order, it is evident that the leadership process is the initial one, the personnel that provide their services in the organization must direct the system with the implementation of this process, the result at their exit will be the authorizations, the assignment of the authority of each process and its responsibilities, from all processes receives information (Fig. 13.1).

The leadership process personnel represented by the process authority will emit the disposal that the planning process personnel design the system in general, will

Table 13.8 Relationship between mission and policy

(Mission) Functional area	(Policy) system elements
Leadership	Leadership process
Planning	Planning process
Administration	Administration process
Production	Production process
Sales	Sales process
Check	Verification process
Improvement	Improvement process

Fig. 13.1 Interrelation of the leadership process with the other processes

indicate the need to gather external information for the design of the system, will also define the result or exit from the system, must indicate that the stated purpose is fulfilled; each process will be assigned its authority; in general, all processes must be interrelated (Fig. 13.2).

The vision will be defined with the design of the interrelation of the processes, its structure is carried out by applying the PDVA cycle, called the "Deming" cycle. When the leadership process indicates the beginning of the operation, the planning process defines the period of production, quantities of products and/or services to be generated, this represents the letter "P", indicates "Plan"; administration, production,

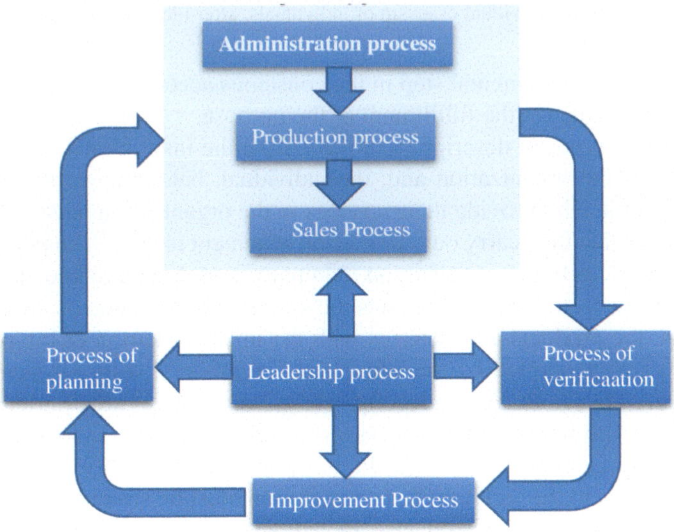

Fig. 13.2 Interrelation of the organization's system processes

and sales must implement the plan, represent the letter "D", means "Do", following the mission's specifications, the processes will be implemented.

The set of processes "D" must have complied with the settled down of the mission, having performed the activities indicated in the plan; it is verified if it was successful comparing the results obtained in "D" with what was planned in "P", this is done by the personnel that is responsible for the verification process, corresponds to the letter "V"; if there is any disagreement, the root cause is corrected or the root cause is sought to eliminate it. The improvement process personnel make a plan for this purpose, deliver it to "D" for its implementation, then proceed with "V", if the result of the verification is positive, the result of "D" corresponds to "P", the improvement process "A" is passed, the authorizations will be drawn up; being authorized will be sent to the planning process "P" to start the next cycle.

The vision of the organization implies knowledge of the system, its processes, its interrelations, its entrances and exits; it is important that all personnel know her; they will see the organization as a whole, it will allow them to visualize in the scheme the place within the process where they provide their services, they will infer how it contributes to the achievement of the purpose by participating in its performance; will identify the relevance of their work, will deduce the consequences of the deficiency of their work, this will allow the "introjection" of the system in the personnel, will unconsciously assimilate the ideas of the relevance of their work, the consequences of their deficiencies, will emerge in them a feeling of belonging to the organization, they will make it their own, they will worry about constantly improving their work, introjection is defined as (RAE 2019).

Introjection	"… unconscious process by which a subject incorporates activities, ideas, beliefs, etc., of an individual or group of individuals, after identifying with them"

Introjection is a fundamental step in the conscious acceptance of a commitment, an essential condition for the fulfillment of the purpose.

There are two types of description of the vision: the institutional, corresponding to the vision of the organization and, the individual, belonging to the vision that each of the people who provide their services in the organization have of their individual activities that they carry out. The vision statement is this: "The personnel that provides its services to the organization (company name), having knowledge of the client's and legal requirements, operates the system whose elements are processes; will generate the products and services with the implementation of the production and sale process, supported by the administration process, with a system that has plans authorized by another of leadership, where your changes are decided, and with the verification and improvement processes to fully guarantee, satisfaction of your customers' requirements, achieving profitability, with sustainability, concerned about system improvement through time", its description must be structured (Table 13.9).

The introjection of the vision also includes the personnel who provide their services in or for the organization, those who have hired, external personnel who perform some work for it, both together constitute the collective intelligence of the organization, their mental process coordinated by the leadership process. The human

Table 13.9 Institutional vision

Order of elements of purpose	Elements
I	The personnel who provide their services
II	In and for organization (name)
III	Having knowledge of customer requirements and
IV	The legal
V	Operates a system consisting of processes
VI	Hat generate the products and services with the implementation of the production and sales process, supported by the administration process
VII	With an authorized system generator planning process
VIII	By another of leadership, where their changes are decided
IX	And with the verification and improvement processes to fully guarantee
X	Satisfaction of your customers' requirements
XI	Achieving profitability
XII	With sustainability
XIII	Concerned about system improvement
XVI	Through time

intellect is the capital of greater value in the organization, with the introjection when making the system their own, the personnel will worry about their work, will commit themselves to seek the continuous improvement of the activities that they carry out. An example of a vision for a shoe manufacturing company would be described with this statement: "The staff that provides their services in and for the shoe manufacturing company La Favorita SA de CV, having knowledge of the need to protect the feet of their customers with comfort, low prices, variety, diversity of sizes, ease of acquisition determined by the client and the applicable tax, environmental and security legal provisions; operates the shoe manufacturing manufacturing system for men and women older than one year of the Central Valley region, Oaxaca; with the implementation of the processes of: production, sale, administration, planning, leadership, verification and improvement; fully guarantee the satisfaction of their clients' requirements, sustainability, profitability and continuous improvement over time".

The individual vision also needs to be developed by each of the people who work in the organization, at all levels, one way to describe individual vision it can be written in the following way; "Me, (name of person), having knowledge of the company system (name or name of the organization), where I provide my services like (indicate the name of the position) in the process (write process name), performing my activities of (include the list of activities developed), I contribute to satisfy the needs of (indicate the needs you satisfy with the result of your work in another process), of the activities of (indicate the name of the activities carried out in the other process) (write the name of the process) by the personnel that holds the position of (write the name of

Table 13.10 Individual vision (one for each person)

Order of elements of purpose	Elements
I	Me, (name of person), having knowledge
II	Of the company system (name or name of the organization)
III	Where I provide my services (indicate the name of the position)
IV	In the process (write process name)
V	Performing my activities of (include the list of activities developed)
VI	I contribute to satisfy the needs of (indicate the needs you satisfy with the result of your work in another process)
VII	Of the activities of (indicate the name of the activities carried out in the other process) (write the name of the process)
VIII	By the personnel that holds the position of (write the name of the position)
IX	With the result of my performance I contribute the value of (indicate what the work done is useful)
X	In case I had any deficiencies in my work
XI	I generate a problem in the activities of (name of the activities) of the process (indicate the name of the process)
XII	Which will have difficulty in the development of its activities

the position), with the result of my performance I contribute the value of (indicate what the work done is useful) in case I had any deficiencies in my work I generate a problem in the activities of (name of the activities) of the process (indicate the name of the process) which will have difficulty in the development of its activities"; this declaration of individual vision has the structure indicated in Table 13.10.

Each process will need to be documented later, there are standards that can serve as auxiliary for both the development of the system and for the documentation of each of the processes, ISO 9001, referring to quality management systems will serve as the basis for such effect. In the international standard ISO 9001, the use of the Deming cycle is recommended, having the leadership process as a starting point, the mentioned cycle does not include it. There will be a process for planning, then others intended to implement the planned; with the intervention of the process aimed at ensuring effective performance, the verification process, who passes the information to the improvement process, identifies the deficiencies, makes plans for its correction, finds the root cause of the deficiencies, elaborate how to eliminate them, send the plans for their implementation in the process where the product and/or service is generated. There will be a reprocess in case of making corrections. When what has been done has been in accordance with what was planned, there is the awareness of observing where you can have some intervention in order to improve the system, these opportunities are passed to have a more efficient and effective system to the process of improvement who will be responsible for making the corresponding plan, will be sent to the leadership in order to authorize the changes, ending the cycle, and starting another with the changes made previously to the system.

The vision can be designed with the quality management system model (IMNC 2015)(IMNC 2015); the administrative strategy must be designed through processes, developing the sequence of them in their implementation for their greatest understanding (Wiedemann 2013).

13.7 Culture

The vision indicates the responsibility of the personnel to operate the system of the organization, its behavior is an extremely important factor, it is the difference between the effective fulfillment of the purpose or its partial fulfillment.

The idiosyncrasy of the organization will be determined by the values to be culminated in developing (Ortiz Ordaz et al. 2010) the functions by its personnel, conceptualizing the culture as:

Culture	Way of acting by custom (habit) of the personnel of an organization induced by its valuation attitude of important aspects common to all

It is clear the need to identify the values to cultivate, hence the valuation attitude is born, it gives a value to certain ways of acting concerning different important aspects for the organization. Values are the key to maintaining a beneficial organizational culture, it will foster obedience to the provisions established in the organization by the personnel. Organizational culture influences worker behavior and varies from organization to organization (Pfister 2009). The values of a company are part of the organizational culture, they induce the behavior of employees; the exercise of values in the development of activities by workers provides them with a basis for creating the way they will feel in the organization (Wallkiser 2005).

The initial value to instill in the personnel is “self-training”, they themselves must read the manuals, procedures, observe the technology and subsequently receive training from competent personnel; the personnel will collaborate by having sensitized to the development of their individual vision, their repetition to defined intervals will ensure their internalization, knowledge and skills needs will be identified, the training will be effective, each worker will know what aspects should be strengthened.

“Attention” is another value to consider; after the personnel is trained, it requires diligently observing from the preparation of materials, services, and technology, to obtaining the results, also the regulatory provisions established in their activity within their process and applying them 100%.

The personnel should have in mind their responsibility to fulfill the purpose, they will know that it belongs to everyone, they will achieve it by fulfilling their mission with the operation of the system; now the value of “synergy” is introduced, it will make them more efficient, they will work as a team, each one will be a leader in their own activity, they will collaborate in those that require additional support.

Notwithstanding the existence of a verification process of the entire system and, of the inclusion in each process of a component of activities with personnel dedicated to their verification, each worker must verify their performance ensuring an effective result; this will be achieved with the introduction of the "assurance" value of individual efficiency and effectiveness.

In exceptional cases of the existence of deficiencies and inefficiencies detected by the same worker in his developed activities, he himself will co-correct himself based on his experience, guided by tacit or documented procedures, will take as reference the standards, the norms and instructions of work or any other normative document applicable to its tasks; after correcting or eliminating the root cause, the deficiencies will have been corrected, the improvement will be assessed.

In situations where there have been no deficiencies and the result was effective, the value of "improvement" will encourage workers to provide ideas to increase the measure of performance variables in case their result is ascendant. expected: greater resistance, greater measurement accuracy, etc., if the value of the variable is descending, for example, less time, less waste, less energy consumption; the personnel is the one who knows best about their activities, is the authority in them in their process, will look for the way to be more productive, together they can have meetings aimed at finding ways to exceed customer expectations, basic condition for the survival of the organization over time will generate loyal customers to the product and/or service.

To recap, the values that will be cultivated will be: self-training, attention, synergy, assurance, and improvement (Fig. 13.3).

Inherent habits of the social nucleus will then be identified, from the cultural environment from which each person who causes inefficiencies or inefficiency comes from; they will be counteracted with the values to be established within the organization such as: punctuality, order, cleanliness, and silence. In the operation of a system, organizational culture leads to the achievement of self-reliance results in the direction of attitudes; each worker will know what to do and how to do it; a harmonious organizational climate is generated, it keeps the personnel of all the hierarchical levels happy, it makes the organization more productive, the purpose will be achieved continuously over time by constantly fulfilling the mission of all the personnel who provide their services in or for the organization, given its knowledge of the developed system, it will generate effective results, leaving customers fully satisfied, making

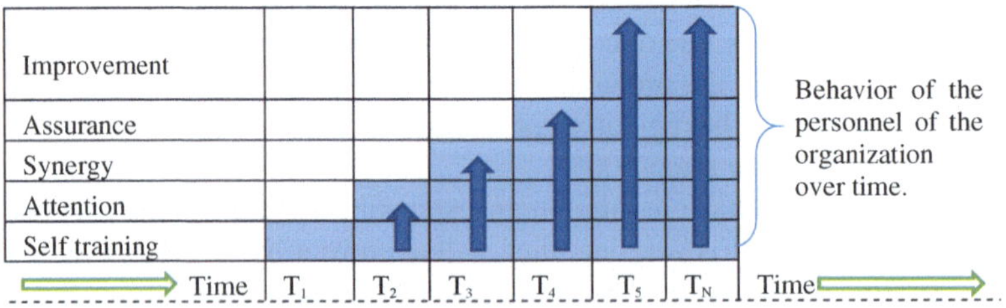

Fig. 13.3 Concatenation of values

the business profitable, operating with sustainability and, complying with the legal precepts.

Each person will feel comfortable with the organizational climate, there will be motivation in the workers; the well-being offered by the organization toward its personnel will generate an attitude of loyalty toward the organization, they will be satisfied and proud to provide their services in the organization.

It is the task of the authority of each process to request resources from the administrative process, which shows its commitment expressed in the organization's policy.

Two streams of information will guide the operation of the organization, the purpose and the policy, the provisions established in these will guide the fulfillment of the mission with the implementation of the system, personnel developing the functions determined for each process, with a beneficial culture for the achievement of productivity (Fig. 13.4).

The information emerges in cascade, the purpose is generated for the fulfillment of the mission, from this the functions with which the system implicit in the vision is designed, the policy provides data conducive to the choice of values, will give rise to personnel behavior; this genetic information will provide the basic elements required for an effective design of the organization's system

The participation of all the workers in the elaboration of their individual vision, commits them to their active participation, the investors and the director of the organization having developed the policy, will understand the needs of obtaining resources for the efficient implementation of the system, that efficiency will generate the expected profitability. Personnel behavior through instilled values will contribute to operational efficiency by reducing production costs in the implementation of organizational system activities.

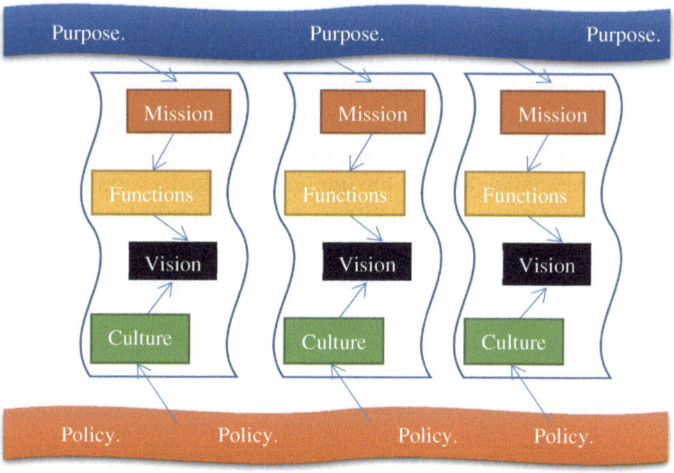

Fig. 13.4 Genetic information of the organization

A good design of the information for the gestation of the organization has the risk of failing having a poor organizational culture, a poor design of the system can be improved with the cultivation of the values determined to be considered in the conduct guide from the workers.

Personnel turnover will be avoided, there will be empathy, the feeling of belonging to the organization will be aroused, the personnel will make yours the mission, fight to improve; human capital with a favorable culture for the performance of its activities will be self-stimulated for its development within the company generating a virtuous circle. With the search for exceeding customer expectations in satisfying their needs and, of all their requirements, the company will be generating quality results, guaranteeing their permanence over time, hence the importance of having a consistent genetic information, with cohesion and appropriate to its environment, constitutes the basis of permanent success.

The introjection of genetic information by personnel who provide their services in and for the organization is necessary and indispensable for their survival over time, it must be operationalized, it can be done in many ways depending on the experience, for example, notebooks with the formats of purpose, mission, functions, politics, institutional, and individual vision and values may be delivered to each worker, leaving values of limited space with a horizontal line where each person will write the ideas that complete the concept of each nucleus of the genetic information, must do it with their handwriting.

A technical design of the genetic information of an organization is a fundamental pillar in its operation with profitability, sustainability, complying with the legal provisions, making it last through time operating efficiently and effectively.

13.8 Conclusions

The technically developed information corresponding to the gestation of the genetics of an organization contributes to increasing its resilience to changes that provoke endogenous or exogenous factors, both gradual and radical, due to the flexibility provided by the synergy of its processes, for the introjection of the organizational philosophy of his personnel, and by having an individual vision of their work area with a continuous improvement mentality each service provider to the organization.

The inclusion of the process approach will keep the organization constantly monitoring its environment so that, when identifying any opportunity, a prospecting is carried out to determine the suitability of its use, in case of identifying a menace it will determine the degree of risk that it will represent for the organization, with the intention of to evaluate the possibility of doing a project to eliminate the potential effects or reduce them.

The adequate design of the information that gives origin to the gestation of an organization represents a competitive advantage with respect to the other existing organizations in the industry to which it belongs, it ensures success in obtaining its results, makes it keep operating indefinitely through time.

For both, for profit and non-profit public, private or social organizations, the information on genetic gestation will be correctly formulated using the proposed models for its development, with its implementation, will make them exceed the expectations of their clients and/or users, who will be the best promoters of its efficiency and effectiveness.

The information of the genetic gestation of an organization correctly structured will convert it into a leading organization, represents its DNA that gives it its way of being to act efficiently, having an effective performance.

References

Baldoni J (2011) Lead with purpose. Giving your organization a reason to believe in itself. American Management Association

Barraza HJ (2019) What is the difference between Purpose, Mission and Vision?, Dictionary entrepreneur

Corvo HS (2019) Business policy: features, types and examples. Lifeder.com. administration and finances

Darbi WPK (2012) Mission and vision statements and their potential impact on employee behavior and attitudes: the case study of a public but profit oriented tertiary institution. Int J Bus Soc Sci 3(14)

Economic Encyclopedia. (2020). Vision. Administration. https://enciclopediaeconomica.com/vision/

IMNC (2015) Quality management systems. Basics and vocabulary. NMX-CC-9000-IMNC-2015

IMNC (2015) Quality management systems. Requirements. NMX-CC-9001-IMNC-2015

Lozano O, Gómez H, Rositas I (2020) What is organizational DNA?. Forbes México

Mendoza I (2013) The basic functions of the company according to Henry Fayol. UTEL Publisher

Muñoz Paredes A (2020) Purpose of the organization. Wolters Cluwer. https://www.guiasjuridicas.es/Content/Documento.aspx?params=H4sIAAAAAAAEAMtMSbF1jTAAASNTY0tLtbLUouLM_DxbIwMDS0NDQ3OQQGZapUt-ckhlQaptWmJOcSoA_WJgEzUAAAA=WKE

Ortiz OF, Camargo T, Isis A (2010) Proposal of values for an organizational culture in sustainable tourism. Tourism and national patriotism magazine. Pasos 8(1)

Pfister JA (2009) Managing organizational culture for effective internal control. From practice to theory. Physica-Verlag a Spring Company Pubisher

RAE (2019) Spanish language dictionary

Snayder NH, Dowd JJ, Houghton DM (1994) Vision, values & courage. Leadership for quality management". The Free Press

Strichow HJ (2013) Our ultimate purposein life. Balboa Publisher

Taiwo AA, Lawal FA, Agwu PE (2016) Vision and mission in organization: myth or heuristic device?. Int J Bus Manage 4(3). March

Wallkiser A (2005) Successful branding. A critical examination of customer experience management for personal international. Diploma thesis

Wiedemann L (2013) Business strategies and value chain management". Author house publisher

Wu T, Wu W, Guo S, Chen R (2006) Studies of enterprises¨ modularization descompsition. Research and practical issues of enterprise information systems. Spring Publisher

Ziegler RJ (1966) Business policies & decision making. Appleton-Century-Crofts Publisher

Chapter 14
Work Engagement as Result of Leadership and Employer Branding: A Case of a Mexican Manufacturing Plant

María Teresa de la Garza-Carranza, Eugenio Guzmán-Soria, Jorge Armando López-Lemus, and Quetzalli Atlatenco-Ibarra

Abstract This chapter proposes a structural equation modeling to analyze three latent variables: employer branding, servant leadership, and work engagement. For testing the dependent variable work engagement, three different models are proposed and seven hypotheses are tested. The model is validated with data obtained through a questionnaire that was administered to 147 workers on a manufacturing setting settled in the central part of Mexico. The hypotheses are tested using parameters of goodness of fit using maximum likelihood procedures. The results indicate that work engagement can be accomplished using strategies of employer branding and servant leadership.

Keywords Work engagement · Structural equation modeling · Mexican workers

14.1 Introduction

Turnover intentions of employees are an increasing problem to be addressed by the human resource managers in many manufacturing plants. The relationship the employee has with his environment and his leaders constructs an identity of a social group that is conceptualized through actions and strategies that the company designs, to keep the productivity levels required to promote the financial results of a company. This is the case of many global manufacturing companies settled in Mexico that have become part of the economy and development of the country. These companies have taken advantage of the trade global system of Mexico, which includes agreements with other countries as the European Union, Canada, the USA, and others.

M. T. de la Garza-Carranza (✉) · E. Guzmán-Soria
Departamento de Ciencias Económico Administrativas, Instituto Tecnológico de Celaya. Av, García Cubas 600, Celaya, Guanajuato, Mexico
e-mail: teresa.garza@itcelaya.edu.mx

J. A. López-Lemus · Q. Atlatenco-Ibarra
Departamento de Estudios Multidisciplinarios, Universidad de Guanajuato, Campus Irapuato-Salamanca, Sede Yuriria, Yuriria, Guanajuato, Mexico

J. L. Garcia Alcaraz et al. (eds.), *Trends in Industrial Engineering Applications to Manufacturing Process*, https://doi.org/10.1007/978-3-030-71579-3_14

The purpose of this paper is to identify key elements that are embedded in the workplace in manufacturing plants that take relevance for the engagement of the employee. According to Steffens et al. (2018), a "growing body of research shows that health in the workplace is impacted by the sense of identity that employees derive from membership in social groups (p. 2)". This social group is affected by the leading practices and by the strategies to retain the employees at work, among others.

For the objective of our paper, we decided to work the strategies derived from recent approaches to involve workers such as employer branding. This theory derived from marketing allows the attraction of employees to their own company using diverse tactics that promote loyalty. In general, the concept is used to create a social group with values that are shared by the employees that include in some cases the worker's family. On the other hand, we used the servant leadership theory to facilitate the enclosure of the worker to the environment of the company. Servant leadership promotes the inclusion of the employee as a member of a group where the values of equity, empowerment, and development are included. The results of our study conclude that there is a strong relationship between the studied variables. Further, we suggested strategies for implementing actions to rise work engagement between workers for increasing the retention of employees.

14.2 Literature Review and Hypotheses Statement

14.2.1 Work Engagement

According to Schaufeli et al. (2002), the definition of work engagement (WE) or "involvement with work" is as follows: It is a positive state of satisfaction and in relation to the mental study of work characterized for vigor, dedication, and absorption. More than a momentary and specific moment, the WE refers to a cognitive–affective state that prevails and does not focus on a particular object or event of individual behavior. In general, this concept has three factors: vigor, dedication, and absorption. The vigor is characterized by high levels of energy and resilience while working that makes one invest effort in the work and persists despite the difficulties. Dedication refers to being strongly involved in work with a sense of meaning, enthusiasm, inspiration, pride, and challenge. Absorption refers to being fully concentrated and happily absorbed in work in such a way that time passes quickly, and the workers have difficulty getting rid of it. On the opposite side, the separation and multiple defenses of people through behaviors that demonstrate both physical and emotional absence, generating passive and incomplete performance, and this phenomenon is known as personal disengagement (Kahn 1990, p. 701). In this sense, when a worker disengages from work, it would not be as productive as he could be (Pech and Slade 2006). Personal disengagement is the simultaneous disconnection, discouragement in the employee behavior that promote a lack of connection with peers and managers that affects his level of performance.

The work engagement (WE) is defined as a positive and satisfactory state of mind characterized by vigor, as well as high levels of energy while working, desired to invest effort in work, and persistence in the face of difficulties (Schaufeli et al. 2002). It is assumed that the WE is the opposite of burnout or "Burning Syndrome", and the concept was developed by Maslach (1993). Burnout is characterized by exhaustion (emptiness of mental energy), cynicism (a negative attitude toward work), and reduction of professional effectiveness (the belief that one is no longer effective in fulfilling job responsibilities). Other authors refer to exhaustion and cynicism as the relevant components of burnout, which is illustrated by the relatively low correlations of professional efficacy with the other two components. In contrast to burnout, the WE at work is defined as compliance, a positive state, related to the work of the mind. WE is characterized by vigor (high levels of energy during work, the willingness to invest effort in work, and persistence in the midst of difficulties), dedication (feeling of enthusiasm, inspiration, pride, and challenge), and absorption (being happily absorbed in the work), so time goes by quickly and one has difficulty in disengage. Vigor and dedication are considered as WE scoring dimensions, while absorption resembles flow, an optimal state of experience, and seems to act as a consequence of the other two variables.

In México, Maldonado-Macías et al. (2015) identified that in Mexican manufacturing plants located in the northern part of Mexico burn out exists due to high levels of stress experimented at work. Also, due to the excessive load of work in Mexican manufacturing plants, Macias-Velasquez et al. (2019) validated the burnout the inventory and found that such phenomena exist in middle and senior managers. Valadez-Torres et al. (2017) also contributed to study burnout in Mexico and related it to physical pain.

Demerouti et al. (2001) proposed a model where it was shown that disengagement is related to factors associated with work resources such as feedback, rewards, participation, job security, and supervisor support. Jubert and Roodt (2019) mention that there are factors that help or stop employees' WE. The factors that help the development of the WE are the organization, the resources, the opportunities of advancement, the safety in the work, the efficiency of the administrator, and the leadership. The factors that do not help the creation of the WE are workload, job insecurity, family conflicts, unclear goals, and intentions to resign. These same authors propose that we have obligations to see at three levels: individual, team, and organizational.

A study conducted by Coetzer and Rothmann (2007) in South Africa stated that employees will be more engaged if organizational factors are supported and developing opportunities are provided. Their study was done in a manufacturing setting, and they concluded that work engagement was best predicted by organizational support and growth opportunities in the job. Job demands showed a weak relationship with work engagement. The energy or vigor that a person plays in the organization is a concept that has been studied by some work researchers (Yanget al. 2019). This vigor can be achieved through interaction with people since energy is a mechanism of influence of social interaction because it is spread among people (Owens et al.

2016). This contagion of energy and vigor influences the performance of people in the organization.

A study conducted by Carter et al. (2018) found that there is an important positive relationship between self-efficacy, with WE and organizational performance. Similarly, Alua et al. established a relationship between employee self-efficacy and job embeddedness. In relation to leadership, Tims et al. (2011) concluded that transformational leadership has a positive effect on employees' WE through some factors such as self-efficiency and optimism. Bailey et al. (2017) conducted a review of various aspects of WE. Their conclusion is that there is evidence that high levels of WE seen from an integral point of view are beneficial for both the individual and the employees. In support of previous results, Saxena and Srivastava (2015), established a relationship between employee engagement, organizational culture, and its impacts on organizational performance in terms of productivity and profitability in manufacturing plants in India,and in their study, they used white- and blue-collar workers. However, there are still gaps for study between the direct effects of leadership and the WE due to the direct influence that a supervisor can exert.

In a study by Shimazu et al. (2015), they sought to distinguish two types of employees, the first with a high investment in work known as "workaholism" (WH) and the second with work engagement. WH is defined as a tendency to work excessively hard and to be obsessed with work, which manifests itself as a compulsive worker. WE is defined as a satisfactory state, related to work with a mental state characterized by vigor, dedication, and absorption. WH is driven by an internal momentum that they cannot resist, and employees who are engaged are intrinsically motivated. According to Crawford et al. (2010), WE is a mediator between perceived organizational support and task performance. In a deep study performed by Wang et al. (2017) among manufacturing workers in China, they suggest that companies should increase strategic ways to develop programs to increase inner positive resources of their workers to enhance mental health and improve organizational performance.

14.2.2 Employer Branding

The employer branding (EB) concept was conceptualized in the USA at the end of 1996 as a response of companies to the decline in birth rates of population and, therefore, to the need to find valuable young people. It is Spain that began to know in the early years of the twenty-first century. Today, it has become generalized as a personnel retention strategy because organizations need to have a prestigious brand that attracts talent and loyalty. It is a technique that applies marketing tools to people management and is based on levers such as processes, communication, values, training, or new technologies (Jiménez 2015). According to with Kunerth and Mosley (2011, p. 19), employer branding is "The package of functional, economic and psychological benefits provided by employment, and identified with the employing company."

Because the retention of human talent for the achievement of organizational objectives is vital in an organization, the search and development of future employees become a strategic activity (Martin et al. 2011). According to Jain and Bhatt (2015), EB helps create a positive image of the organization by sending the right message to current and potential employees. It is also important to develop an understanding of the key attributes of the organization that helps employees who are cohesive to the organization giving a reason to return. The term EB was coined by Ambler and Barrow (1996) as a way to combine brand technique with human resources practices. The EB has expanded massively in research and in the number of companies that investigated it (Barrow and Mosley 2005)

According to Backhaus and Tikoo (2004), the human resources literature describes the concept of EB as a three-step process. First, the company develops a "value position" that has to be incorporated into the brand. Using this information of the organizational culture, its management style, the qualities of current employees, the image of current employees, and the practices of the product or service developed, a concept of particular value that the company offers to employees is developed. The intention is to create a realistic representation of what the company offers to employees, and the "value position" provides a central message that is covered by the brand.

Once the "value position" has been developed, the firm markets its employees with potential objectives, hiring agencies, recruitment fairs, etc. The external marketing of "employer branding" is basically designed to attract a target population, but also to support the product. This is why EB's external strategy must also be compatible with other company brand efforts. The internal marketing of employer branding is the third aspect. It is important because it undertakes the promise made to the employees recruited in the organization and incorporates it as part of its organizational culture. The objective of internal marketing or "internal branding" is to develop a workforce committed to organizational values and goals. They must be congruent with the product, but they have elements that differentiate: (1) It is specifically oriented to the employee and characterizes its identity, and (2) it must be oriented to internal and external audiences and should be different from the company's general *marketing strategy*.

The EB in recent years has achieved prominence in the midst of human resources managers as a significant strategy to create sustainable growth of the organization. Building a reputation brand and positive employer can be considered as vital strategies to create a distinctive and attractive identity with which current employees and potential job seekers can recognize each other. As an example, Coca-Cola Hellenic—the second-largest company of the Coca-Cola system—started with this HR strategy ten years ago and has increased the levels of WE. Also, it has improved talent management and invested in employee development, and the strategy has contributed to the perceptions of an organization as a great place to work (Kunerth and Mosley 2011).

Brand distinguishes products from each other. For example, the sports market is highly competitive like brands as Adidas, Nike, etc. A branding development, from a marketing point of view, includes branding orientation, brand equity, and the brand value chain model, and this concept is compatible with an HRM strategy to develop EB. In a recent study, Tumasjan et al. (2016) conducted a study to link the EB strategy to firm performance. In their study, they included manufacturing industries. According to their results, they found a positive effect between EB and the incumbent of employees. They did not find a positive relationship between EB and recruitment efficiency. In this sense, HR managers need to develop efforts in two sense: internal EB through the enrichment of organizational climate and external EB through the selection of best people in the market. Also, Ferreira (2018) conducted a study in a multinational manufacturing plant of cosmetics to find a relationship between EB and WE. They found that "innovation and growth" were very important, and a rewarding environment contributed to the engagement of workers.

Miles and Mangold (2004) developed an employee branding process that includes four steps: sources of messages, perceptions, interpretations, and consequences. The stage of sources of messages relates to the internal elements that comprise the human resource management systems, the public relation system but also informal influences like organizational culture and coworkers and the managers or leader´s behavior. This stage also is influenced by external sources such as advertising and customer feedback. The perception phase or second stage is related to the psychological contract that is created when people join the organization. This contract establishes the perceptions that new employee has in relation to his new job. This understanding includes that the employee must do his best, but also, the worker expects the organization to be fair, to be caring, and to give him the opportunities he deserves. In this stage, it is very important that human resource management develops an employee brand image for conducting the ideal behavior of employees. The internalization of the mode occurs in the third phase. In this step, the objective is that the employees feel a high level of trust in the organization for which they work. The psychological contract that was made in the previous stage has to be settled in the mind of workers and corroborated in the time. Finally, the consequences of the previous phases are evaluated. In this final stage, the human resource managers must evaluate employee satisfaction and performance and service quality but also if the rotation of personnel is being reduced by the use of the EB strategy.

As a consequence of the EB strategy and the WE process in organizations, we state the following hypothesis:

H1. The employer branding strategy is related to the WE dedication of the employee in a manufacturing setting.
H2. The employer branding strategy is related to the WE absorption in a manufacturing setting.
H3. The employer branding strategy is related to the WE vigor in a manufacturing setting.

14.2.3 Servant Leadership

Researchers have discussed leadership since the beginning of the last century. There have been attempts to integrate all these theories (Derue et al. 2011) into leader traits and behaviors. However, new leadership theories are being developed during the time. Leadership styles have changed according to the needs and requirements of modern organizations. The new organizational forms that have been created as a result of technological changes have led to new ways of understanding between leaders and subordinates. Robert Greenleaf (1904–1990) developed the concept of servant leadership or servant leadership in 1970 as follows:

> [Servant leadership] begins with the natural feeling that one wants to serve, to serve first. Then conscious choice brings one to aspire to lead. . . . The difference manifests itself in the care taken by the servant—first to make sure that other people's highest priority needs are being served. The best test . . . is: do those served to grow as persons; do they, while being served, become healthier, wiser, freer, more autonomous, more likely themselves to become servants? And, what is the effect on the least privileged in society; will they benefit, or, at least, will they not be further deprived? (Greenleaf 2002, p. 15)

Van Dierendonck (2011) proposed the following six characteristics of service leadership after an exhaustive review of the literature.

Empowerment and development of people. The approach is to enable people to be proactive, trust themselves, and develop their personal power. These actions encourage people to develop. Service leadership seeks the value of each individual to recognize him, as well as to identify his abilities so that the person can continue learning.

Humility is the second characteristic of service leadership. It refers to putting personal talents and achievements in perspective. The leader is sought to recognize that he can learn from others. It is about putting the interests of the subordinates first, facilitating their performance, and providing them with support activities that help their development.

Being authentic is the ability of a person to "be her /himself" consistent with her/his own thoughts and feelings. This characteristic is linked to integrity and a code of perceived moral conduct. This characteristic is manifested in several aspects: doing what is promised, visibility with the organization, being honest, and vulnerable, that is, being a person before assuming a professional role.

Interpersonal Acceptance. It is the ability to understand and experience the feelings of others and where people come from. It includes empathy, which focuses on being able to cognitively adopt a psychological perspective of the other and experience feelings of warmth, compassion, and forgiveness in terms of concern for others even in situations where confrontation is made for mistakes, arguments, and mistakes.

Provide address. It ensures that people know what is expected of them, of benefit for the company and for them. The leader must design the work based on the skills and needs of the people. In this sense, management is being accessible that supports a high interpersonal relationship.

Career. It is the desire to take responsibility for a large institution and to serve instead of controlling and seeking one's own interests. It seeks to assume an example role for others. The characteristics of a good server are responsibility, loyalty, and teamwork.

Liden et al. (2008) proposed seven dimensions to identify the leadership of service, and the above was based on previous studies conducted by Barbuto and Wheeler

(2006). The indicators are the following: Emotional healing (EH): the act of showing sensitivity to others by a personal situation. Creating value for the community (CV): Genuine and conscious concern about helping the community. Conceptual skills (CH): Possess the knowledge of the organization and its tasks in such a way that it can guide and assist others, especially close followers. Empowerment (E): To encourage and facilitate others, especially those immediately followed to identify and solve problems, as well as to determine how and where to complete the tasks. Help subordinates develop and succeed (HS): Demonstrate a genuine concern for the careers of others and provide support and mentoring. Subordinates first (SF): Use actions and words to clarify others (especially subordinates) to meet their needs at work as a priority and willing to interrupt personal work by helping others. Ethical behavior (EB): Interact openly, fairly, and honestly with others.

Although service leadership is an emerging theory, Eva et al. (2019) carried out a review of the concept and related it to previous studies that showed the relationship between various attitudinal and performance factors. In relation to behavior, studies confirm a positive relationship with citizenship behavior. The most representative studies on this relationship are those of Linden et al. (2008) and de Zhao et al. (2016). Service leadership has also been associated with positive attitudes of followers such as organizational commitment (Van Dierendonck et al. 2014) and satisfaction (Miao et al. 2014). Choudhary, Akhtar and Zaheer (2013) related service leadership to organizational performance with a mediating variable of organizational learning. Service leadership has also been related as a precursor to customer value creation (Hsiao et al. 2015). A few studies of servant leadership have been done in manufacturing organizations. However, Chiniara and Bentein (2016) analyzed a big manufacturing company in Canada in relation to servant leadership, employee satisfaction, organizational citizenship behavior, and performance; they found a relationship between these factors.

The impacts of service leadership can be measured at three levels: individual level, team level, and organizational level. Relations of positive behavior in the organization have been studied, and its relationship with performance has been demonstrated (Cameron and Caza 2004). Positive behaviors (Searle and Barbuto 2011) at the individual level (hope, self-efficacy, resilience, optimism, creativity, etc.), those at the team level (psychological capital, collective self-efficacy, etc.) and at the group level (virtuosity, forgiveness, gratitude, organizational citizen behavior, etc.). These behaviors promote better results in the organization. Authors such as Newman et al. (2017) and Gillham et al. (2015) studied these types of relationships in different environments, finding positive effects influenced by service leadership. Even, Dutta and Khatri (2017) propose that it can help reduce the intention to quit employment.

Taking in account the revision of the theory above mentioned, we propose the following hypothesis:

H4: There is a relationship between EB and servant leadership in a manufacturing setting.

H5: There is a relationship between servant leadership and the WE (dedication) in a manufacturing setting.

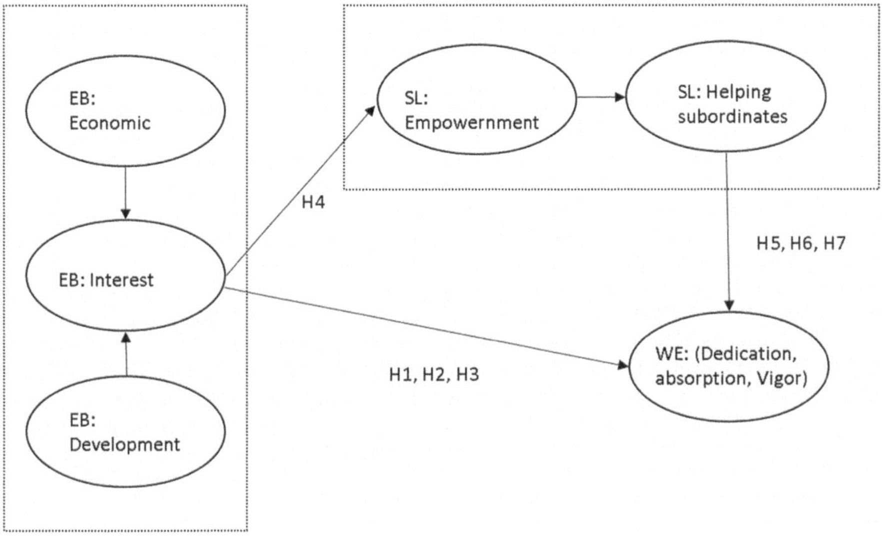

Fig. 14.1 Hypothetical model

H6: There is a relationship between servant leadership and WE (vigor) in a manufacturing setting.
H7: There is a relationship between servant leadership and the WE (absorption) in a manufacturing setting.

The hypothetical model to be proved is as follows (Fig. 14.1).

14.3 Methodology

The following sections describe the methodology followed to test the hypothesis presented in Fig. 14.1.

14.3.1 Sample

The sample of workers was integrated by 147 cases of a manufacturing plant ubicated in the Guanajuato State (México). The plant is located in an industrial cluster where many industries are competing for the best human resources available. Most of the participants agreed to participate, but a few cases did not answer the questionnaire (less than 10%). The plant started operations in 2013 and its part of a car industry global manufacturer. The data was raised between the floor shop workers so the majority (85%) were men and 16% are women. The ages vary according to Table 14.1.

Table 14.1 Age of participants of the study

Age (years)	Amount	%
18–22	56	38
23–27	46	31
28–32	30	21
36–36	8	5
More than 37	7	5

As is shown in Table 14.1, most of the employees are part of the millennial generation. The distribution of education of participants is as follows: 16% has basic studies (9 years minimum), 39% has a high school (at least 12 years), 27% has a technical degree, and finally, 17% has at least an undergraduate degree. Working experience considering the years out of the current job is less than 3 years 45%, between 4–7 years, 32%, and more than 8 years 23%. Most of the participants (80%) have less than 3 years working in the plant, and 20% have 4 years or more. The reason why many of the personnel have a few years working in the plant is due to high rotation rates, and also, the plant has only a few years in operation.

14.3.2 Measures and Descriptive Analysis of the Items

The following questionnaires were used to test the hypothesis.

For work engagement (WE), the short questionnaire developed by Schaufeli et al. (2006) was used. The questionnaire has been used extensively in literature as in Xanthopoulou et al. (2007) and Bakker and Schaufeli (2008). In Latin America, the scale was validated by Chavarría et al. (2017). The scale used was a Likert (0 never to 6 every day). The following table shows the items, the mean, and the standard deviation of the sample. The latent variables, in this case, are dedication, absorption, and vigor. Table 14.2 shows the basic statistics of the analysis.

For employer branding (EB), the questionnaire developed by Bethon et al. (2005) was used. Only three variables of the instrument were taken to assess the EB effectiveness "interest," "economic," and "development." The scale used was a nine-point scale (1 strongly disagree to 9 strongly agree). These three variables were suggested to be considered in the study by the HR department of the organization. The questionnaire has been used also by Schlager et al. (2011) in a multinational insurance company. Table 14.3 shows the results of the descriptive analysis of the data.

In the case of servant leadership, the measures of Liden et al. (2008) were utilized. In this case, we only evaluated two sub-scales as recommended by the HR department of the organization, due that the scales of "empowerment" and "helping subordinates grow and succeed" were in accordance with the values of the company. This scale has

Table 14.2 Descriptive statistics work engagement scale

Item (observable variable)	Latent variable	μ	Standard deviation
WE1. At my work, I feel bursting with energy	Vigor	4.88	1.34
WE4. At my job, I feel strong and vigorous	Vigor	4.96	1.27
W8. When I get up in the morning, I feel like going to work	Vigor	4.5	1.69
W12. I can continue working for very long periods at a time	Vigor	4.25	1.59
WE 2. I find the work that I do full of meaning and purpose	Dedication	4.92	1.28
WE5. I am enthusiastic about my job	Dedication	5.10	1.25
WE7. My job inspires me	Dedication	4.66	1.47
WE14. To me, my job is challenging	Dedication	4.68	1.54
WE3. Time flies when I am working	Absorption	4.65	1.52
WE9. I feel happy when I am working intensely.	Absorption	4.78	1.50
WE14. I get carried away when I am working	Absorption	4.81	1.41
WE16. It is difficult to detach myself from my job	Absorption	4.09	1.75

Table 14.3 Descriptive statistics employer branding scale

Item (observable variable)	Latent variable	μ	Standard deviation
EB4. Feeling good about yourself as a result of working for a particular organization	Development	7.40	1.90
EB5. Feeling more self-confident as a result of working for a particular organization	Development	7.37	1.87
EB6. Gaining career-enhancing experience	Development	8.02	1.64
EB10. Working in an exciting environment	Interest	7.08	2.11
EB11. Innovative employer -novel work practices	Interest	7.62	2.15
EB12. The organization both values and makes use of your creativity	Interest	7.01	2.23
EB13. The organization produces high-quality products and services	Interest	7.82	1.98
EB15. Good promotion opportunities within the organization	Economic	6.87	2.44
EB22. Job security within the organization	Economic	7.66	1.85
EB23. Hands-on inter-departmental experience	Economic	7.13	2.13
EB25. An above average basic salary	Economic	7.19	2.04
EB 26. An attractive overall compensation package	Economic	7.28	2.04

Table 14.4 Descriptive statistics servant leadership scale

Item (observable variable)	Latent variable	μ	Standard deviation
SL13. My manager gives me responsibility to make important decisions about my job	Empowerment	5.86	2.66
SL14. My manager encourages me to handle important work decision on my own	Empowerment	5.79	2.75
SL15. My manager gives me the freedom to handle difficult situations in the way I feel is best	Empowerment	5.63	2.76
SL17. My manager makes my career development a priority	Helping subordinates grow and succeed	4.20	2.55
SL18. My manager is interested in making sure that I achieve my career goals	Helping subordinates grow and succeed	4.41	2.66
SL19. My manager provides me with work experiences that enable me to develop new skills	Helping subordinates grow and succeed	5.02	2.90
SL 20. My manager wants to know about my career goals	Helping subordinates grow and succeed	4.08	2.69

been used widely by researchers as Hu and Liden (2011) and recently by Lemoine and Blum (2019). The scale used was a nine-point scale (1 strongly disagree to 9 strongly agree). The results are shown in Table 14.4.

14.4 Results

In this section, the steps to support the hypothesis are shown. First, the validation of the scales is presented, and then from the construction and validation of the structural model derived from the hypothesis and in the last step, the hypothesis is tested.

14.4.1 Validity of the Scales

For assessing the reliability of the scales, we used Cronbach's alpha. This statistic parameter measures the unidimensional extent to which the scale measures one underlying factor. For evaluating the validity of the questionnaires, exploratory factor analysis was used through the SPSS software. This analysis allows us to identify the factors or latent variables associated with the questionnaire. In factor analysis, we seek to reduce the underlying dimensions by searching which variables seem to cluster together in a meaningful way (Field 2013). To ensure the validity of the analysis, three associated types of tests are required with this analysis: The Kaiser Meyer and Olkin indicator (KMO) and Bartlett's sphericity test and the variance explained by the questionnaire. The KMO test contrasts the partial correlations between variables, which can be calculated for individual or multiple variables, and relates the square of the correlation between the variables with the square of the partial relationship between the variables, in such a way that it is a measure of the adequacy of the sample; the closer the value of 1 is obtained, the better the KMO test. This implies that the correlation patterns are relatively compact; values greater than 0.60 are considered acceptable. Bartlett's sphericity test evaluates the applicability of the factorial analysis of the variables studied so that finding an index lower than 0.05 rejects the hypothesis that the correlation matrix is an identity matrix to determine if the correlations are significant among the variables, and, finally, the explained variance tells us how much the questionnaire explains the phenomenon studied. In our case, we studied each latent variable independently because the questionnaires used were proved by other authors, and we want to find out if the latent variable contributes to our general study, related to the model presented above. Table 14.5 presents the statistics of the factor analysis of the latent variables studied.

14.4.2 Structural Equation Modeling

Structural equation modeling (SEM) was used to test our model. SEM is a statistical method that helps to understand models where several variables intervene in a model that goes beyond multiple linear regressions. The objective of statistical modeling through SEM has answered complex questions in relation to latent variables (Vargas 2019). According to Fox (2002), SEM is a variant of the traditional multivariate models, where a model of structural equations is a system of multiple regressions that may seem as a predictor of another or through mediations of others. The idea of structural equation models is to represent causal relationships between two or

Table 14.5 Validity and reliability analysis

Latent variable	Cronbachα	KMO	Bartlett's test	Explained variance (%)
WE: vigor	0.834	0.759	$p < 0.00$	67
WE: dedication	0.849	0.759	$p < 0.00$	69
WE: absorption	0.780	0.754	$p < 0.00$	61
EB: development	0.802	0.664	$p < 0.00$	71
EB: interest	0.884	0.775	$p < 0.00$	74
EB: economic	0.818	0.790	$p < 0.00$	58
SL: empowerment	0.934	0.734	$p < 0.00$	88
SL: helping subordinates	0.918	0.829	$p < 0.00$	80

more variables simultaneously. Russell et al. (1998) confirmed that SEM is useful to evaluate the mediation effects of the intervention program as in our case because as shown in the hypothetical model, we have to consider interest, empowerment, and helping subordinates as mediators of the work engagement variable. The analysis of structural equations was carried out using the AMOS V 22 software. As a first step of the analysis, a Pearson correlation analysis was conducted as a basic way to evaluate the interaction between the variables. The result is shown in Table 14.6.

As a second step, we evaluated each one of the three models that we proposed to test the hypothesis. There were established three models according to the dependent variable studied. The first model evaluated was the one related to dedication as presented in Fig. 14.2.

As shown in the model, all the loadings are significant. For testing the model, the following indexes were calculated: 1) the chi-square statistic χ^2, 2) the root of the average quadratic residue (RMSEA), 3) the comparative goodness of fit index (CFI), and 4) the Tucker-Lewis index (TLI). In Table 14.7, we show the recommended parameters according to with Rigdon (1996). Model 1 accomplishes the standards of validation of the model.

For model two where the dependent variable is absorption (Fig. 14.3), the following parameters of goodness of fit were obtained $\chi^2 = 404$, degrees of freedom 218, RMSEA = 0.076, TLI = 0.902, CFI = 0.915. Thus, the model is acceptable considering the established criteria explained before.

For model three (dependent variable vigor), the following parameters of goodness of fit were obtained $\chi^2 = 420$, degrees of freedom 218, RMSEA = 0.079, TLI = 0.897, CFI = 0.911. Thus, the model is acceptable considering the stablished criteria explained in Table 6. The model is shown in Fig. 14.4.

Finally, Table 14.8 shows the standardized regression weights of the items and the latent variables used in the model. As shown, all the items explain the latent variables and contribute to the explanation of the model.

Table 14.6 Pearson correlation of the latent variables

Variable	Media	Std. Dev.	1	2	3	4	5	6	7
1. WE (vigor)	4.65	1.20							
2. WE (dedicate)	4.84	1.57	0.84**						
3. WE (absorption)	4.58	1.20	0.72*	0.77*					
4. EB (develop)	7.60	1.54	0.28*	0.37*	0.35*				
5. EB (interest)	7.38	1.83	0.48*	0.51*	0.55*	0.50*			
6. EB (economic)	7.22	1.60	0.22*	0.26*	0.26*	0.49*	0.51*		
7. SL (empower)	5.76	2.56	0.41*	0.42*	0.42*	0.16	0.30*	0.54*	
8. SL (helpings)	4.43	2.42	0.52*	0.47*	0.47*	0.12	0.28*	0.25*	0.69*

* means that the value is statistically significant $p<0.05$
** means that the value is statistically significant $p<0.01$

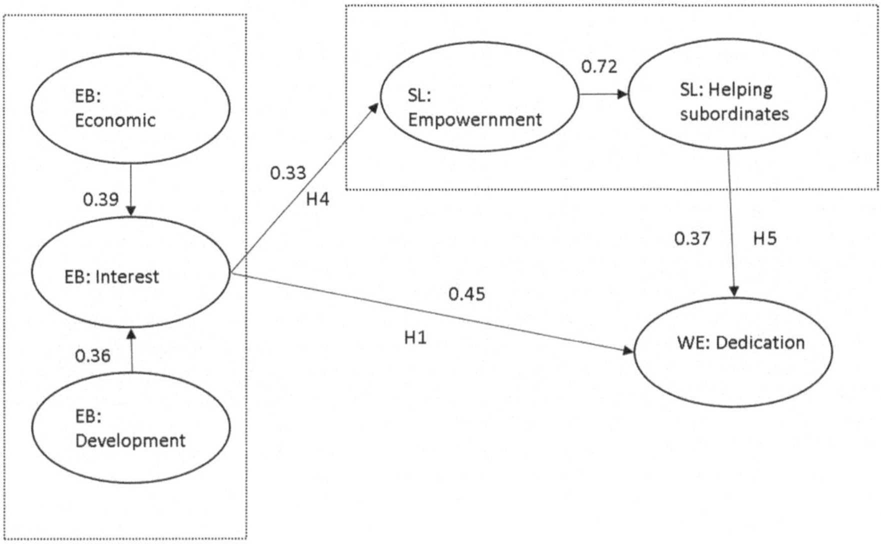

Fig. 14.2 Results of the model 1 for work engagement (dedication). *Note* The model shows the standardized regression weights

Table 14.7 Goodness-of-fit parameters of the model

Model 1: Obtained values		Acceptance criterium
χ^2	435 Degrees of freedom = 239 $p = 0.000$	$p <= 0.05$
RMSEA	0.075	$0.05 \leq \text{RMSEA} \leq 0.08$
CFI	0.917	>0.90
TLI	0.904	>0.90

Source Adapted from Hooper et al. (2008)

14.4.3 Corroboration of Hypothesis

As explained above, all the hypotheses in the model are accepted, although there are some differences between the models that could promote the work engagement of workers in this case. In Table 14.9, we compare the three models to show the variances and further do the pertinent recommendations to promote HR better practices.

As it is observed in the table above that summarizes the findings that each one of the three dependent variables of work engagement, it is confirmed that in the statistical model, the EB and SL variables contribute significatively. The results increase the importance of the leader and the human resources strategies in the floor of production. There has been evidence in other industries that this is an explanation of why EB occurs as a result of organizational behavior variables (Santhanam and

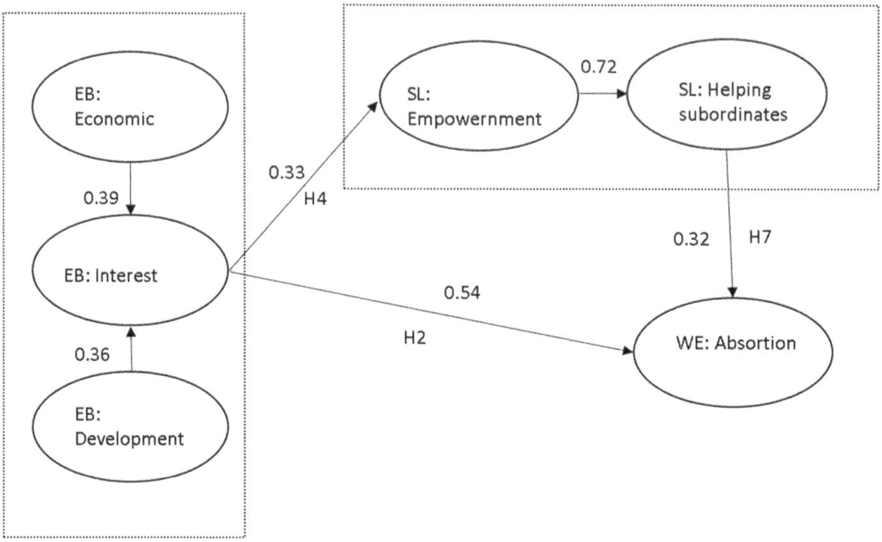

Fig. 14.3 Model 2: Results of the model for work engagement (absorption). *Note* The model shows the standardized regression weights

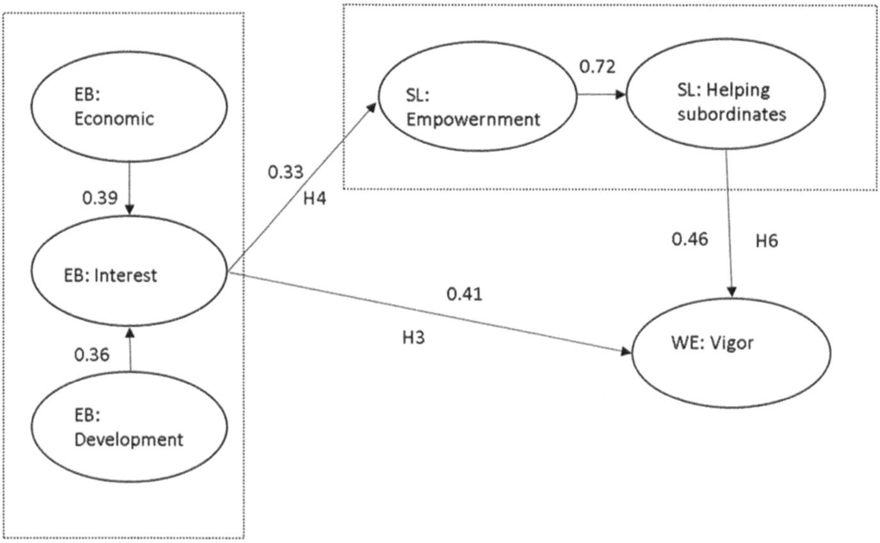

Fig. 14.4 Model 3: Results of the model for work engagement (vigor). *Note* The model shows the standardized regression weights

Table 14.8 Weights of the items of the latent variables

Item	Variable	Standardized regression weights	R^2	*p*
EB26	Economic	0.700	0.49	***
EB25	Economic	0.583	0.34	***
EB23	Economic	0.667	0.44	***
EB22	Economic	0.651	0.42	***
EB15	Economic	0.698	0.49	***
EB13	Interest	0.844	0.71	***
EB12	Interest	0.847	0.72	***
EB10	Interest	0.813	0.66	***
EB11	Interest	0.806	0.65	***
EB6	Development	0.597	0.35	***
EB5	Development	0.792	0.63	***
EB4	Development	0.901	0.81	***
SL13	Empowerment	0.885	0.78	***
SL15	Empowerment	0.883	0.80	***
SL14	Empowerment	0.963	0.93	***
SL17	Helping subordinates	0.885	0.78	***
SL18	Helping subordinates	0.930	0.87	***
SL19	Helping subordinates	0.777	0.60	***
SL 20	Helping subordinates	0.856	0.73	***
WE 10	Dedication	0.606	0.37	***
WE13	Dedication	0.672	0.45	***
WE2	Dedication	0.695	0.48	***
WE5	Dedication	0.766	0.59	***
WE7	Dedication	0.938	0.88	***
WE1	Vigor	0.799	0.64	***
WE4	Vigor	0.841	0.70	***
WE8	Vigor	0.772	0.60	***
WE12	Vigor	0.509	0.26	***
WE3	Absorption	0.757	0.57	***
WE9	Absorption	0.707	0.50	***
WE14	Absorption	0.757	0.57	***
WE16	Absorption	0.622	0.39	***

Note *** $p < 0.01$

Table 14.9 Hypothesis testing

Model	Hypothesis	Standardized regression weight	Criteria
Model 1	H1 (EB-Dedication)	0.456***	Accepted
	H4 (EB-SL)	0.326***	Accepted
	H5 (SL-Dedication)	0.387***	Accepted
Model 2	H2 (EB-Absorption)	0.542***	Accepted
	H4(Interest-SL)	0.328***	Accepted
	H7 (SL-Absorption)	0.328***	Accepted
Model 3	H3 (EB-Vigor)	0.408**	Accepted
	H4 (EB-SL)	0.327***	Accepted
	H6 (SL-Vigor)	0.455***	Accepted

Note *** $p < 0.01$

Srinivas 2019). Also, there is empirical work that demonstrates that engagement or workers are a key element in diverse industries and different cultures (Coetzer and Rothmann 2007; Choo et al. 2013; Kunte and Rungruang 2018). This paper contributes to the literature in the study of work engagement in the Latin American environment. There have been some approaches in the hospitality industry (Ocampo Bustos et al. 2015) and in the gas industry (Contreras-Quevedo 2015). However, our study is restricted only to one organization, and the results have limitations in relation to generalization to other industries.

14.5 Conclusions and Practical Implications

In any business strategy, the inclusion of the human factor is essential. For the purpose of this book, any strategy to increment productivity as lean manufacturing, quality control, total productivity management, etc., should consider that the work associated is a key element for keeping the workshop working smoothly and synchronized. Workers are persons that behave in different ways according to the environment and the stimulus they receive from their leaders and the environment. This study shows statistically how certain strategies for human resources can develop work engagement. In this specific case of study, we investigate WE in relation to EB and SE in a Mexican manufacturing plant.

The first strategy that we studied was the employer branding (EB) strategy to ensure that the worker's values are in accordance with the ones of the corporation. It has been mentioned before that this strategy is successful and should be applied externally and internally to engage new workers. When applied internally, we promote strategies that ensure worker development, economic support for keeping interested in the organization. The strategies studied in this paper are not the only ones that EB promotes, but they are key issues in constructing a nourished environment to develop a social structure within the organization. One of the recent developments within the literature of leadership is "servant leadership." According to our results, two strategies of these theoretical constructions have to be developed among the supervisors or managers of the organization under study: empowerment and helping subordinates to grow and succeed.

In this sense, each organization must develop their self-strategies for EB that could make workers contribute to a healthy environment where leadership could be easily embedded. Leadership can contribute to the engagement of workers, but it has to be naturally dependent on the human resources strategy. Then the work of the managers of human resources is not only designed for the loads of work and the payroll of employees, but his contribution also goes far beyond these activities. They should be responsible for constructing an atmosphere where the worker feels fulfilled and engaged with the company and its values. As we can observe in the validated models, WE is the dependent variable that will contribute to retaining employees at work.

As leadership is a very complex phenomenon and it depends on the attributions of the leader and the perception of the followers, the recommendation is that as there are not two equal organizations, and the leadership style of the supervisor should attend the unique circumstances. Servant leadership is a new development in literature and apparently adapts to many circumstances related to work. The trend is to appreciate the characteristics of humility, service, and empathy for others. These characteristics are appreciated by the millennial generation.

However, for further research, the study of different variables such as burnout and work attitudes should be included to understand the perspective of Mexican and Latin American workers. The emerging economies are employing more qualified workers that are sophisticated, and the understanding of the need for these kinds of workers could be a clue issue in the success of organizations. Many international car industries have manufacturing plants in México and Brazil. Also, aircraft production has moved from their original places to other countries due to cost. But not only the supply chains need competent workers, but this could also be extended to the food supply, the energy supply, etc.

For further studies related to work engagement in a manufacturing setting, there are variables that could be considered that could contribute to the explanation of the employee work's behavior. The variables that could be studied are national culture, work satisfaction, organizational citizenship behavior, and satisfaction with supervision. In this sense, it could be a contribution to study the millennial generation compared with other older generations. Organizations are trying to find solutions to the turnover, and researchers should see more in factors related to work–family conflict, hedonism, etc.

The limitations of our study are that we only studied one organization because of the nature of the project. As it was shown in the demographic data, the majority of employees belong to the millennial generation. This was an advantage for researchers because there are a few studies related to this generation in Latin America. But on the other hand, we considered a few cases of older workers who have devoted many years at the workshop and have more experience. In general, our concern is about the expectative of older and younger workers about the working conditions of the future and its relationship with the economy, technology, and political aspects.

Acknowledgements We want to thank the "Puerto Interior of the State of Guanajuato" for giving us the opportunity in working in this project.

References

Ambler T, Barrow S (1996) The employer brand. J Brand Manag 4:185–206. https://doi.org/10.1057/bm.1996.42

Backhaus K, Tikoo S (2004) Conceptualizing and researching employer branding. Career Dev Int 9(5):501–517. https://doi.org/10.1108/13620430410550754

Bailey C, Madden A, Alfes K, Fletcher L (2017) The meaning, antecedents and outcomes of employee engagement: A narrative synthesis. Int J Manage Rev 19(1):311–353. https://doi.org/10.1111/ijmr.12077

Bakker AB, Schaufeli WB (2008) Positive organizational behavior: engaged employees in flourishing organizations. J Organ Behav: Int J Ind Occup Organ Psychol Behav 29(2):147–154. https://doi.org/10.1002/job.515

Barbuto JE Jr, Wheeler DW (2006) Scale development and construct clarification of servant leadership. Group Organ Manage 31(3):300–326. https://doi.org/10.1177/1059601106287091

Barrow S, Mosley R (2005) Bringing the best of brand management to people at work. Wiley, New York

Bethon P, Ewing M, Hah LL (2005) Captivating company: dimensions of attractiveness in employer branding. Int J Advertising 24(2):151–172. https://doi.org/10.1080/02650487.2005.11072912

Cameron KS, Caza A (2004) Introduction: contributions to the discipline of positive organizational scholarship. Am Behav Sci 47(6):731–739. https://doi.org/10.1177/0002764203260207

Carter WR, Nesbit PL, Badham RJ, Parker SK, Sung LK (2018) The effects of employee engagement and self-efficacy on job performance: a longitudinal field study. Int J Hum Res Manage 29(17):2483–2502. https://doi.org/10.1080/09585192.2016.1244096

Chavarría CAP, Hinestroza MPG, Oliva EJD (2017) Propiedades de la Utrecht Work Engagement Scale (UWES-S 9): análisis exploratorio con estudiantes en Ecuador. Innovar: Revista de ciencias administrativas y sociales 27(64):145–155. https://doi.org/10.15446/innovar.v27n64.62374

Chiniara M, Bentein K (2016) Linking servant leadership to individual performance: Differentiating the mediating role of autonomy, competence and relatedness need satisfaction. Leadersh Q 27(1):124–141. https://doi.org/10.1016/j.leaqua.2015.08.004

Choo LS, Mat N, Al-Omari M (2013) Organizational practices and employee engagement: a case of Malaysia electronics manufacturing firms. Bus Strategy Series 14(1). https://doi.org/10.1108/17515631311295659

Choudhary AI, Akhtar SA, Zaheer A (2013) Impact of transformational and servant leadership on organizational performance: a comparative analysis. J Bus Ethics 116(2):433–440. https://doi.org/10.1007/s10551-012-1470-8

Coetzer CF, Rothmann S (2007) Job demands, job resources and work engagement of employees in a manufacturing organisation. South Afr Bus Rev 11(3):17–32

Contreras Quevedo CA (2015) Determinación del nivel de Engagement laboral en trabajadores de una planta de producción de petróleo y gas costa afuera en México. Ciencia Trabajo 17(52):37–42. https://doi.org/10.4067/S0718-24492015000100008

Crawford ER, LePine JA, Rich BL (2010) Linking job demands and resources to employee engagement and burnout: a theoretical extension and meta-analytic test. J Appl Psychol 95(5):834. https://doi.org/10.1037/a0019364

Demerouti E, Bakker AB, Nachreiner F, Schaufeli WB (2001) The job demands-resources model of burnout. J Appl Psychol 86(3):499–512. https://doi.org/10.1037/0021-9010.86.3.499

Derue DS, Nahrgang JD, Wellman NED, Humphrey SE (2011) Trait and behavioral theories of leadership: An integration and meta-analytic test of their relative validity. Pers Psychol 64(1):7–52. https://doi.org/10.1111/j.1744-6570.2010.01201.x

Dutta S, Khatri P (2017) Servant leadership and positive organizational behaviour: the road ahead to reduce employees' turnover intentions. On the Horizon 25(1):60–82. https://doi.org/10.1108/OTH-06-2016-0029

Eva N, Robin M, Sendjaya S, van Dierendonck D, Liden RC (2019) Servant leadership: a systematic review and call for future research. Leadersh Q 30(1):111–132. https://doi.org/10.1016/j.leaqua.2018.07.004

Ferreira P (2018) Employer brand building from the inside-out: how employer values contribute to employee engagement. 11th Annual conference of the euromed academy of business http://repositorio.uportu.pt:8080/jspui/bitstream/11328/2392/4/Employer%20brand%20building%20from%20the%20inside-out.pdf

Field A (2013) Discovering statistics using IBM SPSS statistics. Sage, London

Fox J (2002) Structural equation models appendix to an R and S-PLUS companion to applied regression

Gillham A, Gillham E, Hansen K (2015) Relationships among coaching success, servant leadership, cohesion, resilience and social behaviors. Int Sport Coaching J 2(3):233–247. https://doi.org/10.1123/iscj.2014-0064

Greenleaf RK (2002) Servant-leadership: a journey into the nature of legitimate power and greatness. Paulist Press, Mahwah

Hooper D, Coughlan J, Mullen MR (2008) Structural equation modelling: guidelines for determining model fit. Electron J Bus Res Methods 6(1):53–60

Hsiao C, Lee YH, Chen WJ (2015) The effect of servant leadership on customer value co-creation: a cross-level analysis of key mediating roles. Tourism Manage 49:45–57. https://doi.org/10.1016/j.tourman.2015.02.012

Hu J, Liden RC (2011) Antecedents of team potency and team effectiveness: an examination of goal and process clarity and servant leadership. J Appl Psychol 96(4):851. https://doi.org/10.1037/a0022465

Jain N, Bhatt P (2015) Employment preferences of job applicants: unfolding employer branding determinants. J Manage Dev 34(6):634–652. https://doi.org/10.1108/JMD-09-2013-0106

Jiménez A (2015) Employer branding: 14 preguntas y una conclusión. Capital Humano 302:84–91

Joubert M, Roodt G (2019) Conceptualising and measuring employee engagement as a role-related, multi-level construct. Acta Commercii 19(1):1–17. https://doi.org/10.4102/ac.v19i1.605

Kahn WA (1990) Psychological conditions of personal engagement and disengagement at work. Acad Manage J 33(4):692–724. https://doi.org/10.5465/256287

Kunerth B, Mosley R (2011) Applying employer brand management to employee engagement. Strategic HR Rev 10(3):19–26. https://doi.org/10.1108/14754391111121874

Kunte M, Rungruang P (2018) Work engagement in manufacturing sector in Thailand. AJMI-ASEAN J Manage Innovation 5(1):100–112. https://doi.org/10.14456/ajmi.2018.8

Lemoine GJ, Blum TC (2019) Servant leadership, leader gender, and team gender role: testing a female advantage in a cascading model of performance. Pers Psychol 1–26. https://doi.org/10.1111/peps.12379

Liden RC, Wayne SJ, Zhao H, Henderson D (2008) Servant leadership: Development of a multidimensional measure and multi-level assessment. The leadership quarterly 19(2):161–177. https://doi.org/10.1016/j.leaqua.2008.01.006

Macias-Velasquez S, Baez-Lopez Y, Maldonado-Macías AA, Limon-Romero J, Tlapa D (2019) Burnout syndrome in middle and senior management in the industrial manufacturing sector of Mexico. Int J Environ Res Publ Health 16(8):1467. https://doi.org/10.3390/ijerph16081467

Maldonado-Macías A, Camacho-Alamilla R, Torres SGV, Alcaraz JLG, Limón J (2015) Determination of burnout syndrome among middle and senior managers in manufacturing industry in Ciudad Juarez. Proc Manuf 3:6459–6466. https://doi.org/10.1016/j.promfg.2015.07.927

Martin G, Gollan PJ, Grigg K (2011) Is there a bigger and better future for employer branding? Facing up to innovation, corporate reputations and wicked problems in SHRM. Int J Hum Res Manage 22(17):3618–3637. https://doi.org/10.1080/09585192.2011.560880

Maslach C (1993) Burnout: a multidimensional perspective. In: Schaufeli B, Maslach C, Marek T (eds) Professional burnout, Taylor & Francis, Washington, pp 19–32

Miao Q, Newman A, Schwarz G, Xu L (2014) Servant leadership, trust, and the organizational commitment of public sector employees in China. Publ Adm 92(3):727–743. https://doi.org/10.1111/padm.12091

Miles SJ, Mangold G (2004) A conceptualization of the employee branding process. J Relat Mark 3(2–3):65–87. https://doi.org/10.1300/J366v03n02_05

Newman A, Schwarz G, Cooper B, Sendjaya S (2017) How servant leadership influences organizational citizenship behavior: the roles of LMX, empowerment, and proactive personality. J Bus Ethics 145(1):49–62. https://doi.org/10.1007/s10551-015-2827-6

Ocampo Bustos RM, Juárez García A, Arias Galicia LF, Hindrichs I (2015) Factores psicosociales asociados a engagement en empleados de un restaurante de Morelos, México. Liberabit 21(2):207–219

Owens BP, Baker WE, Sumpter DM, Cameron KS (2016) Relational energy at work: implications for job engagement and job performance. J Appl Psychol 101(1):35. https://doi.org/10.1037/apl0000032

Pech R, Slade B (2006) Employee disengagement: is there evidence of a growing problem? Handb Bus Strategy 7(1):21–25. https://doi.org/10.1108/10775730610618585

Rigdon EE (1996) CFI versus RMSEA: a comparison of two fit indexes for structural equation modeling. Struct Eqn Model: A Multi J 3(4):369–379. https://doi.org/10.1080/10705519609540052

Russell DW, Kahn JH, Spoth R, Altmaier EM (1998) Analyzing data from experimental studies: a latent variable structural equation modeling approach. J Couns Psychol 45(1):18. https://doi.org/10.1037/0022-0167.45.1.18

Santhanam N, Srinivas S (2019) Modeling the impact of employee engagement and happiness on burnout and turnover intention among blue-collar workers at a manufacturing company. Benchmarking: Int J. https://doi.org/10.1108/BIJ-01-2019-0007

Saxena V, Srivastava RK (2015) Impact of employee engagement on employee performance–Case of manufacturing sectors. Int J Manage Res Bus Strategy 4(2):139–174

Schaufeli WB, Salanova M, González-Romá V, Bakker AB (2002) The measurement of engagement and burnout: A two sample confirmatory factor analytic approach. J Happiness Stud 3(1):71–92. https://doi.org/10.1023/A:1015630930326

Schaufeli WB, Bakker AB, Salanova M (2006) The measurement of work engagement with a short questionnaire: A cross-national study. Educ Psychol Meas 66(4):701–716. https://doi.org/10.1177/0013164405282471

Schlager T, Bodderas M, Maas P, Cachelin JL (2011) The influence of the employer brand on employee attitudes relevant for service branding: an empirical investigation. J Serv Mark 25(7):497–508. https://doi.org/10.1108/08876041111173624

Searle TP, Barbuto JE Jr (2011) Servant leadership, hope, and organizational virtuousness: a framework exploring positive micro and macro behaviors and performance impact. J Leadersh Organ Stud 18(1):107–117. https://doi.org/10.1177/1548051810383863

Shimazu A, Schaufeli WB, Kamiyama K, Kawakami N (2015) Workaholism vs. work engagement: The two different predictors of future well-being and performance. Int J Behav Med 22(1): 18–23. https://doi.org/10.1007/s12529-014-9410-x

Steffens NK, Yang J, Jetten J, Haslam SA, Lipponen J (2018) The unfolding impact of leader identity entrepreneurship on burnout, work engagement, and turnover intentions. J Occup Health Psychol 23(3):373. https://doi.org/10.1037/ocp0000090

Tims M, Bakker AB, Xanthopoulou D (2011) Do transformational leaders enhance their followers' daily work engagement? Leadersh Q 22(1):121–131. https://doi.org/10.1016/j.leaqua.2010.12.011

Tumasjan A, Kunze F, Bruch H (2016) Linking employer branding and firm performance: testing an integrative mediation model. In: Academy of management proceedings, Academy of Management, New York. https://doi.org/10.5465/ambpp.2016.173

Valadez-Torres SG, Maldonado-Macías AA, García-Alcaraz JL, Camacho-Alamilla MDR, Avelar-Sosa L, Balderrama-Armendariz CO (2017) Analysis of burnout syndrome, musculoskeletal complaints, and job content in middle and senior managers: Case study of manufacturing industries in Ciudad Juarez, Mexico. Work 58(4):549–565. https://doi.org/10.3233/WOR-172642

Van Dierendonck D (2011) Servant leadership: a review and synthesis. J Manage 37(4):1228–1261. https://doi.org/10.1177/0149206310380462

Van Dierendonck D, Stam D, Boersma P, De Windt N, Alkema J (2014) Same difference? Exploring the differential mechanisms linking servant leadership and transformational leadership to follower outcomes. Leader Quarter 25(3):544–562. https://doi.org/10.1016/j.leaqua.2013.11.014

Vargas-Chanes D (2019) Aspectos metodológicos para la investigación social: modelos de ecuaciones estructurales. UNAM, México

Wang Z, Liu H, Yu H, Wu Y, Chang S, Wang L (2017) Associations between occupational stress, burnout and well-being among manufacturing workers: mediating roles of psychological capital and self-esteem. BMC Psychiatry 17(1):364. https://doi.org/10.1186/s12888-017-1533-6

Xanthopoulou D, Bakker AB, Demerouti E, Schaufeli WB (2007) The role of personal resources in the job demands-resources model. Int J Stress Manage 14(2):121. https://doi.org/10.1037/1072-5245.14.2.121

Yang F, Liu J, Wang Z, Zhang Y (2019) Feeling energized: a multilevel model of spiritual leadership, leader integrity, relational energy, and job performance. J Bus Ethics 158(4):983–997. https://doi.org/10.1007/s10551-017-3713-1

Zhao C, Liu Y, Gao Z (2016) An identification perspective of servant leadership's effects. J Manag Psychol 31(5):898–913. https://doi.org/10.1108/JMP-08-2014-0250

Chapter 15
The DMAIC Methodology as a Tool for Process Improvement: The Case of a Mexican Manufacturing Company

Arturo Realyvásquez-Vargas, Jorge Luis García-Alcaraz, Guadalupe Hernández-Escobedo, Karina Cecilia Arredondo-Soto, Joel Eduardo García-Ortíz, Julio Blanco-Fernández, and Emilio Jiménez-Macías

Abstract Quality allows manufacturing companies to remain competitive, yet some manufacturers still struggle with a high percentage of defective products. In this chapter, we introduce the case study of a Mexican manufacturing company experiencing problems in the manufacture of radio frequency (RF) and optical fiber amplifiers as a result of defective T3 transformers and optical fibers. Consequently, the company's First Time Yield (FTY) index is lower than 97.5%, the minimum acceptable value. The main goal of this chapter is to implement a method for reducing the defect rates and increasing FYT using the define–measure–analyze–improve–control (DMAIC) methodology, Pareto charts, and the Ishikawa diagram. Our findings revealed that by using these three tools, the number of defective T3s and optical fibers decreased from 90 to 14 (83.3%) and 56 to 12 (21.4%), respectively, during

A. Realyvásquez-Vargas (✉) · G. Hernández-Escobedo · J. E. García-Ortíz
Departamento de Ingeniería Industrial, Tecnológico Nacional de México/IT Tijuana, Calzada del Tecnológico S/N, Col. Tomás Aquino, Tijuana, Baja California, Mexico
e-mail: arturo.realyvazquez@tectijuana.edu.mx

J. L. García-Alcaraz
Department of Industrial Engineering and Manufacturing, Universidad Autónoma de Ciudad Juárez, Av. Del Charro 450 Norte, Col. Partido Romero, Juárez, Chihuahua, Mexico

Division of Research and Postgraduate Studies, Tecnológico Nacional de México/Instituto Tecnológico de Ciudad Juárez. Av. Tecnológico, 1340, Fuentes del Valle, 32500. Ciudad Juárez 32500, Chihuahua, México

K. C. Arredondo-Soto
Faculty of Chemical Sciences and Engineering, Universidad Autónoma de Baja California, Tijuana, Baja California, Mexico

J. Blanco-Fernández
Department of Mechanical Engineering, University of La Rioja. Edificio Departamental—C/San José de Calasanz, 31, Logroño, La Rioja, Spain

E. Jiménez-Macías
Department of Electrical Engineering, University of La Rioja, Edificio Departamental—C/San José de Calasanz, 31, Logroño, La Rioja, Spain

J. L. Garcia Alcaraz et al. (eds.), *Trends in Industrial Engineering Applications to Manufacturing Process*, https://doi.org/10.1007/978-3-030-71579-3_15

the January–May 2019 period. Similarly, FYT increased from 87.7% to 97.5% in that same period.

Keywords Manufacturing industry · Defects · FTY · DMAIC · Pareto chart · Ishikawa diagram · Damaged T3 · Damaged optical fibers

15.1 Introduction

In manufacturing systems, customer satisfaction is a key factor for competitiveness. Consequently, manufacturers try to take into account customer needs from the moment a product is being designed (Wang 2013), while simultaneously ensuring the highest possible quality (Abolhassani et al. 2018). Conversely, defective products lead to unhappy customers, customer loss, higher production costs, decreased sales, and few economic benefits (Galbreath 2002; Saeidi et al. 2015).

In order to offer high-quality products and avoid losses, manufacturing systems follow the lean manufacturing philosophy (Sartal et al. 2017; Kumar et al. 2018; Journal of Applied Sciences 2018), which relies on a set of production process improvement tools (Liu et al. 2016), such as the Deming or plan–do–check–act (PDCA) cycle (Dudin et al. 2015), process flow charting (PFC) (Bunce et al. 2008), the eight disciplines (8Ds) method (Cheng and Chen 2010), six sigma, and the define–measure–analyze–improve–control (DMAIC) method (Chakravorty and Shah 2012), among others.

Concisely, lean manufacturing seeks to increase earnings while adding value to products by reducing waste. By doing so, lean manufacturing ensures on-time deliveries (Botti et al. 2017), reduces waste along the production process (Kumar et al. 2018; Journal of Applied Sciences 2018), allows implementing improvement strategies, ensures defect-free products, and standardizes work (Marodin et al. 2018). In manufacturing settings, waste is a kind of process inefficiency that adversely affects timely deliveries, product quality, and production/product costs (Botti et al. 2017). There are seven types of wastes: overproduction, inventory, transport, waiting, motion, over processing, and defects (Jadhav et al. 2018). Defects are a great struggle in many manufacturing systems (Sreedharan et al. 2018), because they are the main cause of damages in product components (Zhou and Gosling 2018). Consequently, defects prevent many manufacturers from complying with timely delivery times (Guiras et al. 2018).

To estimate to what extent they ensure quality along the production process, manufacturing companies measure indicators such as process capability (Cpk), process performance (Ppk), In Process Yield (IPY), Rolled Throughput Yield (RTY), and First Time Yield (FTY), among others (Wang et al. 2019). Specifically, the FTY index is a quality metric used to indicate the percentage of product units completed without needing any rework (Brown et al. 2006). Similarly, FTY represents the likelihood for a product to be successfully approved from the beginning (Pérez-Urrego et al. 2014). FTY is calculated with Eq. (15.1) as follows (Brown et al. 2006):

$$FTY = \left(1 - \frac{defective\ units}{total\ units}\right) \times 100 \quad (15.1)$$

Many manufacturing companies seek to reach an FTY higher than 97.5%; otherwise, it is an indicator that a problem exists in the process and must be identified from the root cause. To identify problems, companies implement lean manufacturing, which encompasses a range of tools and techniques, such as the DMAIC process.

15.2 The DMAIC Methodology

DMAIC is described as a method for reducing variation in the production process, and it has demonstrated its effectiveness as a quality improvement tool across various sectors (de Mast and Lokkerbol 2012), such as manufacturing (Hakimi et al. 2018) and health care (Gupta et al. 2018), to name but a few. DMAIC focusses on one process at a time, finds the inconsistencies, adjusts them, and standardizes work swiftly (Gupta et al. 2018). Each phase of the DMAIC method occurs subsequently and depends on the results from the previous phase. According to Leaphart et al. (2012), in an improvement project, the define (D) phase involves the project team creating a project charter to outline the project's focus, goals, success measures, constraints, risks, and timeline. At the measure (M) phase, the team must determine the start point or baseline of the process to be improved. Also, the team collects and reviews data to determine the root causes of process waste, including variations and defects. The analyze (A) phase seeks to help the team find out what is causing the issue that they are trying to fix. To this end, team members review their charts and graphs and use their observations to uncover the interrelations between the causes and effects of each problem harming the process. The improve (I) phase implies identifying potential changes or improvement actions that could remove the problem from its root cause. At this stage, the team pilots process changes and ultimately sets new standards for the baseline (Leaphart et al. 2012). Finally, at the control (C) phase, improvements are sustained while monitoring the process, thus generating feedback for new project implementations (Leaphart et al. 2012).

The DMAIC methodology is widely used across different sectors. Table 15.1 summarizes our review of the literature reporting cases of successful DMAIC implementation. However, for further information, readers can consult the work of (Srinivasan et al. 2016), who found that DMAIC can increase production capacity in manufacturing systems.

15.2.1 DMAIC Techniques and Tools

The DMAIC methodology relies on multiple techniques to identify the root causes of problems arising in production systems. Two of these techniques are Pareto charts and the Ishikawa diagram. A Pareto chart is a special type of graph that contains bars and a line graph, and each bar represents a different problem (Joiner Associates

Table 15.1 Literature review: success cases of DMAIC implementation

References	Problem	Results following DMAIC implementation
(Tong et al. 2004)	Defects in the solder joint of surface-mounted printed circuit boards (PCBs) lead to circuit failure. Capability of the screening process is under 1.33—the company's requirement	Improved sigma level of the screening process in the manufacture of surface-mounted PCBs
(Hwang 2006)	Insufficient data/misunderstanding of the DMAIC methodology causes project failure in manufacturing systems	Improved process performance and capability, cycle time, RTY, and operating costs Implementation of the DMAIC methodology quickly and precisely
(Gijo et al. 2011)	Defects in the fine grinding process in an Indian automotive manufacturer.	Reduction of defects in the fine grinding process from 16.6% to 1.19% Implementing the DMAIC methodology had a significant impact on the company's profitability in terms of reduced scrap costs, man-hour saving on rework, and increased output Annual savings of approximately 2.4 million USD were reported from this project
(Chang et al. 2012)	The effectiveness and efficiency of a company's information systems were below the company's standard levels, which caused poor production performance	Improved performance of the production planning procedure through information system integration
(Jonny and Christyanti 2012)	PT BBI, an asbestos supplier in Indonesia faced quality problems of asbestos roofing	Increased sigma level, from 4.91 to 5.02
(Antony et al. 2012)	A high-precision and critical process in the manufacture of automotive products had a low performance—85%, which is below the company's requirement. Moreover, the company received averagely 34 customer complaints monthly	Production process performance increased to 99.4% Implementation of the DMAIC methodology had a significant impact on the company's profitability Annual savings increased to 70,000 USD, and quality on returns and sales increased, too

(continued)

Table 15.1 (continued)

References	Problem	Results following DMAIC implementation
(Leaphart et al. 2012)	Hyponatremia occurs among renal transplant recipients, and no efficient method exists to prevent it	The authors sought to identify the main causes of hyponatremia and found that 0.45% sodium chloride administered to patients during renal transplantation is the most likely cause. Hence, administering 0.9% % sodium chloride intraoperatively resulted in normal serum sodium levels postoperatively in 59 of 64 patients (92%)
(Srinivasan et al. 2014a)	In a small furnace manufacturing company, the shell and tube heat exchanger—a critical component of furnaces—had low efficiency and affected the overall quality of the furnaces	Shell and tube heat exchanger effectiveness was enhanced from 0.61 to 0.664, and annual monetary savings were of 0.34 million Rs
(Srinivasan et al. 2014b)	A leading shock absorber manufacturing company faced problems in the shock absorber spray painting process, namely peel off and blisters. As a result, overall product quality and customer satisfaction decreased	The sigma level was enhanced from 3.3 to 4.5, thus improving the performance of the manufacturing process
(Tenera and Pinto 2014)	A Portuguese telecommunication company suffered the adverse effects of the economic crisis, which increased the demand for profitable solutions to gain a competitive advantage	The authors identified the company's main project management problems and corresponding root causes and then selected the causes to be first attended. Also, the authors systematically addressed the actions and solutions to be implemented in order to keep the long-term continuous improvement of the company's project management processes
(Desai et al. 2015)	An Indian food-processing industry experienced variation in the weight of milk powder pouches, which ultimately caused production and earning losses	The weight variation problem was satisfactorily addressed, and the rejection rate decreased to 50%. The annual financial benefits projected as a result of the project were 800,000 Rs

(continued)

Table 15.1 (continued)

References	Problem	Results following DMAIC implementation
(Cunha and Dominguez 2015)	The billing process in a Portuguese car dealer group showed deficiencies, since the financial metrics in use did not control compliance standards for car brands	The company set new financial metrics (% of time compliance to do the service and bill it, % time compliance reception, and % time to find defective part in an audit). These metrics generated time benefits and consequently a more controlled cash flow
(Pugna et al. 2016)	A car manufacturing company in Romania implemented quality improvement initiatives (e.g., total quality management, ISO, agile and lean manufacturing) which, in practice, seemed to be neither time efficient nor profitable in terms of quality	Quality was remarkably improved in reasonable time
(Gupta et al. 2018)	Waist circumference (WC) is a superior surrogate marker of central obesity than the body mass index. However, WC is not measured routinely in pediatric clinics	Application of an evidence-based DMAIC protocol led to remarkable improvements in terms of both assessing central obesity in ambulatory clinic practice and appropriate counseling regarding cardiometabolic risk reduction among children and adolescents with central obesity during an eight-month period
(Hakimi et al. 2018)	A food-processing company experienced quality issues in the plain yogurt production process as a result of the acidity of the yogurt. The factors affecting plain yogurt acidity were unknown	The authors found that incubation time and fat percentage were significant factors on pH values of plain yogurt. The optimum settings for these factors were defined as 12 h for incubation and 1.5% of fat
(Smętkowska and Mrugalska 2018)	A manufacturing company experienced bottlenecks along its production process	Machinery efficiency was increased

(continued)

Table 15.1 (continued)

References	Problem	Results following DMAIC implementation
(Srinivas and Sreedharan 2018)	In the car manufacturing supply chain, automotive components are packed and distributed to customers and retail units; however, various other operations performed during distribution may contribute to component failure	The packing method was identified as the root cause of the problem The new packing method, consisting of a cushion between the packing material and the spare part, proved to be more economic and required making minor changes to the original packing method
(Khajuria et al. 2018)	A thermoforming setup struggled with a high rate of defective pieces due to numerous reasons	The number of defective parts was brought down, and the sigma level increased to 4.56
(Sharma et al. 2018)	In the aluminum alloy wheel machining process, defects were caused by unclean alloy wheel, surface finish deterioration, and non-conformance to geometrical and dimensional specifications, among others	Cp, Cpk, and Cpm showed signs of improvement from an initial value of 0.66, -0.24, and 0.27 to a final value of 4.19, 3.24, and 1.41, respectively
(C.R. and Thakkar 2019)	A high number of rejections in the manufacturing of telecommunication cabinet doors	Sigma level increased from 3.49 to 3.67 Defects per million opportunities (DPMO) were brought down from 23,271 to 15,873 Rework costs were reduced from 56,666,662.55 to 38,650,779 Rs
(Gaikwad et al. 2019)	A medical device manufacturing company experienced a rejection rate of 0.76% in spring supports, an important component of the tissue dissector device	Rejection rate was brought down to 0%
(Guo et al. 2019)	In an air-conditioner manufacturer in China, profits from single air conditioners declined 19% due to higher manufacturing costs	The company achieved monthly profits of 797,051 RMB
(Gandhi et al. 2019)	A casting unit manufacturing cylinder blocks experienced a rejection rate as high as 30% due to formation of blowholes, other than the surface	The problem was significantly reduced from 28,111 to 9,708 parts per million, which represents net annual savings of 1,256,640 INR

(continued)

Table 15.1 (continued)

References	Problem	Results following DMAIC implementation
(Wang et al. 2019)	In a manufacturing company, defects in touch panels were caused by pollution, which still accounts for 30% of the total variation	Abnormal proportion of pollution decreased from 0.35% to 0.13% (improvement rate reached 63%) It is estimated that the entire product can generate 3 million USD of performance in one year

1995). The heights of the bars represent frequency or costs, and they are arranged in descending order, that is with the longest bars on the left and the shortest bars on the right. Also, the vertical axis on the left can be expressed in terms of quantity or amount, and the vertical axis on the right can be expressed in terms percentage from 0% to 100% (Joiner Associates 1995; Beheshti et al. 2018). Finally, the line graph drawn from left to right represents the cumulative total. Table 15.2 summarizes our

Table 15.2 Use of Pareto charts: success cases

Authors	Applied to	Findings
(Nabiilah et al. 2018)	Manufacturing	Bits were the most frequent defect in the electrodeposition process
(Acharya et al. 2018)		Defects such as sand inclusions, gas porosity, and hard metal contributed more than 50% of total defects in furan no-bake casting
(Sharma and Suri 2017)		The most common defects in low voltage panel boards were poor welding, paint thickness/shade, loose connection, and sticker missing
(Rahman Ansar et al. 2018)		The most common defects in garment manufacturing were broken stitch, skipped stitch, and open stitch (23%)
(Chokkalingam et al. 2017)		The root causes of the shrinkage defect in an automobile body casting were larger volume section in the cope, insufficient feeding of riser, and insufficient poured metal in the riser
(Kholif et al. 2018)	Dairy laboratories	The highest percentage of errors was from contamination of ultra-high-temperature milk samples, followed by non-homogeneous sample errors and insufficient volume of sample errors
(Rahman et al. 2018)	Apparel industry	The most common defect categories in a garment manufacturing company were broken stitch, skipped stitch, open seam, pleat, and down stitch

review of the literature reporting successful implementation of Pareto charts across multiple sectors.

The Ishikawa diagram, or fishbone diagram, is another tool used when implementing the DMAIC methodology. It is a quality management/improvement tool and was first developed by Kaoru Ishikawa during the 1960s (Doshi et al. 2012). The Ishikawa diagram is used to graphically represent cause–effect relationships, helping teams identify, classify, and display the root causes of a specific problem, especially a quality-related problem. The first step when implementing an Ishikawa diagram is to analyze the problem or effect and create a problem statement that is a statement that explicitly and clearly explains what the problem (effect) is. The next step involves identifying the potential causes of the problem. Traditionally, causes are grouped into six categories: measurement, materials, people, environment, methods, and machines. Once the categories are set, team members can list out all the individual causes. The Ishikawa diagram allows us to identify the relationships between customer requirements and the quality of a final product, thus helping companies better detect product properties (Gawdzińska 2007). Nowadays, Ishikawa diagrams are used in a range of sectors, including manufacturing, marketing, research, and services (Doshi et al. 2012).

15.3 Research Problem and Goal

This research is conducted in a Mexican manufacturing company based in the city of Tijuana. The company produces radiofrequency and fiber optic amplifiers using electronic boards. Lately, these boards have experienced quality problems, causing the company's FTY to decrease to 88.3%, even though the standard has been set to FTY $\geq$ 97.5%. The company thus struggles with numerous customer complaints that have even resulted in returns of entire batches of defective amplifiers. Consequently, the company's reputation has been adversely affected, and the costs of production are increasing as a result of greater logistic and administrative expenses. The most common defects in the electronic boards of the amplifiers include damaged T3 transformer (cut wire), damaged optical fiber, solder bridges, and wrong components.

The goal of this research is to implement the DMAIC methodology along with both Pareto charts and Ishikawa diagrams to identify the quality problems causing a low FTY in the company to increase its value. The remainder of this article is structured as follows: Sect. 15.2 discusses the methodology, Sect. 15.3 introduces the results of the case study, Sect. 15.4 discusses the overall results, and Sect. 15.5 presents the research conclusions, limitations, and future work recommendations.

15.4 Methodology

The DMAIC methodology comprises five phases—define, measure, analyze, improve, and control. The following subsections discuss how each of these phases was followed in this research and how the corresponding tasks were performed.

15.4.1 Phase 1. Define

We identified the current situation of the FTY index at the AR subassembly area. From July 2018 to February 2019, we collected data on the failures having occurred in the production flow control system. We also formed an improvement team. Then, we began planning the project by defining its goal and the tasks to be performed. Next, each team member was assigned a specific role based on the tasks for which they were responsible. Finally, the team defined the project's scope and constraints, deadlines per stage, and outcomes. The planning and definition phase lasted from January to February 2019.

15.4.2 Phase 2. Measure

We collected data on the different defects in the amplifier circuit boards, such as damaged T3s (cut wire), damaged optical fiber, solder bridges, and wrong components, among others. The data were collected from the production flow control system, which reports the failures occurring in the assembly area. Next, we designed a Pareto chart to arrange the defects with respect to their frequency. We applied the 80/20 rule, thus considering that 80% of the defects could arise from 20% of the causes. Consequently, by removing 20% of the causes, 80% of defects could be avoided.

15.4.3 Phase 3. Analyze

We analyzed the possible causes of those defects that contributed the most to the problem (i.e. low FTY). To this end, we used Ishikawa diagrams.

15.4.4 Phase 4. Improve

In this phase, we brainstormed about the possible solutions that could eliminate each of the defect causes. Once the solutions were determined, we performed a root cause analysis (RCA) on them to determine their suitability. To this end, we used a special form (see appendix 1). The RCA weights six criteria—factor, direct cause, solution, feasibility, measurability, and costs)—on a 1–3 scale (1 = totally disagree, 2 = neutral, 3 = totally agree) (Flower et al. 2019). The following questions were formulated to respond to the six aforementioned criteria (Instituto de Productividad Empresarial Aplicada 2019):

1. Does this factor lead to the problem?
2. Is this a direct cause of the problem?
3. Does removing this cause solve the problem?
4. Is there a feasible solution?
5. Can we measure the impact of the solution?
6. Is the solution low cost?

We implemented the solution having the highest weighting. The improvement proposals were tested and validated in one production line. The validation involved analyzing both the FTY results obtained gradually during the implementation of the proposals and the number of defective products. Once the improvement proposals were validated, we standardized the workflow to generate a new production process.

15.4.5 Phase 5. Control

In this phase, we implemented the techniques necessary to control the new process. To this end, we monitored process behavior using control graphics, Pareto charts, and worksheets. Similarly, we visually supervised the production lines.

15.5 Results

This section discusses the results obtained following the implementation of the DMAIC methodology.

15.5.1 Phase 1 Results: Define

After analyzing the data collected on the company's production process, we found that FTY lay at low levels, i.e., below 97.5%, during the entire period (July 2018–February 2019). Figure 15.1 illustrates such results; the red line indicates the

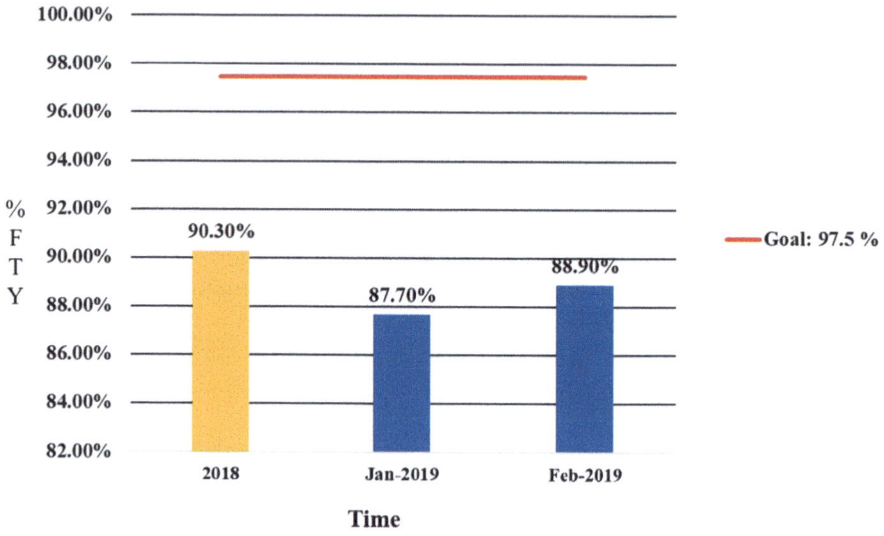

Fig. 15.1 FTY values

minimum FTY level required by the company, whereas the percentages over the graph bars indicate the average FTY values during July-December 2018, January 2019, and February 2019. As can be observed, the low percentages are an indicator of problems with the amplifiers, which are used to manufacture the nodes in the following process. The problem ultimately impacts time deliveries, since the nodes are sent to external customers.

15.5.2 Phase 2 Results: Measure

After measuring the collected data, we found that during the July 2018–February 2019 period, 602 defects, classified into 10 types, were reported in the circuit boards used for the amplifiers. The Pareto chart from Fig. 15.2 visually represents such data. As can be observed, the most common defects included damaged T3 and damaged optical fiber, with 283 and 177 occurrences, respectively. In other words, together the two accounted for 76.4% of the total defects, 47 and 29.4%, respectively. Such results indicate that damaged T3s and damaged optical fibers are the main causes of low FTY values.

Figure 15.3 allows for the comparison of two T3s, one with a cut wire and one in good condition (no cut end). As previously mentioned, damaged T3s are the leading cause of defective circuit boards. From a similar perspective, Fig. 15.4 illustrates an example of a damaged (polluted) optical fiber and that of an optical fiber in good condition (not polluted).

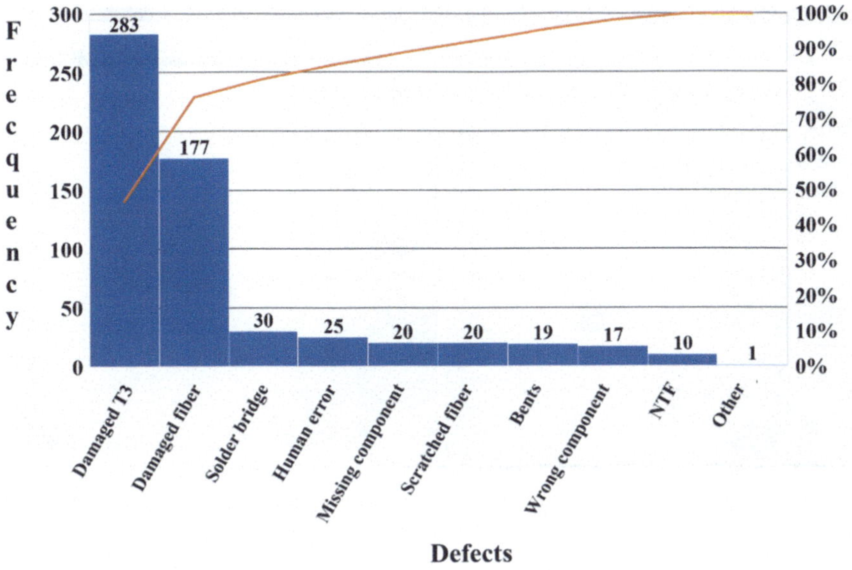

Fig. 15.2 Defects and frequency of occurrence

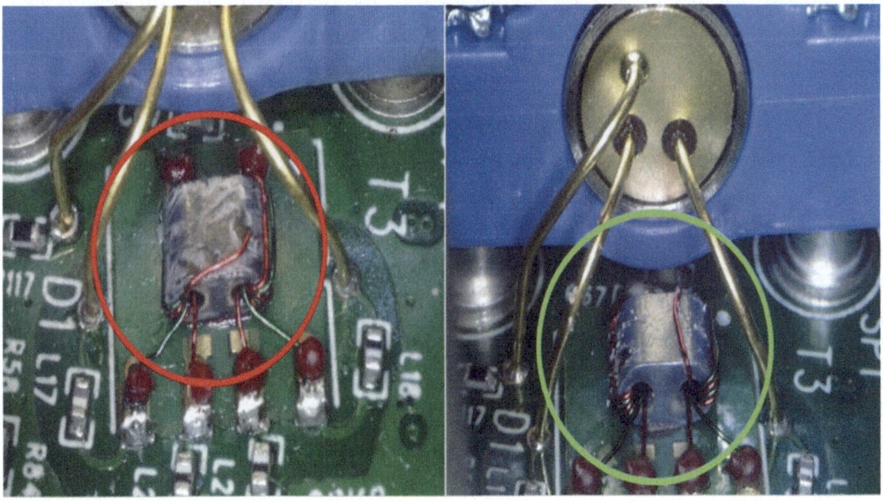

Fig. 15.3 Damaged T3 (defect) vs T3 in good condition

Following such results, the goal of the project was to reach a FTY equal to or higher than 97.5% in the AR subassembly area.

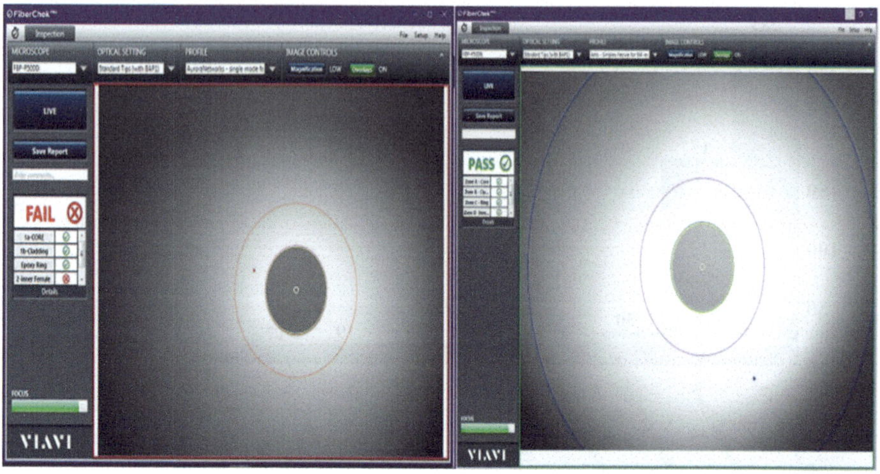

Fig. 15.4 Damaged/polluted optical fiber (defect) vs optical fiber in good condition

15.5.3 Phase 3 Results: Analyze

In this phase, we performed a RCA for each of the two most common defects, that is damaged T3s and damaged (polluted) optical fibers. The following paragraphs discuss the results of such analyses.

15.5.3.1 Root Cause Analysis of Damaged T3 Defect

Figure 15.5 illustrates the Ishikawa diagram designed to study the problem of damaged T3s in circuit boards. Some of the causes of this problem include miscalibrated equipment, defective supplier material, inefficient process supervision, AR temperature, and lack of a standardized alignment method, among others.

Table 15.3 summarizes the results of the RCA performed for defect type damaged T3s. As can be observed, the leading causes of this problem (i.e., those reporting the highest frequency as reported in the TOTAL column) are an unstandardized alignment method and an inappropriate alignment fixture, both causes reported a weighting of 17. For all the causes of this type of defect, we defined the possible solutions (see SOLUTIONS column).

According to the RCA, the damaged T3 defect is chiefly due to the absence of a standardized alignment method (manual manipulation) and a wrong alignment fixture. Consequently, we implemented improvement actions to eliminate these factors.

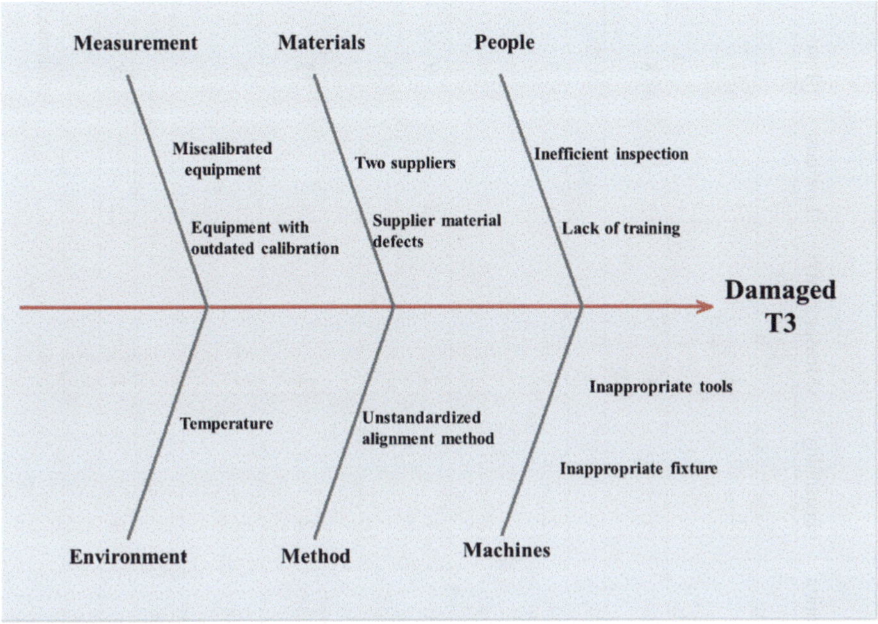

Fig. 15.5 Ishikawa diagram for damaged T3s defect

15.5.3.2 Root Cause Analysis of Damaged Optical Fiber

Figure 15.6 illustrates the Ishikawa diagram designed to study the occurrence of damaged optical fibers in the circuit boards. Two of the main causes of this defect include an unstandardized method for inspecting the fibers and the presence of dirt.

Table 15.4 summarizes the results of the RCA for the problem of damaged optical fibers. As can be observed, the leading cause of the problem is the set of fiber inspection parameters entered on the Aurora profile, which is a piece of software used internally in the company to inspect optical fibers (see Fig. 15.7 and Tables 15.5 and 15.6).

Following the RCA results, we concluded that the main factor contributing to damaged optical fibers on the circuit boards was the set of inspection parameters. The second leading cause of this defect involved the absence of a standardized inspection method. Consequently, we implemented solutions destined to remove such causes.

Table 15.6 shows the parameters of the IEC 61300-3-35 international standard, a common set of global requirements for optical fiber inspection (Lee 2016). The IEC 61300-3-35 standard sets requirements for allowable surface defects such as scratches, pits, and debris, which may affect optical performance. As can be observed from the table, the requirements of IEC 61300-3-35 are more flexible than those set by Aurora. For instance, in terms of cladding (1b), IEC 61300-3-35 accepts any number of removable particles smaller than 2 μm but no particles larger than > 5 μm (Table 15.7).

Table 15.3 Root cause analysis for damaged T3 defect

Causes	Solutions	Criteria						
		Factor		Direct cause	Solution	Feasibility	Measurability	Low cost	Total
Measurement									
Miscalibrated equipment	Provide training on equipment calibration	3		1	1	2	1	3	11
Equipment with outdated calibration	Check calibration status during TPM	2		1	1	1	1	2	8
Material									
Two suppliers	Consider the possibility of working with one supplier, only	1		1	1	2	1	1	7
Defective supplier material	Inform suppliers of quality issues with their materials	2		1	1	1	2	2	9
People									
Inefficient inspection	Provide training on inspection tools	3		2	1	2	2	3	13
Lack of training	Design and implement a worker training plan	2		1	1	1	2	1	8
Environment									
Temperature	Reduce testing time	2		1	1	1	1	2	8
Method									

(continued)

Table 15.3 (continued)

Causes	Solutions	Criteria						
		Factor	Direct cause	Solution	Feasibility	Measurability	Low cost	Total
Unstandardized alignment method	Design and implement a plan for standardizing alignment method	3	3	3	3	2	3	17
Machines								
Inappropriate alignment fixture	Use right alignment fixture	3	3	3	3	2	3	17
Inappropriate tool	Provide appropriate tool	3	3	1	2	1	2	12

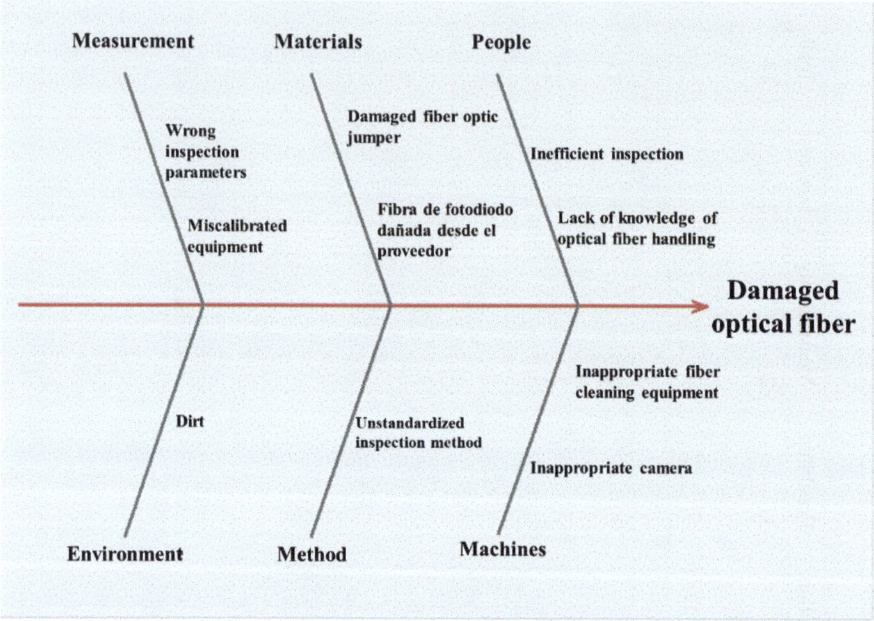

Fig. 15.6 Cause–effect diagram of the damaged optical fibers defect

Considering our findings at this phase, we created a new optical fiber inspection profile, which is further discussed in the following section.

15.5.4 Phase 4 Results: Improve

15.5.4.1 Damaged T3

Figure 15.8 depicts the original fixture used in the company to manually align the AR subassemblies. Note that the cavity through which technicians insert the tools to manipulate the transformer windings is relatively narrow (i.e., 1 × 1.5 cm, see red rectangle). This reduces the technicians' visibility when manipulating the subassembly, which in turn increases the likelihood of damaging the T3.

Figure 15.9 depicts the improvement proposal implemented to remove this first cause of the T3 defect. The fixture cavity was made larger (2 × 4 cm, see green rectangle) for technicians to have more visibility of the T3 transformer and better access the windings.

Another improvement action implemented to reduce the T3 defect was to request the company' suppliers T3 transformers in which the widely visible part of the coil wire winded around the transformer's balun core remains on the upper part of the core, not on the side, as shown in Fig. 15.10. This decision emerged from

Table 15.4 Root cause analysis for damaged optical fiber defect

Causes	Solutions	Criteria						
		Factor		Direct cause	Solution	Feasibility	Measurability	TOTAL
Measurement								
Wrong inspection parameters entered in Aurora	Create profile for Arris and change parameters	3	3	3	3	2	3	17
Miscalibrated equipment	Provide training on equipment calibration	2	2	1	2	1	2	10
Material								
Defective supplier materials	Inform supplier	2	1	1	1	2	2	9
People								
Inefficient inspection	Provide training course on inspection tools	3	1	1	2	2	3	12
Lack of knowledge for optical fiber handling	Plan training program on optical fiber handling	3	3	2	2	2	2	14
Environment								
Dirt	Design preventive maintenance plan	3	3	2	2	1	2	13
Method								
Unstandardized method for optical fiber inspection	Standardize method for optical fiber inspection	3	3	3	3	2	3	17
Machines								
Inappropriate camera	Evaluate feasibility of purchasing new camera	2	2	2	2	2	2	12
Inappropriate cleaning equipment	Review cleaning equipment	2	3	1	1	2	1	10

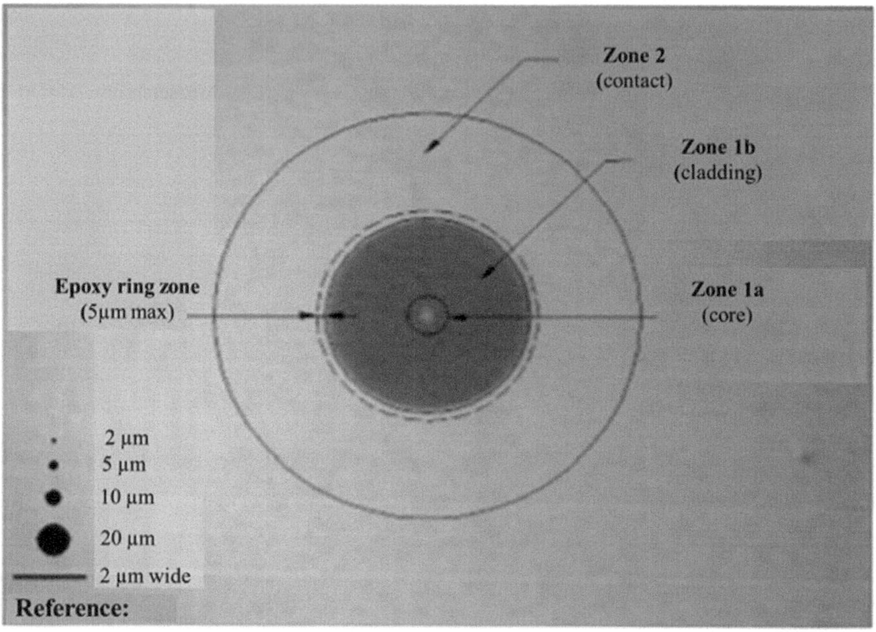

Fig. 15.7 Visual inspection of fiber

Table 15.5 Internal fiber inspection parameters in Aurora

Area	Description	Defects
A: (1a)	Core	None
B: (1b)	Cladding	None
C: (Epoxy ring)	Adhesive	≤5 µm
D: (2)	Contact	≤5 µm

Table 15.6 Specifications of the IEC 61300-3-35 standard

Area	Description	Defects
A: (1a)	Core	None
B: (1b)	Cladding	No limit <2 µm/5 from 2 µm to 5 µm/None >5 µm
C: (Epoxy ring)	Adhesive	No limit
D: (2)	Contact	None ≥10 µm

Table 15.7 Parameters of the Arris simplex profile

Area	Description	Defects
A: (1a)	Core	None
B: (1b)	Cladding	<0.8 µm
C: (Epoxy ring)	Adhesive	≤5 µm
D: (2)	Contact	≤5 µm

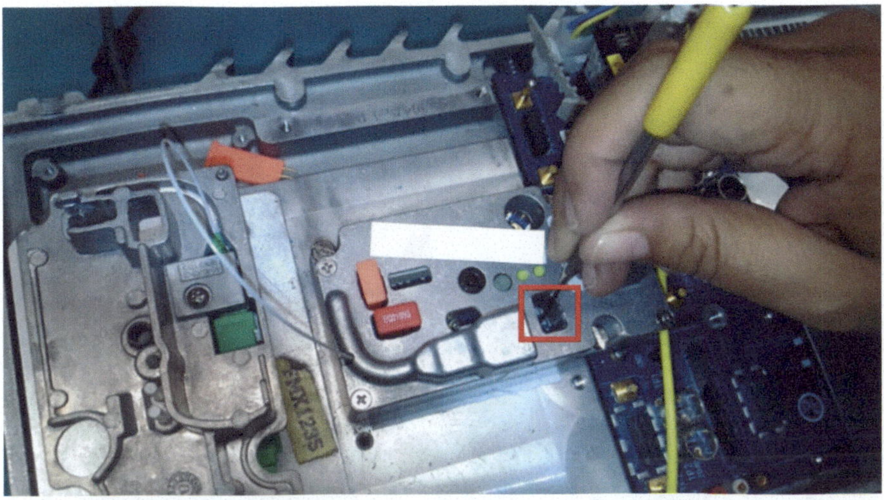

Fig. 15.8 Original alignment fixture with narrow cavity

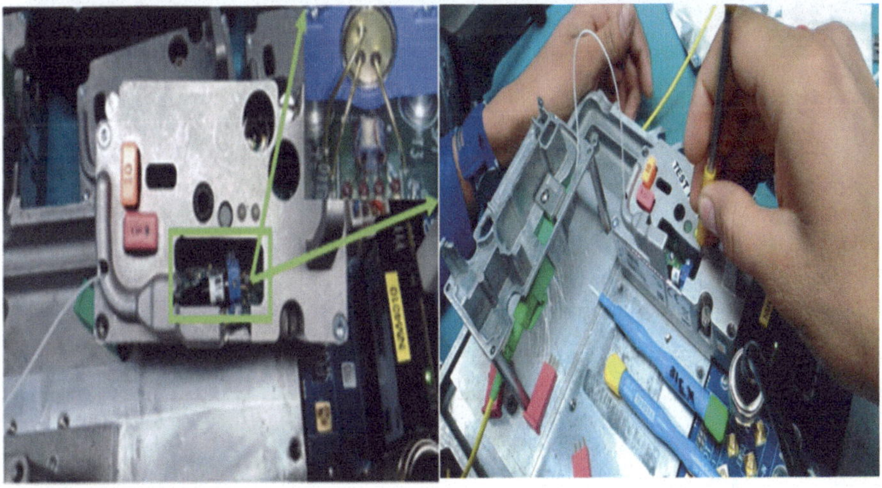

Fig. 15.9 Improvement action implemented for T3 defect

the results of a previous test, conducted with 30 samples of non-defective transformers. While conducting the test, we realized that all these good transformers had the aforementioned winding characteristic.

To test the improvement actions, we conducted a fitness test with 30 product samples, analyzing three variables—*Gain, Slope,* and *Flatness*—with respect to the behavior of the T3 transformer. *Gain* is the amplified difference between the input and output signals of an amplifier, *Slope* is the relative difference of radio frequency

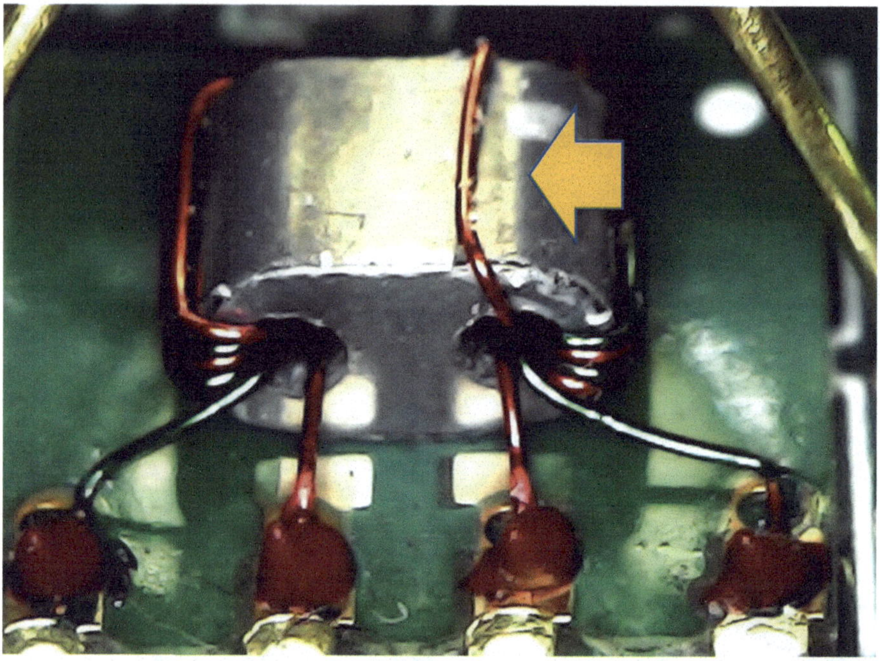

Fig. 15.10 Sample requested to suppliers

(RF) signal levels due to the coaxial cable, and *Flatness* refers to the flatness of an RF signal with respect to the *Slope*. The three variables are expressed in decibels (dB). The results—visually depicted in Figs. 15.11, 15.12, and 15.13—revealed that the three variables had values within the acceptable ranges during the test. Note that the intervals were calculated based on the product's test specifications. Acceptable

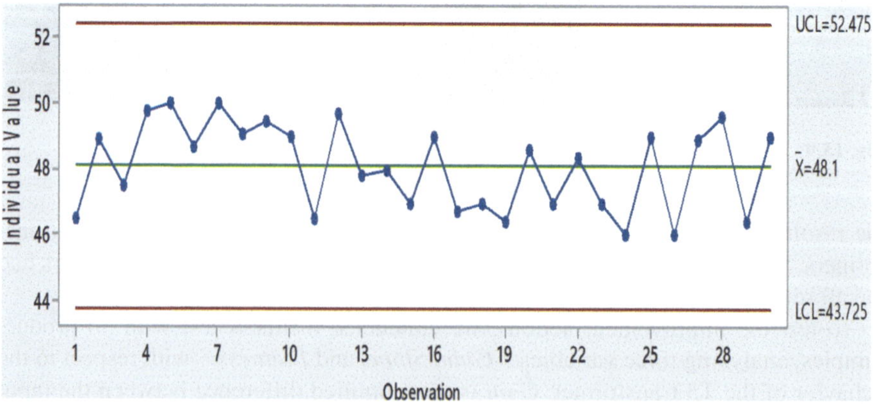

Fig. 15.11 *Gain* results with 30 samples

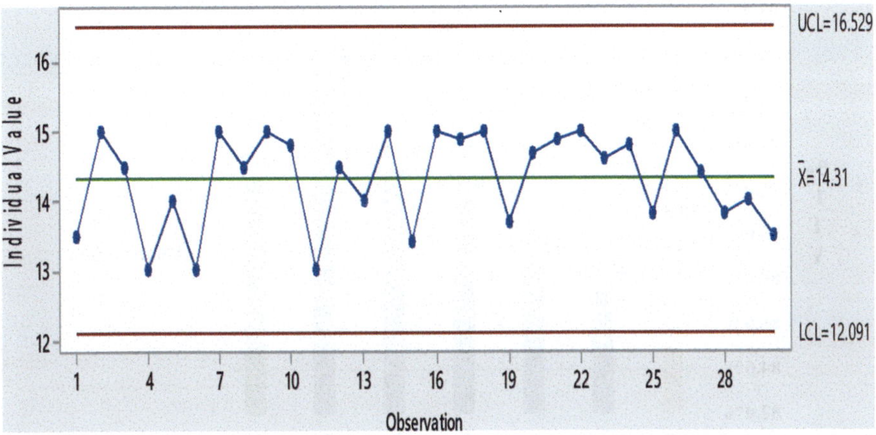

Fig. 15.12 *Slope* results with 30 samples

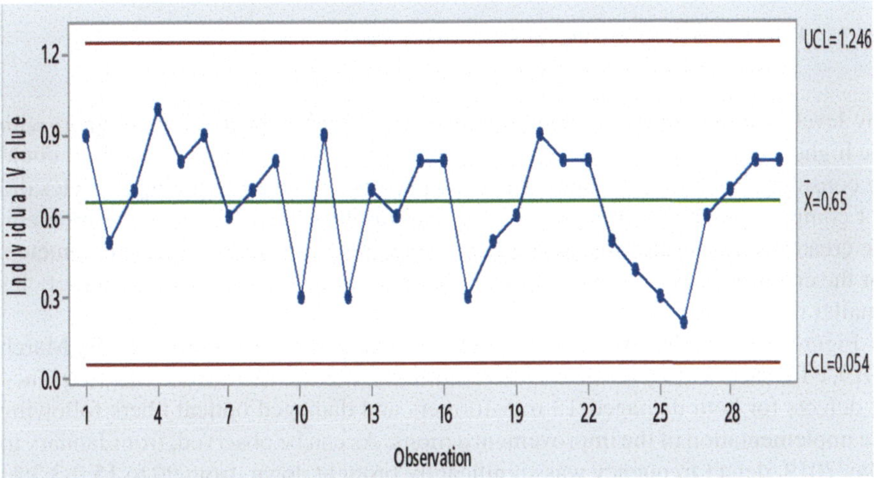

Fig. 15.13 *Flatness* results with 30 samples

ranges were 43.725–52.475 dB for *Gain* (median = 48.1 dB), 12.091–16.529 dB for *Slope* (median = 14.31 dB), and.054-1.246 dB for *Flatness* (0.065 dB). According to such results, the improved process is stable in terms of these three variables.

15.5.4.2 Damaged Optical Fiber

To measure pollution in optical fibers, the company follows a supervision process using Aurora. As previously mentioned, Aurora's inspection profile does not tolerate

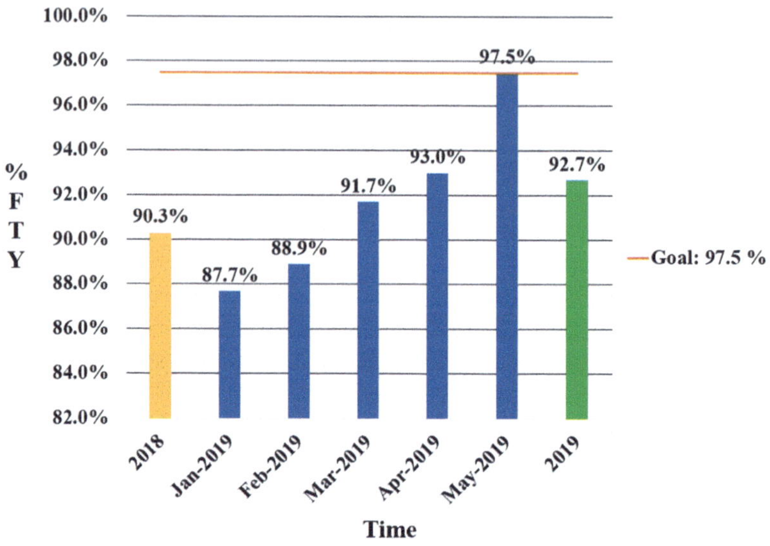

Fig. 15.14 Evolution of FTY in the AR subassembly area

any level of pollution in the cladding area (1b), which happens to be the part with the highest rejection rate (107 of 177). Interestingly, some of these rejections could be considered as false rejections, since the presence of small polluting particles did not compromise the performance of the final product itself. To solve the problem, we created a new inspection profile (see Table 15.7) with more flexible parameters for the cladding area (1b) of the optical fiber, thus allowing the presence of particles smaller than 0.8 μm.

Figure 15.14 depicts the evolution of FTY in the AR subassembly area. By March 2019, FTY successfully reached 97.5%. Similarly, Fig. 15.15 illustrates the frequency of defects for both damaged T3 transformers and damaged optical fibers following the implementation of the improvement actions. As can be observed, from January to May 2019, defect frequency was significantly brought down, from 90 to 15 (83.3%) in the case of T3 transformers, and from 56 to 12 (21.4%) in the case of optical fibers.

15.5.5 Phase 5 Results: Control

Following the encouraging results obtained in the improvement phase, we proceeded to control the new, enhanced process through standardization. Figure 15.16 illustrates the visual aid designed to support technicians during the T3 electrical testing process. The chart considers the improved fixture, which has a wider cavity for technicians to better manipulate the transformers. Also, it mentions and highlights all the necessary tools to perform the test.

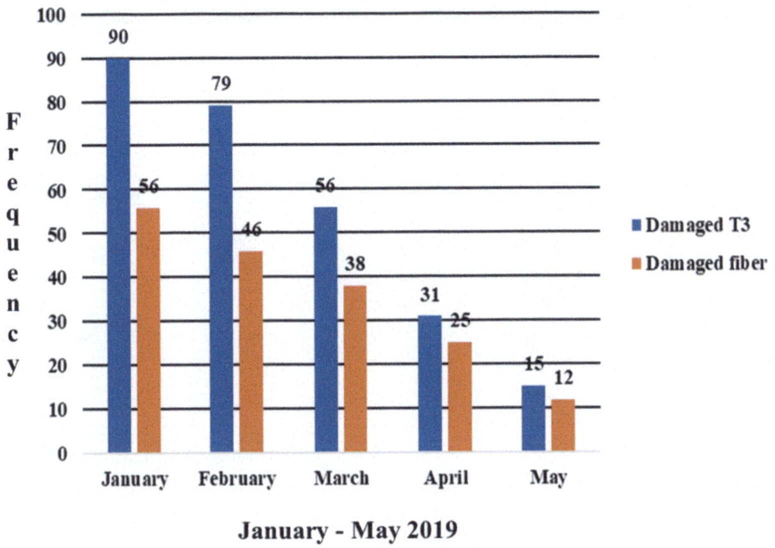

Fig. 15.15 Evolution of defect occurrence in 2019 for damaged T3 and damaged optical fiber

Fig. 15.16 Visual aid for T3 electrical testing

Figure 15.17 depicts the visual aid designed to guide technicians during the optical fiber inspection process. The chart includes a workflow diagram that orderly details the inspection steps. The diagram arrows indicate the sequence of the process.

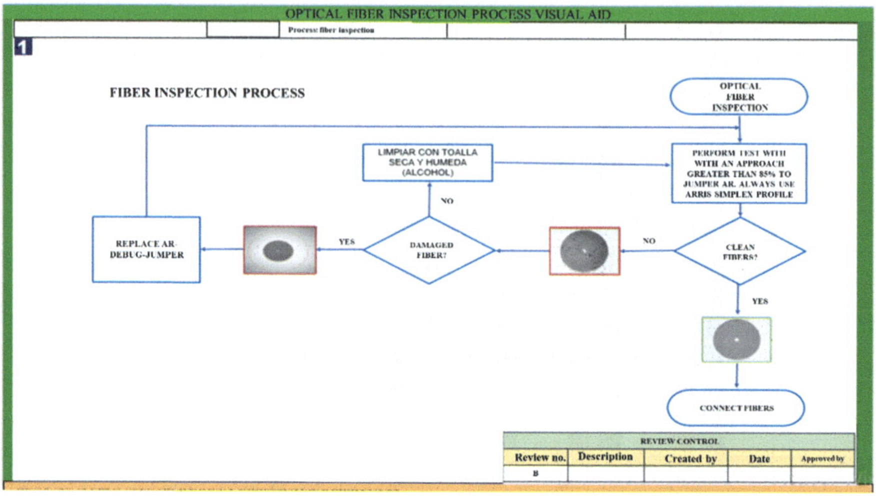

Fig. 15.17 Visual aid for optical fiber inspection

15.6 Conclusions and Recommendations

The DMAIC methodology can have a remarkably positive impact on the overall performance of manufacturing systems. Nowadays, large manufacturing companies—especially in developed countries—manage to remain competitive thanks to their investment in and implementation of advanced manufacturing technology. However, DMAIC is a relatively simple methodology that can bring significant benefits regardless of the company's degree of technological specialization. Also, in developing countries, such as Mexico, DMAIC is an essential tool for manufacturing companies to maintain a competitive status. The methodology can be implemented in an array of different projects to increase economic savings and comply with timely deliveries.

This research demonstrates that the DMAIC approach, along with quality management tools such as Pareto charts and the Ishikawa diagram, can help significantly improve product quality, consequently increase customer satisfaction, minimize costs and production time, and decrease variability in the production process. In our case study, the three tools—DMAIC, Pareto charts, and the Ishikawa diagram—were pivotal in finding both the specific problems (defects) affecting amplifier manufacturing and the causes of such problems. As a result, the company managed to increase FTY in the AR assembly area until reaching the standard (i.e., 97.5%). Our results are consistent with those reported by Tong et al. (2004), Hwang (2006), Gijo et al. (2011), Chang et al. (2012), Gandhi et al. (2019), Gaikwad et al. (2019), and (C.R. and Thakkar 2019), among others. Finally, it is noteworthy mentioning that the case study reported in this article is one of the many cases of successful DMAIC implementation in Mexican manufacturing systems, yet it is also one of the few that become reported and scientifically published.

References

Abolhassani A, Harner J, Jaridi M, Gopalakrishnan B (2018) Productivity enhancement strategies in North American automotive industry. Int J Prod Res 56:1–18. https://doi.org/10.1080/00207543.2017.1359700

Acharya SG, Sheladiya MV, Acharya GD (2018) An application of PARETO chart for investigation of defects in FNB casting process. J Exp Appl Mech 9:33–39

Antony J, Gijo EV, Childe SJ (2012) Case study in six sigma methodology: manufacturing quality improvement and guidance for managers. Prod Plan Control 23:624–640. https://doi.org/10.1080/09537287.2011.576404

Beheshti MH, Hajizadeh R, Dehghan SF et al. (2018) Investigation of the accidents recorded at an emergency management center using the pareto chart: a cross-sectional study in Gonabad, Iran, During 2014–2016. Heal Emergencies Disasters Q 3:. https://doi.org/10.29252/NRIP.HDQ.3.3.143

Botti L, Mora C, Regattieri A (2017) Integrating ergonomics and lean manufacturing principles in a hybrid assembly line. Comput Ind Eng 111:481–491. https://doi.org/10.1016/J.CIE.2017.05.011

Brown CB, Collins TR, McCombs EL (2006) Transformation From Batch to Lean Manufacturing: The Performance Issues. Eng Manag J 18:3–14. https://doi.org/10.1080/10429247.2006.11431689

Bunce MM, Wang L, Bidanda B (2008) Leveraging Six Sigma with industrial engineering tools in crateless retort production. Int J Prod Res 46:6701–6719. https://doi.org/10.1080/00207540802230520

C.R. A, Thakkar JJ (2019) Application of Six Sigma DMAIC methodology to reduce the defects in a telecommunication cabinet door manufacturing process: A case study. Int J Qual Reliab Manag IJQRM-12-2018-0344. https://doi.org/10.1108/IJQRM-12-2018-0344

Chakravorty SS, Shah AD (2012) Lean Six Sigma (LSS): an implementation experience. Eur J Ind Eng 6:118–137. https://doi.org/10.1504/EJIE.2012.044813

Chang S-I, Yen DC, Chou C-C et al (2012) Applying Six Sigma to the management and improvement of production planning procedure's performance. Total Qual Manag Bus Excell 23:291–308. https://doi.org/10.1080/14783363.2012.657387

Cheng H-R, Chen B-W (2010) A case study in solving customer complaints based on the 8Ds method and Kano model. J Chinese Inst Ind Eng 27:339–350. https://doi.org/10.1080/10170669.2010.495508

Chokkalingam B, Raja V, Anburaj J et al (2017) Investigation of shrinkage defect in castings by quantitative ishikawa diagram. Arch Foundry Eng 17:174–178. https://doi.org/10.1515/afe-2017-10032

Cunha C, Dominguez C (2015) A DMAIC Project to improve warranty billing's operations: a case study in a portuguese car dealer. Procedia Comput Sci 64:885–893. https://doi.org/10.1016/J.PROCS.2015.08.603

de Mast J, Lokkerbol J (2012) An analysis of the six sigma DMAIC method from the perspective of problem solving. Int J Prod Econ 139:604–614. https://doi.org/10.1016/J.IJPE.2012.05.035

Desai DA, Kotadiya P, Makwana N, Patel S (2015) Curbing variations in packaging process through Six Sigma way in a large-scale food-processing industry. J Ind Eng Int 11:119–129. https://doi.org/10.1007/s40092-014-0082-6

Doshi JA, Kamdar JD, Jani SY, Chaudhary SJ (2012) Root cause analysis using ishikawa diagram for reducing radiator rejection. Int J Eng Res Appl 12:684–689

Dudin MN, Frolova EE, Gryzunova NV, Borisovna ES (2015) The deming cycle (PDCA) concept as an efficient tool for continuous quality improvement in the agribusiness. Asian Soc Sci 11:239–246. https://doi.org/10.5539/ass.v11n1p239

Flower KB, Higginbotham LB, Jamison SD et al (2019) Alignment of preventive medicine physicians' residency training with professional needs. Am J Prev Med 56:908–917. https://doi.org/10.1016/J.AMEPRE.2019.01.012

Gaikwad LM, Sunnapwar VK, Teli SN, Parab AB (2019) Application of DMAIC and SPC to improve operational performance of manufacturing industry: a case study. J Inst Eng Ser C 100:229–238. https://doi.org/10.1007/s40032-017-0395-5
Galbreath J (2002) Twenty-first century management rules: the management of relationships as intangible assets. Manag Decis 40:116–126. https://doi.org/10.1108/00251740210422794
Gandhi SK, Sachdeva A, Gupta A (2019) Reduction of rejection of cylinder blocks in a casting unit: a six sigma DMAIC perspective. J Proj Manag 4:81–96. https://doi.org/10.5267/j.jpm.2019.1.002
Gawdzińska K (2007) Application of the Pareto chart and Ishikawa diagram for the identification of major defects in metal composite castings. Arch Foundry Eng 11:23–28
Gijo EV, Scaria J, Antony J (2011) Application of six sigma methodology to reduce defects of a grinding process. Qual Reliab Eng Int 27:1221–1234. https://doi.org/10.1002/qre.1212
Guiras Z, Turki S, Rezg N, Dolgui A (2018) Optimization of two-level disassembly/remanufacturing/assembly system with an integrated maintenance strategy. Appl Sci 8:1–25. https://doi.org/10.3390/app8050666
Guo W, Jiang P, Xu L, Peng G (2019) Integration of value stream mapping with DMAIC for concurrent lean-kaizen: a case study on an air-conditioner assembly line. Adv Mech Eng 11:168781401982711. https://doi.org/10.1177/1687814019827115
Gupta N, Lteif A, Creo A et al (2018) Improved utilization of waist-to-height ratio in cardiometabolic risk counselling in children: Application of DMAIC strategy. J Eval Clin Pract 1–6. https://doi.org/10.1111/jep.13055
Hakimi S, Zahraee SM, Mohd Rohani J (2018) Application of Six Sigma DMAIC methodology in plain yogurt production process. Int J Lean Six Sigma 9:562–578. https://doi.org/10.1108/IJLSS-11-2016-0069
Hwang Y-D (2006) The practices of integrating manufacturing execution systems and six sigma methodology. Int J Adv Manuf Technol 31:145–154. https://doi.org/10.1007/s00170-005-0164-0
Instituto de Productividad Empresarial Aplicada (2019) Diagrama Causa—Efecto, o de Ishikawa. Resolución de problemas. In: Diagr. Causa—Efecto, o Ishikawa. Resolución Probl. https://www.ipeaformacion.com/resolucion-de-problemas/diagrama-causa-efecto-ishikawa/. Accessed 13 Jun 2019
Jadhav PK, Nagare MR, Konda S (2018) Implementing lean manufacturing principle in fabrication process—a case study. Int Res J Eng Technol 5:1843–1847
Joiner Associates I (1995) Pareto charts: plain & simple. Oriel Incorporated, Madison
Jonny Christyanti J (2012) Improving the quality of asbestos roofing at pt bbi using six sigma methodology. Procedia—Soc Behav Sci 65:306–312. https://doi.org/10.1016/J.SBSPRO.2012.11.127
Journal of Applied Sciences (2018) Special issue: applied engineering to lean manufacturing production systems. https://www.mdpi.com/journal/applsci/special_issues/aelmps. Accessed 3 Oct 2018
Khajuria A, Raina A, Singh MP (2018) Implementation of DMAIC six sigma principle in thermoforming for improving rate of production. Int J Res Mech Civ Eng 4:13–22
Kholif AM, Abou El Hassan DS, Khorshid MA et al (2018) Implementation of model for improvement (PDCA-cycle) in dairy laboratories. J Food Saf 38: https://doi.org/10.1111/jfs.12451
Kumar M, Vaishya R, Parag (2018) Real-time monitoring system to lean manufacturing. Procedia Manuf 20:135–140. https://doi.org/10.1016/J.PROMFG.2018.02.019
Leaphart CL, Gonwa TA, Mai ML et al (2012) Formal quality improvement curriculum and DMAIC method results in interdisciplinary collaboration and process improvement in renal transplant patients. J Surg Res 177:7–13. https://doi.org/10.1016/J.JSS.2012.03.017
Lee B (2016) Conector de fibra óptica. Higiene, Texas
Liu AX, Liu Y, Luo T (2016) What drives a firm's choice of product recall remedy? the impact of remedy cost, product hazard, and the CEO. J Mark 80:79–95. https://doi.org/10.1509/jm.14.0382

Marodin G, Frank AG, Tortorella GL, Netland T (2018) Lean product development and lean manufacturing: testing moderation effects. Int J Prod Econ 203:301–310. https://doi.org/10.1016/J.IJPE.2018.07.009

Nabiilah AR, Hamedon Z, Faiz MT (2018) Improving quality of light commercial vehicle using PDCA approach. J Adv Manuf Technol 12:525–534

Pérez-Urrego ML, Peláez-Zúñiga JS, Carrión-García A (2014) La capacidad de procesos como métrica de calidad para características cualitativas. In: IX Encuentro Internacional de Investigadores de la Red Latinoamericana de Cooperación Universitaria. Cali

Pugna A, Negrea R, Miclea S (2016) Using six sigma methodology to improve the assembly process in an automotive company. Procedia—Soc Behav Sci 221:308–316. https://doi.org/10.1016/J.SBSPRO.2016.05.120

Rahman Ansar A, Chowdhury-Shaju SU, Kumar-Sarkar S et al (2018) Application of six sigma using define measure analyze improve control (DMAIC) methodology in Garment Sector. Indep J Manag Prod 9:810–820. https://doi.org/10.14807/ijmp.v9i3.732

Rahman M, Dey K, Kapuria TK, Tahiduzzaman M (2018) Minimization of sewing defects of an apparel industry in bangladesh with 5S & PDCA. Am J Ind Eng 5:17–24. https://doi.org/10.12691/ajie-5-1-3

Saeidi SP, Sofian S, Saeidi P et al (2015) How does corporate social responsibility contribute to firm financial performance? the mediating role of competitive advantage, reputation, and customer satisfaction. J Bus Res 68:341–350. https://doi.org/10.1016/J.JBUSRES.2014.06.024

Sartal A, Llach J, Vázquez XH, De Castro R (2017) How much does Lean Manufacturing need environmental and information technologies? J Manuf Syst 45:260–272. https://doi.org/10.1016/j.jmsy.2017.10.005

Sharma G, Rao PS, Babu BS (2018) Process capability improvement through DMAIC for aluminum alloy wheel machining. J Ind Eng Int 14:213–226. https://doi.org/10.1007/s40092-017-0220-z

Sharma H, Suri NM (2017) Implementation of quality control tools and techniques in manufacturing industry for process improvement. Int Res J Eng Technol 4:1581–1587

Smętkowska M, Mrugalska B (2018) Using six sigma DMAIC to improve the quality of the production process: a case study. Procedia-Social Behav Sci 238:590–596. https://doi.org/10.1016/j.sbspro.2018.04.039

Sreedharan R, Rajasekar S, Kannan SS et al (2018) Defect reduction in an electrical parts manufacturer: a case study. TQM J 30:650–678. https://doi.org/10.1108/TQM-03-2018-0031

Srinivas SS, Sreedharan R (2018) Failure analysis of automobile spares in a manufacturing supply chain distribution centre using Six Sigma DMAIC framework. Int J Serv Oper Manag 29:359. https://doi.org/10.1504/IJSOM.2018.089828

Srinivasan K, Muthu S, Devadasan SR, Sugumaran C (2016) Six Sigma through DMAIC phases: a literature review. Int J Product Qual Manag 17:236–257. https://doi.org/10.1504/IJPQM.2016.074462

Srinivasan K, Muthu S, Devadasan SR, Sugumaran C (2014a) Enhancing effectiveness of shell and tube heat exchanger through six sigma DMAIC Phases. Procedia Eng 97:2064–2071. https://doi.org/10.1016/J.PROENG.2014.12.449

Srinivasan K, Muthu S, Prasad NK, Satheesh G (2014b) Reduction of paint line defects in shock absorber through six sigma DMAIC phases. Procedia Eng 97:1755–1764. https://doi.org/10.1016/J.PROENG.2014.12.327

Tenera A, Pinto LC (2014) A lean six sigma (LSS) project management improvement model. Procedia—Soc Behav Sci 119:912–920. https://doi.org/10.1016/J.SBSPRO.2014.03.102

Tong JPC, Tsung F, Yen BPC (2004) A DMAIC approach to printed circuit board quality improvement. Int J Adv Manuf Technol 23:523–531. https://doi.org/10.1007/s00170-003-1721-z

Wang C-H (2013) Incorporating customer satisfaction into the decision-making process of product configuration: a fuzzy Kano perspective. Int J Prod Res 51:6651–6662. https://doi.org/10.1080/00207543.2013.825742

Wang C-N, Chiu P-C, Cheng I-F, Huang Y-F (2019) Contamination improvement of touch panel and color filter production processes of lean six sigma. Appl Sci 9:1893. https://doi.org/10.3390/app9091893

Zhou X-Y, Gosling PD (2018) Influence of stochastic variations in manufacturing defects on the mechanical performance of textile composites. Compos Struct 194:226–239. https://doi.org/10.1016/J.COMPSTRUCT.2018.04.003

Chapter 16
Adaptability of the Lean-Sigma Methodology for Operations in a Multicultural Workplace

Noé Gaudencio Alba-Baena and Aldo Salcido-Delgado

Abstract Overseas operations, as beneficial as there are, present challenges such as the adoption of the organizational culture, working methodologies, and practices. Different strategies have been exercised to successfully reach the expected benefits of these operations and take advantage of the local resources. Managers have faced local resistance and miscommunication barriers, moreover, the cultural differences have impacted the operations reducing in general, the efficiency of these operations. The need for integration in the cultural component has been addressed by different researchers; however, there is limited information on methodologies used for such integration. The present work proposes a methodology for the integration of the national cultural characteristics for creating a local workplace culture. The described methodology is accompanied by an example of implementation in a multicultural scenario. The case study shows successful implementation of the Lean-Sigma methodology in response to a problematic situation in the operation site of Juarez (Mexico). A study of the local culture helps in the selection of the tools used during this implementation and to exceed the customer expectative and compete for larger projects with this and other customers.

Keywords Lean-Sigma · Workplace culture · Overseas operations

16.1 Introduction

In searching to maintain a competitive business in the growing globalized market through growth in a globalized market, companies find several benefits in expanding to international customers and outsourcing parts or their complete production systems.

Such strategies include partnerships, the merging with other players, or the acquisition of companies in different regions around the world. The potential for increasing

N. G. Alba-Baena (✉) · A. Salcido-Delgado
Department of Industrial Engineering and Manufacturing, Universidad Autónoma de Ciudad Juárez, Av. Del Charro 450 Norte, Col. Partido Romero, Ciudad Juárez, Chihuahua, Mexico
e-mail: nalba@uacj.mx

J. L. Garcia Alcaraz et al. (eds.), *Trends in Industrial Engineering Applications to Manufacturing Process*, https://doi.org/10.1007/978-3-030-71579-3_16

the assets because of these movements have been limited in part due to communication issues and cultural differences between the different actors. The benefits, economic value, and competitive advantages expected from these movements have overseen the challenges occurring due to the cultural aspects given in each region. Experience has shown that it is common to find resistance and confrontation in adapting policies and working methods when local employees and foreign management test them during the day-to-day operations. Managers in charge of foreign origin face administrative challenges, especially for those managers who are not familiar with local culture and common customs (Omar and Urteaga 2010).

In their report, Berson et al. (2005) argue that organizational culture is one of the fundamental characteristics of strong organizations and that organizations can be influenced by the values of their leaders permeated through the organizational practices. In consequence, long-lasting companies develop their own cultures and tend to hire and attract employees with similar culture and behavior characteristics, including their moral values (Johns 2006). Moreover, in developing a culture, the organizations have included systemic influences from their immediate environment (Findler et al. 2007). At the time of taking the organizational culture to overseas operations, their practices, and cultural values can influence the national or regional culture in the recipient environment. House et al. 2004 remarks that if the organizational culture is strong, their values permeate and transfer from the organization to the overseas culture. However, Berson et al. (2005) and De Hilal (2006), observed that in overseas operations the interaction will be as reactive as the differences in values are with the overseas values and the resistance will count for each to the challenged cultural points. At the same time, Inglehart (2004) considers that the work culture of an organization can be influenced by the values, sociopolitical environment, and geographical location of the region where it is located in foreign operations.

The organization itself can absorb characteristic features of external environments such as the culture of the workers. Generally, this phenomenon is given by the contribution of the adopted values towards efficiency, value-added, and guided by the increase of operational profits. This adoption not only reflects flexibility and value of the elements in a business, moreover, this phenomenon also influences the response of employees to management strategies adaptation regardless of whether they come from abroad or are local practices.

The success in taking overseas advantages has been studied for the last decades with contrasting results. Aycan et al. (2000) reported the variations in the distance to power after the implementation of different strategies used by human resources managers (ARH) in ten countries. The authors conclude that a common resistance and reactivity of the employees toward the power (directives) is resultant of the lack of enrichment or empowerment toward downstream the organization. For a specific case, Kim et al. (2004) show how the organizational culture had impacted the overall performance at several operations in Singapore. Concluding that there is a strong correlation between the cultural values with the performance metrics affecting the outcomes, productivity counts for the manufacturing, annual premiums growth, and the sum insured by insurance companies. Moreover, the influence of cultural values affects directly personnel retention, as shown by Kerr and Slocum (2005).

To weight this effect, the authors compared an organizational culture and their values with its national staff (both from the USA), measuring the assimilation of the organization values, how the culture influences the "loyalty" of the staff and how impacts the employee turnover cycle. When moving operations abroad, the challenges are more complex than the mere adaptability of the organizational goals and values. The challenges include the cultural strives but also the methodological contrasts, the analysis, and understanding of the context will help in the decision criteria used for defining the methods and management tools used for running overseas operations (Williams et al. 1984).

A successful cultural exchange depends also on the staff's fast integration into the organizational culture and the synergy between the local culture and the organizational values. To have more possibilities in succeeding, managers can influence the staff selection and search for the more suitable to the organizational culture and values. Managers need to have a clear understanding of the local culture and be able to make this link (Naor et al. 2014). The efficiency of this understanding is reflected directly in the enthusiasm, commitment, and loyalty of employees (Marc and Farbrother 2003). Then, the clear measurement of the local culture and the later comparison to the organizational culture will provide the tools for maximizing the human factors in operations abroad. Other factors must be considered if an organization is looking for keeping its operations competitive in the XXI century. Productivity, quality, costs, and deliverability among others are critical factors to address in their strategic considerations. For facing these challenges, managers have several world-class manufacturing strategies (WCMS), such as Lean Manufacturing, Six-Sigma, Lean- Sigma, and other more specialized approaches. Modern competitive challenges require organizations to have lean, efficient, cost-effective, and flexible manufacturing practices. However, in general, the implementation of these methodologies has been reported low rates of success (See Fig. 16.1).

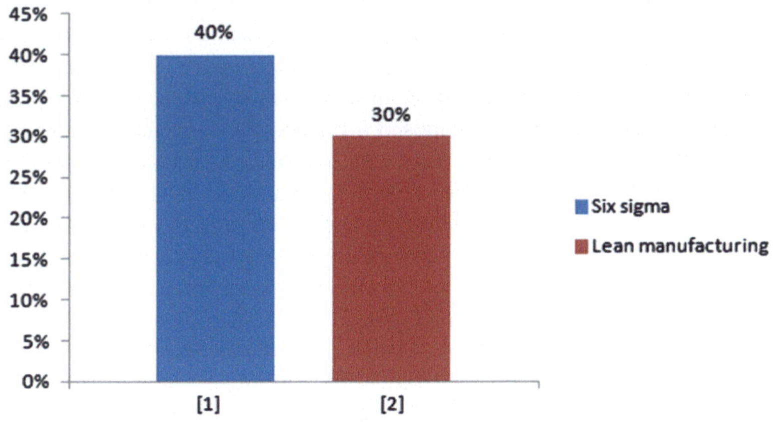

Fig. 16.1 Failure rates after implementing WCMS [1] Six-Sigma and [2] Lean Manufacturing (Roser, 2017)

The success rates have been reported in unexpected levels of 40% for Six-Sigma projects and 30% for Lean Manufacturing projects (Where Lean Went Wrong—A Historical Perspective 2017). One of the main reasons can be attributed to the influence of the cultural factors mentioned above. Under the described conditions, different studies and strategies have been tested for increasing the world-class methodologies' success rates; Estrada-Orantes et al. (2019) proposed the use of the E-strategy as guidance for solving problems in organizations using a Lean-Sigma approach in overseas operations.

This approach defines a series of steps interactive for validating the decisions, increasing with different tools communication (in Horizontal Organizations), and reducing the impact of avoidances of the local staff answering to the management requirements in vertical organizations. The success in the adoption of these WCMS and methodologies is also critical from the customer side, as the implementation of world-class strategies, influences the perception and confidence of the intermediate and final customers. In the same approach, this strategy can help to keep a market advantage, besides the detailed management, planning, budgeting, training, and efficiency, it is necessary to respond to the customer needs and expectations. A case shown by Salcido-Delgado et. al (2019), presents a process behaving as expected but with the need to respond to a market competitive requirement and a customer expectative. behavioral challenges in organizations, keep hidden from most managers' scope, and are valued in large by experienced managers. For managers abroad, this behavioral component is a dark region that delays the expected results.

To increase the possibility for implementation of WCMS with success rates in overseas operations then, it is necessary to consider and measure the cultural factors of the region/nation. The cultural analysis can be key guidance for determining the chosen organizational structure at the given site, in the implementation of a world-class methodology for the selection of tools and activities or used for the selection between a horizontal and vertical local organization, will also help in meeting the expectations from the staff and workers at the site and the original organization. Helping also in determining the tool kit with the integration or distancing expected culturally, for example, a consensus tool is more efficient in a horizontal structure and a directive instruction more efficiently in a vertical organization. For measuring the cultural aspects, Hofstede (1980) proposed a method for measuring it for different nations divided as follows: distance to power, resistance to uncertainty, femininity, masculinity, individualism, and collectivism. The data coming from this study has been the guide for many studies addressing and comparing the cultural aspects of organizational integration. However, the social integration and globalization from the last decades have impacted people in different ways and some regions have developed unique multicultural regional cultures, like the one at the border region between Mexico and the USA. Alba-Baena et al (2019) described this region as highly competitive, with investments from North American countries (the USA and Canada), from Europe (France, Italy, and others) and Asia (Japan, South Korea, and others); divided into more than 3,700 factories, with more than 1,300,000 workers. After six decades of interaction, this border region has developed a unique culture and social behavior, which is different from the Mexican and USA working and social cultures. This scenario is suitable for testing a methodology for integrating the cultural aspects of projects based on world-class methodologies.

This chapter includes a methodology for integrating the cultural aspects to the tool selection during the adaption of a world-class methodology such as Lean-Sigma in a factory located in Juarez, Mexico, a city with more than 320 factories employing more than 275,00 workers. The steps include the measure of the cultural aspects of the site for implementation (following the Hofstede (1980) survey or the later version for example), the analysis of the adaptability of the organizational values to the local values, the selection among the possible tools to be adapted to the given scenario, and finally the measurement of the level of success in the adaptation of the tools and the adoption of the methodology by measuring the changes in the process outcomes. The following sections present the regional culture analysis, the list of tools integrated for solving the case, and presents the steps followed for implementing the Lean-Sigma methodology and solving a specific problem in this facility. For reference, this facility has management of French origin and vertical organization is observed as an operative structure. The problem to solve is a cosmetic defect in the stamp-printing process.

16.2 Project Preparation

Firstly, the work-culture analysis is based on the survey proposed by Wu et al. (2001) which considers the six dimensions of culture proposed by Hofstede (1980). The questionnaire was translated to Spanish and the 24 questions were divided into groups of four questions for evaluating the proposed dimensions. After gathering the data, coded results are compared to the scores obtained by Hofstede for the USA, Mexico, Italy, and France (management origin) for the surveyed groups. Surveys in two manufacturing operations were gathered (see Table 16.1 for the data characteristics). The instrument representativity was validated using a random sample of 30 and evaluated using the Cronbach's alpha coefficient to estimate the reliability of this instrument under the given conditions according to Tavakol and Dennick (2011). The Cronbach coefficient shows a result of 0.837 in case 1 (of French management) and 0.712 in the second case of Italian management indicating that the reliability of the applied questionnaires. Data resultant is then compared to the indicators proposed by Hofstede as seen in Fig. 16.2.

As expected for a multicultural site, the large scores in several dimensions are the product of the influence of the international cultural values merging in this region (see Fig. 16.2). As shown in Fig. 16.2a, top values are registered for the cases in

Table 16.1 Characteristic parameters of the personnel surveyed in two locations

	# of respondents	Average age (years)	% Female staff	% Male staff
Case 1	108	30	37	63
Case 2	140	31	44	56

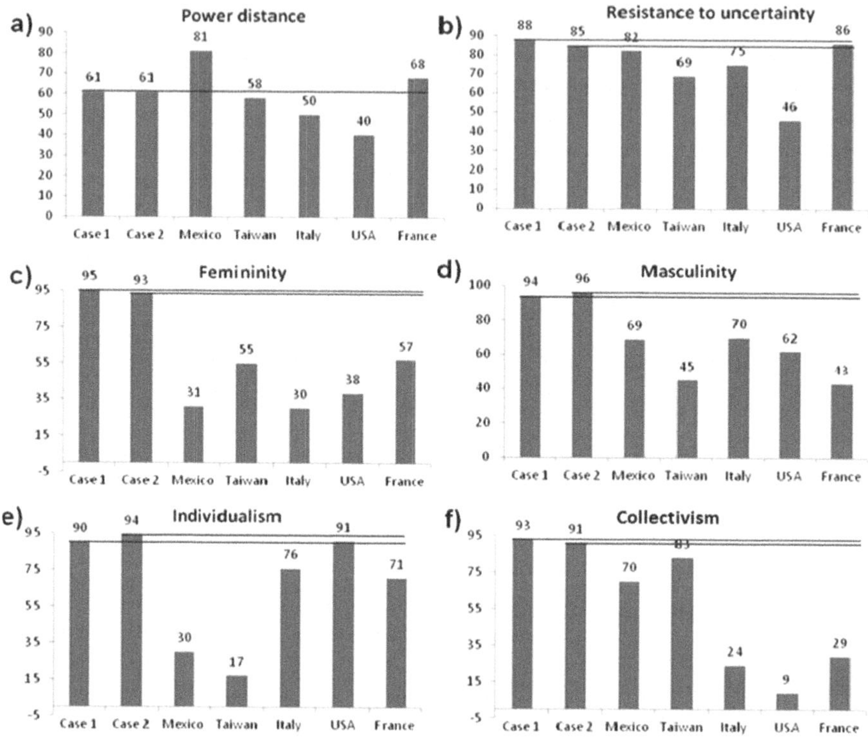

Fig. 16.2 Comparisons between the scores of Mexico, Italy, France, and the USA against the scores obtained in cases 1 and 2 for the dimensions **a** distance to power, **b** resistance to uncertainty, **c** femininity, **d** masculinity, **e** individualism, and **f** collectivism respectively

Juarez except for those values for the distance to power (or the extent to which the less powerful members of organizations and institutions accept unequally) in both cases of Juarez of 61. These scores compared to those for Mexico and France, are showing that the staff attitude is more equalitarian than those expected for the national character. Knowing this, managers will find more efficient the combination of quantitative and qualitative tools for communicating and make decisions in this environment. Two other values are also remarkable in Fig. 16.2e and f, is shown a combination of attitudes that promotes individualism and collectivism at the same time. Such a combination of scores is reflected in the worker's decision to satisfying his personal needs (individualism) and keep efforts towards the higher benefit (collectivism). Using this information, managers will know that the decisions on economic goals or to design strategies must show benefits to both the workers and the organization.

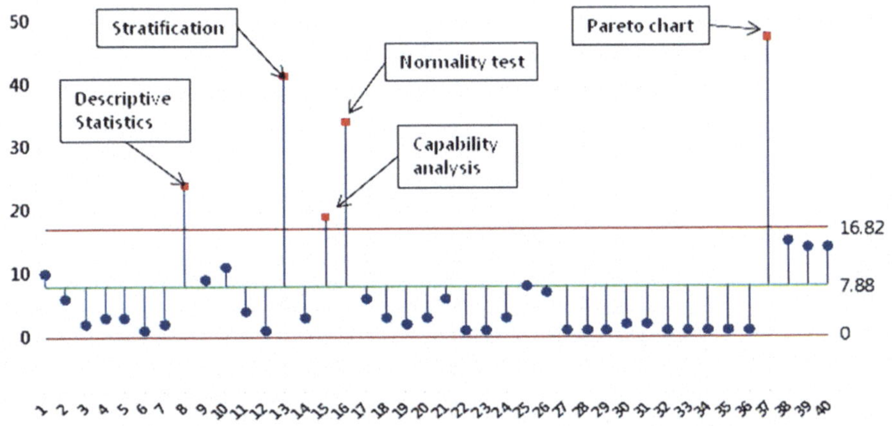

Fig. 16.3 Graph comparing means of data values from the tools used in the analyzed projects

16.2.1 Set of Tools for the Case Study

Considering that the use of a given tool reflects a cultural comfort, then, it is possible to measure and compare the use of a tool over others and determine the tendency to use a specific type of tool. In this case, reports in a repository at the autonomous university of Ciudad Juárez is used to identify industrial projects implementing world-class methodologies in their solutions.

A meta-analysis is used for filtering the projects following the four-phase information flow in the PRISMA statement (Moher et al. 2009). The filtered projects are used for identifying the tools used, data then is quantified and coded. With the list of tools a statistical evaluation is carried out using a Means Analysis, the results of this analysis are shown in Fig. 16.3. 40 of the top used tools are shown. From these five are the relevant: descriptive statistics, stratification, normality test, capability tests, and the Pareto charts. These results show that for the Juarez case, the preferred tools are related to statistics and quantitative values, an approach, favoring the distance to the power and decisions that require less communication among ranks, which is expected in the Mexican work culture.

16.2.2 Adapting a Methodology

The described list of tools can be organized in the scope of a WCSM, Six-Sigma for example. In Mexico, The National Council for Standardization and Certification of Working Competencies (CONOCER, www.conocer.gob.mx), certifies the expertise and use of Six-Sigma as a problem-solving methodology with the norm EC0264 (problem-solving through the Six-Sigma methodology level I). In the same context, the same council certifies the expertise in the use of the methodology. WoW-Vation

Table 16.2 List of tools suggested for use in each stage for solving problems (EC0264)

Stages Norm 0264	DMAIC available tools
D	Stratification, Pareto, quality function deployment (QFD)
M	Descriptive statistics, repeatability, and reproducibility studies (R & R), Pareto, stratification, point chart, bar chart, pie chart, data collection sheet
A	Brainstorm, stratification, 5 whys, Ishikawa diagram, descriptive statistics, histogram, normality test
I	Experiment design, main effects chart, Pareto, stratification, XR chart, XS chart, I-MR chart, Z-MR chart, means analysis, point chart, box chart, proportion chart, P chart, NP chart, a graph of points, bar graph, pie graph, two-sample t, one-sample t
C	Failure mode analysis (FMA), control plan, visual instructions

for solving problems with inventive (EC1074), the use of a method of root-cause for incidents analysis (EC0479). The use of the 5s methodology for continuous improvement is certified under the (EC0491 norm). The case of the EC0264 certification is based on the DMAIC methodology (see Table 16.2) which includes tools shown in Fig. 16.3, which are the base for the adaption of Lean-Sigma.

16.3 Integration into a Lean-Sigma Project

The resultant cultural scores for Juarez, Mexico, and the organizational culture from the company are now matched for selecting the tools for the implementation of Lean-Sigma. For the adoption of a Lean-Sigma methodology, the tools are now matched in the steps of the Lean-Sigma: (a) Identify and measure the problem, (b) conduct a root cause analysis, (c) develop a solution, (d) verify the solution, and have (e) a control plan, as described by Estrada-Orantes et al. (2014). However, the tools' selection requires a larger definition of the activities, then, the steps are divided into 21 activities as is shown in Fig. 16.4. Then, the tools are selected by the combination of local and organizational cultures under the activity requirement. The main goal of integrating the cultural data (in Fig. 16.2) into the selection process is to increase the potential for implementation success. Is expected to increase efficiency as the team members are "playing comfortably in common grounds." In this case study, the distance to power plays a distinctive role in the tools' selection, for example, the tools that promote the numeric evaluations and decisions, such as the Pareto charts, will be preferred over more interactive tools such as the Delphi method.

Following the sequence presented in Fig. 16.4, Table 16.3 shows in part the tools selected during this integration. In these tables, the column "Step" (of the implemented methodology) is related to one step from the methodology to adapt, the "Tool" column shows the selected tools, which are based on the described criteria, this according to the cultures' comparison, finally, the column "Workplace culture adaptation" gives a brief description of the reasons to select a specific tool at that stage were selected at a given stage.

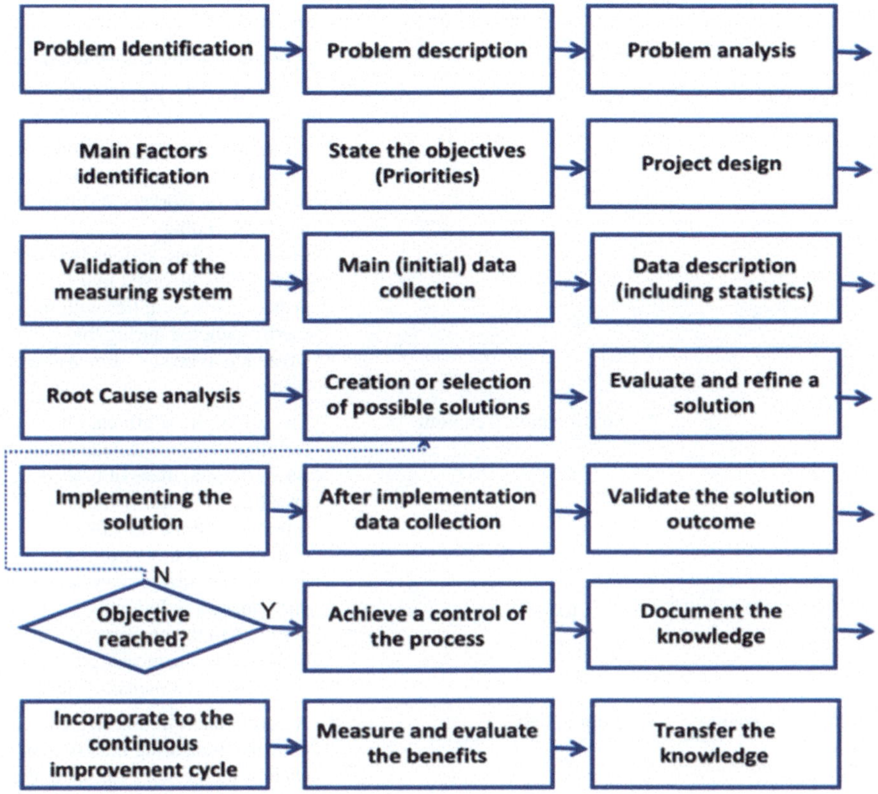

Fig. 16.4 Twenty-one activities chosen for adapting Lean-Sigma tools to the workplace culture (Alba-Baena unpublished data 2020)

Table 16.3 and its continuations highlight different reasons for using some tools, giving the reader an idea of the selection process. For example, a Pareto chart can be used in the different steps of the adaptation process, however, the use of this tool in the present case helps in giving a visual and a numerical comparison instead of promoting interaction, a Pareto chart is an efficient tool when the team has a vertical structure and members of different ranks may conflict in a decision talking process (for example in the first row in Table 16.3). By using this tool, the discussion is mainly focused on the quantitative values and their effects on the outcomes. The same occurs when the implementation includes a root cause analysis in a step (row 8 in Table 16.3). These tools help in reducing the subjectivity of the decisions and qualitative evaluation. The brainstorming tool is used in the "Develop a solution" step (row 7 in Table 16.3). Finally, other statistical tools are used under the same logic for integrating the Lean-Sigma methodology and its adaptation to this cultural environment.

Table 16.3 Relationship between problem-solving tools and dimensions of work-culture

Step	Tool	Workplace culture adaptation
Identify and measure the problem	Pareto chart	In this step, the Pareto chart helps in visualizing the information concisely, allowing a short time for interaction between the workers at different levels. Avoiding discussions-interactions between participants involved in a project. This complies with the marks gathered during the "distance to power" dimension values
	Stratification technique	The use of the stratification technique helps in categorizing a series of data, these of objective, or subjective values, and separates them to identify patterns. This tool requires a level of socialization among the participating individuals involved in the project and can be linked to the measured "collectivism dimension" level
Conduct a root cause analysis	Pareto chart	A Pareto chart in this step can be used for comparing the effect of each of the primary causes in the outcome values and determine the expected impact of the activities proposed for solving the problem and as mentioned can be related to the "distance to power" dimension
	Stratification technique	Combined with the Pareto chart, the stratification technique helps in the evaluation of the causes including the quantitative and qualitative considerations for reaching a robust solution, while the team have limited interaction, making the decision process to be based on numeric considerations

(continued)

Table 16.3 (continued)

Step	Tool	Workplace culture adaptation
	Normality test	Normality tests and further descriptive statistics help in the characterization of the collected data in the measuring step. This is a critical evaluation that determines the statistical treatment of the initial and outcome values. The calculations and interpretation required training and the ability to describe the outcomes to the team. Then, this tool can be associated with the "resistance to uncertainty" dimension and even to the "distance to power" values obtained in this example
Develop a solution	Pareto chart	The Pareto chart allows, for example, to track the effects and impact of the changes occurring during the implementations. The concise presentation and monitoring help the team to keep the levels of "distance to power" as expected
	The brainstorming technique	The brainstorming technique promotes team interaction, ideas about a specific issue, or problem which are proposed. This technique is associated with the "collectivism" dimension (especially in the same level team) as it requires the open participation of the members involved in the project
	The Ishikawa diagram	The Ishikawa diagram visually assists the identification of the root cause by the interaction of the team. More efficient is as combined with the brainstorming technique. This combination is suitable to the high levels graded for the "Collectivism" dimension in this example

(continued)

Table 16.3 (continued)

Step	Tool	Workplace culture adaptation
	The 5 whys analysis	The 5 whys analysis tool is based on the integration of consecutive questions for the searching of the causes and effects of the different aspects in the process conditions. Here, the team must interact and, however, is more efficient in the same hierarchy rank members. Then, this tool can be linked to the "collectivism" dimension levels
Verify the solution	Pareto chart	The rearranged effects of the remaining causes can be visualized and compared (in this step) by using a Pareto chart
	Descriptive statistics	Descriptive statistics show the characteristics of the data obtained in a quantitative way and on this step can be used as the base for the statistical comparison and evaluation of the changes and effects of the project or implementations in the production process
	Validation using a two-sample test	As for the descriptive statistics, the statistical evaluation provided by this test is based on quantitative grounds. However, the interpretation of the results is of a qualitative understanding. These tools are more than useful for teams working within a vertical organization but value a collectivist approach, which is the case of the present study

16.3.1 Using the Adopted Methodology in the Case Study

After, in the case study, the decision is to use the adapted Lean-Sigma methodology tested previously as a reliable methodology for short-term projects and simple problems, but also a powerful methodology for long-term projects and/or complex problems.

Thanks to its adaptability, fast solving time, Lean-Sigma can include statistical analysis that provides long-term reliability of the solutions (Estrada-Orantes and Alba-Baena 2014). Each step of this methodology is flexible, and it is open to use the required the tool from the engineering, management, or statistical areas (Estrada-Orantes et al. 2019). The five basic steps of the Lean-Sigma methodology were originally proposed by George et al. (2005) and for this case, the tools for each step are selected following the criteria of the workplace culture for facilities located in Juarez, Mexico, similarly to the mentioned and partially described in Table 16.3). During the implementation the adaptation follows the five Lean-Sigma steps previously mentioned (see Estrada-Orantes et al. 2014) and enlisted with their goals:

- *Identify* and measure the problem. How big is it?
- *Root cause analysis*: What is the root cause of the problem?
- *Develop a solution*: Identify the alternative solution that best solves the problem
- *Verify the solution*: Make sure that the problem is eliminated by the proposed solution
- *Control Plan*: Make a quick and effective plan, so that the previous situation does not come back

16.3.2 Case Description

This project is implemented in a facility dedicated to the automotive industry. The focus is on a stamping process of a label that identifies rubber hoses (final product). A pad printing technology (tampography process as shown in Fig. 16.5) is used for transferring the designs as painted patterns to the product. The hoses must be stamped according to the customer specifications, tampography is a common method for printing images on non-planar or other non-flatten surfaces such as spherical, conical, cylindrical, and other curved objects and uses a deformable pad that receives images from a flat cliché plate and transfers the images to the curved surface which is to be printed as seen in Fig. 16.5. Typically, an inverted cup containing a quantity of printing ink is used to apply the ink to the plate. Also to apply a new coating of ink to the plate, and the ink cup and a plate are moved relative to each other following each ink transfer operation (Dietz et al. 2001). However, such a process has several disadvantages such as the use of lots of consumables, many maintenance time, several cleaning periods, and set up in between, due to changes in the ink viscosity. The stamping process starts when the operator takes a precut hose from a container serving as a supermarket using an average of 1 min/pc. The plate setup (of 3 min/lot) is finished by placing a plate engraved, the paint container is filled talking an average of 3 min/pc. The first imprinting is made on a hose, and the distance to

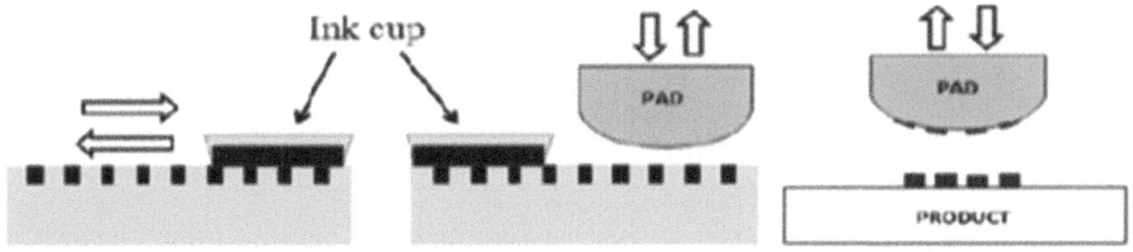

Fig. 16.5 Schematic of a tampography process

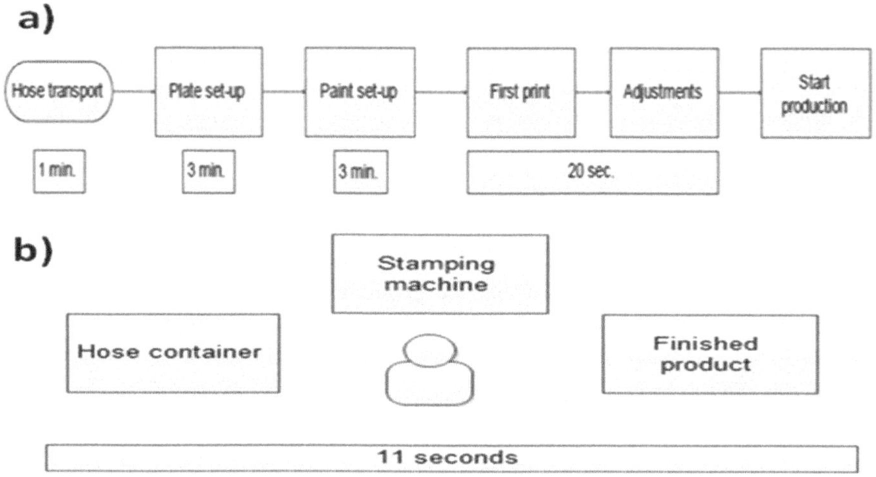

Fig. 16.6 **a** Initial setup of the tampography process, **b** Processing method used for stamping the product

the edge of the hose is measured, after making the proper adjustments (3 min/lot), the production time begins (see Fig. 16.6a). During the production process, the steps are as follow: the operator takes a hose and place this on a fixture with a specific orientation, then activates the stamping machine thru a pedal, after, removes the stamped hose and finally allocates the hose in a container. The layout and time for this process are shown in Fig. 16.6b showing a measured average of 11s/pc.

However, this time is not including the daily adjustments (55 times per shift) to keep the product under the specifications, causing the reduction of the productivity to an average rate of 75 pcs/hr. The customer has documented a complaint because of these productivity levels and delivery failures. At the same time, this customer is launching a new product, and biddings are called for the new project. At the moment, the facilities located in Juarez are the main vendor of this product, and the complaint risks not only the continuation of this business but the opportunity for growing sales to this customer. In consequence, regaining the confidence of the customer is critical not only for keeping the business with this customer but for competing with better chances in the new product bidding. The manager has assembled a multifunctional team for having an improvement and respond strongly to the customer's complaint, and prepare for the bidding process.

16.3.3 Implementation Plan

For using the adapted methodology, the description is divided into three sections: diagnosis, solution, and improvement. For using the Lean-Sigma approach in this case, the "develop a solution" step includes both, the solution and improvement sections of the project.

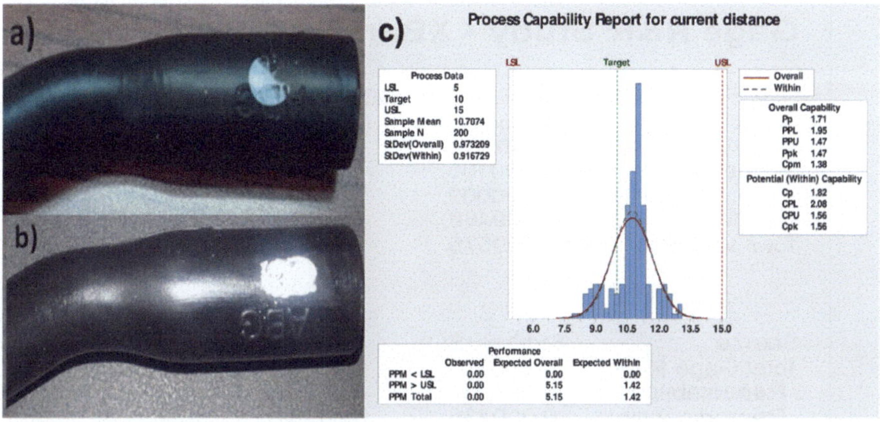

Fig. 16.7 Examples of cosmetic defects due to **a** lack of paint, and **b** excess of paint. **c** Cpk and data of the edge to the printed stamp

16.3.4 Identify and Measure the Problem Step

To identify the problem, the customer requirement (or complaint) is translated to a quantitative value. An initial diagnosis is used to identify the main variables (such as Productivity, Cost, and Quality), data then are arranged in tables and control charts for visualization and initial analysis. The statistical description of the obtained data then is used for making statistical comparisons to the product specifications or standardized values of the process which is used to determine the size of the problem.

In this case, the processing times are distributed as follow: 3% for the initial set-up, 75% in adjustments related to the operation, and 22% in production time, resulting in the mentioned average rate of 75 pcs/hr, which are below the required 400 pcs/hr. Also, quality defectives are about 4%, cosmetic defects are detected as the main causes as seen in the examples in Fig. 16.7a and b. From a quantitative point of view, the design specification requires a distance to the edge of the hose to be around 10 mm with a tolerance of 5 mm. To eliminate recurrent causes, initial actions include the implementation of programmed maintenance activities, after, statistical data is gathered and compared to the expected limits as seen in Fig. 16.7c), here is possible to see that the process Cpk is 1.56 (4 ppm), increasing the productivity to 78 pcs/hr and the cost to 5.33 IEU (internal economic units)/pc.

16.3.5 Root Cause Analysis Step

According to Estrada-Orantes et al. (2019) in finding the root cause, variations from the measuring system (a recurrent cause) and the process need to be analyzed before moving into deeper causes. In this case, The measuring system is based on a metallic

Gage R&R Study - XBar/R Method

Source	VarComp	%Contribution (of VarComp)
Total Gage R&R	0.0000106	0.10
Repeatability	0.0000105	0.10
Reproducibility	0.0000002	0.00
Part-To-Part	0.0109468	99.90
Total Variation	0.0109575	100.00

Source	StdDev (SD)	Study Var (6 × SD)	%Study Var (%SV)
Total Gage R&R	0.003262	0.019575	3.12
Repeatability	0.003233	0.019399	3.09
Reproducibility	0.000436	0.002615	0.42
Part-To-Part	0.104627	0.627763	99.95
Total Variation	0.104678	0.628068	100.00

Number of Distinct Categories = 5

Fig. 16.8 Results for the gage R&R test on the edge distance for the stamping process

ruler and the worker's visual readings. A Gage Repeatability & Reproducibility (G R&R) study shows that the first is not contributing to the causes of variation as the variation caused by the process (see the part to part values in Fig. 16.8). A combination of tools such as brainstorming, stratification, and cause and effect analysis helps the team moving deeper from the symptoms to conditions, and causes of variation in the process. The analysis is shown in the partially filled cause and effect (Ishikawa) diagram (see Fig. 16.9). Actual equipment conditions and the analysis results in three possible solutions for increasing the productivity and competitiveness as seen in Table 16.4) to upgrade to newer technology, replace the equipment with the same technology, and replace the required components.

16.3.6 Develop a Solution Step

Up to this point, the majority of the tools used are of quantitative nature, were selected as described previously, and used as exemplified in Table 16.3. Before the implementation of the solution and the change of technology, a containment plan is implemented to increase the productivity to 16,000 pcs/wk. With these actions is expected to reach rates to comply with the customer requirements. This containment plan requires extra efforts from the production staff testing their commitment by increasing to three regular shifts and two weekend overtime shifts causing also an immediate increase in costs. With these actions, the production time moves to 176

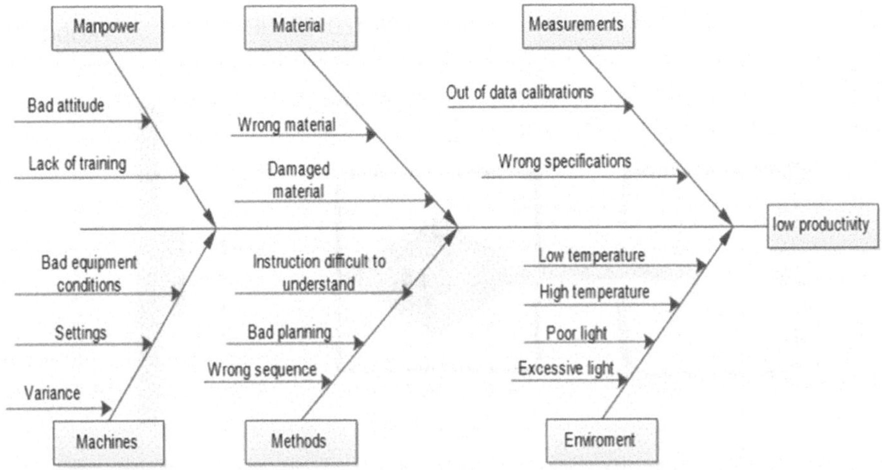

Fig. 16.9 Partially filled cause and effect (Ishikawa) diagram used in this case

Table 16.4 Options for solving the pad printing problem

Ideas	Advantage	Disadvantage	Classification
Upgrading to a newer technology	Reduction of ink consumption, reduction of preparation time, reduction of preparation time of stamping plate	Training on new technology, cost of technology	1
Replacing the equipment with the same technology	With new machinery print set, times are reduced	Cost of machinery, implications of getting rid of old machinery	3
Replacement of the required components	With the replacement of components, the reduction of production times is achieved	Components must be replaced periodically to maintain productivity	2

production hours/wk. From the problem analysis (see the previous section and Table 16.4), three options were suggested after exercising with the Nominal Group Technique (NGT). Table 16.4, shows the solutions (ideas), the decision factors (advantages and disadvantages columns), and the order for its viability (classification). During this decision-making process, the team considers the advantages and disadvantages of each possible solution, moreover, the evaluation includes the return over investment (ROI), productivity, reliability, quality, and adaptability of the solutions for this and other projects. The next step is to use the brainstorming technique for selecting among the available. The team decides for upgrading to a newer technology, which is cheaper and more reliable than the second of them, (to keep the same equipment but with new functional parts), or to acquire replacement equipment keeping the same technology. Options for the selected technology are offered by different

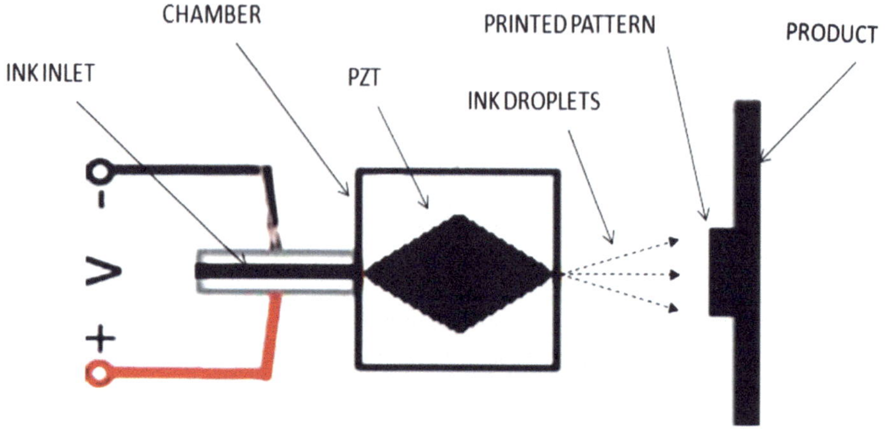

Fig. 16.10 Schematic of the selected inkjet stamping process proposed in the solution step

vendors weighing them for the mentioned and equipment characteristics. An inkjet technology is selected (see schematic in Fig. 16.10) because of its productivity, ROI, and quality characteristics. The advantages of using this technology include quicker changeovers, less and lower time for maintenance, cleaner stamping results (as compared to the actual process) and the more efficient use of the ink. In the same way, the price of the supplies and time for adapting this technology to the process are considered among the disadvantages. In the same way, the price of the supplies and time for adapting this technology to the process are considered among the disadvantages. After the proper implementations, adjustments, and training, initial data shows the potential of the updated process to speed up the production process and meeting the customer requirements. During the adaptation of the proposed technology, a vision system is included to synchronize the stamping distance with the paint jet drive and for comparing the stamp characteristics with the main patterns, helping in this way in measuring the quantitative values and feedback the operator during the visual (or qualitative) inspections. With the sensing arrangements including in the vision system, data is collected to a database in the intranet system, preparing in this way the process with Industry 4.0 characteristics. Also, this data is helping in monitoring the productivity data and the quality characteristics of the operations from the management position site and remotely from the headquarters. Results of the implementation show a viable production rate of 600 pcs/hr, a cycle time of 6 sec./pc is measured and a cost below 1.0 IEU/pc. Comparing data from the initial conditions to the adapted process, it can be seen that the process capability (Cpk) moves from 1.56 to 2.06 (see Fig. 16.7c and Fig. 16.11 respectively) an improvement that is reflected in an expected quality level or Six- Sigma level of 0.002 ppm. The return over the investment (ROI) is calculated using the latest unit cost of 0.9 IEU/pc. A value that is expected to adjust as the process is reaching its stabilization and the learning curve of the workers reaches its plateau. Considering the total investment the ROI then is calculated to reach its point of equilibrium at the production point of

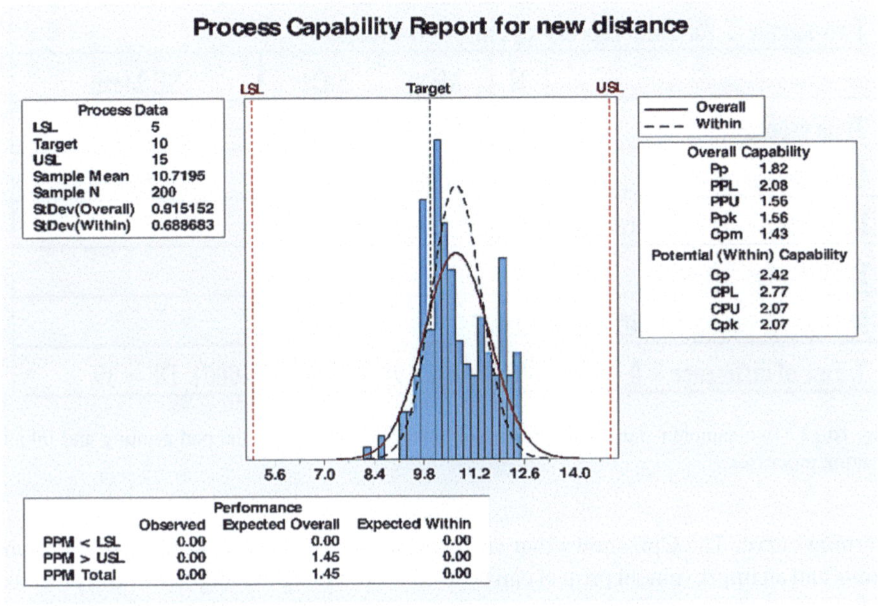

Fig. 16.11 Process capability study results after implementing the solution

615,000 pieces which, with the new production rate (600 pcs /hr) is manufactured in approximately two weeks.

16.3.7 Verify the Solution

For the organization and the local management is important to keep the success definition as quantitative as possible, then, during the validation step, statistical data is gathered to validate the efficiency of the implementation and the solution. Firstly, a cycle time comparison (Fig. 16.12) is used between the two processes. Forty random sample times are taken from the stamping process, using the pad printing (identified as Time process) and from the inkjet printing process (Time new process in Fig. 16.12). The statistical comparison shows a P-value of 0.0001 which is lower than the 0.05 for considering a statistically valid difference among the two samples, in this case, there is a change in the cycle times. In the same Fig. 16.12, it is observed that the mean values changed in 3 minutes, about 50% time reduction, despite the increase in variation measured by the standard deviation reduction that increases about ten times. However, the variation in the cycle time is expected to be reduced with the

Two-sample T for Time process vs Time new process				
	N	Mean	StDev	SE Mean
Time process	40	8.912	0.126	0.02
Time new process	40	5.98	1.15	0.18
Difference = μ (Time process) - μ (Time new process)				
Estimate for difference: 2.931				
95% lower bound for difference: 2.622				
T-Test of difference = 0 (vs >): T-Value = 15.98 P-Value = 0.0001 DF = 39				

Fig. 16.12 Two-sample t statistical comparison for cycle times using the pad printing and inkjet printing processes

learning curve. The Cpk shows that such variation is still between the specification limits and an improvement project can be used for reducing the inkjet printing process variation.

The statistical comparison for the quality characteristics uses the Two-Sample t tool data from both processes are gathered, observations of the cosmetic defects expected in the product do not appear (up to the writing of this report) when using the inkjet printing process. In the distance to the edge readings the printing distance data shows to be similar. The samples' comparison in Fig. 16.13 shows statistical data validation. Two hundred random samples are used from each stamping process. The pad printing (identified as current distance) and the inkjet (seen as new distances) stamped comparison shows a P-value larger than 0.05 indicating that the averages of the distances measured remain statistically the same.

From the results, it is possible to say that with the use of Lean-Sigma methodology in combination with the work culture analysis, the potential for succeeding and reaching the goals increases. More studies are required for confirming this approach,

Two-sample T for current distance vs new distance				
	N	Mean	StDev	SE Mean
current distance	200	10.707	0.973	0.069
new distance	200	10.72	0.915	0.065
Difference = μ (current distance) - μ (new distance)				
Estimate for difference: -0.0122				
95% CI for difference: (-0.1979, 0.1735)				
T-Test of difference = 0 (vs ≠): T-Value = -0.13 P-Value = 0.897 DF = 396				

Fig. 16.13 Statistical comparison study for printing variation before and after the solution

however, the improvements and reliability of the implemented changes in this production process are supported by the measured outcomes. In this case, the process is now able to compete in strong terms for this product and has now the flexibility for bringing more business. The increase in capability reduces the operation's time and frees one-third of the regular production capability and is possible to accept new projects at a lower cost (of 0.9 IEU/pc) with quality levels expected from a WCMS.

16.3.8 Control Plan

Finally, the Lean-Sigma methodology requires a control plan and the possibility to transfer the acquired knowledge to the organization in general. Two actions and tools are highlighted from this step, the use of a new Failure Mode & Effects Analysis (FMEA) and the adjustments to the improvement process plan. Changes include the documentation to identify potential failure modes in the system and their causes and effect due to the use of the newer equipment and failure modes by using this technology. Adjustments, documentation, changes, and additions of the functional elements of quality control are included, and the further institutionalization of the modifications will help in preventing or be prepared for future system failures.

16.4 Conclusions

The adaptation of the organizational culture to overseas operations will help in the creation of a workplace culture that connects to the organization and the people at the operations' location. This chapter presented a methodology for increasing the success potential when a world-class methodology, such as Lean-Sigma, is implemented. The case study shows an example of the use of this methodology, cultural differences and characteristics were evaluated during the adaptation of this methodology. After conducting a survey and using the classification and scores of the cultural aspects proposed by Hofstede the local scores for Juarez, Mexico were obtained. Of the cultural characteristics, the local staff shows low scores for the cultural characteristic of "distance to power" and high scores for the "collectivism" aspect as expected for a Mexican operation work culture. However, in the Juarez city case, other scores show that the local culture is different from the national (Mexico), the vicinity (USA), or the organizational (of French origin) culture showing characteristics of a multicultural environment. Under this scenario, with the cultural differences and organizational need for communicating among the different levels, the communication between the members of the organization may be difficult, the implementation of methodologies is also compromised. For example, a transversal approach for communication may confuse the local staff. Then, including the cultural component for implementing a methodology increase the possibilities for a successful implementation. The methodology presented show that the use of statistical tools allows members of both cultures to discuss and communicate on common grounds.

The case discusses the use of the steps of Lean-Sigma, dividing the activities into 21 to disaggregate the steps and identify the potential tools and methods to use in each activity then, choosing the ones that adapt better to the culture of both the organization and the local one. This report shows the results of selecting a different technology and improvements in the process, that increases the productivity rates from 178 to 600 pc/hr, the quality levels increased by 50% reaching a statistical Cpk value of 2.06, and the operational cost reduction from 5.33 to 0.9 IEU/pc. Aside from the ROI in two weeks, the adaptation is considered a success in the adaptation of the Lean-Sigma methodology. Finally, this case and methodology serve as a guide for organizations facing these challenges, while expanding operations around the world. Several implementations show that is possible to increase the success rates in adopting methodology, by adapting the tools and methods to the working site culture.

References

Alba Baena N (2020) Lean-sigma sinergy-conference notes, unpublished data

Alba-Baena N, Estrada-Orantes F, Valenzuela-Reyes C (2019) Use of lean-sigma as a problem-solving method in a restrictive environment. Manag Innovation Highly Restrictive Environ 35

Aycan Z, Kanungo R, Mendonca M, Yu K, Deller J, Stahl G, Kurshid A (2000) Impact of culture on human resource management practices: a 10-country comparison. Appl Psychol 49(1):192–221

Berson Y, Oreg S, Dvir T (2005) Organizational culture as a mediator of ceo values and organizational performance. Paper presented at the Academy of Management Proceedings

De Hilal AVG (2006) Brazilian national culture, organizational culture and cultural agreement: findings from a multinational company. Int J Cross Cult Manage 6(2):139–167

Dietz V, Grob H, Fastje KU (2001) U.S. Patent No. 6,276,266. U.S. Patent and Trademark Office, Washington, DC

Estrada-Orantes FJ, Alba-Baena NG (2014) Creating the lean-sigma synergy. In: Lean manufacturing in the developing world. Springer, Cham, pp 117–134

Estrada-Orantes FJ, García-Pérez AH, Alba-Baena NG (2019) The E-strategy for lean-sigma solutions, Latin American case study in a new product validation process. In: Best practices in manufacturing processes. Springer, Cham, pp 297–322

Findler L, Wind LH, Barak MEM (2007) The challenge of workforce management in a global society: modeling the relationship between diversity, inclusion, organizational culture, and employee well-being, job satisfaction and organizational commitment. Adm Soc Work 31(3):63–94

George ML, Rowlands D, Price M, Maxey J (2005) The lean six sigma pocket tool book, 2005, Ma-Graw-Hill, ISBN 0-07-144119-0

Hofstede insights, https://www.hofstede-insights.com/

Hofstede G (1980) Culture and organizations. Int Stud Manage Organ 10(4):15–41

House RJ, Hanges PJ, Javidan M, Dorfman PW, Gupta V (2004) Culture, leadership, and organizations: the GLOBE study of 62 societies: Sage publications

Inglehart R (2004) Human beliefs and values: a cross-cultural sourcebook based on the 1999–2002 values surveys: Siglo XXI

Johns G (2006) The essential impact of context on organizational behavior. Acad Manage Rev 31(2):386–408

Kerr J, Slocum JW Jr (2005) Managing corporate culture through reward systems. Acad Manage Perspect 19(4):130–138
Kim J, Lee S, Yu K (2004) Corporate culture and organizational performance. J Manage Psychol 19(4):340–359
Marc D-L, Farbrother S (2003) Changing organization culture, one face at a time. PM. Public Management 85(9):14–17
Moher D, Liberati A, Tetzlaff J, Altman DG (2009) Preferred reporting items for systematic reviews and meta-analyses: the PRISMA statement. Ann Intern Med 151(4):264–269
Naor M, Jones JS, Bernardes ES, Goldstein SM, Schroeder R (2014) The culture-effectiveness link in a manufacturing context: a resource-based perspective. J World Bus 49(3):321–331
Omar A, Urteaga F (2010) El impacto de la cultura nacional sobre la cultura organizacional. Universitas Psychologica 9(1):79–92
Salcido-Delgado A, Zhou L, Alba-Baena NG (2019) Lean-Sigma for Product Improvement Using the VoC for Enhancing the Product. Oper Manage Syst Eng: Sel Proc CPIE 2018:69
Tavakol M, Dennick R (2011) Making sense of Cronbach's alpha. Int J Med Educ 2:53
Where Lean Went Wrong—A Historical Perspective (2017) Retrieved may September 11, 2018, https://www.allaboutlean.com/where-lean-went-wrong/
Williams JC, Calás de Birriel MC (1984) Conducta organizacional: Iberoamérica
Wu M-Y, Taylor M, Chen M-J (2001) Exploring societal and cultural influences on Taiwanese public relations. Publ Relat Rev 27(3):317–336

Chapter 17
Using Simulation for Facility Layout Problems: A Case Study of an Ecuadorian Handcraft Chocolate Company

Diego Viteri-Viteri, Andreé Rodríguez-Luna, Sonia Avilés-Sacoto, and Galo Mosquera-Recalde

Abstract During the last years, the global confectionery market has experienced an increase in their demand, being headed by chocolate due to the high consumption around the world. This performance and the expected growth represent a challenge for the confectionery market, especially for participants along the chocolate supply chain. This is the case of Cacao Company a growing handcraft chocolate company based in Ecuador, which seeks to expand to international confectionery market through the exportation of its products and whose current capacity is not enough to satisfy this expected demand. Therefore, a new facility layout design is required to cover the new demand, and it also should maintain the artisan aspects of the production process. Currently, solutions for facility layout problem (FLP) focus on design algorithms that use a from-toward table, which includes more information about the material handling system (MHS). A mixed integer programming problem (MIP) is also needed, because it finds the most efficient arrangement of the production departments within the new facility, minimizing the material handling cost (MHC). Several design alternatives generated by the MIP are evaluated through a discrete events simulation model in order to consider additional factors as relative locations of equipment, capacities, transport speeds and statistical distributed processing times at the workstations. As a result of this study, the most suitable layout alternative is selected, and its economic analysis is presented.

Keywords Chocolate · Facility layout problems · Material handling · Mixed integer programming · Simulation

---

D. Viteri-Viteri · A. Rodríguez-Luna · G. Mosquera-Recalde
Departamento de Ingeniería Industrial, Colegio de Ciencias e Ingenierías, Universidad San Francisco de Quito (USFQ), Diego de Robles y Vía Interoceánica, 170901 Quito, Ecuador

S. Avilés-Sacoto (✉)
Departamento de Ingeniería Industrial, Colegio de Ciencias e Ingenierías, Instituto de Innovación en Productividad y Logística CATENA-USFQ, Universidad San Francisco de Quito (USFQ), Diego de Robles y Vía Interoceánica, 170901 Quito, Pichincha, Ecuador
e-mail: svaviless@usfq.edu.ec

J. L. Garcia Alcaraz et al. (eds.), *Trends in Industrial Engineering Applications to Manufacturing Process*, https://doi.org/10.1007/978-3-030-71579-3_17

17.1 Introduction

During the last years, the global confectionery market has experienced an increase in their demand, being headed by chocolate due to the high consumption around the world. Chocolate represents one of the most popular forms of snacks, and it is broadly recognized as an affordable indulgence by consumers. Thus, it represents 55% of the global confectionery market and is above the sugar confectionery (31%) and chewing gum (14%) (Thomas 2017). According to Grand View Research (2018), in 2016, the global confectionery market was valued at US $123.7 billion and is expected to grow 2.5% in the coming years. This performance and the expected growth represent a challenge for the confectionery market, especially for participants along the chocolate supply chain. However, the requirements of chocolate customer are constantly changing. It is known that consumers are choosing to spend more on occasional chocolate purchases of premium products and want to experience moments of happiness from small-scale artisan production methods (Jewett 2017). This is complemented by the fact that the chocolate industry is under consumer pressure, because they prefer chocolate where crops, sourcing and production processes present sustainable sourcing techniques (Jewett 2017). This is the case of Cacao Company, an Ecuadorian company that focuses on the production of premium and exclusive boxes of chocolates, through the development of a gourmet product with emblematic aromas and flavors of Ecuador.

The global confectionary market is highly competitive due to the increasing demand and growing popularity, where several companies seek to gain a major market share. Usually, these companies make long-term investments with the aim of repositioning themselves in the market, whether the decision is expanding the current capacity, implementing new information systems, new facilities design, marketing or research and development (R & D) (Morris 2014). These types of strategies intend to generate a base that allows to achieve long-term achievements, which in the future develop the desired levels of productivity and profitability (Morris 2014).

According to Yang et al. (2011), facilities planning and material handling impact the company's productivity and profitability. In addition, manufacturing and service companies spend a significant amount of time and money designing or redesigning their facilities (Yang et al. 2011). Singh and Yilma (2013) found that 20–50% of the total costs within manufacturing are related to material handling and an effective facility planning can reduce these costs by 10–30%. This is the main challenge that any organization faces, even the companies of the global confectionery market.

17.2 Literature Review

Several factors are addressed in the literature in order to solve facility layout problems, such as the design of a material handling system, workshop characteristics and resolution approaches. These factors are detailed below.

17.2.1 Facilities Planning Problems

According to Hosseini-Nasab et al. (2018), facilities planning problems can be divided into four categories, which are facility layout problem, facility location problem, facility location-allocation problem and mobile facility routing problem. The facility layout problem (FLP) seeks the most efficient arrangement of elements on the factory floor subject to different constraints in order to meet one or more objectives (Hosseini-Nasab et al. 2018). Regarding the Facility Location Problem, it is a branch from Operations Research, which aim is to locate a new facility among several existing facilities, whose purpose is to minimize or maximize at least one objective function (like cost, profit, revenue, travel distance, service, waiting time, coverage and market shares) (Farahani et al. 2010). The facility location-allocation problem considers a set of new facilities so that, the transportation cost from facilities to customers is minimized and it determines the location of one or more facilities in an area of interest in order to satisfy the demand (Azarmand and Neishabouri 2009). Finally, the mobile facility routing problem (MFRP) seeks to determine routes for a fleet of mobile facilities to maximize the amount of demand serviced in a continuous-time plan-routes for a fleet of mobile facilities to maximize the amount of demand serviced in a continuous-time planning horizon (Halper and Raghavan 2010). Companies need a facility layout that is well-designed, in order to operate efficiently besides planning their activities and establish many operational procedures (Hosseini-Nasab et al. 2018).

Hosseini-Nasab et al. (2018) provide a general overview of the classification of the FLP, which is composed of four main branches (i.e., layout evolution, workshop characteristics, problem formulation and resolution approaches).

17.2.2 Workshop Characteristics Impacting Layout

The facility layout is directly influenced by specifications of the production system and factors such as variety of products and production volumes (Hosseini-Nasab et al. 2018). Tompkins et al. (2010) classify facility layout planning according to the volume-variety of the product, which can be divided into four categories, by product, fixed location of materials, family of products and processes. *By product* consists of high production volumes and low variety of products. Usually, in these systems, the demand is high and remains stable over time. It is ideal for standardized products, since workstations are arranged sequentially to obtain a product. However, the problem is that they are not very flexible to changes in product specifications. *By processes* is characterized by low production volumes and high variety of products. In this case, stations are grouped when similar processes perform or that uses resources of the same type. *Fixed location of materials*, when there are low production volumes and low variety of products. It is characterized by low and sporadic demand, since the products are usually large and difficult to move. Thus, the product remains fixed,

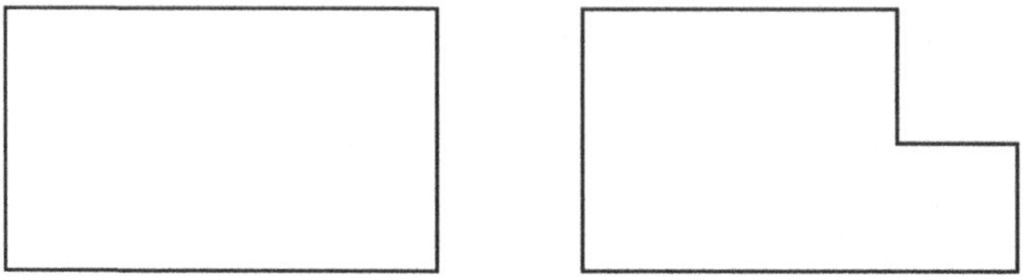

Fig. 17.1 Regular and irregular shapes (authors elaboration)

and the machines go to it. *Family of Products*, the production volumes and the variety of the product remain in medium levels, since the products are similar.

Other factor to be considered involves the facility shape and dimension. According to Drira et al. (2007), two different facility shapes are often distinguished by regular, like rectangular shape and irregular, like polygons containing at least a 270° angle. Figure 17.1 introduces the graphical representation of these shapes.

However, out of 186 articles reviewed by Hosseini-Nasab et al. (2018), 144 publications suggest to use regular facility shapes to generate feasible layout alternatives. Additionally, a facility can have given dimensions, defined by a fixed length of facility i (L_i) and a fixed width of a facility i (W_i) (Drira et al. 2007), and this is known as fixed or rigid block. Likewise, Hosseini-Nasab et al. (2018) expressed that a facility may be also represented by its area or its aspect ratio. The aspect ratio (α_i) is defined as shown in Eq. (17.1):

$$\alpha_i = \frac{L_i}{W_i} \tag{17.1}$$

where α_i is between an upper bound α_{iu} and a lower bound α_{il} ($\alpha_i \in [\alpha_{il}, \alpha_{iu}]$). If $\alpha_i = \alpha_{il} = \alpha_{iu}$, it corresponds to the case of fixed-shape blocks (Hosseini-Nasab et al. 2018) but, out of 186 articles reviewed, only 55 used the aspect ratio.

17.2.3 Material Handling System

Drira et al. (2007), Tompkins et al. (2010) and Hosseini-Nasab et al. (2018) agreed that the material handling system (MHS) is an important factor to consider in facilities planning. According to Frazelle (1986), material handling occupies 25% of all employees, 55% of all factory space and 87% of production time. Material handling is the art and science of moving, saving, protecting and controlling the material through its manufacturing, distribution, consumption and waste processes. It consists on providing the right amount of the right material in the right place, at the right position, at the correct sequence and at the ideal cost, using the right method (Tompkins et al. 2010). Tompkins establishes six steps for the design of a material handling system:

1. Define the objectives and scope of the material handling system.
2. Analyze the requirements to move, store, protect and control materials.
3. Generate design alternatives that meet the requirements of the material handling system.
4. Evaluate material handling design alternatives.
5. Select the most convenient design to move, store, protect and control materials.
6. Implement the chosen material handling design.

This model seeks to find the most efficient arrangement of department areas and equipment inside a facility.

17.2.4 Resolution Approaches

According to Hani et al. (2007), the facility layout problem is known to be non-deterministic polynomial time hardness (NP-Hard), and it is usually formulated as a quadratic assignment problem (QAP). QAP assigns a set of facilities to a set of locations in order to minimize the total allocation cost (the flow between the facilities and the distance between the locations of the facilities), based on an initial distribution. However, there are numerous resolutions approaches for FLPs. Recently, a full description of classification of FLP and resolution approaches was made by Hosseini-Nasab et al. (2018). This paper describes the resolution approaches starting by classifying them into two main categories single-criterion FLPs and multi-criterion FLPs. Regarding multi-criterion FLPs, ideally FLPs must consider qualitative and quantitative factors. Therefore, there are various approaches for solving multi-criterion FLPs, such as weighted sum method, global criterion method, fuzzy multi-objective programming, analytic hierarchy process and analytic network process (Hosseini-Nasab et al. 2018). However, according to Hosseini-Nasab et al. (2018), of 186 articles reviewed, only 57 fall in the field of multi-criterion FLPs, and therefore, they concluded that different criteria simultaneously are ignored in order to provide appropriate layouts. Therefore, it is suggested to solve FLP's problems by single criterion. Single-criterion FLPs seek to minimize the material handling cost (MHC) (quantitatively) or maximize the closeness between the facilities (Qualitatively). For solving the single-criterion FLP, several resolution approaches are presented (exact approaches, approximated approaches and stochastic approaches).

Exact approaches are useful to find optimal solutions for small-sized FLP (Hosseini-Nasab et al. 2018). For example, the studies of (Amaral 2006), Meller et al. (2007) and Tavakkoli-Moghaddam et al. (2007) solve FLP using exact approaches like branch and bound methods to find an optimum solution (Singh and Sharma 2006). Other formulation can be applied such as mixed integer programming (MIP) that was developed in previous studies by Konak et al. (2006), Amaral (2006) and

Leno et al. (2015). However, Madhusudanan Pillai et al. (2011) found that optimization methods are not capable of solving problems with 15 or more facilities in a reasonable amount of time, and therefore, there is a need for approximated approaches or heuristic methods that provide good sub-optimal solutions.

Approximated Approaches, according to Hosseini-Nasab et al. (2018) these approaches are classified in construction algorithms, improvements algorithms and meta-heuristics algorithms. Regarding to construction algorithms, a solution is constructed from the scratch. Construction algorithms are the simplest and oldest heuristic approaches to solve the QAP from a conceptual and implementation point of view (Singh and Sharma 2006). The quality of the solutions may be far from the optimal, since it only yields one layout solution. Examples of construction algorithms are computerized relationship layout planning (CORELAP), automated layout design program (ALDEP) and programming layout analysis and evaluation technique (PLANET) (Hosseini-Nasab et al. 2018). Improvement algorithms start with a feasible solution and try to improve it by interchanging the locations of the facilities (Singh and Sharma 2006). The swap continues until the solution cannot be improved any further. However, the solution is very sensitive of an initial solution. Example of these methods is pairwise exchange, insertion neighborhood, Lin–Kernighan neighborhood, computerized relative allocation of facilities technique (CRAFT) and computerized facility aided design (COFAD) (Hosseini-Nasab et al. 2018). Finally, meta-heuristics approaches include genetic algorithm (GA), tabu search (TS), simulated annealing (SA) and ant colony optimization. Yang et al. (2011) used GA to solve the facility layout problem, considering the handling cost, the facility moving cost and the facility setup cost. Mazinani et al. (2012) proposed a GA to solve a FLP based on flexible bay structure. Samarghandi and Eshghi (2010) developed an efficient TS algorithm for the single row facility layout problem. McKendall and Hakobyan (2010) used the TS to improve the solution of the boundary search construction algorithm to solve a FLP with unequal areas. Dong et al. (2009) proposed a shortest path-based SA for dynamic FLP under dynamic business environment. Guan and Lin (2016) developed a hybrid algorithm based on variable neighborhood search and ant colony optimization to solve the single row FLP.

Stochastic Approaches, according to Pourvaziri and Pierreval (2017), these algorithms solve problems that imply non-deterministic inputs, generating near-optimal solutions with high probability (Hosseini-Nasab et al. 2018). Orderly material flows can be solved by regular exact approaches of FLPs, but a dynamic volume of material flow occurring in specifics periods of time can affect significantly the overall efficiency of the layout, requiring modifications (Vitayasak et al. 2017). Therefore, to evaluate the manufacturing system, the problem is modeled as an open queueing network (OQN) that allows formulating the layout problem as an analytical discrete event simulation (Pourvaziri and Pierreval 2017).

17.3 Methodology

The following sections describe the methodology depicted in Sect. 17.2.3. However, the scope in this case of study includes steps 1 through 5.

17.3.1 Define the Objectives and Scope of the Material Handling System

The objective and scope of a material handling system are to find the most efficient arrangement of department areas and equipment. There are two main resolution approaches to find an efficient arrangement:

- Quantitatively, minimizing the total material handling cost (MHC) between the production departments.
- Qualitatively, maximizing the closeness of the production departments.

Hosseini-Nasab et al. (2018) found that the most common approach is to minimize the MHC, because most of the research in the literature is focused on this criterion as a factor for deciding the suitability of a layout in terms of cost and distance. However, Hillier and Liberman (2015) established that, before implementing the system, an agreement between the facility design analysts and the customer is necessary, in order to make sure that the model solution is accurately translated to an operating procedure and to rectify any possible flaws uncovered by the quantitative solution. Therefore, qualitative requirements were considered to determine the functionality of the final model. As a result, the arrangement provided would be the safest and most satisfactory.

17.3.2 Analyze the Requirements to Move, Store, Protect and Control Materials

In order to analyze the requirements to move, store, protect and control materials, the equation of the material handling system is used. Figure 17.2 introduces a graphical representation of the material handling equation.

This equation provides the structure to identify solutions for any material handling system. According to Tompkins et al. (2010), *what* defines the type of materials that are going to be moved; *where* and *when* establish the requirements of place and time as when the material is delivered, received and stored; *how* and *who* indicate the methods of material handling. Analyzing these requirements provides alternatives for the material handling system. Finally, the system which is preferred and has convenient cost is selected.

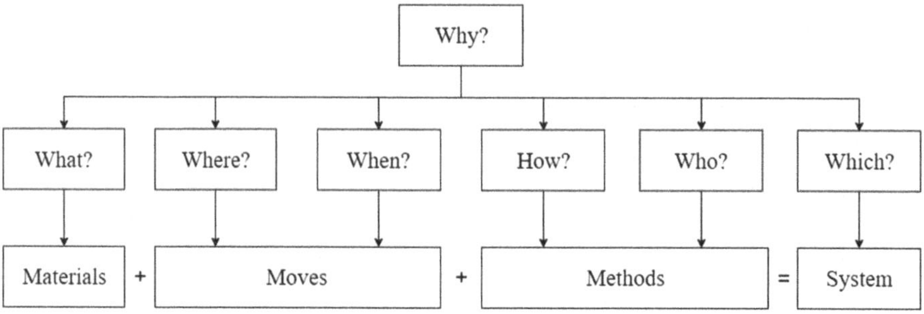

Fig. 17.2 Material handling equation (Tompkins et al. 2010)

17.3.3 Generate Material Handling Design Alternatives that Meet the Process Requirements

Design alternatives are generated by defining the scope and analyzing the requirements of the material handling system. According to Tompkins et al. (2010), the correct arrangement of the production departments can be done based on their closeness ratings (through a relationship table) or material flow intensities (through a from-toward table). The relative location of departments based on their closeness ratings or material flow intensities can be reduced to an algorithmic process Tompkins et al. (2010). However, the current trend is toward design algorithms that use a from-toward table, which generally requires more time and effort to prepare, but offer more information about the material handling system. Emami and Nookabadi (2013) determined that the most significant indicator of the efficiency of a layout is the material handling cost (MHC). Since, from 15 to 70% of the total manufacturing cost of a product is attributed to the material handling cost, it is suggested to minimize this indicator Mohamadghasemi and Hadi-Vencheh (2012).

17.3.4 Evaluate Material Handling Design Alternatives

When evaluating design alternatives obtained by the algorithms, it is important to consider the qualitative requirements. Most algorithms do not include closeness ratings in their resolution. Therefore, these aspects are considered in this phase, when the client or an expert in the process evaluates the design alternatives obtained through the algorithms. This guarantees that the final solution provided by the mathematical algorithm is aligned with the operational processes and thus rectify any possible defect of the quantitative solution (Hillier and Liberman 2015).

17.3.5 Select the Most Convenient Design to Move, Store, Protect, and Control Materials

Once each design alternative is evaluated, the alternative that minimizes more the sum of material flow in terms of cost and distance is chosen. It is important to consider that, when generating and evaluating design algorithm, the distance between each production department is measured through the rectilinear distance between their centroids. So, the relative locations of machines, shelves or any other material necessary to move, store, protect and control materials are not considered.

Alemany et al. (2018) suggests that, in order to select the most convenient layout, it is important to analyze factors related to production and use emerging technologies that facilities the estimation of the future. Once the complete layout design is obtained, Khusna et al. (2010) suggest the use of simulation software to evaluate the functionality of the layout, because it allows to conduct the design before finally building the facility. Therefore, a discrete events simulation software, known as FlexSim, was used to select the most appropriate design, since this software allows to model, analyze and include factors such as

- Capacities of each workstation and transportation methods.
- Experimentation with the model.
- Exploration of new scenarios.
- Statistical distributed processing times at the workstations.
- Visualization of results using statistical graphs simulated in real time.

17.4 Results

The following sections describe Cacao Company case study, a growing handcraft chocolate company based in Ecuador, which seeks to expand to international confectionery market through the exportation of its products and whose current capacity is not enough to satisfy this expected demand.

17.4.1 Case Study

Cacao company is dedicated to the production of exclusive premium chocolates. It has the objective to develop gourmet products stuffed with emblematic Ecuadorian aromas and flavors; this situation entails expand its facilities to reach such objective. Due to its recent performance and limited storage space in the current location, this company wants to establish the new facility that offers sustainable sourcing techniques, maintain its artisanal production processes, which differ from the other chocolate companies and maximize the benefits of the growing chocolate market. Therefore, the objective of this case study is to develop a prefeasibility study for the design of a new facility of handcraft chocolates.

17.4.1.1 Production Process

The chocolate production process is divided into four main areas: raw material, filling, molding and chocolate dipped and Packaging. Each involves specific operations that are distributed throughout all departments. This allows the development of a wide variety of type of chocolates, each with different qualities and finishes. Appendix 17.1 introduces a graphical representation of the activities that are carried out within each of the areas.

17.4.1.2 Demand

According to Anecacao (National Association of Cocoa Exporters—Ecuador), the Ecuadorian cocoa industry has presented an annual production increase of 6% since 2011, reaching 300 thousand tons in 2017 (Anecacao 2019). However, only 10% of this production represents semi-processed products, such as cocoa beans, cocoa liquor, cocoa butter and all kinds of chocolates with milk or fruit (Anecacao 2019). Cacao Company sells its chocolates through the HORECA channel, which means hotels, restaurants and coffee shops. Despite this, other niche markets that the Company has are the cruise chains that are managed within the Galapagos Islands, super-markets and stores located in different airports in the country. Figure 17.3 presents a graphic representation of the distribution of the market, according to the demand presented by each of these sectors.

Due to the market increment in recent years, Cacao Company expects its demand to increase over the next 3 years. Therefore, three possible scenarios are established, and each represents a different expected growth of its current demand. These scenarios described below were established together with the company's CEO.

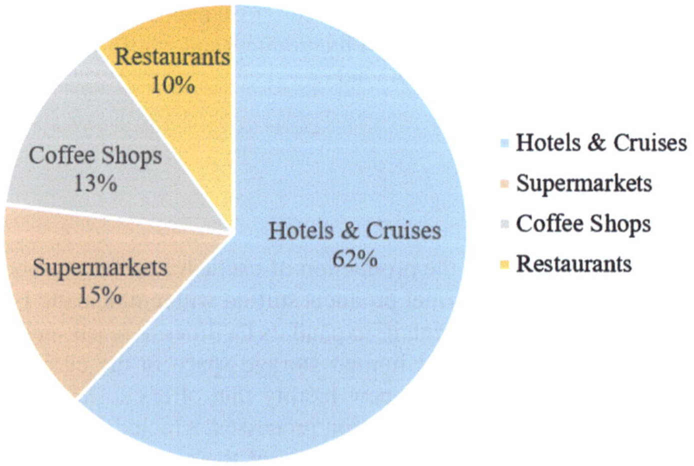

Fig. 17.3 Market distribution 2019 (authors elaboration)

- Scenario No. 1: This scenario implies that the expected demand of the company maintains a continual growth, between 10 and 13%, as it has been doing during the last years. This scenario involves the design of a new facility with an efficient arrangement, capable of satisfying the current demand of the company plus this margin of growth.
- Scenario No. 2: This scenario considers a growth of 30% compared to the current demand. Because Cacao Company could expand its market to Peru, a country consumes 70% more chocolate per year per person (Lira 2019), than in Ecuador. The chocolate market in Peru has shown a growth of 9% during 2018, and it is expected to increase in 11% in 2019 (La República 2018). The distribution channel within this new market is through the branches of hotel chain. This will be carried out through a 3-year contract with two hotels.
- Scenario No. 3: The last scenario includes a growth of Cacao Company within the country, considering the impulse of entities such as the Cámara de Comercio Franco Ecuatoriana (CCIFEC) toward companies that process Ecuadorian cocoa and generate a product with added value. This is expected to expand the consumption of national and international chocolate by 30% increasing its export to European countries (Anecacao 2019). The CEO of Cacao Company expects that the development of the Ecuadorian cocoa industry will allow the company to export their products, which would imply a growth of 60–100% of the current demand in the next 3 years. This is also considering the possibility of achieving the agreements described in the second scenario.

17.4.1.3 Infrastructure

Its current infrastructure consists of a kitchen of 24 m^2, with two storage spaces of an area smaller than 6 m^2 each. Its small size allows it to maintain only the storage of two days of production, limiting the use of large production batches. The current facility is the size of a standard kitchen, which allows precise temperature control in this space, an essential factor in chocolate making. However, keeping the entire space within the same environment limits the minimum temperature with 18 °C, increasing the cooling time and therefore the total production time. Appendix 17.2 shows an illustration of the current location.

17.4.1.4 Capacity

Cacao Company provided data for 2018 and the first quarter of 2019, which show production levels of 20,000–30,000 units per month. Currently, the company is dedicated to the production of four product families, which are Truffles, Chocolate plate, Molds, and Be-Expert. Each product family is characterized by the type of chocolate and the market segment to be attained. Table 17.1 shows the monthly production volume.

Table 17.1 Company production 2018–2019 [units]

Year	Months	Molds	Chocolate plate	Truffles	Be-expert	Total
2018	January	6320	3600	7786	1878	19,584
	February	12,244	7486	2700	1968	24,398
	March	10,129	5318	1300	2306	19,053
	April	6840	6421	2245	2068	17,574
	May	10,592	8860	5342	592	25,386
	June	8560	4819	1589	3025	17,993
	July	10,096	11,250	2807	1997	26,150
	August	12,895	6986	6277	870	27,028
	September	11,518	8749	4064	1310	25,641
	October	11,192	7782	10,230	0	29,204
	November	9537	4225	1534	1975	17,271
	December	13,420	7351	7401	0	28,172
2019	January	12,300	9971	2505	1827	26,603
	February	9576	6945	3632	1946	22,099
	March	9596	6075	3674	1686	21,031
	April	10,892	8776	3911	318	23,897

Production levels in August, October and December are the highest for all types of chocolates. However, each of the types of chocolates shows specific demand patterns, linked to cultural events, such as Christmas and Valentine's Day. The company reaches these levels of production, by increasing the workload of its regular operators, between 2 and 4 h daily and not hiring more people (Fig. 17.4).

17.4.2 Restrictions

17.4.2.1 Available Space

Cacao Company owns a plot of land of 550 m^2 on the outskirts of the city of Quito of which, only 230 m^2 are available for the construction of the new facility. The first phase of the project is carried out in the West of the building space, and Appendix 17.3 shows a graphical representation of the available space for the design of the new facility. Within this area, enough production capacity is estimated to meet the demand forecast for the next 3 years. This rectangular space involves an area of 132 m^2, which should include the productive areas, omitting space for administrative departments.

Once the maximum capacity of the facility has been reached, it has 98 m^2 available to continue expanding. This is phase 2 of the project and implies a monthly production requirement, greater than 40,000 units per month, which will not be included in this analysis.

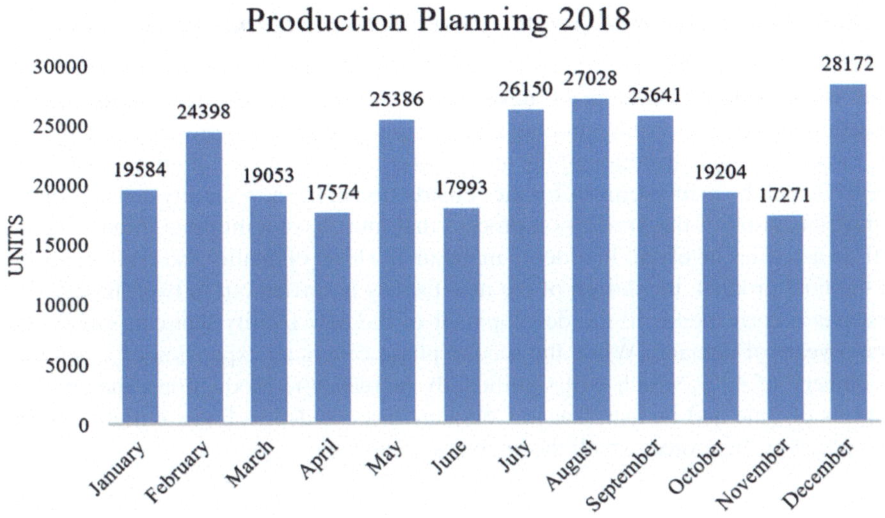

Fig. 17.4 Production planning 2018 (authors elaboration)

17.4.2.2 Fixed Positions Requirements

Within the new facility, there exists production departments that, according to Cacao Company's requirements, must be assigned to a specific place. In this case, there are four departments that need to be in a fixed position, those are receiving zone, packaging material warehouse, wet and dry-cleaning area and fruit pulping area. In this case, these areas must be in the peripheral areas of the facility, considering their impact on food safety and product quality.

17.4.2.3 Closeness and Product Safety Requirements

The final arrangement of department areas and equipment presents a closeness requirement, where specific production departments are undesirable to be close. In this case, the plate chocolate making area and painting area of chocolate molds cannot be located next to the filling chocolate area, because this department has temperatures above 18 °C, which affects the process of cooling the chocolate during the preparation of chocolate plates and decorating molds.

In addition, the design of a new food production facility must include an analysis of the type of materials and machines that will be used within it, with the aim of preserving food safety.

17.4.3 Objectives and Scope of the Material Handling System

This study seeks to develop the design of a chocolate production facility of the Ecuadorian company, contemplating all the requirements that it presents and finding the most efficient distribution of the areas and equipment of each production department. One of the main premises for the construction of the new facility consists in the industrialization of the production process that must have a minimal impact on the artisan activities involved. In order to maintain the level of quality and craftsmanship of the final product, the design of the new facility is carried out in two phases. The first phase corresponds to the development of the new facility that can satisfy the first 3 years of demand. While the second phase covers an expansion of space and machinery of 42%, which will significantly increase the production capacity. For reasons of time and budget, Cacao Company has decided to focus initially on the first phase of the project, so all the analyzes are made.

17.4.4 Requirements to Move, Store, Protect and Control Materials

In order to design a new facility, it is necessary to understand the material handling equation, because it allows to identify the requirements of movement, place, time, space and quantity. Therefore, the objective is to understand each question in Fig. 17.2 in relation to Cacao Company.

- What → What types of materials are going to move? Being a food production facility, most of the objects mobilized are organic products. However, there is also a wide variety of kitchen utensils and containers that will be transported along it.
- Where → Where is the material stored, delivered and received? All products received in the reception and dispatch area are classified upon entry, in order to locate them in their respective storage areas. This minimizes the possibility of cross contamination within the facility.
- When → When should the material be moved? The materials within the facility are mobilized according to the production requirements. Therefore, production plans are executed twice a week, which are updated daily. The receipt of raw materials is carried out twice a week, while the delivery of finished product is carried out daily.
- How → How is the material moved, stored or delivered? The material is stored within spaces with the controlled temperature, which depending on the material, these can be 4 and 18 °C. When handling small production batches, the products are manually mobilized throughout all departments.
- Who → Who handles the material? Within the facility, operators must rotate weekly throughout the different process activities. All operators can transport all materials within the facility.

- Which → Which type of material handling should be considered? The type of material handling is completely manual, due to the use of small production batches, which limits the loading of material mobilized simultaneously by a person.

Therefore, when understanding the requirements of movement, place, time, space and quantity, for this material handling system; two results are obtained that serve as input data to generate design alternatives, which are the flow matrix and the space requirements.

17.4.4.1 Flow Matrix (From-Toward Table)

The material flow is measured quantitatively in terms of the amount of material transferred between the production departments. Therefore, the most used method to record these flows is through a from-toward table. To create the from-toward table, it is necessary to establish an equivalent flow measurement, since the transferred materials have different sizes, weights, shapes and other attributes. Thus, it is necessary to establish a common unit of measure so that the quantities recorded in the from-toward table represent the appropriate relationship of movement volumes between each production department (Tompkins et al. 2010). In this case study, the unit of measure was defined as an individual unit of chocolate of 10 g, since it is the unit of sale of the product and the company manages its performance indicators in units.

According to the production process, the following production departments were determined

- A1: Receiving Zone
- A2: Raw Material Warehouse
- A3: Package Material Warehouse
- A4: Filling Chocolate Area
- A5: Plate Chocolate Making Area
- A6: Cold Room (0–4 °C)
- A7: Chocolate Crystallization Area
- A8: Chocolate Decoration Area (after bathing and sealing the chocolate)
- A9: Intermediate Product Warehouse (18 °C)
- A10: Packaging Area

- A11: Area of Finished Product (18 °C)
- A12: Painting Area of Chocolate Molds
- A13: Wet and Dry-Cleaning Area
- A14: Fruit Pulping Area.

When analyzing the production process and the established flow measurement, the flow volumes are recorded in the from-toward table. The monthly amount of 10-g chocolate units transferred between each production department for each scenario is shown in Appendices 17.4, 17.5 and 17.6.

17.4.4.2 Space Requirements

Each production department needs specific tools and machines to perform the processes assigned as well as containers to transport the products through the facility. Considering that the material handling system is completely manual, the size of the containers is limited by the loading capacity of the operators. In addition, the composition of the material for each equipment is considered. Ahmad and Ibrahim (2007) recommend that the materials used for food contact must be 316 and 304 stainless steel. The advantage of this material is that it does not affect the texture and taste of the chocolate itself, because it is a safe material and clean that takes care of the safety of the product (Ahmad and Ibrahim 2007). Therefore, these types of steel are the most suitable for chocolate processing. Finally, in order to select each equipment that will be used in the facility, the company's production process and the material handling equation were analyzed. Subsequently, the materials selected to store, move and control 10-g chocolate units were validated with chocolate experts and operators. Therefore, it is guaranteed that the selected equipment is aligned with the operational processes. Appendix 17.7 details the production and loading capacity of each equipment required for the facility.

The dimensions of each department are obtained from the maximum flow of products through each department for one day. This implies that each department must handle a maximum amount of daily product, with the exception of storage departments such as raw materials warehouse, packaging material warehouse, cold room (0–4 °C), intermediate products warehouse (18 °C) and finished product area (18 °C). Space requirements vary according to the type of scenario selected. Scenario No. 1 generates 10% more of the current production, which implies that its average monthly production is 21.000 units. This means an increase in the space in the storage areas. This extension requirement is carried out within the three scenarios, but with different magnitude. However, the variation of machines and tools within Scenario No. 2 allows it to meet an average demand of 33.000 units. Likewise, this last scenario, the increase in capacity for Scenario No. 3, is due to the addition of tools for handling materials and cooling space, since a production of 40.000 units must be achieved. Appendix 17.8 details the materials required for each production department and the area required for each scenario.

17.4.5 Generation of Design Alternatives

The design of alternatives is toward algorithms that use a from-toward table, because it includes more information about the material handling system and allows to develop a design that minimizes the sum of material flow in terms of cost and distance. Since the complexity of Cacao Company case study is within the limits that Madhusudanan Pillai et al. (2011) suggest, that is, if the design of a facility does not exceed 15 or more production departments, then exact approaches are useful to find an efficient arrangement. Therefore, facility layout problems can be formulated as a mixed integer programming (MIP) problem, if all departments are assumed to have a regular facility shape (rectangular) (Tompkins et al. 2010). This model based on mathematical programming is considered a construction and improvement algorithm because it finds an optimal arrangement for the specified departments and does not require an initial arrangement. Tompkins et al. (2010) established the dimensions of each department as decision variables, so that the mathematical model is formulated as follows:

Parameters

f_{ij}: denotes the flow from department i to department j (expressed in the amount of unit loads displaced by unit time)
c_{ij}: represents the cost from department i to department j
B_x: is the length of the layout (measured along the x coordinate)
B_y: is the length of the layout (measured along the y coordinate)
P'_i: lower perimeter of department i
P''_i: upper perimeter of department i
L'_i: is the length lower limit of the department i
L''_i: is the length upper limit of the department i
W'_i: is the width lower limit of the department i
W''_i: is the width upper limit of the department i
M: it is a big number.

Decision variables

α_i: is the x coordinate of the centroid of the department i
β_i: is the y coordinate of the centroid of the department i
x'_i: is the x coordinate of the left (west) side of department i
x''_i: is the x coordinate of the right (east) side of department i
y'_i: is the y coordinate of the upper part (north side) of department i
y''_i: is the y coordinate of the lower part (south side) of department i
z^x_{ij}: is equal to 1 if department i is strictly east of department j and 0 otherwise
z^y_{ij}: is equal to 1 if department i is strictly north of department j and 0 otherwise.

The definitions of the parameters and variables form the following model:

$$\text{Minimize}\, z = \sum_i \sum_j f_{ij} c_{ij} \left(\alpha_{ij}^{+} - \alpha_{ij}^{-} + \beta_{ij}^{+} - \beta_{ij}^{-} \right) \tag{17.2}$$

Subject to:

$$L_i' \leq \left(x_i'' - x_i' \right) \leq L_i'' \quad \text{for all}\ i \tag{17.3}$$

$$W_i' \leq (y_i'' - y_i') \leq W_i'' \quad \text{for all}\ i \tag{17.4}$$

$$P_i' \leq (x_i'' - x_i' + y_i'' - y_i') \leq P_i'' \quad \text{for all}\ i \tag{17.5}$$

$$0 \leq x_i' \leq x_i'' \leq B_x \quad \text{for all}\ i \tag{17.6}$$

$$0 \leq y_i' \leq y_i'' \leq B_y \quad \text{for all}\ i \tag{17.7}$$

$$\alpha_i = 0.5 x_i' + 0.5 x_i'' \quad \text{for all}\ i \tag{17.8}$$

$$\beta_i = 0.5 y_i' + 0.5 y_i'' \quad \text{for all}\ i \tag{17.9}$$

$$\alpha_i - \alpha_j = \alpha_{ij}^{+} - \alpha_{ij}^{-} \quad \text{for all}\ i \text{ and } j,\ i \neq j \tag{17.10}$$

$$\beta_i - \beta_j = \beta_{ij}^{+} - \beta_{ij}^{-} \quad \text{for all}\ i \text{ and } j,\ i \neq j \tag{17.11}$$

$$x_i'' \leq x_i' + M \left(1 - z_{ij}^{x} \right) \quad \text{for all}\ i \text{ and } j,\ i \neq j \tag{17.12}$$

$$y_i'' \leq y_i' + M \left(1 - z_{ij}^{y} \right) \quad \text{for all}\ i \text{ and } j,\ i \neq j \tag{17.13}$$

$$z_{ij}^{x} + z_{ji}^{x} + z_{ij}^{y} + z_{ji}^{y} \geq 1 \quad \text{for all}\ i \text{ and } j,\ i < j \tag{17.14}$$

$$\alpha_i, \beta_i \geq 0 \quad \text{for all}\ i \tag{17.15}$$

$$x_i', x_i'', y_i', y_i'' \geq 0 \quad \text{for all}\ i \tag{17.16}$$

$$\alpha_{ij}^{+}, \alpha_{ij}^{-}, \beta_{ij}^{+}, \beta_{ij}^{-} \geq 0 \quad \text{for all}\ i \text{ and } j,\ i < j \tag{17.17}$$

$$z_{ij}^{x}, z_{ij}^{y} \epsilon \{0, 1\} \quad \text{for all}\ i \text{ and } j,\ i < j \tag{17.18}$$

According to Tompkins et al. (2010), the equations represent the following:

- Equation (17.2) represents the objective based on distances, which seeks to minimize the cost per unit time per movement of materials between departments.
- Equations (17.3) and (17.4) are constraints that ensure that the length and width of each department are within the established limits.
- Equation (17.5) is a constraint that represents the area requirement of each department. According to Tompkins et al. (2010), the area of a department can be controlled through its perimeter, which results in a linear function of its length and width. Where P_i' and P_i'' represent the lower and upper perimeter of department i, respectively (Tompkins et al. 2010). This reduces the possibility of not obtaining feasible solutions when fixed departments are present (Tompkins et al. 2010) which include the following Receiving Zone (A1), Packaging Material Warehouse (A3), Wet and Dry-Cleaning Area (A13) and Fruit Pulping Area (A14).
- Equations (17.6) and (17.7) are constraints that ensure that the sides of the departments determined correctly (rectangular shape) and are located within the limits of the facility in the directions of x and y.
- Equations (17.8) and (17.9) are constraints that represent the x and y coordinates of the centroid of each department.
- Equation (17.12) is a constraint that ensures that $x_j'' \leq x_i'$, that is, if $z_{ij}^x = 1$, department i is located strictly east of department j, otherwise department i is located west of department j.
- The constraint represented by Eq. (17.13) works in the same way as the constraint in Eq. (17.12), but in the direction of y, that is, if $z_{ij}^y = 1$, the department i is located strictly north of department j, otherwise the department i is located south of department j.
- Equation (17.14) is a constraint that guarantees that two departments do not overlap in the east–west or north–south direction.
- Equations (17.16) and (17.17) represent non-negativity constraints. Finally, Eq. (17.18) is a constraint that designates the binary variables.

Therefore, objective function finds the most efficient arrangement of the areas and equipment of each production department, by minimizing the sum of material flow in terms of cost and distance. Considering the flow of materials and the space requirements for each production department of each scenario, the design alternatives are generated for each one. It should be noted that the common unit of measure of material flow is a chocolate unit of 10 g and the same method of transport (people) is used throughout the facility. Thus, it is possible to establish a standard cost (c_{ij}) of \$1, which represents the cost of moving a 10-g chocolate unit from department i to department j. Therefore, the objective function minimizes the distance traveled, which implies a lower material handling cost, hence, an efficient design. If there are several transportation methods in a facility, such as people, forklifts, pallet trucks, conveyor belts or other methods of cargo transport, this cost may vary. However, for this case study, it was not necessary to calculate this parameter.

To solve the mathematical model presented by Tompkins et al. (2010), the AMPL IDE modeling language was used, and since it is a linear model, CPLEX solver was needed. The model was solved in a computer with Intel Core i7 4.6 GHz processor and a 16 GB RAM of LPDDR3 RAM at 2133 MHz. The processing time to obtain the solution was 30 s. Appendices 17.9, 17.10 and 17.11 show the results obtained by this software. These results represent the coordinates of the vertices of each production department of each scenario. In this way, it is possible to control the dimensions of the departments, considering the flow of materials and the space requirements. The results obtained for each scenario are plotted.

Figure 17.5 shows the optimal arrangement obtained for each scenario, which minimizes the material handling cost (MHC) through the mathematical model, considering the requirements of the available space and fixed positions of some departments. From the three resulting scenarios, it can be observed that the optimal arrangement does not vary significantly in scenarios 1 to 3. However, the department arrangement from scenario 1 suggests that departments A5, A8, and A12 should be in a different location, compared to scenarios 2 and 3, which suggest a similar department arrangement, given that the space requirements of these two scenarios are similar. However, considering scenario 3, in which a growth of 60–100% of its current demand is expected in the next 3 years, this arrangement is the alternative most efficient in the long term. Therefore, by using design algorithms to find the best arrangement within a facility, it is possible to determine that, by increasing the

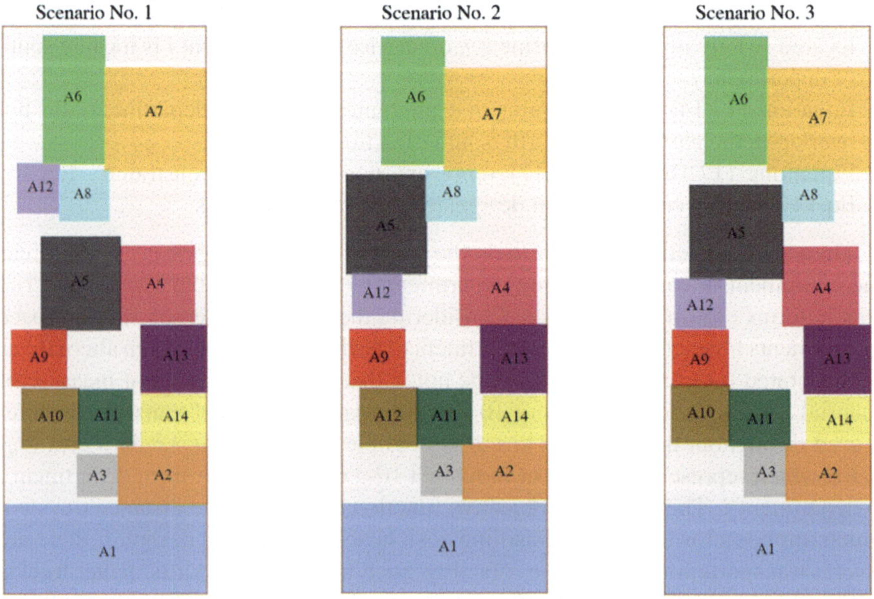

Fig. 17.5 Arrangement of production departments inside the facility for each scenario (authors elaboration)

flow of materials from one department to another by a growth factor, in this case, 30% (scenario 2) and 60–100% (scenario 3), the positions may vary, but the essence of the arrangement provided is maintained, as long as the space requirements for the production departments do not vary significantly. In this way, the arrangement obtained by scenario 3 is the one selected to be evaluated.

17.4.6 Evaluation of Design Alternatives

When evaluating the design of a material handling system, represented by the arrangement obtained by the design algorithms, it is important consider that, before implementing the system, an agreement between the facilities design analysts and the client is necessary. This guarantees that the solution proposed is aligned with the operational processes and thus rectifies any possible defect of the quantitative solution (Hillier and Liberman 2015). Although the MIP algorithm does not mathematically consider closeness ratings, these aspects are considered in this phase. Once the proposed arrangement obtained from scenario 3 was presented to the client, the product's proximity and safety requirements were analyzed to determine if the proposed solution guarantees that the operational processes do not affect the quality of the product. Considering these requirements, two final alternatives were established (Fig. 17.6).

Therefore, once the changes required by the client have been made, each design alternative is evaluated through the Eq. (17.19):

$$\text{Minimize}\, z = \sum_{i}\sum_{j} f_{ij}c_{ij}d_{ij} \tag{17.19}$$

The objective is to minimize the sum of the flow of materials in terms of cost and distance. Where f_{ij} denotes the flow from department i to department j (expressed in the amount of unit loads displaced by unit time), c_{ij} represents the cost of moving a unit load in a unit of distance from department i to department j, and d_{ij} is the distance between department i to department j (Tompkins et al. 2010). Finally, when evaluating both alternatives through Eq. (17.19), by considering the flow of materials from scenario 3, the following results were obtained

- Alternative 1 = 2,196,591.651
- Alternative 2 = 2,199,975.684.

17.4.7 Select the Most Convenient Design

Each design alternative was evaluated through Eq. (17.19). Mathematically, the option that minimizes the sum of the flow of materials in terms of cost and distance should be chosen. Therefore, Alternative 1 should be selected, because it minimizes more the material handling cost within the facility. However, before selecting this

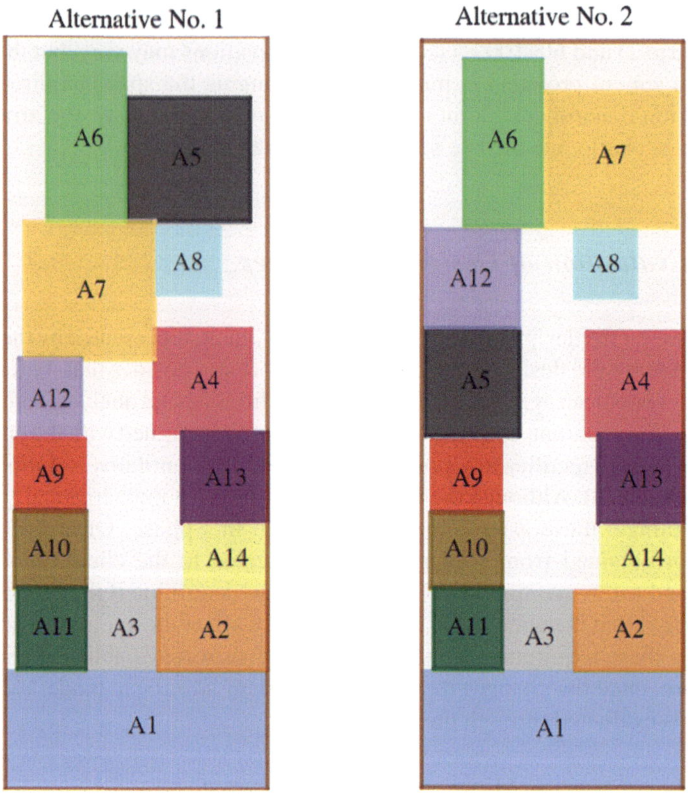

Fig. 17.6 Final alternatives (authors elaboration)

alternative, other factors should be considered by using a simulation software. Since Scenario No. 3 is the one that expects the highest growth, 60–100% of its current demand in the next 3 years, Cacao Company wants to identify the adequate amount of resources (operators) that needs to perform their processes in the new facility, in order to reach this expected demand and minimize daily operational costs. Therefore, a comparison of both alternatives through simulation is necessary, before selecting the most convenient design.

17.4.7.1 Assumption for the Development of the Simulation Model

Prior to developing the simulation of the model, assumptions were established regarding the operation of Cacao Company and the data provided, given that in a simulation model, not all events can be modeled. Considering the production process and the capacity of each equipment selected, the following assumptions and considerations at a general level were considered, for the simulation of the future location:

- As it was mentioned before the company produces four types of products, but for this simulation, it is considered chocolate molds, plates and truffles.
- It is considered that the company handles two types of finished products boxes that contain chocolate molds and boxes that contain truffles.
- Regarding the boxes that contain both the molds as well as the truffles, different presentations have been considered. For example, the molds can have 4 presentations that is, they can come 2 units per box, 3 units per box, 9 units per box, or 25 units per box. On the other hand, the truffle has a single presentation, this is 10 units per box.
- Human or machine failures are not considered. Likewise, any active pause or additional rest for operators is not contemplated in the simulation. Only a 15-min rest in the morning and an hour for lunch is considered from 13:00 to 14:00 p.m. Additionally, operators work from Monday to Friday from 8 to 17:30 p.m. and on Saturdays from 8 to 14:00 p.m.
- It is considered that, for producing each type of chocolate, there is enough inventory at the raw material warehouse.

17.4.7.2 Model Development

Once the production process has been analyzed and each workstation was identified in both layout design alternatives, it is necessary to take collect data of production processes times, so that in the simulation model, the variability of each activity is represented through statistical distributions. However, due to the physical restrictions of the current facility, considering that the production of all the types of chocolate is not simultaneous, but based on a weekly production planning and that, in some activities, processing time exceeds 24 h, a sample of 50 data was collected for each detailed activity in Appendix 17.1. Therefore, to find the statistical distribution that best fits the data set collected from each activity, the statistical complement of FlexSim, ExpertFit is used, because it automatically figures out which statistical distribution best represents a data set. In Appendix 17.12, it is possible to appreciate that every activity is properly adjusted to a given distribution, according to the chi-square statistical test and a statistical significance of 95%.

By using the 3D potential of the FlexSim software, the simulation model for both alternatives was developed, which included all the specifications of the case, such as the processing times of the resources, their speeds, storage capacities and % product demand, among other considerations that should be implemented to reflect as closely as possible, the operations of Cacao Company. Additionally, it is important to mention that the performance indicator to compare both alternatives is the total throughput of units monthly.

17.4.7.3 Experimental Design

Since Cacao Company wants to identify the adequate amount of resources (operators) that needs to perform their processes in the new facility, in order to reach the expected demand proposed by Scenario No. 3, an experimental design is necessary to find the optimal number of operators required to produce between 40,000 and 45,000 units of chocolates monthly. Therefore, a screening (or pilot test) was carried out in the simulation model of Alternative 1, since this option minimizes more the material handling cost within the facility. It is important to mention that a screening or pilot test is useful to determine the factors or levels to evaluate. Once the levels have been determined, both layout designs are compared under the same conditions, and thus, the most suitable design is selected.

Cacao Company works with four operators, who allows it to produce the quantities previously exposed. Since the new facility is bigger compared to the current facility, most likely the company needs to hire additional operators. Therefore, to find the answer to this issue, the screening test is necessary. Initially, the following levels of operators are proposed:

- Scenario 1: 6 operators
- Scenario 2: 8 operators
- Scenario 3: 10 operators
- Scenario 4: 12 operators
- Scenario 5: 14 operators.

After defining the levels for the pilot test, it was decided to perform 15 runs per scenario to find differences between the levels initially raised. Appendix 17.13 shows the results obtained by the pilot test. Once the pilot test was completed, by analyzing the results of the test, it was found that from 10 to 14 operators met the requirements of Scenario No. 3 and from 6 to 10 operators satisfy the requirements up to Scenario No. 2. However, since the objective is to find the number of operators required to meet the requirements of Scenario No. 3, levels 6 and 8 operators are discarded. Thus, it was determined that both layout alternatives are evaluated using the following levels of the factor operators:

- Scenario 1: 10 operators
- Scenario 2: 12 operators
- Scenario 3: 14 operators.

This experiment is known as one-way or single-factor analysis of variance (ANOVA) model because only one factor is investigated (Montgomery 2012). Then, in order to determine the number of replicates for each scenario, it is necessary to calculate the power curve for the experiment, since this statistical tool is often helpful in selecting a sample size to use in an experiment (Montgomery 2012). The power curve of a one-way ANOVA is the probability that the test will determine that, if the maximum difference between a group of means is statistically significant, when that difference truly exists (Montgomery 2012).

Thereby, the power curve allows to find the number of replicates required to find the number of operators required to produce between 40,000 and 45,000 units of chocolates. Appendix 17.14 shows a graphical representation of the power curve for this one-way ANOVA analysis. This evidence that, with a power of 84.18%, 100 replications per scenario are needed, in order to find if the difference between the means of 10 – 12 – 14 operators is statistically significant. Therefore, 100 replicates were performed for each design alternative simulation model. Finally, Appendix 17.15 shows the results obtained for Alternative 1, and Appendix 17.16 shows the results obtained for Alternative 2.

Since the objective of Scenario No. 3 is to produce more than 40,000 units, each proposed scenario (number of operators) for each alternative is feasible, because with a 95% confidence interval, the total throughput is fulfilled. Therefore, in order to determine which throughput means statistically differ, it is necessary to test the differences between all pairs of operators for Alternative 1 and 2. In this way, Tukey's multiple (pairwise) comparison test is ideal for this case, because it controls the experimentwise or "family" error rate at the selected level (α), when the interest focuses on pairs of means (Montgomery 2012). Table 17.2 shows the Tukey's procedure results.

Therefore, with a 95% confidence, the null hypothesis of equality of throughput means is rejected since the P value for this factor is lower than the established level of significance ($\alpha = 0.05$). So, there exists at least one scenario for each alternative, where its total throughput mean is significantly different. In this way, Tables 17.3 and 17.4 show Tukey's procedure results for each design alternative.

The results obtained show that for both alternatives, scenario 1 (10 operators) is statistically different compared to scenario 2 and 3. In addition, scenario 2 and 3 are statistically equal. Therefore, in order to minimize the daily operational cost, scenario 3 (14 operators) is discarded for Alternative 1 and 2. However, since scenario 1 and 2 produce more than 40.000 units monthly and are statistically different in both design alternatives, these options are economically evaluated, in order to determine which option results more profitable for Cacao Company.

Table 17.2 Tukey's procedure *P*-value

	P-value
Alternative 1	0.000
Alternative 2	0.000

Table 17.3 Tukey's procedure Alternative 1

Factor	Alternative 1		
	N	Mean	Grouping
Scenario 3	100	41,507.6	A
Scenario 2	100	41,350.6	A
Scenario 1	100	40,728.2	B

Table 17.4 Tukey's procedure Alternative 2

Factor	Alternative 2		
	N	Mean	Grouping
Scenario 3	100	41,583.3	A
Scenario 2	100	41,525.1	A
Scenario 1	100	41,134.0	B

17.4.7.4 Model Proposed

Mathematically, Alternative 1 should be chosen, since it minimizes the sum of the flow of materials in terms of cost and distance. From the previous analysis, mathematically, Alternative 1 should be chosen, because it minimizes the sum of the flow of materials in terms of cost and distance. However, considering all the additional factors that simulation software allows, the results show that Alternative 2 can be feasible too. Therefore, additional analysis should be carried out. It is here, where economic analysis is suitable to get further information to select the best alternative between them.

Economic Analysis

The economic analysis of the design alternatives relies in the comparison of the profit generated by each one in a monthly period. Within the calculation of the monthly profit of each scenario, the operational costs (fixed and variable) are involved, as well as the taxes that must be covered for the operation of the facility.

Table 17.5 shows the monthly profitability for each alternative and the scenarios evaluated. The results obtained evidence that Scenario No. 1 (10 operators) of Alternative 2 has the highest monthly profitability among all options, generating an annually net profit of $129,102.36. It is important to emphasize that even though there are scenarios that include a high number of operators, which leads to higher production levels, these scenarios do not generate a large profit margin. The financial profitability of the project is evaluated from two financial indicators, the return on investment rate (ROI) and the net present value (NPV). For this, the cash flow of the next three years is simulated, and the calculation is executed with an initial investment of $56,427.00 (cost of assembly). Once the calculations have been made, a ROI of 17% is obtained, which indicates a greater return to that presented by the Central Bank of Ecuador equal

Table 17.5 Monthly profitability for each alternative

	Scenario	Profit
Alternative 1	No. 1: 10 operators	$10,758.53
	No. 2: 12 operators	$10,221.73
Alternative 2	No. 1: 10 operators	$10,981.83
	No. 2: 12 operators	$10,317.43

to 8.7% (ECB 2019). In addition, an NPV of $82,917.25 was obtained, which allows establishing that the project is economically feasible considering all the assembling costs and resources required to satisfy the expected demand.

Once Alternative 2 has been defined as the design of the new facility, it is necessary to determine that the cost of executing this alternative is within the budget allocated for the project, and in addition, it generates enough profitability to consider sustainable. The economic evaluation process begins with the determination of assembly costs, and these include all the tools and machines that are required to produce chocolate within the quality parameters established by Cacao Company. However, this analysis does not consider land costs and construction of the space where the facility will be built, because the company's directors established the construction of the facility as a different project, the same that is not included within the scope of the case study. The assembly costs of the facility reach $56,427. The design proposal was presented to the company's board of directors, in June 2019. It was subsequently presented and accepted by the Municipio de Quito, the regulatory entity for land use, in August 2019.

Cacao Company agreed to build the facility with an Ecuadorian construction company, which, due to confidentiality agreements, cannot be disclosed. This construction company estimates that the completion of the construction will be in February 2020, and that it can begin operation in April of the same year.

Cacao Company Layout Design

Finally, to facilitate the understanding and visualization of the model proposed, a graphic representation of Alternative 2 was translated into an architectural layout in AutoCAD and a 3D model. These illustrations can be found in Appendices 17.17, 17.18 and 17.19.

17.5 Conclusions, Limitations and Recommendations

Cacao Company case study is based on the use of mathematical algorithms to determine the most efficient arrangement of the production departments in the new facility. However, before implementing the system, an agreement with the customer is necessary, in order to make sure that the model proposed is accurately translated to an operating procedure and to rectify any possible flaws uncovered by the quantitative solution. Therefore, qualitative requirements are important to determine the functionality of the final model. An example of this is the food safety requirements, the same ones that altered the alternatives obtained. Another factor to consider was the temperature of each department, considering that the chocolate must reach a temperature below 18 °C to crystallize and be properly decorated. As a result, the arrangement provided for Cacao Company is the safest and most satisfactory. Mathematically, Alternative 1 must have been selected, since it minimizes the sum of the flow of materials in terms of cost and distance. However, considering all the additional factors that a simulation software can analyze, because it allows to conduct the design before finally building the facility, Alternative 2 is selected. Additionally, thanks to the one-way ANOVA analysis, it is possible to identify the optimal amount of resources (number of operators) that Cacao Company will need to operate in the new facility, and this number will also allow the company to achieve the desired monthly throughput. Finally, FlexSim simulation software allows to represent the material handling system designed through a 3D model, which generates an added value for the directors of Cacao Company. The economic evaluation establishes that Cacao Company new facility design is an economically feasible project by evaluating the return on investment rate (ROI) and the net present value (NPV).

Some limitations presented during the development of this case study are related to the current infrastructure of the company, which made it difficult to gather information, especially in data collection of process time for the development of the simulation model. In addition, the determination of the scenarios, because the expected growth was determined mostly by assumptions and experience of Cacao Company's directors. It should be noted that these growth percentages play a fundamental role in calculating space requirements and determining the flow of materials between the production departments.

Finally, according to our findings, some recommendations can be proposed

- A larger sample data of production processes, in order to represent better the variability of the system.
- Develop a more detailed market study to determine a more accurate growth percentage for the scenarios.
- Conduct a sensitivity analysis, since the expected growth is quite high. This allows to determine the range of demand under which the selected alternative is profitable.
- Run a benchmarking with other handcraft chocolate companies, in order to compare their material handling system.
- Once the facility begins its operations, consider the implementation of good manufacturing practices (GMP) and hazard analysis and critical control points (HACCP) standards, within critical production processes. These certifications are important for the exportation of food products.

Appendix 17.1: Chocolate Production Process Flowchart

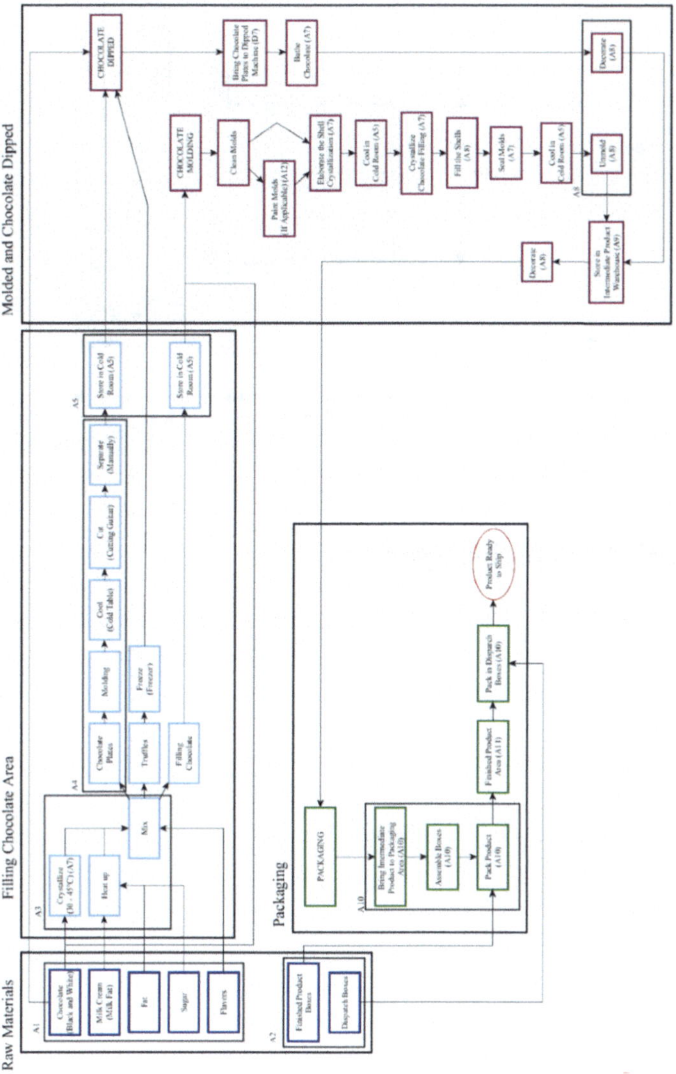

Appendix 17.2: Current Facility Layout

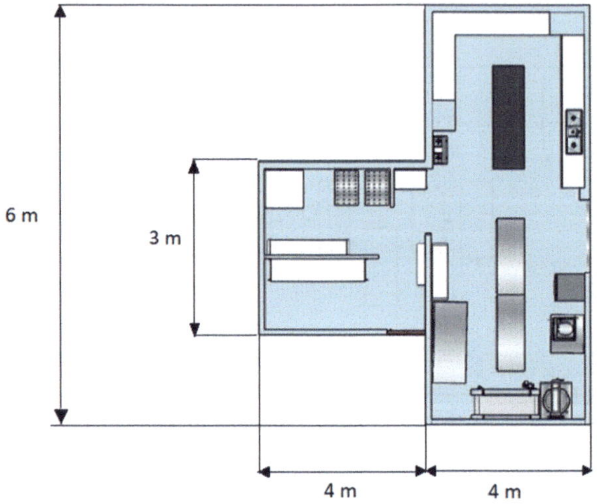

Appendix 17.3: Available Space

Appendix 17.4: Scenario No. 1 Flow Matrix

	A1	A2	A3	A4	A5	A6	A7	A8	A9	A10	A11	A12	A13	A14
A1	0	23,810.3	43,897.2								19,255.6			4700
A2		0		19,708.4			22,117.6							
A3			0							19,509.9				
A4				0	3313.8	10,860.2	5534.4							
A5					0	3313.8								4700
A6						0	14,174	4284.1				2861.97		
A7							0	15,941.8						
A8								0	19,255.6					
A9									0	19,255.6				
A10										0	19,255.6			
A11											0			
A12												0		
A13													0	
A14														0

Appendix 17.5: Scenario No. 2 Flow Matrix

	A1	A2	A3	A4	A5	A6	A7	A8	A9	A10	A11	A12	A13	A14
A1	0	35,715.4	65,845.8								28,883.5			7050
A2		0		29,562.6			33,176.4							
A3			0							29,264.8				
A4				0	4970.82	16,290.2	8301.6							
A5					0	4970.82								7050
A6						0	21,261	6426.15				4292.95		
A7							0	23,912.7						
A8								0	28,883.5					
A9									0	28,883.5				
A10										0	28,883.5			
A11											0			
A12												0		
A13													0	
A14														0

Appendix 17.6: Scenario No. 3 Flow Matrix

	A1	A2	A3	A4	A5	A6	A7	A8	A9	A10	A11	A12	A13	A14
A1	0	59,525.67	109,743								48,139.12			11,750
A2		0		49,271.08			55,294.04							
A3			0							48,774.67				
A4				0	8284.7	27,150.38	13,836							
A5					0	8284.7								11,750
A6						0	35,435.08	10,710.25				7154.92		
A7							0	39,854.42						
A8								0	48,139.12					
A9									0	48,139.12				
A10										0	48,139.12			
A11											0			
A12												0		
A13													0	
A14														0

Appendix 17.7: Equipment Capacity

Equipment	Capacity	Unit	Description
Stainless steel rack	700	kg	3 levels and loads up to 235 kg each
Spieth rack for light load	1500	Box	Store 1500 boxes
Granite kitchen counter (large)	5	Bowls	Each bowl loads up to 10 kg
Fridge	13	Bucket	A bucket loads 3–5 kg
Kitchen	32.75	kg	Produce up to 32,750 units daily
Chocolate temper	6.5	kg	Able to crystallize 1300 units daily
Pot	10	kg	Produce 1950 daily units
Small kitchen counter	6	Tray	Handle 3000 units simultaneously
Cutting guitar	2	kg	Cut 1550 units per hour
Stainless steel tables (mobile)	4	Tray	Handle 1060 units per table
Tray holder	55	Tray	Store 5583 units for each tray holder
Stainless steel rack	900	kg	Store bowls, trays and other containers
Black chocolate dipped machine	12.6	kg/h	Melt and dip 1260 units per hour
White chocolate dipped machine	3.75	kg/h	Melt and dip 375 units per hour
Stainless steel tables (mobile)	9	Tray	Handle 2550 units in trays per table
Intermediate product rack	690	kg	3 levels, store 687 units. at each level
Plastic drawer	30	kg	Store 300 units in each drawer
Stainless steel tables (mobile)	9	Tray	Handle 1700 units for packing per table
Finished product rack	1340	Tray	Store 33,500 units in each rack (4 levels in each rack)
Decorating materials rack	150	kg	Store decoration material and a compressor
Stainless steel tables (mobile)	9	Tray	Handle 1460 units in trays per table
Electric dishwasher			–
Industrial stainless-steel oven	4	kg	Produce filling for 400 units per hour
Stainless steel tables (mobile)	9	Tray	Handle 1460 units in trays per table

Appendix 17.8: Area Requirements for Each Scenario

	Departments	Workstation	No. 1	No. 2	No. 3
A1	Receiving zone	Receive and dispatch area	21.00	21.00	21.00
A2	Raw material warehouse	Ingredient rack	2.72	2.72	2.72
A3	Package material warehouse	Storage space	4.56	4.56	4.56
A4	Filling chocolate area	Kitchen counter	5.28	5.28	5.28
A5	Plate chocolate making	Fridge	2.92	2.92	2.92
		Kitchen	1.86	1.86	1.86
		Temperator	2.94	2.94	2.94
		Pot	2.31	2.31	2.31
		Small kitchen counter	1.10	1.10	1.10
		Cutting guitar	0.30	0.30	0.30
A6	Cold room (0–4 °C)	Stainless steel tables	5.92	8.88	11.84
		Tray holder	5.98	8.97	9.97
A7	Chocolate crystallization	Materials rack	6.53	6.53	6.53
		Black chocolate dipped machine	13.00	13.00	13.00
A8	Chocolate decoration	White chocolate dipped machine	5.60	5.60	5.60
		Stainless steel tables	5.92	5.92	5.92
A9	Intermediate product warehouse (18 °C)	Intermediate product rack	4.10	4.10	4.10
A10	Packing area	Plastic drawer	4.57	8.22	9.14
		Stainless steel tables	5.92	5.92	5.92
A11	Finished product (18 °C)	Finished product rack	3.26	3.26	3.26
A12	Painting of chocolate molds	Decorating materials rack	1.52	3.05	3.05
A13	Wet and dry-cleaning area	Stainless steel tables	5.92	5.92	5.92
		Electric dishwasher	4.18	4.18	4.18
A14	Fruit pulping area	Small oven	1.28	1.28	1.28
		Kitchen counter	5.28	5.28	5.28
		Total [m^2]	123.98	153.11	139.98

Appendix 17.9: AMPL IDE Solution for Scenario No. 1

```
CPLEX 12.9.0.0: optimal integer solution; objective 806646.9289
322 MIP simplex iterations
0 branch-and-bound nodes
absmipgap = 1.16415e-10, relmipgap = 1.4432e-16
ampl: display x1,x2,y1,y2;
:     x1      x2      y1      y2      :=
1     0       4       0       1.88
2     2.06    4       1.83    2.89
3     1.23    2.06    1.88    2.71
4     2.3     3.76    5.28    6.74
5     0.75    2.3     5.11    6.91
6     0       1.2     8.34   10.8
7     2       4       8.2    10.2
8     0.285   1.105   7.38    8.34
9     0.165   1.225   4.05    5.11
10    0.36    1.47    2.865   3.975
11    1.47    2.53    2.89    3.95
12    1.105   2.085   7.22    8.2
13    2.71    4       3.89    5.2
14    2.71    4       2.89    3.89
;
```

Appendix 17.10: AMPL IDE Solution for Scenario No. 2

```
CPLEX 12.9.0.0: optimal integer solution; objective 1245101.788
160 MIP simplex iterations
0 branch-and-bound nodes
absmipgap = 2.32831e-10, relmipgap = 1.86997e-16
ampl: display x1,x2,y1,y2;
:     x1      x2      y1      y2      :=
1     0       4       0       1.88
2     2.06    4       1.83    2.89
3     1.23    2.06    1.88    2.71
4     2.31    3.79    5.52    6.98
5     0.1     1.65    6.21    8.11
```

```
6    0       1.2     8.34    10.8
7    2       4       8.2     10.2
8    0.215   1.175   5.39     6.21
9    0.165   1.225   4.05     5.11
10   0.36    1.47    2.865    3.975
11   1.47    2.53    2.89     3.95
12   1.65    2.63    7.22     8.2
13   2.71    4       3.89     5.2
14   2.71    4       2.89     3.89
;
```

Appendix 17.11: AMPL IDE Solution for Scenario No. 3

```
CPLEX 12.9.0.0: optimal integer solution within mipgap or absmipgap; objective 2059716.428
261 MIP simplex iterations
0 branch-and-bound nodes
absmipgap = 182.283, relmipgap = 8.84991e-05
ampl: display x1,x2,y1,y2;
:     x1      x2      y1      y2      :=
1    0       4       0       1.88
2    2.06    4       1.83    2.89
3    1.23    2.06    1.88    2.71
4    2.3     3.76    5.62    7.22
5    0.5     2.3     6.11    7.91
6    0       1.2     8.34   10.8
7    2       4       8.2    10.2
8    0.215   1.175   5.15    6.11
9    0.165   1.225   4.05    5.11
10   0.36    1.47    2.865   3.975
11   1.47    2.53    2.89    3.95
12   2.3     3.28    7.22    8.2
13   2.71    4       3.89    5.2
14   2.71    4       2.89    3.89
;
```

Appendix 17.12: Statistical Distribution for Each Production Process

Process	Statistical distribution	Critical value (chi-square)	Score	Parameters	
Cool	Log Laplace	3.2	99.22	Location	0.000000
				Shape	7.970000
				Scale	10.956716
Filling process	Log logistic (E)	10.8	98.39	Location	5.344654
				Shape	16.466309
				Scale	9.914652
Truffle making	Beta	8.8	87.5	Lower endpoint	4.668469
				Upper endpoint	15.787534
				Shape #1	5.243192
				Shape #2	10.588883
Unmold chocolate molds	Beta	19.2	85	Lower endpoint	1.105469
				Upper endpoint	1.245889
				Shape #1	3.772009
				Shape #2	5.621652
Fill dark chocolate	Johnson SB	70.8	82.14	Lower endpoint	0.007075
				Upper endpoint	0.080039
				Shape #1	−0.527481
				Shape #2	1.402037
Fill white chocolate	Beta	94.8	80.68	Lower endpoint	0.007073
				Upper endpoint	0.075923
				Shape #1	6.387951
				Shape #2	3.233468
Mold fruit	Triangular	7.2	99	Min	212
				Max	240
				Mode	228
Chocolate plate making	Beta	10.4	99.22	Lower endpoint	14.947124
				Upper endpoint	32.757234
				Shape #1	1.734645
				Shape #2	2.811234
Chocolate plate making with fruit	Log logistic	17.6	95.16	Location	0.000000
				Shape	22.436642
				Scale	13.079730
Black chocolate crystallization	Weibull (E)	40.4	85.58	Location	0.634675
				Shape	0.123160

(continued)

(continued)

Process	Statistical distribution	Critical value (chi-square)	Score	Parameters	
				Scale	6.431569
White chocolate crystallization	Johnson SB	5.6	95.97	Lower endpoint	8.825640
				Upper endpoint	15.317296
				Shape #1	−0.023583
				Shape #2	1.280530
Painting of chocolate molds	Johnson SB	29.2	89.17	Lower endpoint	0.253458
				Upper endpoint	0.425993
				Shape #1	0.615591
				Shape #2	2.047847
Decoration of truffles	Beta	55.6	93	Lower endpoint	0.037740
				Upper endpoint	0.119699
				Shape #1	2.561288
				Shape #2	4.564534
Decoration of chocolate molds	Weibull (E)	56	85.23	Location	0.018358
				Shape	0.055636
				Scale	3.696733
Packing box of 2 units	Log logistic	12.4	88	Location	0.000000
				Shape	0.331116
				Scale	19.424458
Packing box of 25 units	Beta	4.8	89.58	Lower endpoint	0.232180
				Upper endpoint	0.377150
				Shape #1	2.554675
				Shape #2	1.597855
Packing box of 3 units	Johnson SB	11.6	86.96	Lower endpoint	0.003785
				Upper endpoint	0.362995
				Shape #1	0.478220
				Shape #2	3.030693
Packing box of 9 units	Weibull	16.4	84.78	Location	0.024983
				Scale	0.129263
				Shape	4.688172
Packing box of truffles	Beta	20.8	86.54	Lower endpoint	0.173294
				Upper endpoint	0.435848
				Shape #1	15.090556
				Shape #2	15.983751

Appendix 17.13: Pilot Test Results

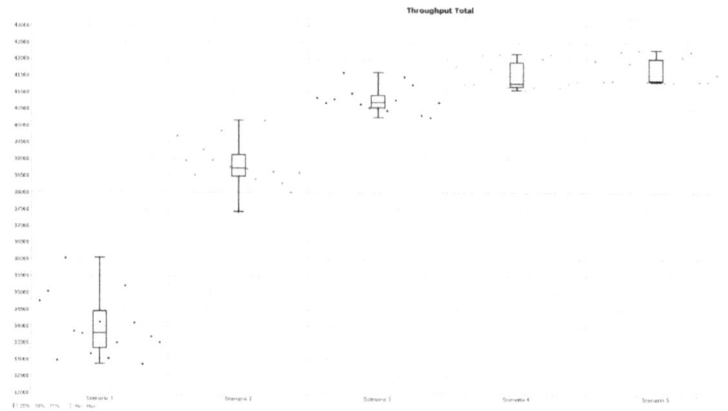

Total Throughput

	Mean (95% Confidence)					Sample Std Dev	Min	Max
Scenario 1	33553	<	33972	<	34390	920	32860	36035
Scenario 2	38493	<	38817	<	39141	712	37415	40154
Scenario 3	40609	<	40788	<	40966	393	40241	41600
Scenario 4	41327	<	41518	<	41709	420	41071	42150
Scenario 5	41479	<	41658	<	41837	393	41316	42266

Appendix 17.14: Power Curve

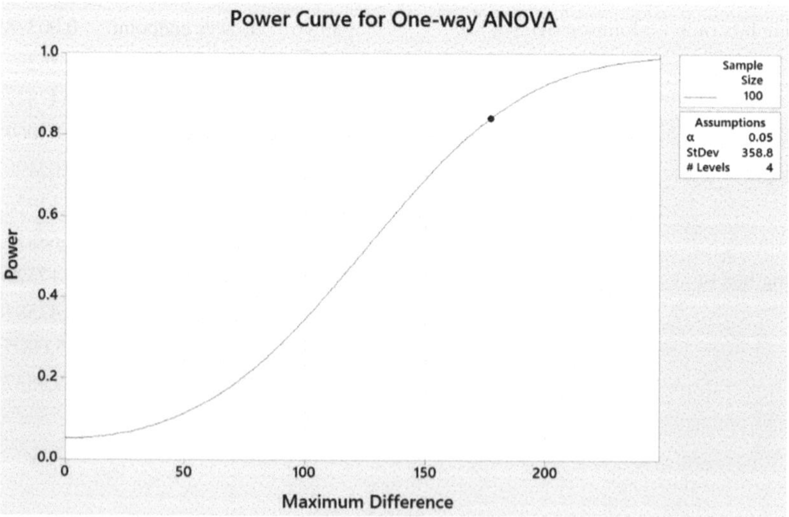

Appendix 17.15: Simulation Results for Alternative 1

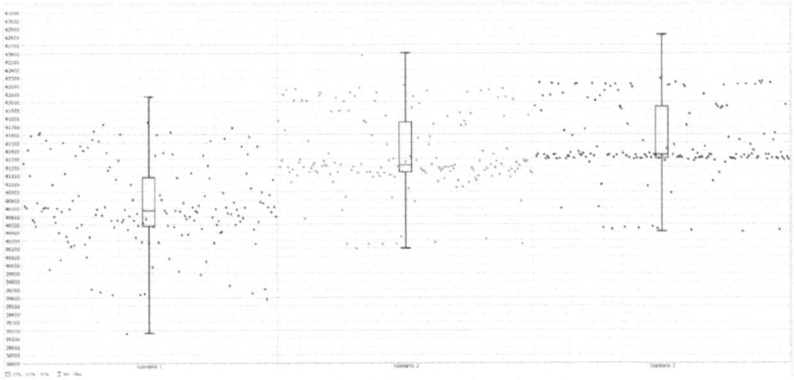

Total Throughput

	Mean (95% Confidence)					Sample Std Dev	Min	Max
Scenario 1	40661	<	40728	<	40796	525	39159	42057
Scenario 2	41289	<	41351	<	41413	480	40197	42599
Scenario 3	41448	<	41508	<	41568	464	40409	42823

Appendix 17.16: Simulation Results for Alternative 2

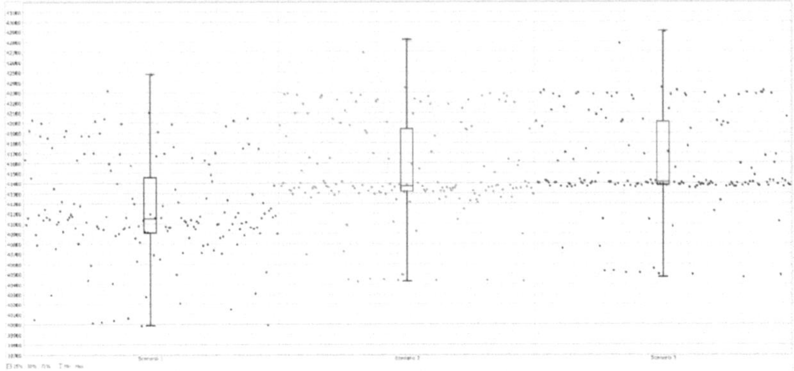

Total Throughput

	Mean (95% Confidence)					Sample Std Dev	Min	Max
Scenario 1	41070	<	41134	<	41199	501	39986	42476
Scenario 2	41463	<	41525	<	41587	480	40428	42819
Scenario 3	41521	<	41583	<	41644	475	40467	42901

Appendix 17.17: Cacao Company AutoCAD Layout

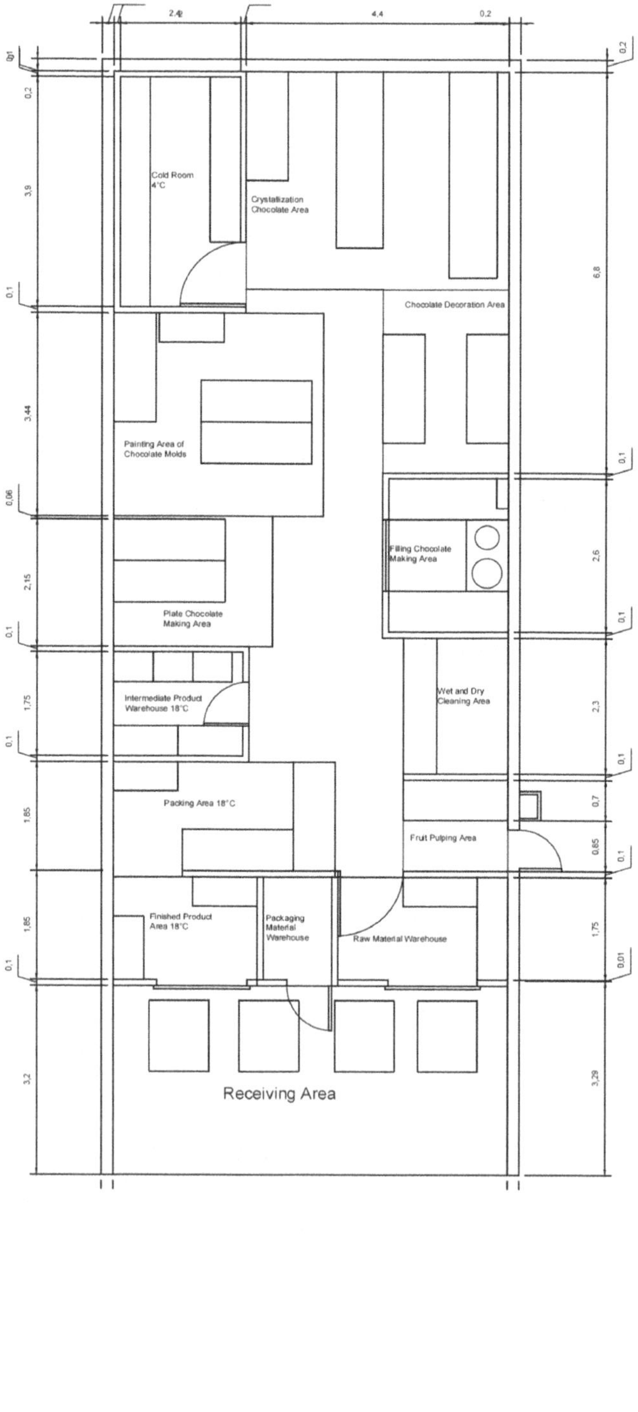

Appendix 17.18: Cacao Company 3D Model Top View

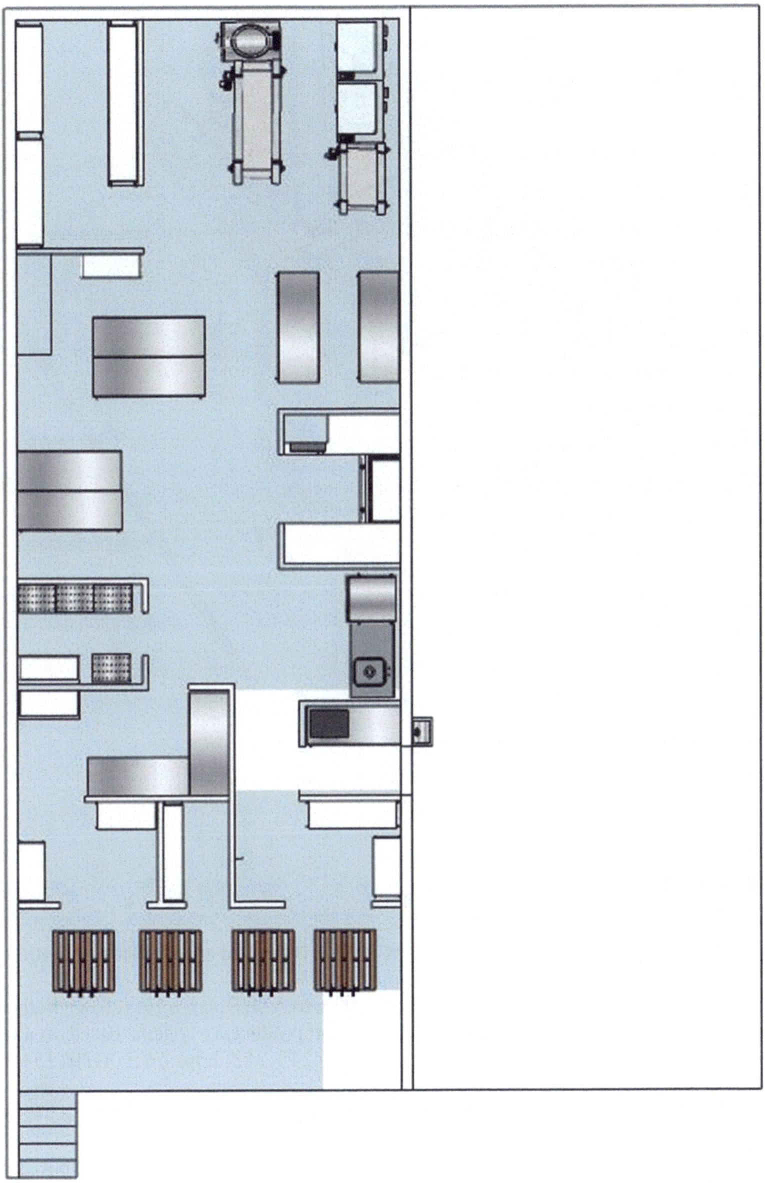

Appendix 17.19: Cacao Company 3D Model Side View

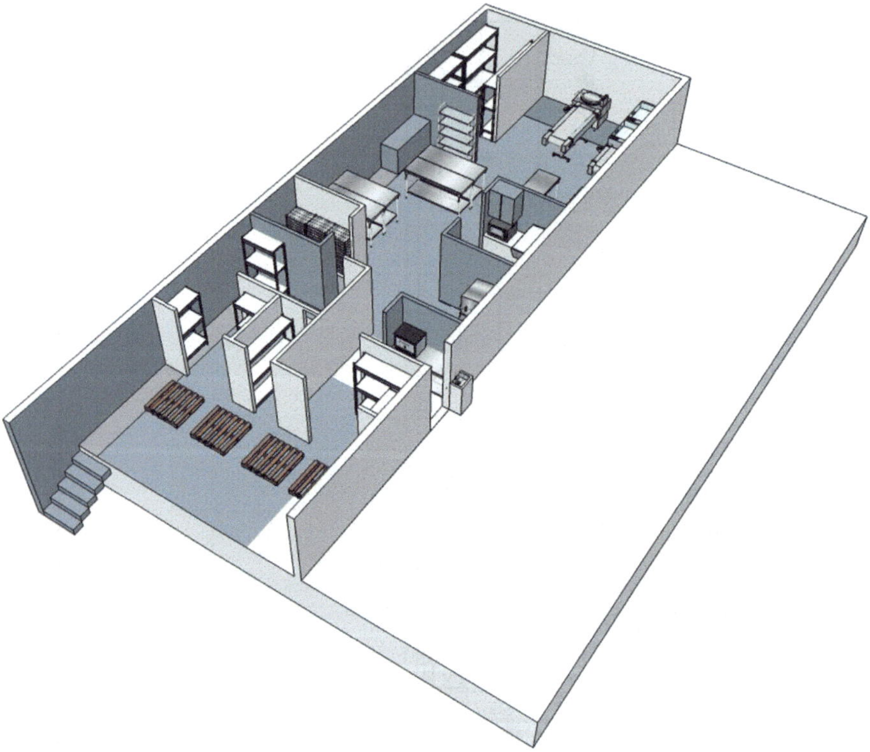

References

Ahmad MM, Ibrahim N (2007) Development of 3 in 1 chocolate filling machine with variable volume using semi or full automatic process, pp 65–73

Alemany MDMe, Peluffo DH, Torres JC, Alpala LO, Rosero AM, Bolaños FA (2018) Methodology for the design and simulation of industrial facilities and production systems based on a modular approach in an "industry 4.0" context. Dyna 85(207):243–252. https://doi.org/10.15446/dyna.v85n207.68545

Amaral A (2006) On the exact solution of a facility layout problem. Eur J Oper Res 173:508–518. https://doi.org/10.1016/j.ejor.2004.12.021

Anecacao (2019) Sector Exportador de Cacao 2019. Asociación Nacional de Exportadores de Cacao—Ecuador. Extracted on 14 Oct 2019 from https://www.anecacao.com/index.php/es/estadisticas/estadisticas-actuales.html

Azarmand Z, Neishabouri E (2009) Location allocation problem. In: Facility location. Physica, Heidelberg, pp 93–109. https://doi.org/10.1007/978-3-7908-2151-2_5

Dong M, Wu C, Hou F (2009) Shortest path based simulated annealing algorithm for dynamic facility layout problem under dynamic business environment. Expert Syst Appl 36(8):11221–11232. https://doi.org/10.1016/j.eswa.2009.02.091
Drira A, Pierreval H, Hajri-Gabouj S (2007) Facility layout problems: a survey. Annu Rev Control 31:255–267. https://doi.org/10.1016/j.arcontrol.2007.04.001
ECB (2019) Tasas de Interés Efectivas Vigentes. Banco Central del Ecuador [Ecuadorian Central Bank]. Retrieved from https://contenido.bce.fin.ec/documentos/Estadisticas/SectorMonFin/TasasInteres/TasasVigentes102019.htm. Date: October 10, 2019 at 4:00 pm
Emami S, Nookabadi AS (2013) Managing a new multi-objective model for the dynamic facility layout problem. Int J Adv Manuf Technol 68:2215–2228. https://doi.org/10.1007/s00170-013-4820-5
Farahani RZ, SteadieSeifi M, Asgari N (2010) Multiple criteria facility location problems: a survey. Appl Math Model 34(7):1689–1709. https://doi.org/10.1016/j.apm.2009.10.005
Frazelle EH (1986) Material handling: a technology for industrial competitiveness. Material Handling Research Center, Georgia Institute of Technology
Grand View Research (2018) Chocolate confectionery market size, share & trends analysis report by product (boxed, countlines, molded bars, seasonal chocolates, straightlines, others), by type (milk, dark, white), by region and segment forecasts, 2018–2025. Retrieved from https://www.grandviewresearch.com/industry-analysis/chocolate-confectionery-market
Guan J, Lin G (2016) Hybridizing variable neighborhood search with ant colony optimization for solving the single row facility layout problem. Eur J Oper Res 248(3):899–909. https://doi.org/10.1016/j.ejor.2015.08.014
Halper R, Raghavan S (2010) The mobile facility routing problem. Transp Sci 45(3):413–434. https://doi.org/10.1287/trsc.1100.0335
Hani Y, Amodeo L, Yalaoui F, Chen H (2007) Ant colony optimization for solving an industrial layout problem. Eur J Oper Res 183:633–642. https://doi.org/10.1016/j.ejor.2006.10.032
Hillier F, Liberman G (2015) Introduction to operations research, 10th edn. McGraw-Hill Education
Hosseini-Nasab H, Fereidouni S, Fatemi Ghomi SMT, Fakhrzad MB (2018) Classification of facility layout problems: a review study. Int J Adv Manuf Technol 94(1–4):957–977. https://doi.org/10.1007/s00170-017-0895-8
Jewett S (2017) Artisan chocolate making. In: Beckett's industrial chocolate manufacture and use, pp 456–478. https://doi.org/10.1002/9781118923597.ch18
Khusna D, Dawal SZM, Jamasri, Aoyama H (2010) A proposed study on facility planning and design in manufacturing process. In: Proceedings of the international multi-conference of engineers and computer scientists, vol 3, no 1, pp 1640–1645
Konak A, Kulturel-Konak S, Norman BA, Smith AE (2006) A new mixed integer programming formulation for facility layout design using flexible bays. Oper Res Lett 34(6):660–672. https://doi.org/10.1016/j.orl.2005.09.009
La Republica (2018) Consumo de Chocolate en Perú, Crecimiento del Mercado de Chocolates. Extracted on 7 July 2019 from https://larepublica.pe/economia/1286290-consumo-chocolate-peru-bajos-america-latina/
Leno IJ, Sankar SS, Ponnambalam SG (2015) MIP model and elitist strategy hybrid GA–SA algorithm for layout design. J Intell Manuf 29(2):369–387. https://doi.org/10.1007/s10845-015-1113-x
Lira J (2019) Cacao Peruano. Gestión. Extracted on 7 July 2019 from https://gestion.pe/economia/cacao-peruano-consumo-per-capita-239008
Madhusudanan Pillai V, Hunagund IB, Krishnan KK (2011) Design of robust layout for dynamic plant layout problems. Comput Ind Eng 61(3):813–823. https://doi.org/10.1016/j.cie.2011.05.014
Mazinani M, Abedzadeh M, Mohebali N (2012) Dynamic facility layout problem based on flexible bay structure and solving by genetic algorithm. Int J Adv Manuf Technol 65(5–8):929–943. https://doi.org/10.1007/s00170-012-4229-6

McKendall AR, Hakobyan A (2010) Heuristics for the dynamic facility layout problem with unequal-area departments. Eur J Oper Res 201(1):171–182. https://doi.org/10.1016/j.ejor.2009.02.028

Meller RD, Chen W, Sherali HD (2007) Applying the sequence-pair representation to optimal facility layout designs. Oper Res Lett 35(5):651–659. https://doi.org/10.1016/j.orl.2006.10.007

Mohamadghasemi A, Hadi-Vencheh A (2012) An integrated synthetic value of fuzzy judgments and nonlinear programming methodology for ranking the facility layout patterns. Comput Ind Eng 62(1):342–348. https://doi.org/10.1016/j.cie.2011.10.004

Montgomery DC (2012) Design and analysis of experiments, vol 2, 8th edn. Wiley. https://doi.org/10.1198/tech.2006.s372

Morris JA (2014) KPMG—a taste of the future. In: Consumer markets, June 2014 (KPMG), 20. https://doi.org/10.1038/nbt1004-1203

Pourvaziri H, Pierreval H (2017) Dynamic facility layout problem based on open queuing network theory. Eur J Oper Res 259(2):538–553. https://doi.org/10.1016/j.ejor.2016.11.011

Samarghandi H, Eshghi K (2010) An efficient tabu algorithm for the single row facility layout problem. Eur J Oper Res 205(1):98–105. https://doi.org/10.1016/j.ejor.2009.11.034

Singh SP, Sharma RRK (2006) A review of different approaches to the facility layout problems. Int J Adv Manuf Technol 30(5–6):425–433. https://doi.org/10.1007/s00170-005-0087-9

Singh AP, Yilma M (2013) Production floor layout using systematic layout planning in Can manufacturing company. In: 2013 international conference on control, decision and information technologies, CoDIT 2013, pp 822–828. https://doi.org/10.1109/CoDIT.2013.6689649

Tavakkoli-Moghaddam R, Javadian N, Javadi B, Safaei N (2007) Design of a facility layout problem in cellular manufacturing systems with stochastic demands. Appl Math Comput 184(2):721–728. https://doi.org/10.1016/j.amc.2006.05.172

Thomas J (2017) The global chocolate confectionery market. In: Beckett's industrial chocolate manufacture and use, pp 654–674. https://doi.org/10.1002/9781118923597.ch27

Tompkins JA, White JA, Bozer YA, Tanchoco JMA (2010) Facilities planning, 4th edn. Wiley, New Jersey

Vitayasak S, Pongcharoen P, Hicks C (2017) A tool for solving stochastic dynamic facility layout problems with stochastic demand using either a Genetic Algorithm or modified Backtracking Search Algorithm. Int J Prod Econ 190:146–157. https://doi.org/10.1016/j.ijpe.2016.03.019

Yang CL, Chuang SP, Hsu TS (2011) A genetic algorithm for dynamic facility planning in job shop manufacturing. Int J Adv Manuf Technol 52(1–4):303–309. https://doi.org/10.1007/s00170-010-2733-0

Chapter 18
A Framework for the Formulation of an Operations Strategy in Manufacturing Systems

Jorge A. Vivares, William Sarache, and Jorge E. Hurtado

Abstract The objective of operations strategy (OS) is to contribute to the consolidation of lasting competitive capabilities in organizations, from the standpoint of production/operations system (POS) enhancement. The OS permanence formulation process remains an open problem in the scientific literature, as there are several possible *frameworks*, and no single focus universally fits all companies. Consequently, the objective of the present investigation is to develop and apply an OS formulation *framework*. Said *framework* consists of two phases: (1) production/operations system evaluation, based on the measurement of its current maturity (baseline), and (2) formulation of strategy, based on a group of projects oriented toward the maturity level maximization using stochastic optimization model and several investment scenarios. By way of an action research process, the *framework* was applied in a manufacturing company located in the Colombian coffee region. The results obtained confirmed the feasibility of the OS formulation framework. This investigation makes both academic and practical contributions, as, on the one hand, it contributes to enhance the field of knowledge, by providing a structured novel approach that may be replicated in the future investigations. On the other, it provides a practical solution to address the complex OS formulation process in the real world.

Keywords Operations strategy · Manufacturing strategy · Maturity model · Strategy formulation · Action research

---

J. A. Vivares (✉)
Escuela de Ciencias Básicas, Tecnología e Ingeniería, Ingeniería Industrial, Universidad Nacional Abierta y a Distancia, Dosquebradas, Colombia
e-mail: jorge.vivares@unad.edu.co

W. Sarache
Departamento de Ingeniería Industrial, Facultad de Ingeniería y Arquitectura, Universidad Nacional de Colombia, Manizales, Colombia

J. E. Hurtado
Departamento de Ingeniería Civil, Facultad de Ingeniería y Arquitectura, Universidad Nacional de Colombia, Manizales, Colombia

J. L. Garcia Alcaraz et al. (eds.), *Trends in Industrial Engineering Applications to Manufacturing Process*, https://doi.org/10.1007/978-3-030-71579-3_18

18.1 Introduction

Operations strategy (OS) is a long-term plan for production systems (POS), with which lasting competitive abilities are developed to support organizations' general strategies. As a field of study, OS emerged during the 1960s, via the work of Professor Wickham Skinner, who positioned POS as a strategic competitive weapon for organizations, beyond the mere technical character assigned thereto by managers of the time (Skinner 1966, 1969). Traditionally, OS has been analyzed from two perspectives (Leong et al. 1990): content and process. Vivares and Sarache (2019) recently proved the existence of a type of "missing link" in this division and promoted the concept of manufacturing's strategic role as an integrating element, given that studies have found that companies with the highest performance levels (World Class Manufacturers (WCM)) are characterized by their simultaneous consideration of content and process (Brown et al. 2007).

OS content encompasses a group of decisions oriented toward POS intervention for strategy development. This has typically been studied from two aspects (Leong et al. 1990): competitive priorities and strategic decision areas. Competitive priorities are measured with strategic POS performance. The literature review indicates that the most relevant competitive priorities are as follows: cost, quality, flexibility, deliveries, service, and environmental protection (Avella et al. 2011). Hill (1985) divides competitive priorities into winning criteria and order qualifier criteria. The formers are those with which the company wishes to differentiate itself from its strongest competitor, and the latter are those with which they hope to achieve performance at the same level as their competition. For Choudhari et al. (2010), competitive priorities must be understood from both external and internal perspectives. On the external level, these represent performance variables perceived by the client. On the internal level, these should translate into abilities to be developed in the POS.

Strategic decision areas refer to the group of subsystems that conform the POS, and which are the object of intervention, to improve performance. Miltenburg (2005) coined the term "manufacturing levers" to refer to strategic decision areas, as these are subsystems which must be adjusted to support organizations' general strategies. Said author proposes that manufacturing levers include human resources, organization structure and controls, production planning and control, sourcing, process technology, and facilities.

In order to design an OS, a coherent and organized formulation process must be implemented. To this end, a *framework* (Miltenburg 2008), methodology (Hill 2000), or procedure (Dangayach and Deshmukh 2001) that facilitates this being put into practice must be adopted. In other words, a pattern of reasoning that permits structured decision making is required for POS intervention. The strategy formulation process is carried out in function of the content or is based on the identification of the competitive priorities upon which the POS will focus to gain a distinctive advantage. Based thereupon, the manufacturing levers to be used are defined, with strategic manufacturing decisions.

Despite the fact that the literature has been emphatic in its indication of the due linkage between process and content (Hill 2000) to achieve satisfactory results, Vivares and Sarache's (2019) study indicates that manufacturing's strategic role is a relevant component that, unfortunately, has remained absent in the necessary linkage between process and content. Manufacturing's strategic role seeks to establish POS positioning as a competitive weapon for the company, on the strategic level, and has two classical references from the conceptual point of view that, despite their age, have yet to be investigated in depth: Hill (1983) and Hayes and Wheelwright (1984). For example, Hayes and Wheelwright (1984) proposed a four-stage model in which said role may be found: (1) internally neutral, (2) externally neutral, (3) internally supportive, and (4) externally supportive. However, on the topic of said contribution, Barnes and Rowbotham (2003, p. 613) affirmed that "… little practical application of the model is reported in the literature." This same situation was reported by Vivares and Sarache (2019).

Despite the number of contributions that have addressed OS study, the majority of these have focused on content, ignoring the formulation process (Vivares et al. 2015). Said tendency was corroborated in the literature reviews performed by Chatha et al. (2018), Boyer et al. (2005), and Dangayach and Deshmukh (2001). In these contributions, it is confirmed that the OS formulation process requires further investigation, for two main reasons: on the one hand, in OS development, many *frameworks* are possible, and there is no single framework that is best for all company types (da Silveira 2005; Miltenburg 2009). On the other hand, the complexities of each company and competitive environment impede the establishment of a single OS formulation route (Platts et al. 1998; Schroeder and Flynn 2001).

Another challenge in the OS formulation process is uncertainty management. As proposed by Kaplan and Norton (1997), a strategy is a group of causal hypotheses in which it is impossible to have complete certainty about the effects generated by each decision. Effectively, given that the strategic proposal carries with it decisions that require risky long-term investments (e.g., facility capacity increases, product design changes, process automatization, facility relocation, etc.), selecting a POS project intervention portfolio in an uncertain situation is a necessary condition for OS formulation. This condition, however, has rarely been addressed in previous investigations.

In the literature, several contributions have been identified which are oriented toward the OS formulation process and permit the identification of research opportunities. Jia and Bai (2011) performed a case study in which they propose a focus based on quality function deployment (QFD) for OS development. However, not all uncertainty content variables were included or considered for decision making. Lindström and Winroth (2010) proposed a methodology to select the level of OS machine automatization. However, no formulation process was addressed in the strictest sense. Miltenburg's (2005, 2008, 2009) research proposed a framework for OS deployment, which facilitates the alignment between strategic manufacturing decisions (manufacturing levers) and performance dimensions associated with competitive priorities (manufacturing outputs). However, this same author recognizes that the proposed framework lacks empirical validation and suggests research in that vein.

Generally, literature reviews permit the identification of OS formulation process research opportunities. Concretely, the few advances identified present two main weaknesses: (1) the majority are limited to conceptual proposals that lack empirical validation, and (2) they fail to incorporate uncertainty, as a matter of relevance, in OS formulation.

As such, the objective of the present investigation is to develop and validate a *framework* for OS formulation. The proposed framework incorporates a group of procedures and mathematic treatment to guide decision making in uncertain conditions. This occurs in two phases. In the first, the POS maturity level is evaluated, in terms of its performance in competitive priorities, manufacturing lever capabilities, and the strategic role that management assigns to the manufacturing area. In this phase, the maturity model proposed by Vivares et al. (2018) was employed. In Phase 2, a group of intervention (improvement) decisions was deployed from the manufacturing levers, based on a group of strategic projects selected in function of uncertainty (potential impact on manufacturing levers) and budget limitations. All this was done to improve POS performance levels.

In order to validate its feasibility and utility, the *framework* was applied in a manufacturing company, via an action investigation process. The action investigation focus has been promoted for various decades by Meredith et al. (1989) and Westbrook (1995), as an investigative paradigm that permits the achievement of useful contributions to the solution of real problems (Avella and Alfaro 2014). According to Zhang et al. (2015), there are very few studies performed in OS using this investigative perspective. For this process, it was necessary participate in an eight-month immersion in the company, and to interact with the employees and management in an organized fashion, by way of workshops, meetings, and interviews.

The validation process with company personnel indicates that the framework proposed is applicable, and that as such, it may constitute a useful tool with which to support the OS formulation process. Thus, a double contribution occurs: a contribution is generated for the OS formulation process, by way of an empirically validated framework that may also be replicated in the future investigations, and a useful, practical tool, which may be applied in manufacturing companies, is provided.

The remainder of this chapter is organized as follows: In Sect. 18.2, the structure and components of the proposed framework are described. Section 18.3 presents the results of its validation in a manufacturing company. In Sect. 18.4, results are discussed, and finally, in Sect. 18.5, the most relevant conclusions and future topics for investigation are presented.

18.2 Proposed Framework

In this section, the structure of the proposed framework is presented, and certain methodological clarifications, which were considered for its application, are made.

18.2.1 Framework Description

The *framework* developed in the present study is presented in Fig. 18.1 and is composed of two phases: (1) POS evaluation, based on current maturity level, and (2) strategy formulation within the framework of a continuous improvement process. Phase 1 is supported by Vivares et al. (2018), a contribution derived from Miltenburg's research (2005, 2008, 2009). In order to establish the baseline, three groups of variables associated with OS content and process must be studied: competitive priorities (CP), manufacturing levers (ML), and manufacturing's strategic role (RO).

Of the competitive priorities, the following were considered: cost, quality, flexibility, innovation, deliveries, service, and environmental protection. In the case of the two manufacturing levers, human resources, organizational structure and culture, supply and distribution, planning and production control, process, facility, and management and support subsystem technology were considered. In terms of the strategic role of manufacturing, the degree of protagonism that upper management assigned to the POS was evaluated. Based on these three groups of variables, the POS may be classified in one of the five established maturity levels (see Table 18.1): preinfantile, infantile, industry average, adult, and World Class Manufacturing (WCM). A detailed description of these levels may be found in Vivares and Sarache (2019).

In order to classify the POS on any of these levels, a maturity index must be calculated for the time of k planning (MI_k) (see Eq. 18.1). Similarly, this maturity index is added from the calculation of three sub-indices: the competitive priorities index (MI_{CP}), manufacturing lever index (MI_{ML}), and manufacturing's strategic

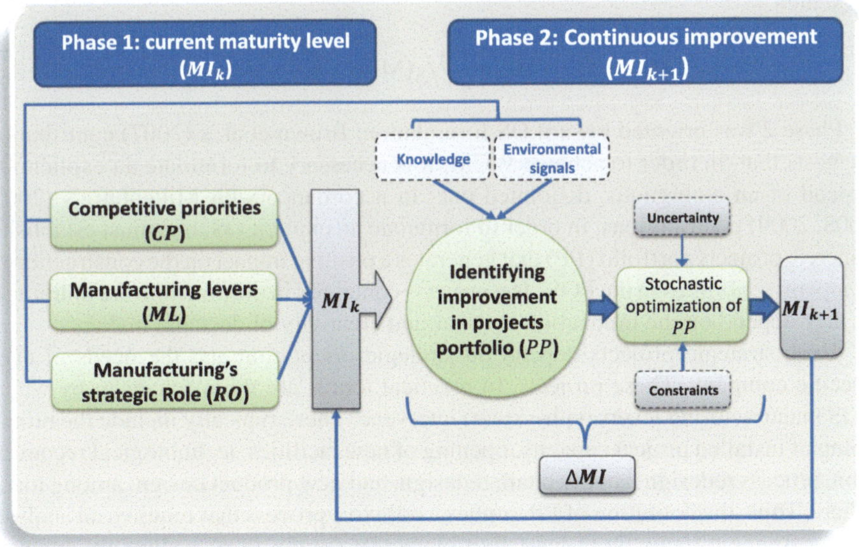

Fig. 18.1 Proposed framework for OS formulation

Table 18.1 Maturity levels for POS evaluation (Phase 1)

Level	Description
Preinfantile	Very low. POS with many structural weaknesses, including problems complying with legal provisions, chaotic system
Infantile	Low. POS is poorly developed, with a low technological level. Complies with some legal provisions
Industry average	Average. POS similar to the industry average, sufficient to survive, but not to stand out
Adult	Good. POS performance is slightly superior to the industry average and/or direct competition, achieving certain prestige in the market
World Class Manufacturing	Excellent. The POS stands out, owing to its innovative character, following technological advances. The company stands out among the best in the world

Table 18.2 POS evaluation levels (Phase 1)

Maturity level	Rank
Preinfantile	$0 \leq MI_k \leq 30$
Infantile	$30 < MI_k \leq 50$
Industry average	$50 < MI_k \leq 70$
Adult	$70 < MI_k \leq 90$
World Class Manufacturing	$90 < MI_k \leq 100$

role (MI_{RO}). In Table 18.2, the conditions for POS maturity index classification are presented.

$$MI_k = {}^1/_3(MI_{CP}) + {}^1/_3(MI_{ML}) + {}^1/_3(MI_{RO}) \tag{18.1}$$

Phase 2 was oriented toward OS formulation. Brown et al.'s (2007) contribution suggests that, in order to achieve WCM, it is necessary to formulate an explicit OS instead of an ambiguous, disjointed one. In accordance with Miltenburg's (2005, 2008, 2009) contributions, in order to formulate an explicit OS, one must establish a concrete projects portfolio (PP) that generates a positive impact on the construction of competitive abilities in the POS. The projects contained in the portfolio are unlimited, as they depend on the innovation abilities and creativity of decision makers.

Thus, strategic projects depend on strategic orientation and the needs of each specific company. These projects, in practical terms, are the mechanism by which POS manufacturing levers (subsystem) intervene. These typically include the broadening of installed project capacity, opening of new facilities, technological reconversion, process redesign, supply chain redesign, and new product design, among many others. Thus, the definition of PP implies a reflexive process that requires an analysis with a strategic focus on at least four elements: (a) the POS evaluation obtained, by way of MI_k, (b) the theoretical knowledge (education), and practical knowledge (experience) of management (decision makers), (c) environmental signals (threats and opportunities), and (d) company policies.

Theoretically, the successful execution of PP must provoke a change in maturity index (ΔMI), or the POS′ competitive capacity. Initially, it is expected that the change be positive. However, the effect of a project may not be that which is expected, owing to uncertainty. This is normal in long-term decisions. Strategic-level projects generally imply large investments. Owing to uncertainty, however, the real costs may be different than those planned. A more expensive project, then, does not necessarily generate greater impact on POS performance than a less expensive project. Further, the company may have financial restrictions for the execution of all PP projects. In order to address uncertainty and financial restrictions, the *framework* developed involves a stochastic optimization model, which is described below.

The maturity index change (ΔMI) occurs because each project P_i affects e_i system maturity improvement. However, the effect is uncertain, for which reason a random component may be incorporated into the coefficients (denoted in this study as a function of ω), such that the effect of project P_i on maturity would be given by the function $e_i(\omega)$. Thus, an improved maturity index could be expected $k + 1$ (MI_{k+1}) following strategy implementation (see Eq. 18.2). As such, the cost of a project i (c_i) has an uncertainty burden that, as in the case of the effects, may be modeled by incorporating a random component to the coefficients (ω), such that the cost of a project P_i would be given by the function $c_i(\omega)$.

$$\mathrm{IM}_{k+1} = \mathrm{IM}_k + \Delta\mathrm{IM} = \mathrm{IM}_k + \Sigma_i\, e_i(\omega) \tag{18.2}$$

Using this logic, OS formulation seeks the maximization of ΔIM through the prioritization of a project portfolio that minimizes uncertainty and is adjusted to a limited budget.

The equation system shown below presents the summarized version of the stochastic optimization model associated with the *framework* proposed. The decision variable is P_i ($P_i = 1$ if project P_i is selected for OS formulation, and $P_i = 0$ in the opposite case). The objective function presented in Eq. (18.3) and its restrictions are described below.

$$\text{Objective function: Max } Z = \Delta\mathrm{MI} \tag{18.3}$$

The first restriction is related to the available budget (Eq. 18.4), in which the company has an investment threshold K for strategy implementation. Each project P_i has a cost $c_i(\omega)$ that is a function of a random cost, to model uncertainty.

$$\Sigma_i\, P_i \times c_i(\omega) \le K \tag{18.4}$$

Equation (18.5) represents an exclusion restriction, given a hypothetical case in which there are various alternatives or projects to meet an objective, such that EXC is a group of ordered pairs of projects $P(a, b)$ such that project P_a and project P_b are mutually exclusive (one of the two must be chosen). In other words, if one chooses to implement one of them, the other is discarded.

$$P_a + P_b \leq 1, \quad \forall(a, b) \in \text{EXC} \tag{18.5}$$

Equation (18.6) describes a dependence relationship between any two projects. In other words, it is assumed that, in certain cases, the results of a project depend on the results of another, for example, where a macroproject is concerned. In this scenario, DEP represents a group of ordered pairs of projects $P(m, n)$ such that project P_m depends upon project P_n.

$$P_m \geq P_n, \quad \forall(m, n) \in \text{DEP} \tag{18.6}$$

Equation (18.7) models an inclusion relationship between projects. This occurs when there are interdependence relationships or mutual *feedback* between projects, for which reason, if one projects is selected to be implemented, the other is as well. Thus, INC represents a group of ordered pairs of projects $P(y, z)$ such that projects P_y and P_z are mutually inclusive.

$$P_y = P_z, \quad \forall(y, z) \in \text{INC} \tag{18.7}$$

Lastly, Eqs. (18.8), (18.9) and (18.10) permit the restriction of the value of the decision variable P_i so that it may be binary (zero or one).

$$P_i \geq 0 \tag{18.8}$$

$$P_i \leq 1 \tag{18.9}$$

$$P_i \in N \tag{18.10}$$

The solution to the model is performed via a process with n iterations, such that functions $e_i(\omega)$ and $c_i(\omega)$ assume specific values in each of the iterations, to model the uncertainty of the effect on maturity as well as the uncertainty of project costs. The OS is formulated by way of the group of projects chosen to be implemented, on resolution of the optimization model. Given that the optimization model resolves n times, different combinations of possible projects (or possible strategies) may be obtained. There the deciding party considers the behavior of each possibility encountered. OS formulation concludes with an implementation schedule, in which projects are scheduled over time, thus representing a strategic route for POS improvement.

18.2.2 Framework Validation Strategy

The validation of the proposed framework occurs in accordance with the principles of action investigation. To this end, one must link the company that is the object of study in an early phase of investigation (Avella and Alfaro 2014), such that, from the beginning, it may be confirmed that the scientific problem is simultaneously a practical problem in the real world. Action investigation could be considered a variant of the case study (Westbrook 1995), in which criteria must be defined to select the company that is the object of study. Based on the contribution of Yin (2014), the criteria presented in Table 18.3 were established.

Framework validation occurs on the part of upper management, line managers, and certain middle managers, as necessary, via the evaluation of the results obtained. For this, anonymous surveys are applied, based on a group of evaluation criteria, using a five-level scale (1 = Totally disagree, 5 = Totally agree).

18.3 Results

Based on the criteria in Table 18.3, a manufacturing company located in the so-called Colombian coffee region was selected. The company forms part of the metal-mechanic sector (production of tools for the agricultural and construction sectors), manufactures 11 groups of products, and exports 30% of their sales. *Framework*

Table 18.3 Criteria for company selection

Criteria	Description
Sector	The company must form part of the industrial sector in a relevant subsector for regional and national industry
Organizational structure	The company must have an organizational structure with upper- and middle-management committed to long-term POS planning
Attitude and accessibility	Company with attitude of improvement, willing to strengthen the university-company relationship
Availability and participation	The company must be willing to dedicate time to the process, and involve upper-management, middle-management, and eventually, operative personnel
Permission to reveal results	The company must be willing to permit the dissemination of scientific results, with the guarantee of information confidentiality
Safety and entrance prerequisites	The company must provide the researcher with the safety elements necessary to enter the production plant

application occurred by way of a collaborative process between researchers and a work team selected by the company, and lasted for eight months. During said period, all activities necessary for POS evaluation and OS formulation occurred. To this end, the research team kept in permanent contact with the company team, provided training to strengthen their theoretical OS knowledge, interacted in regular meetings, and performed workshops on strategic thought.

In line with the action investigation focus, the results were divided into two groups: (1) the results of *framework* application in the company that was the object of study (strategy formulation), and (2) *framework* validation on the part of company upper-management and middle-management team. With this scheme, the academic and practical contributions of the framework proposed herein were verified.

18.3.1 Strategy Formulation

On following the procedures and calculations for the implementation of Phase 1 (POS evaluation—baseline), the results presented in Table 18.4 were obtained. The POS scored a maturity index of 68.0, which places it within the industry average level. This evaluation constituted the baseline upon which the OS would be formulated.

Next, a portfolio was identified with 10 projects (PP) that were potentially eligible for OS formulation. Their interrelationships and costs were included as well, in accordance with a uniform distribution function to model $c_i(\omega)$ (see Table 18.5). For example, project P_1 (implementation of a total productive maintenance (TPM) program) has no alternatives or mutually exclusive projects. In accordance with the strategic analysis performed by the company, results with this kind of project depend

Table 18.4 General Phase 1 (POS evaluation) results

Indices	Index description	Value	Maturity level
MI_{CP}	Competitive priorities	68	Industry average
MI_{ML}	Manufacturing levers	67	Industry average
MI_{RO}	Manufacturing's strategic role	69	Industry average
MI_k	Global POS at the time k	68	Industry average

Table 18.5 General information about the project portfolio (PP)

P_i	Project	Relationships between projects			Cost $(c_i(\omega))$ (millions-COP)[d]	
		A[a]	B[b]	C[c]	Min	Max
P_1	Implementation of a total productive maintenance (TPM) program		5, 7, 8, and 9		321	612
P_2	Application of a single-minute exchange of dies (SMED) system for reference changes		5, 9, 1, and 7		45	87
P_3	Technical reinforcement for production planning		5		40.5	54
P_4	Reinforcement of the internal and external institutional image, audiovisually		5		10	15
P_5	Reinforcement of organizational culture			7	60	130
P_6	Environmental performance positioning		5, and 9	10	25	40
P_7	Effective implementation of the eight disciplines			5	40.6	57.2
P_8	Strengthening of communication and synergy between areas		5, and 7		33	52
P_9	Effective implementation of the five S		5	10	35	54
P_{10}	Reinforcement of safety strategy		5, and 7	6, and 9	38	77
Total					648.1	1178.2

Note [a]Type A projects (mutually exclusive). [b]Type B projects (dependent). [c]Type C projects (inclusive). [d]Limits [A, B] for uniform probability distribution on the implementation of projects P_5, P_7, P_8, and P_9. Projects inclusive with P_1, however, were not found. Considering the cost uncertainty present, the company established that the cost of this project could fluctuate between $COP 321 and $COP 612 million. The modeling of the uncertainty of the effect of the projects, on maturity and its respective function $e_i(\omega)$, is presented in Appendix.

On review of the company's budget, upper management proposed four scenarios for the investment threshold (K): 360, 540, and 720 million, and 1.08 billion-COP. Note that, if all projects in the portfolio are implemented, the investment to be made could fluctuate between $COP 648.1 and $COP 1178.2 million (see Table 18.5). In other words, even in the best-case investment scenario, they may not have sufficient funds to implement all projects. Considering the above, the particular stochastic optimization model generated for this company begins with the objective function reviewed in Eq. (18.11), and has its first restriction in Eq. (18.12).

$$\text{Max } Z = \Delta \text{MI} \tag{18.11}$$

$$\Sigma_i\, P_i \times c_i(\omega) \leq K, \quad \forall K \in \{360, 540, 720\,\text{million}, \text{ and } 1.08\,\text{billion}\} \tag{18.12}$$

Equations (18.13)–(18.21) model dependence relationships between projects. Equations (18.22)–(18.24) model the condition of mutually inclusive projects, and Eqs. (18.25)–(18.27) permit the restriction of the decision variable value P_i (zero or one).

$$P_9 \geq P_1 \tag{18.13}$$

$$P_8 \geq P_1 \tag{18.14}$$

$$P_9 \geq P_2 \tag{18.15}$$

$$P_1 \geq P_2 \tag{18.16}$$

$$P_7 \geq P_2 \tag{18.17}$$

$$P_5 \geq P_3 \tag{18.18}$$

$$P_5 \geq P_4 \tag{18.19}$$

$$P_5 \geq P_8 \tag{18.20}$$

$$P_5 \geq P_{10} \tag{18.21}$$

$$P_5 = P_7 \tag{18.22}$$

$$P_6 = P_{10} \tag{18.23}$$

$$P_9 = P_{10} \tag{18.24}$$

$$P_i \leq 1 \tag{18.25}$$

$$P_i \geq 0 \tag{18.26}$$

$$P_i \in N \tag{18.27}$$

where

P_i	Project i may be eligible for OS formulation.
$c_i(\omega)$	random cost function for project P_i.
K	available budget for OS implementation (four scenarios: 360, 540, 720 million, and 1.08 billion-COP).

The model was solved in $n = 10{,}000$ iterations for each one of the four investment scenarios (360, 540, 720 million, and 1.08 billion-COP). In Fig. 18.2, the histograms obtained for the maturity increase (ΔMI) over the 10,000 iterations, in the different scenarios, are presented. The presence of a combed histogram suggests different solutions for OS formulation, which ratifies the information presented in Table 18.6. For example, in Scenario 1, eight possible solutions or project combinations were identified, the first of which repeats 4025 times during the 10,000 iterations and indicates that the company OS must implement projects P_3, P_4, P_5, P_6, P_7, P_9, and P_{10}. With this strategy, projects P_1, P_2, and P_8 would enter into a project bank until investment budget uncertainty was resolved, which would depend on future company cash flow. Lastly, in Fig. 18.3, curves that reflect the probability of obtaining improved maturity indices are presented (MI_{k+1}). In this figure, observe the high probability of achieving slightly superior maturity indices than the current one. It gradually decreases in higher maturity indices.

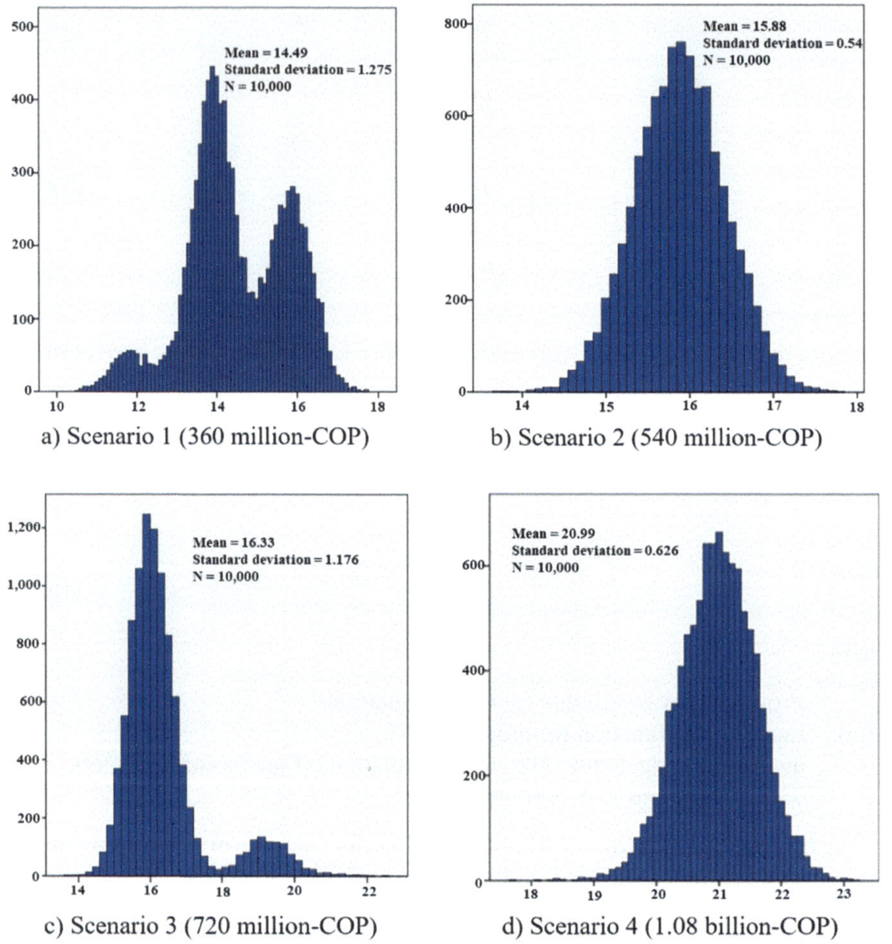

Fig. 18.2 Histograms obtained in the four investment scenarios

The results obtained provide multiple alternatives for OS formulation. Said information was discussed with company management, and it was decided that Scenario 2 would be approved, as it was the most probable (Scenario 1 was pessimistic, and Scenario 3 was optimistic). OS formulation in this scenario implies ruling out the implementation of projects P_1 and P_2. Figure 18.4 presents the sequence and schedule for implementation of the selected portfolio projects. Those in Fig. 18.5 project an improved maturity index (MI_{k+1}) that could be achieved on complete strategy implementation. Thus, if implementation of the selected project occurs successfully, it is expected that the POS maturity index would move from 68 (industry average) to 83 (adult).

Table 18.6 Solutions found for OS formulation in the different investment scenarios

Scenario	Solution	Possible OS formulation solutions										Frequency	Total
		P_1	P_2	P_3	P_4	P_5	P_6	P_7	P_8	P_9	P_{10}		
1 (360 million-COP)	1	0	0	1	1	1	1	1	0	1	1	4025	10,000
	2	0	0	1	0	1	1	1	1	1	1	1523	
	3	0	0	1	0	1	1	1	0	1	1	1199	
	4	0	0	1	1	1	1	1	1	1	1	2227	
	5	0	0	0	0	1	1	1	1	1	1	337	
	6	0	0	1	1	1	0	1	1	0	0	580	
	7	0	0	0	1	1	1	1	1	1	1	53	
	8	0	0	0	1	1	1	1	0	1	1	56	
2 (540 million)	1	0	0	1	1	1	1	1	1	1	1	10,000	10,000
3 (720 million-COP)	1	0	0	1	1	1	1	1	1	1	1	7816	10,000
	2	1	0	1	0	1	1	1	1	1	1	326	
	3	1	0	1	1	1	1	1	1	1	1	678	
	4	1	0	0	1	1	1	1	1	1	1	1033	
	5	1	0	0	0	1	1	1	1	1	1	87	
	6	1	1	0	0	1	1	1	1	1	1	9	
	7	1	1	1	0	1	1	1	1	1	1	33	
	8	1	1	0	1	1	1	1	1	1	1	2	
	9	1	1	1	1	1	1	1	1	1	1	16	
4 (1.08 billion-COP)	1	1	1	1	1	1	1	1	1	1	1	9853	10,000
	2	1	1	1	0	1	1	1	1	1	1	88	
	3	1	0	1	1	1	1	1	1	1	1	59	

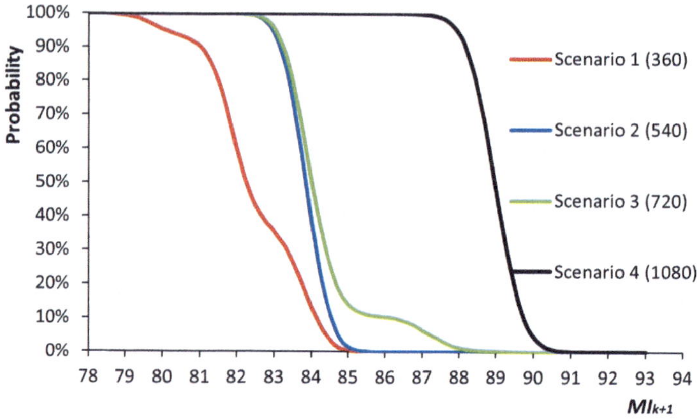

Fig. 18.3 Probability curves by scenario

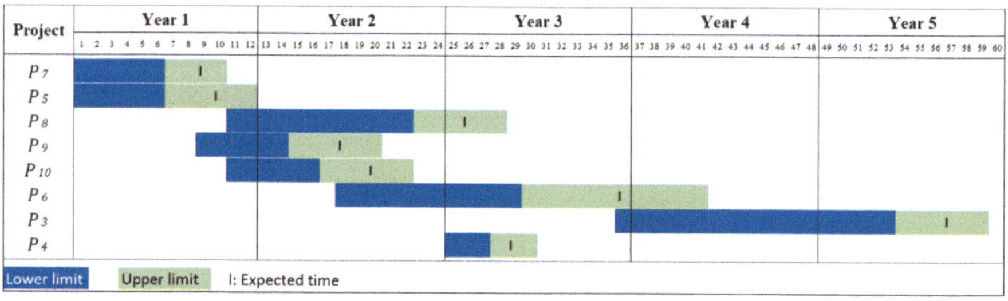

Fig. 18.4 Implementation schedule for strategic projects that support OS

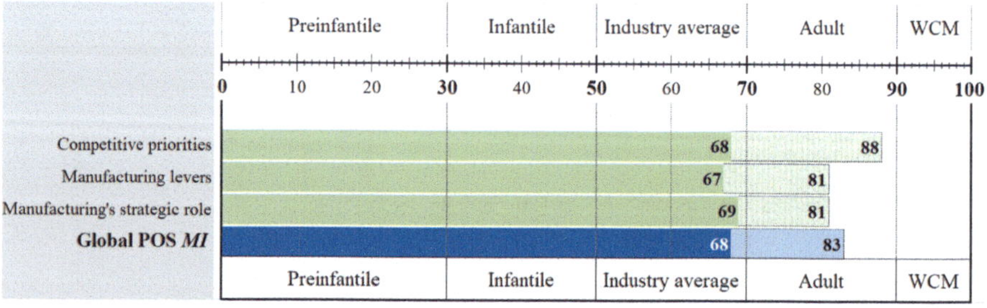

Fig. 18.5 Current and expected (MI_k) maturity indices (MI_{k+1})

18.3.2 Framework Validation

The results presented in the previous section show the specific OS formulation for the company participating in the project, which corroborates the feasibility of framework application in the real world. Given that its application requires a process, the performance of workshops, routines, and a mathematical model for the obtention of results, in this section, a framework validation, based on participant perceptions, is presented. As such, 16 statements were submitted to evaluation (anonymously) by employees and managers who participated in the process. Said statements were divided into two groups: evaluation of the specific application results in the company, and knowledge transfer actions performed during the action investigation process. Participants indicated their level of agreement on a scale of one to five (1 = Totally disagree, 5 = Totally agree). With the scale used, a score of three (3) meant a neutral opinion about the developed framework for OS formulation, while a score of 4 was suitable to demonstrate framework quality. Considering that not all participants responded to the survey, a t-test was performed to evaluate the following hypotheses: H_0: $\mu \geq 4.0$, H_1: $\mu < 4.0$. Based on Table 18.7, the framework and knowledge transfer suitability performed on the company throughout the investigation was confirmed.

18.4 Discussion and Further Research

OS formulation has remained an open problem in the literature for decades. On this topic, Platts et al. (1998) proposed that companies needed focuses to be feasible and useful for OS formulation in the real world. The present contribution demonstrates that it is still possible to continue making novel contributions to a field as mature as OS, particularly in the formulation process. Based on these results, four guiding lines may be emphasized to guide the discussion.

Firstly, the replication of Vivares et al.'s model (2018) for POS maturity model evaluation corroborated its utility as a baseline for OS formulation. In accordance with the evaluation performed by participating employees, in terms of the results obtained, the maturity index is comprehensible, adequately reflects the company reality and is worth using in other organizations to evaluate integral POS performance. Considering that, in this investigation, only one company participated, the framework application should be extended by way of new case studies, in order to broaden the sample and improve its robustness.

Secondly, the present study shows that these projects are quintessentially vehicles to formulate and implement OS in practice, given that they permit the materialization of analyses and policies in concrete improvement actions that facilitate the mapping of a strategic route to be followed. However, the different foci available in the scientific literature have rarely addressed this perspective, and when they have, quantitative analyses for its prioritization and selection have been practically null. Future investigation could delve further, theoretically, into this concept and add other variables to improve attributes modeled in this contribution (cost and relationships between projects).

Table 18.7 Result evaluation by company practitioners

Evaluation criterion	Mean	Median	Standard Deviation	Test (H_0: $\mu \geq 4$)	
				t	Decision
Results obtained					
The maturity index reflects the company reality	4.022	4.000	0.746	0.206	H_0
The maturity index is understandable	4.333	4.000	0.440	5.254	H_0
The maturity index may be useful and valuable for the evaluation of POS in other companies	4.422	4.500	0.621	4.709	H_0
The policies that resulted in strategy formulation are accurate and valuable	4.033	4.000	0.777	0.291	H_0
The solutions found for the different investment scenarios are reasonable and useful to support decision making	4.115	4.000	0.497	1.597	H_0
The improved maturity index projection is reasonable and useful to support decision making	4.063	4.000	0.649	0.667	H_0
Global *framework* application was enriching and provided sufficient elements with which to formulate the OS	4.362	4.500	0.508	4.937	H_0
The framework is valuable and has potential to be applied in other companies for OS formulation	4.340	4.500	0.618	3.818	H_0
The researchers fulfilled their roles and effectively supported the company	4.686	5.000	0.450	10.558	H_0

(continued)

Table 18.7 (continued)

Evaluation criterion	Mean	Median	Standard Deviation	Test (H_0: $\mu \geq 4$)	
				t	Decision
Knowledge transfer					
The quality of the knowledge transferred was satisfactory	4.661	5.000	0.472	7.400	H_0
The topics analyzed broadened our view of POS on the strategic level	4.643	5.000	0.427	7.962	H_0
The project prompted me to learn more, in the future, about the topics covered	4.625	5.000	0.571	5.789	H_0
We learned about important concepts for the improvement of strategic POS management	4.554	4.500	0.550	5.325	H_0
The knowledge transferred is important for any company seeking to formulate an OS	4.643	5.000	0.507	6.715	H_0
Researchers had mastered the topics and were competent to transfer said knowledge	4.839	5.000	0.306	14.517	H_0
The proper methodology was employed, generally	4.607	5.000	0.533	6.024	H_0

As a third line of discussion, it may be said that the results presented align with the conclusions of the worldwide macroproject high performance manufacturing, in accordance with which there is no universal strategy for companies. As such, it is necessary for each one to formulate their own strategic route, considering aspects such as their specific situation, country, industry, and corporate strategy, among others (Schroeder and Flynn 2001). This premise is consistent with the results obtained, as in the different investment scenarios, multiple solutions (project combinations) were found for OS formulation. Scenario 3 provided the highest number of possibilities (nine possible solutions or project combinations).

Lastly, the stochastic optimization model yielded good, practical results, in addition to being a contribution that had not been previously employed as an alternative for OS formulation. Practitioners indicated that the solutions encountered in the investment scenarios were reasonable and useful to support decision making. The utility of the improved maturity index was also confirmed (MI_{k+1}) as was its applicability in the framework proposed. For future investigation, it is necessary to incorporate the time required to implement projects in the optimization model, as some companies may consider this a relevant restriction.

The present study made both academic and practical contributions, which is consistent with the action investigation paradigm. From the scientific point of view, the findings presented motivate future studies to replicate and improve the *framework* proposed. From a practical point of view, the participating employees highlight the value and utility of the proposed framework to guide OS formulation in the real world.

18.5 Conclusions

The development of lasting competitive capacities in the production/operations system (POS) requires that companies follow a long-term improvement route, via the formulation of an operations strategy (OS). The OS formulation process has remained open in the literature for various decades because many focuses are possible, and those which exist still provide improvement possibilities. In the present investigation, a framework was developed for OS formulation, which has three distinctive values: (a) it uses a recent development to evaluate the maturity of the POS, (b) promotes the identification of a project portfolio to solidify strategy, and (c) mathematically models the problem of uncertainty, which permits prioritization of projects for OS formulation.

The framework, developed in two phases, proved its utility in the real world. Its focus permitted the conclusion that, in order to formulate an OS, it is vital to establish a project portfolio for POS improvement. Similarly, it was established that the uncertainty inherent to decision making, especially that related to cost and effect on project maturity, may be modeled and incorporated into the OS formulation process. However, the creation of new applications for the proposed framework remains an open field of study.

It is also important to mention that the study was performed within the action investigation paradigm, via a process which implied collaboration between researchers and practitioners and sought a double academic and practical contribution. Action investigation has been promoted for various decades as a legitimate OS paradigm, and its use has been quite rare in previous investigations. However, in this study, its utility was confirmed for investigations with great impact on specific organizational problems.

Appendix: Project Effects (Example with Two Projects)

In the following table, a sample of the effect of projects on maturity, modeling function $e_i(\omega)$ as a triangular probability distribution, is presented. Only the first three projects are presented with their effect based on pessimistic (P), realistic (R), and optimistic (O) values.

Variable structure			MI_k	MI_{k+1}								
Latent variables		Items		P_1			P_2			P_3		
Level 2	Level 1			P	R	O	P	R	O	P	R	O
Competitive priorities	Cost	1	70	72	75	80	70	75	80	75	80	80
		2	66	73	80	85	68	74	78	66	75	80
	Quality	3	76	76	79	85	76	78	80	76	78	80
		4	68	75	75	86	69	73	76			
		5	79									
	Flexibility	6	63	63	66	70	66	74	77	68	73	82
		7	66	66	69	72	66	72	75	70	75	80
	Innovation (products)	8	68				68	70	72			
		9	64	64	66	67	64	65	66			
	Deliveries	10	69	69	73	82	71	75	78	70	76	80
		11	60	62	72	75	63	67	71	65	75	80
	Service	12	68									
		13	70									
		14	65									
	Environmental protection	15	66	67	73	78	68	70	72			
		16	47									
Human resources	Personnel talent and competencies	17	73	74	77	82						
		18	74	76	82	85						
	Personnel satisfaction and climate	19	80	82	84	90				80	81	82
		20	76	80	84	90						

(continued)

(continued)

Variable structure			MI_k	MI_{k+1}								
Latent variables		Items		P_1			P_2			P_3		
Level 2	Level 1			P	R	O	P	R	O	P	R	O
		21	76	80	81	85				76	78	80
	Human resource policies and practices	22	69	40	50	55						
		23	61	65	68	78						
		24	64									
		25	63	63	74	77						
		26	64	64	66	70						
		27	66	68	75	80						
		28	71	71	75	80	70	75	80	71	75	80
	Work health and safety	29	73	76	78	84						
Structure and culture	System organizational structure	30	67									
		31	63	63	66	80	60	69	73	65	75	80
		32	65	65	68	80						
		33	67	70	75	84						
	Organizational culture	34	65	66	71	76						
		35	77	77	80	85						
Supply and distribution	Inventory management	36	52							52	60	65
		37	66				65	70	73	66	70	75
	Vertical integration	38	85									
		39	55									

(continued)

(continued)

Variable structure			MI_k	MI_{k+1}								
Latent variables		Items		P_1			P_2			P_3		
Level 2	Level 1			P	R	O	P	R	O	P	R	O
	Provider management	40	57							60	65	70
		41	75									
	Chain (or network) coordination	42	70							70	72	75
Production planning	Medium-term planning	43	48							54	60	65
	Production activity scheduling	44	60	62	65	68	62	64	68	64	70	75
		45	59	64	71	81	64	77	82	62	65	70
		46	73							75	80	82
		47	60							65	70	75
Process technology	Production capacity	48	100									
	Know-how	49	65	66	80	89				65	67	70
	Machinery and equipment	50	58	58	60	65						
	Work study	51	68	68	73	78						
		52	65	68	73	75	65	69	71			
		53	57	57	60	62						
	Information and communications technology (ICT)	54	62	62	64	72				62	65	70
		55	59							59	62	65
Facilities	Plant infrastructure	56	69	69	75	82						

(continued)

(continued)

Variable structure			MI_k	MI_{k+1}								
Latent variables		Items		P_1			P_2			P_3		
Level 2	Level 1			P	R	O	P	R	O	P	R	O
		57	75	77	82	85						
		58	79									
	Layout	59	58									
	Localization	60	80									
Management and support subsystems	Quality engineering and management	61	52									
		62	62									
		63	59	60	66	80						
	Product development	64	90									
		65	80									
		66	70									
	Maintenance management	67	70	80	85	89						
		68	63	80	85	89				63	65	70
		69	55	80	85	89						
	Environmental management	70	65									
	Performance evaluation	71	64									
Manufacturing's strategic role	Strategic manufacturing environment	72	80	80	82	85						
		73	73	73	75	80						
		74	73									

(continued)

(continued)

Variable structure			MI_k	MI_{k+1}								
Latent variables		Items		P_1			P_2			P_3		
Level 2	Level 1			P	R	O	P	R	O	P	R	O
		75	78									
	Continuous improvement	76	58	59	61	63						
		77	62	62	67	72				62	65	70
		78	72							72	75	80
		79	61				62	65	70			

References

Avella L, Alfaro JA (2014) Spanish University Business Chairs used to increase the deployment of Action Research in Operations Management: a case study and analysis. Action Res 12(2):194–208. https://doi.org/10.1177/1476750314528010

Avella L, Vazquez-Bustelo D, Fernandez E (2011) Cumulative manufacturing capabilities: an extended model and new empirical evidence. Int J Prod Res 49(3):707–729. https://doi.org/10.1080/00207540903460224

Barnes D, Rowbotham F (2003) Developing a questionnaire for the four-stage model of operations strategy. Prod Plan Control 14(7):613–622. https://doi.org/10.1080/09537280310001626205

Boyer KK, Swink M, Rosenzweig ED (2005) Operations strategy research in the POMS journal. Prod Oper Manag 14(4):442–449

Brown S, Squire B, Blackmon K (2007) The contribution of manufacturing strategy involvement and alignment to world-class manufacturing performance. Int J Oper Prod Manag 27(3):282–302. https://doi.org/10.1108/01443570710725554

Chatha KA, Butt I, Sadiq MS, Arshad M (2018) Theoretical developments in empirical quantitative manufacturing strategy literature. Int J Oper Prod Manag 38(1):183–210. https://doi.org/10.1108/IJOPM-08-2016-0486

Choudhari SC, Adil GK, Ananthakumar U (2010) Congruence of manufacturing decision areas in a production system: a research framework. Int J Prod Res 48(20):5963–5989. https://doi.org/10.1080/00207540903164644

da Silveira GJC (2005) Market priorities, manufacturing configuration, and business performance: an empirical analysis of the order-winners framework. J Oper Manag 23(6):662–675. https://doi.org/10.1016/j.jom.2005.01.005

Dangayach GS, Deshmukh SG (2001) Manufacturing strategy: literature review and some issues. Int J Oper Prod Manag 21(7):884–932. https://doi.org/10.1108/01443570110393414

Hayes RH, Wheelwright SC (1984) Restoring our competitive edge: competing through manufacturing. Wiley, New York

Hill TJ (1983) Manufacturing's strategic role. J Oper Res Soc 34(9):853–860

Hill T (1985) Manufacturing's strategy—the strategic management of the manufacturing function. Macmillan, London

Hill T (2000) Manufacturing strategy: text and cases. McGraw-Hill, Boston

Jia GZ, Bai M (2011) An approach for manufacturing strategy development based on fuzzy-QFD. Comput Ind Eng 60(3):445–454. https://doi.org/10.1016/j.cie.2010.07.003

Kaplan RS, Norton DP (1997) El cuadro de mando integral: the balanced scorecard. Gestión 2000, Bogotá

Leong GK, Snyder DL, Ward PT (1990) Research in the process and content of manufacturing strategy. Omega 18(2):109–122. https://doi.org/10.1016/0305-0483(90)90058-H

Lindström V, Winroth M (2010) Aligning manufacturing strategy and levels of automation: a case study. J Eng Tech Manage 27(3):148–159. https://doi.org/10.1016/j.jengtecman.2010.06.002

Meredith JR, Raturi A, Amoako-Gyampah K, Kaplan B (1989) Alternative research paradigms in operations. J Oper Manag 8(4):297–326. https://doi.org/10.1016/0272-6963(89)90033-8

Miltenburg J (2005) Manufacturing strategy: how to formulate and implement a winning plan, 2nd edn. Productivity Press, New York

Miltenburg J (2008) Setting manufacturing strategy for a factory-within-a-factory. Int J Prod Econ 113(1):307–323. https://doi.org/10.1016/j.ijpe.2007.09.001

Miltenburg J (2009) Setting manufacturing strategy for a company's international manufacturing network. Int J Prod Res 47(22):6179–6203. https://doi.org/10.1080/00207540802126629

Platts KW, Mills JF, Bourne MC, Neely AD, Richards AH, Gregory M (1998) Testing manufacturing strategy formulation processes. Int J Prod Econ 56–57:517–523. https://doi.org/10.1016/S0925-5273(97)00134-5

Schroeder RG, Flynn BB (2001) High performance manufacturing. Global perspectives. Wiley, New York

Skinner W (1966) Production under pressure. Harv Bus Rev 44(6):139–146
Skinner W (1969) Manufacturing-missing link in corporate strategy. Harv Bus Rev 47(3):136–145
Vivares JA, Sarache W (2019) Manufacturing's strategic role and management practices: evidence from Colombian companies. In: García Alcaraz JL, Rivera Cadavid L, González-Ramírez RG, Leal Jamil G, Chong Chong MG (eds) Best practices in manufacturing processes: experiencies from Latin America, pp 325–345. https://doi.org/10.1007/978-3-319-99190-0_15
Vivares JA, Castaño LE, Sarache W (2015) Estrategia de operaciones: una revisión sistemática de literatura. In: II Congreso Internacional Industria y Organizaciones—Logística, Innovación y Desarrollo Tecnológico, Bogotá, 4–6 agosto, pp 1–15
Vivares JA, Sarache W, Hurtado JE (2018) A maturity assessment model for manufacturing systems. J Manuf Technol Manag 29(5):746–767. https://doi.org/10.1108/JMTM-07-2017-0142
Westbrook R (1995) Action research: a new paradigm for research in production and operations management. Int J Oper Prod Manag 15(12):6–20. https://doi.org/10.1108/01443579510104466
Yin RK (2014) Case study research: design and methods, 5th edn. Sage, Thousand Oaks
Zhang W, Levenson A, Crossley C (2015) Move your research from the ivy tower to the board room: a primer on action research for academics, consultants, and business executives. Hum Resour Manage 54(1):151–174

Part III
Human Factors and Ergonomics

Chapter 19
The Human Factor as a Central Element in the Design of the Workplace. A Systematic Review

Gabriela Jacobo-Galicia, Carlos Raúl Navarro-González, Mildrend Montoya-Reyes, Ismael Mendoza-Muñoz, and Eusebio Jiménez-López

Abstract Manufacturing workplace design enables products with varying levels of complexity to be manufactured in multiple configurations. However, despite technological advances, the central methodology used continues to be based on applying solutions through the exploration of possibilities rather than implementing rules or algorithms to solve a specific problem. The way the workplace is designed has a profound impact on the staff and the organization's goals, as satisfied workers tend to be more productive than less satisfied ones. This chapter presents a systematic review of workplace design considering health effects and regulations. It also provides guidelines that consider psychosocial risk factors during the design process and new trends in manufacturing workplace design from three perspectives: technology, human factors, and health and safety for the well-being of workers.

Keywords Manufacturing workplace design · Worker's health · Workplace regulations

19.1 Introduction

The essential function of the organization is to provide standardized solutions to recurring problems. The success of this process will depend on the degree to which the entity allows its employees to focus their attention on a way that allows them to complete their tasks correctly (Schabracq 2003). There are different factors from an organization's structure and environment that influence the workers' well-being (ILO and WHO 1986). Among these factors are the firm's general policies, lack of authentic

G. Jacobo-Galicia · C. R. Navarro-González · M. Montoya-Reyes (✉) · I. Mendoza-Muñoz
Department of Industrial Engineering, Universidad Autónoma de Baja California, Blvd. Benito Juárez S/N, Mexicali, Baja California, México
e-mail: mildrend.montoya@uabc.edu.mx

E. Jiménez-López
CIAAM, Southern Sonora Technological University—Northwest La Salle University-IIMM, Calle Dr. Norman Borlaug, Cd. Obregón, Sonora, México

J. L. Garcia Alcaraz et al. (eds.), *Trends in Industrial Engineering Applications to Manufacturing Process*, https://doi.org/10.1007/978-3-030-71579-3_19

consultation, non-participation in decision making, limitations to individual initiative, poor internal communication, low support levels, lack of agreement between the organizational goals, lack of personal development support, ambiguity and conflict of roles, and responsibility for other people (Moreno-Jiménez and Báez-León 2010; ILO 2016; ILO and WHO 1986).

Some aspects of a company's function and culture are incredibly important, such as the organization as a task development environment, the problem-solving ecosystem, and a unique development environment (ILO 2016). It is stated that there is a possibility of an increase in stress levels if there are deficiencies in these aspects (Cox et al. 2005). However, even though human factors have recently assumed a leading role in process design, there is a lack of standard references and structured protocols to evaluate how workers experience industrial practice in an effective and predictive manner (Oldham and Fried 2016; Peruzzini et al. 2018).

Process design is strongly identified with workplace design, which refers to how the activities, responsibilities, and scopes of a job in an organization are configured. This chapter presents the results of a recent literature review in new workplace design trends and how new technologies have supported the development of innovative methodologies and approaches that consider the human factor as a central element of the design process while prioritizing health and safety.

This chapter is divided into two main sections; the first is a review of the literature on workplace design topics considering physical, biomechanical, cognitive, and psychosocial elements and their effects on workers' well-being. It contains an analysis of current workplace design approaches based on the level of maturity of the process. It emphasizes health and safety and proposes applicable standards for the design of the workplace. The second section focuses on emerging workplace design trends, describing more holistic techniques from three perspectives: digital technology use, the human factor, and health and safety in workplace design.

19.2 Literature Review

19.2.1 Manufacturing Process and Workplace Design

Manufacturing process design allows assembling products with different complexity levels in multiple configurations; however, the ability to apply knowledge to a process interactively and intuitively continues to be imprecise, so a heuristic approach is generally used, applying solutions through the exploration of possibilities rather than implementing rules or algorithms to solve a specific problem (Grewal 2011). Expert knowledge of the design process usually resides in technicians' minds, and it mainly refers to the rules that have an internal logical relationship within the general decision-making process and the reasoning that leads to design (Dorst 2011).

Grewal (2011) proposes a series of steps to design a manufacturing process—as shown in Fig. 19.1—where two main stages can be observed: product development

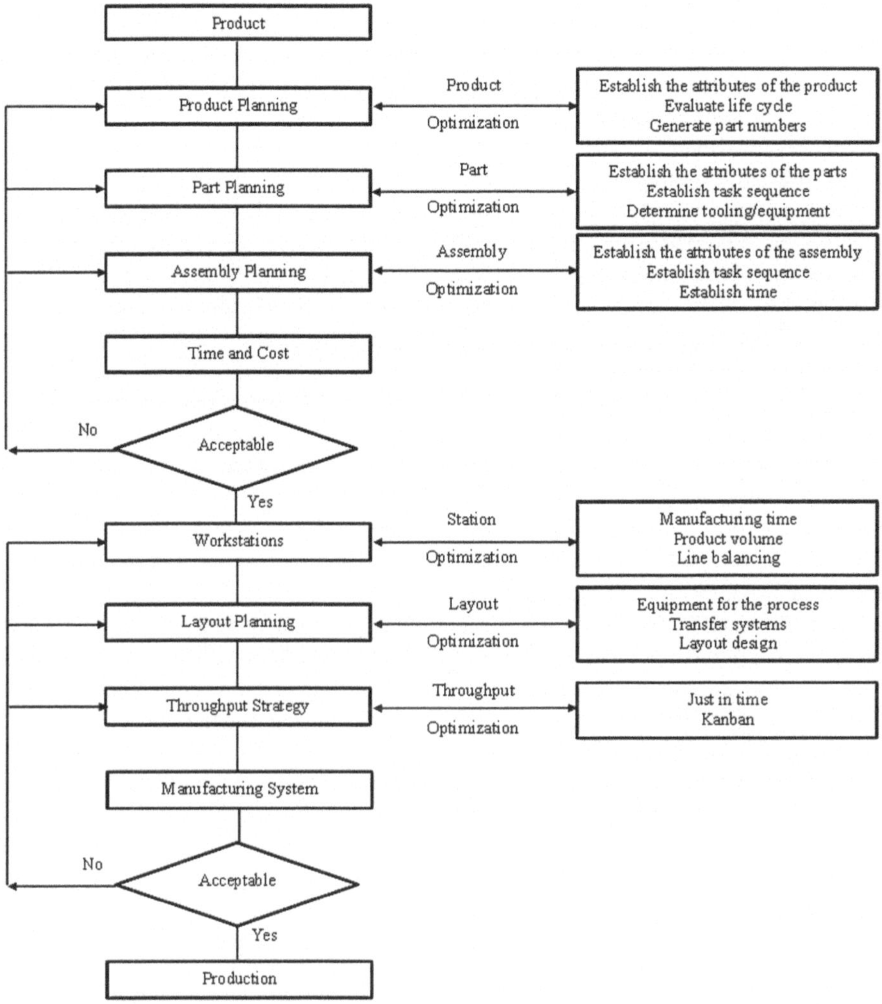

Fig. 19.1 Manufacturing process design stages. *Source* Own creation based on Grewal (2011)

[from product planning to the determination of production time and cost] and manufacturing process [from workstation design to establishing a manufacturing system]. In broad terms, once a product has been designed, and the parts that constitute it have been identified, the required basic processes are determined along with the necessary equipment and the instructions on material handling between workstations, which will be fine-tuned in the new process' testing period, before starting real production (Vaughn 2010).

Once the basic processes have been identified, the production quantities are known, the sequence of the operations has been established, and the equipment and tooling have been selected, tentative time and cost can be established. These will be modified during real production runs (Grewal 2011; Vaughn 2010). However, even though human and ergonomic factors have recently assumed a leading role in process design, there is a lack of common references and structured protocols to evaluate how workers experience industrial practices in an effective and predictive manner (Peruzzini et al. 2018).

The literature on the topic points links between a poorly designed process [such as poorly balanced workloads, poor workstation design, lack of control or autonomy in activities to be performed, poorly defined or ambiguous roles, and physical risk, to name a few] and musculoskeletal issues (Abdullah et al. 2015; Gerr et al. 2014b; Maakip et al. 2016), stress (Barber and Santuzzi 2014; Magnusson Hanson et al. 2014; Unda et al. 2016), and job burnout (Barrios León and Illada 2013; Jiménez 2019) among other issues.

In practice, process design focuses on work design, or the real structure of the tasks that the employee performs as part of the daily activities they do for the organization (Oldham and Fried 2016). The way work is designed profoundly impacts the people, and virtually all the goals that the organization intends to reach [safety, performance, productivity, creativity, motivation] are affected by it (Parker 2015). A case study developed by the cutting department of an electric company evaluated the impact of work design in manufacturing cells over 12 years. The study found that the process transformed into a more functional system with fewer cells, fewer but more advanced machines put to better use, and fewer workers with a higher degree of specialization and a process-oriented approach (Molleman and Slomp 2001). Hence, the understanding that the way this process happens is essential to the success of any business (Waschull et al. 2019).

The literature identifies two emerging aspects of work design: on the one hand, we have the relational perspective that focuses on analyzing how jobs, roles, and tasks are socially embedded since there is increasing interdependence and interaction with coworkers and customers; on the other hand, we have the proactive perspective that highlights the growing importance of employees taking the initiative to anticipate and create changes in how work is performed, based on increasing uncertainty and dynamism (Grant and Parker 2009).

The study of the relational approach has linked the existence of a right work environment [with adequate levels of social interaction, available at work by colleagues and supervisors] with increased productivity, as satisfied workers tend to be more productive than those not so satisfied (Cox et al. 2005; Foldspang et al. 2014). Besides, this perspective generates a higher level of commitment to work when employees feel they have the social support of their peers and supervisors, so it is recommended that relational resources be considered within the workplace's design elements (Freeney and Fellenz 2013).

The proactive approach is exemplified in the application of techniques such as participatory design [PD], which promotes the development of collaborative work environments (Pilemalm 2018). In the context of manufacturing systems, these environments arise from the involvement of multiple stakeholders within the organization, to identify opportunities for improvement and provide a valuable contribution to optimizing a manufacturing system. This technique helps identify all critical requirements that the system must meet and the immediate processing of product-specific requirements, such as most projects developed from traditional production processes (Andersen et al. 2018).

From the operational perspective, one can consider the transition from traditional workstations to lean workstations and appreciate how inherently different they are. Traditional workplaces are designed to facilitate material handler's work, not to increase the operator's value-added. A lean workstation is designed focused on operator concerns, such as safety and ergonomics, and minimal wasted motion, to get parts efficiently and find tools quickly. Assembly materials, tools, or parts should be strategically positioned to allow the operator to reach them instantly, without interfering with their safety and comfort (Weber 2005).

In this context, Granath et al. (2005) state that two main dimensions should be distinguished in workplaces: functionality and usability. Functionality refers to the ability to carry out a regular function and is more concerned by the workstation's characteristics, which do not make the workstation usable individually. The workstation must be designed based on ergonomic and lean principles to improve the functionality dimension. Usability can be defined as the system's capability to achieve a specific goal in a specific context of use.

According to Blakstad et al. (2009), usability can be divided into three consecutive levels: effectiveness, efficiency, and satisfaction. Effectiveness refers to the possibility of providing the desired effect. An organization, and consequently workstations, will only achieve its goals when the employees' social needs are fulfilled. Efficiency expresses how the resources are allocated and organized to increase productivity, and satisfaction is fully achieved during total integration and synchronization within the organization [systems, people, resources, etc.].

Efficiently designed workstations are essential to provide flexibility and significant mass production. Unfortunately, it is common to encounter industrial workstations that have been built without intention. Workplace design should focus on the requirements of users, but also on the requirements of each task, which will allow organizations to optimize their production indicators (less time, space, and cost) and quality levels (Gonçalves and Salonitis 2017) without compromising employee health, safety, and well-being.

19.2.2 Effects of Workplace Design on Workers' Well-Being

At its most basic level, work design refers to the real structure of the employees' tasks or activities (Oldham and Fried 2016). In a broader sense, the concept covers the content and organization of tasks, activities, responsibilities, and interpersonal relationships associated with the work itself, which employees carry out for the organization every day, so the dynamism of the emerging roles and the changes in a work position that are developed as an integrated whole are considered, also taking into account the interdependencies among skills, organization, and technology (Jiménez 2019; Parker 2014; Vinay et al. 2012).

Task design includes several aspects that are risk factors: the low value of work, underutilization of capabilities, lack of activity diversity and repetitiveness of the job, uncertainty, lack of learning opportunities, high attention demand, conflicting demands, and insufficient resources (ILO 2016). In this sense, work design can be a powerful medium for learning and development, but it can also be a resource to maintain and improve physical and mental health, which are important factors given the challenges in today's organizations (Parker 2014).

From the occupational health point of view, workplace design impacts the well-being of people in different ways. Examples are found in the modification of attitudes, behaviors, and health of individual workers, whether they feel motivated and committed, and in their perception of the effort they put in, as well as their safety (Driscoll et al. 2008; Parker and Griffin 2014). The inclusion of these aspects approach to workstation design would be critical. Nafuna (2019) proposes a hazard structure to consider during the design process which includes physical, chemical, biological, psychosocial, mechanical, and ergonomic hazards. However, other authors indicate that musculoskeletal disorders have a higher prevalence over ergonomic risk assessment (Aghilinejad et al. 2016; Beheshti et al. 2018; Hossain et al. 2018). Parker and Griffin (2014) propose a more succinct hazard structure which organizes the elements already mentioned within clearly defined categories. They propose that the following elements must be considered when making decisions about the evaluation of the effectiveness of work design:

- Physical Elements: Aspects of the work environment or context that create physical or physiological demands on the human body, such as chemical hazards, noise, lighting, and vibration.
- Biomechanical Elements: Aspects of the work that include hazardous manual tasks and the biomechanical risk factors that lead to musculoskeletal disorders.
- Cognitive Elements: Aspects of the work that can create demands on the human mental capacity.
- Psychosocial Elements. Social, psychological, and organizational aspects of work that place demands on human capacities.

A distinction of these four elements at work allows making decisions regarding activities aligned with a deep range of hazards and risks, encouraging eliminating or minimizing them to foster a healthy and safe workplace where work design optimizes human performance, productivity, and job satisfaction (Parker and Griffin 2014).

These four categories [physical, biomechanical, cognitive, and psychosocial elements] will be treated as health workplace elements since all of them are linked to the well-being and health of workers. The following sections will go more in-depth into each of these elements.

19.2.2.1 Physical Elements

Numerous factors can make up a working environment. These include noise, light, heat and cold, particulates in the air, gases, air pressures, gravity, etc. (Parsons 2000; Schonfeld and Chang 2017). The effects on humans that are usually considered are those that affect the workers' health, comfort, and performance (Parsons 2000). The workplace's strong physical restrictions represent an aggressive context due to environmental risks, such as activities in environments under extreme temperatures and pressure, humidity, inadequate lighting, and loud noises (Neffa 2015). These factors affect the human senses, overload the nervous system, and negatively affect general health and cause stress (Kralikova et al. 2019).

Most of these factors can be measured objectively and with a certain degree of reliability and validity, so they may be easily controlled. In some cases, norms can be used to regulate exposure to these factors (Cox et al. 2005). However, recent literature indicates that in some professions and contexts, there are barely any regulations to guarantee that work is performed under adequate temperature and humidity conditions (Al Horr et al. 2016; Cheung et al. 2016). Heat stress can result in illnesses, thus increasing the risk of workplace injuries, but it may also reduce brain function of reasoning ability, creating additional hazards (ILO 2017a).

A link between noise levels that are above acceptable and physical issues such as professional hearing loss (Martín Leal and Rojas Sánchez 2014; Pinosova et al. 2015), as well as psychological issues such as stress (Cox et al. 2005), have also been reported. Concerning noise exposure, the most effective way to protect workers is to eliminate the source of noise, avoiding the usage of noisy machines, or moving noisy operations away from other work activities (Lingard et al. 2019). Lighting is mentioned in literature not only as a comfort factor related to visual strain (Al Horr et al. 2016; de Araújo Vieira et al. 2016; Hawes et al. 2012; Leccese et al. 2017) but in other contexts, such as mining, it tends to be a crucial physical safety element (Sammarco et al. 2012).

Some studies have identified a link between musculoskeletal disorders and risk factors due to the work environment (Chanchai et al. 2016; Gerr et al. 2014b; Vandergrift et al. 2011; Widanarko et al. 2014). As these studies indicate, physical factors can significantly risk employees (Kralikova et al. 2019).

There are four principal methods of assessing human response to the environment: subjective methods, where representatives of the user population report on the response to the environment; objective measures, where the occupant's response is directly measured [e.g., body temperature, hearing ability, performance at a task]; behavioral methods, where the behavior of a person or group is observed and related to responses to the environment [e.g., change posture, move away, switch on lights]; and modeling methods. Modeling methods include those where predictions of human response are made from models based on the experience of human response in previously researched environments [empirical models] or rational models of human response to environments that attempt to simulate the underlying system, and hence can be used to relate cause and effect (Parsons 2000).

19.2.2.2 Biomechanical Elements

Biomechanical elements refer to the workplace characteristics such as sustained physical effort, loading, and unloading heavy objects, working extended periods in uncomfortable or unnatural positions, and exposure to vibration. These physical restrictions represent an aggressive context due to situations related to work intensity and frequency or to the duration and configuration of time spent at work, which in time can cause joint pain, lower back pain, and musculoskeletal disorders (Neffa 2015). People are an essential element of every organization, and they are fundamental to deliver the right quality products. This resource becomes even more costly when a worker gets injured or sick, which increases direct costs and the loss of services provided by them (Vinay et al. 2012). Onawumi et al. (2016) show in a health hazard research work that musculoskeletal issues can have a prevalence of 29% of the workforce in the case of metal fabrication workers, which is considered to be significant.

Globally, from all injuries, back pain represents the more vital proportion of years lost to disability [YLD], with neck pain and other musculoskeletal disorders in the top ten of illnesses included in this calculation (Vos et al. 2015). In Mexico, the Mexican Social Security Institute [IMSS for its Spanish acronym, which stands for Instituto Mexicano del Seguro Social] reports that lumbar pain is the most common affliction seen in Physical Medicine and Rehabilitation Clinics and that in 2019, physical therapy personnel provided services to almost 2.5 million beneficiaries, who received 9,125,000 therapies (IMSS 2020).

Ergonomics is the science of designing workplaces that adapt to the capabilities and limitations of the body. Said design aims to prevent injuries by identifying tasks in the workplace that represent a risk to develop musculoskeletal disorders (Inyang et al. 2012). Multiple studies have determined that poor ergonomic design in workstations produces problems such as shoulder, neck, back, and lower back pain, among other issues, which negatively impact productivity in organizations (Allahyari et al. 2016; Gerr et al. 2014a; Vinay et al. 2012). Even though the risk of suffering an injury is different depending on the profession and the location where tasks are performed, manual materials handling is one activity that leads to work-related injuries with greater frequency. It is the most related to musculoskeletal disorders, sprains, and strains (Khanzode et al. 2012).

Musculoskeletal disorders linked to ergonomic issues have the characteristic that discomfort and other issues are not immediately felt; instead, they take a while because these injuries are generally slow to appear and harmless, which causes people to ignore the symptoms until they become chronic and the damage is permanent (Chávez et al. 2017). One of the reasons for this is that there are significant deficiencies in how human factors have been approached in production and manufacturing systems in terms of work methods and interactions with the work environment (Del Rio Vilas et al. 2013). For example, these situations sometimes arise because the human interface is designed with anthropometric data for different countries, leading to unexpected ergonomic issues, especially in tasks and activities related to workstations. Particularly in industries that need to change their manufacturing processes quickly to satisfy the world market's demands, ergonomic hazards caused by incorrect workstation design can be severe (Lin and Chan 2007).

Ergonomic intervention for individual-centered workstations must start with the premise that it is not only the worker utilizing the devices who should be considered for the ergonomic analysis of a work situation, but the specific use of the resource should also be taken into account because it allows finding ways to improve a job (Castillo Martínez 2007).

19.2.2.3 Cognitive Elements

Cognitive elements include information load, processing load, complexity, and duration of the task. These are factors that create demands on human mental capacity (Parker and Griffin 2014). The work schedule is also a cognitive factor since, as ILO (2017b) states, there is a link between long working hours and adverse effects on health, alertness, and performance. From the cognitive ergonomics point of view, the cognitive elements of work correspond to how an operator develops mental processes while performing the task work. These may include different phases and stages of reception, treatment, action, and decision in the face of the different sources of information that constitute the work situation, which require different cognitive supports that are necessary to fulfill the goals established to perform a task (Castillo Martínez 2007).

Cognitive elements are relatively common in workforces and are associated with issues such as restriction of perception, reduced concentration, functional memory disorders, hesitation to make decisions, changes in reasoning, and reduced creativity, which have a significant impact on the self-esteem of the worker and the achievement of the organizational goals (ILO and WHO 1986; Stenfors et al. 2013). The worker's actions must complete answering to the requirements of a work situation [evolutional and complex] evidence the central role of the processes related to problem–solution, building representations, and the processes that allow assessing, deciding, and planning (Castillo Martínez 2007).

Considering the cognitive elements of work has become increasingly important as physical tasks become mechanized. In many occupations, information processing requirements have increased in complexity due to technological evolution, globalization, and other factors that give place to dynamic changes (Parker and Griffin 2014). For example, in the case of manufacturing areas, it has been demonstrated that the growing complexity and demand for manufacturing processes have increased the cognitive load on assembly line workers (Hoedt et al. 2017; Thorvald et al. 2019). Also, it has been identified that increased quantitative work demand also increases cognitive issue complaints (Stenfors et al. 2013). Increased work demands deplete employees' physical and mental resources, which generate reduced energy and possible burnout and contributes to health deterioration (Hu et al. 2017).

On the other hand, it is not rare for people to experience thoughts or behaviors unrelated to the task while performing their work activities, which make them forget about rules and procedures, not pay attention to coworker or customer requests or inadvertently toss valuable documents. In the best-case scenario, these mistakes reduce effective performance at work, but they could lead to disastrous accidents (Wallace and Chen 2005). It is essential to consider that even though excessive workloads could reduce the performance of the operator and increase errors, the opposite is also true: a lack of activities [low workload] could lead to boredom, loss of awareness of the situation, and a reduced state of alertness, which could also cause problems (Rubio-Valdehita et al. 2017). These types of failures are caused by errors in the cognitive process produced in a task's performance that a person would usually execute successfully (Martin 1983).

Also, any electronic or technology device with the ability to gather, store, or send, represents a threat to cognitive aspects in workstation design. However, with mixed effects: while they could help accomplish tasks with a lower cognitive level, they could also be used to increase job demand, information load, learning expectations, and hassles from technology [malfunctions, incompatibility, security issues, needed upgrades, increased skills] (Wang et al. 2020). So, using electronics and technology to "rescue" people from routine work should be taken with caution.

The health effects of poorly balanced cognitive load could also cause musculoskeletal issues (Darvishi et al. 2016; Deeney and O'Sullivan 2017; Haghshenas et al. 2018; Heidarimoghadam et al. 2019), stress (Goh et al. 2015; Gujski et al. 2017; Körner et al. 2019), burnout (Moreno-Jiménez and Báez-León 2010), and poor life satisfaction (Vargas et al. 2012), among other issues.

19.2.2.4 Psychosocial Elements

Psychosocial elements are the factors that generate psychosocial hazards in the workplace, including social, psychological, and organizational aspects that could be challenging for an employee (Parker and Griffin 2014). The term psychosocial work hazard is understood as the incident, situation, or state that is a consequence of work organization, with a high probability of affecting the worker's health and whose consequences could be severe (Moreno-Jiménez and Báez-León 2010). From a broader perspective, psychosocial hazards refer to the potential of psychosocial threats to cause damage, and represent significant economic implications at a social level, affecting organizations of all types and sizes (BSI 2011). Unlike physical hazards, which are directly observable and measurable, psychosocial hazards, being the product of social interactions, cannot be understood outside of the context in which they exist (Jespersen et al. 2016). The primary sources of psychosocial hazards are new forms of employment contracts and job insecurity; aging workforce; work intensification; high emotional demands at work; and low work–life balance (Milczarek et al. 2007). Regarding these types of risks, some key entities, such as the International Labor Organization [ILO], the World Health Organization [WHO], and the European Commission, indicate that they have the potential of generating adverse consequences for organizations, including sickness-related absences, reduced productivity, and human error (Leka et al. 2015). Among the main types of psychosocial hazards, we find work-related stress, workplace violence, workplace sexual harassment, and burnout (Moreno-Jiménez and Báez-León 2010).

The factors that favor the emergence of psychosocial risk can be defined in two aspects, according to ILO: first, according to the interaction between the content of a task, the management and organization of the work, and other environmental and organizational conditions. Second, they can be defined according to the employees' needs and skills, which have a hazardous influence on worker health through their perceptions and experiences (Rosário et al. 2016). When these factors are not properly managed at workplaces, they can represent hazards that could affect the safety, health, and well-being of individuals, with negative effects on the organization [through sickness, absence, reduced productivity, human error] as well as on society through increased disability pensions, health costs, etc. (Leka and Jain 2016).

Psychosocial risk factors are all the conditions experienced by a person concerning the organizational and social environment that are a consequence of work management or organizational, environmental, or individual aspects and may have the potential to cause physical, psychological, or social harm (Charria et al. 2011). These factors could favor quality of life at work by fostering workers' personal development, but they could also damage their health and well-being (Gil-Monte 2012). The factors with the ability to generate psychosocial hazards, according to Cox et al. (2000), can be categorized based on the conditions that define the corresponding hazard (see Table 19.1).

By observing this classification, factors associated with technical aspects can be identified, such as workload or work pace, which can be solved if the interests of the business and the design considerations are included in the configuration of workstations (Gil-Monte 2012). It is important to mention that there is currently a very successful and extensive theoretical and practical foundation to improve the performance of a process by using already-established techniques. However, this is not the case for achieving satisfaction and well-being in the workplace since it depends on both physical and psychological factors (Hollnagel 2014). Recent literature indicates that applying safety principles to process design is a useful technique for generating a safer, sustainable, and economically feasible production plan (Rathnayaka et al. 2014). Without failing to consider that the success or effectiveness of the organizational measures to prevent psychosocial hazards depends on the level of involvement of the company, as well as its values, needs, and the competitive context of its environment, occupational health must play a role in the business strategy of any organization (Grote 2014).

19.2.3 Current Approaches to Workplace Design

The current approach to workplace design changes according to the conditions under which the design process becomes necessary and the designer's point of view. This may happen based on the level of maturity of the production process, from a preventive point of view [before the operational establishment of the production process], or

Table 19.1 Taxonomy of psychosocial hazards

Category	Conditions that define the hazard
Job content	
Environment and equipment	Issues related to equipment or facility reliability, availability, suitability, maintenance, or repair
Task design	Lack of variety and short work cycles, fragmented or meaningless work, underuse of skills, high uncertainty
Work overload/work pace	Work overload or under load, lack of control over pace, high levels of time pressure
Work schedule	Shift work, inflexible work schedules, unpredictable hours, long or unsociable hours
Job context	
Organizational culture and function	Poor communication, low levels of support for problem-solving and personal development, lack of definition of, or agreement on, organizational objectives
Function in the organization	Function ambiguity and conflict, responsibility for other people
Professional development	Career stagnation and uncertainty, under promotion or over-promotion, poor pay, job insecurity, low social value to work
Decision-making autonomy (decision latitude), control	Low participation in decision making, lack of control over workload [control, mainly in the form of participation is also a broader organizational and contextual matter]
Interpersonal relationships at work	Social and physical isolation, poor relationships with superiors, interpersonal conflict, lack of social support
Home–work interface	Conflicting demands of work and home, low support at home, dual-career problems

Source Cox et al. (2000), retrieved from ILO (2016)

a corrective perspective [when the process has been established]. Another approach can arise from the way design improves workers' health and/or safety. A third approach considers the legal requirements, regulations, and norms associated with workplaces. There can also be mixed approaches that involve two or more of the conditions mentioned above. The following sections will present recent examples of these approaches.

19.2.3.1 Workplace Design Based on the Process Maturity Level

Workplace design could be approached from its correlation with the life cycle of the product (Caple et al. 2019; Jacquemin et al. 2012), where two stages could be identified depending on the maturity of the production process. The first stage is set in an early phase, usually associated with the product design stage (Matuszek et al. 2020). While the second one is related to a redesign of the process, looking to improve efficiency, reduce costs (Kovács and Kot 2017), and introduce the latest assembly methods and manufacturing technology (El-Nounu et al. 2018), or to involve worker well-being (Daniels et al. 2017). These stages are shown in Fig. 19.2.

Figure 19.2 shows how original process design and process redesign are correlated with the product life cycle; however, they should be considered independently since both use different techniques and methods for the workstation design process. This separation of design into two stages is also proposed by Jacquemin et al. (2012), who differentiates product life cycle and process life cycle while highlighting the importance of renovating and restoring the process design in a second stage, reviewing the impact of the production process with environmental and industry regulations.

Designing for the product and manufacturing processes could significantly impact manufacturing costs. Matuszek et al. (2020) indicate that while design cost represents 5% of a new product's startup, it impacts up to 70–80% of the manufacturing costs, excluding indirect labor costs. This means that changes in the production project at the right time have the most significant impact on the efficiency and performance of subsequent manufacturing. Machac et al. (2017) emphasize that 80% of nonconformities in manufacturing processes appear in the product design phase. Not only nonconformities, but most of the problems of the process could be detected early. Mourtzis (2016) points out the importance of early virtual engineering in the first

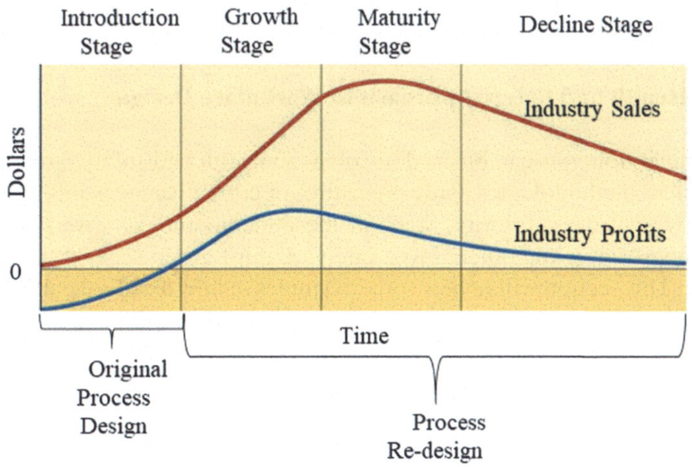

Fig. 19.2 Workplace design stages. *Source* Adapted from Solorzano (n.d.) retrieved from Kuka (2018)

stage of design while implementing discussions about potential problems using "what if" scenarios.

When referring to the original process design stage, there are several traditional techniques that can be used to aid the workstation design process, such as design for X, quality function deployment [QFD], failure mode and effect analysis [FMEA], design for assembly [DFA], and design for manufacturing [DFM] (Matuszek et al. 2020). Butt and Jedi (2020) show that the DFA and DFM techniques could lead to design efficiency and reduced manufacturing costs from 20 to 80% of the original estimate. Also, including the use of more advanced technological devices or software that may include augmented or virtual reality to model and simulate the manufacturing process, like CAD/CAM, enterprise resource planning [ERP], supply chain simulation, or knowledge management, could aid in workstation design by replicating probable ergonomic conditions, and material and process flow in a realistic way (Mourtzis 2020).

Regarding the process redesign stage, changes occurring during this manufacturing design phase could help the company at the growing, maturing, and declining periods of the product's life cycle. According to Kuka (2018), taking the right marketing and product management measures could increase the company benefits from products or end it abruptly, depending on how the control of the product and its processes ensure the whole operation's performance and profitability. Thus, a constant production process redesign should be done. Iuga et al. (2017) suggest that this stage could include principles of circular economy and sustainable development of industrial products such as: reducing consumption of raw materials, and energy and water consumption, usage of fewer toxic materials, and waste generation or emissions to air control.

It is essential to distinguish between these two proposed stages of manufacturing workplace design because they should be applied at different time frames of the product's life cycle, and, as Table 19.2 shows, their differences are significant.

19.2.3.2 Health and Safety Approach to Workplace Design

Different methodologies can be used to solve the health and safety issues faced by workers. These methodologies apply preventive or corrective measures based on the element they consider a priority. Some of these methodologies give preference to the health approach while others favor safety. A third group seeks to balance both approaches. This section will review some examples of how to solve health and safety problems. Classifications that contrast their characteristics will be presented.

In the view of several authors, considering ergonomics and occupational health at process redesign could increase its productivity. Onawumi et al. (2016) say that even small changes in workstation dimensions can severely impact productivity and workers' occupational health and safety. However, these improvements should be measured by an indicator, score, or parameter. For instance, Boulila et al. (2018) show that posture improvement reduces musculoskeletal disorders and accidents and increases productivity and quality using what they call a "Normalized Score". Colim

Table 19.2 Differences between original process design and process redesign stages

	Original process design stage	Process redesign stage
Location within the life cycle of the product	Early introduction stage associated with the product design stage (Matuszek et al. 2020)	At growing, maturity and declining stages (Caple et al. 2019; Jacquemin et al. 2012)
Main goal	Core company's support benefits the new product introduction period of the life cycle (Kuka 2018)	Increase company benefits from product by supporting the growing, maturing, and declining periods of the product's life cycle, and by differentiating process life cycle from product life cycle (Jacquemin et al. 2012; Kuka 2018)
Improvement focus	Efficient product and process introduction by considering new technologies and government regulations while involving worker's well-being and sustainability principles (El-Nounu et al. 2018; Daniels et al. 2017; Iuga et al. 2017)	Productivity and quality enhanced by making changes at workstation and process (Kovács and Kot 2017) Introduce the latest assembly methods and manufacturing technology (El-Nounu et al. 2018) Involve workers well-being (Daniels et al. 2017) Review impact of production process with regulations [environmental, safety, and health] and involve sustainability principles (Iuga et al. 2017)
Impact at production cost, quality, efficiency, and performance from production process	Highest, could significantly impact manufacturing costs (Matuszek et al. 2020) and problems and nonconformities during the process (Machac et al. 2017)	Lowest and more difficult to implement (Machac et al. 2017)
Methods and techniques	DFA, DFM, QFD, FMEA, CAD/CAM solutions, ERP, supply chain simulation, knowledge management, ergonomics, material flow, and process simulation, augmented reality, virtual engineering (Butt and Jedi 2020; Matuszek et al. 2020; Mourtzis 2016, 2020)	Kaizen events, six sigma, lean manufacturing, ergonomic studies, process checklists, use principles of circular economy, sustainable development (Kovács and Kot 2017)

Source Own creation

et al. (2020) also agree that the relationship between productivity and ergonomics is relevant and apply observational methods and questionnaires to promote a collaborative framework to positively impact productivity, flexibility toward jobs, and workstation adjustments rather than replacing workers.

According to Nafuna (2019), a prime concern to regulatory entities [policymakers, stakeholders, and researchers] should be occupational hazards in the workplace in all employment sectors. Hazards also have been introduced over time as workstations change, equipment and tools become worn, maintenance is neglected, or cleaning practices decline. Likewise, a relative control of safety hazards has been paid far less attention by the manufacturing industry, a legal duty is barely fulfilled, and the adoption of general health and wellbeing initiatives is confusing (Lingard et al. 2019).

Some authors conclude that knowledge of the occupational hazards and safety practices that factory workers should have is important to reduce the negative consequences of health effects while reinforcing the proper use of protective measures (Nurwani and Minghat 2018; Tadesse et al. 2016). Nafuna (2019) states that failure to identify or recognize hazards that exist and could be anticipated is one of workplace incidents' root causes. Thus, any adequate safety and health program is dependent on the proactive identification of hazards at the process level, where the empowerment of trained personnel is a critical element. García-Gómez et al. (2020) suggest that each organization should perform a risk assessment looking to highlight the hazards and risk levels while identifying who will be eliminated or reduced through a risk reduction plan.

Sorensen et al. (2018) emphasize as best practice the importance of involving proactive personnel in working condition improvements. Front-end workers should understand the relationships between working conditions, worker safety, and health outcomes, which helps them inform, prioritize, lead, or improve workplace conditions. Those activities could be done through small projects, Kaizen, six sigma, or lean action projects at the production level.

There are five risk-control levels to ensure safety and health in the workplace: elimination, substitution, engineering, administrative, and personal protective equipment (PPE). They are named in descending order of effectiveness, so the fifth-level measures are far less effective while the first-level measures are very effective (Lingard et al. 2019).

Table 19.3 presents a classification of the health and safety proposals from the above authors per stage of workstation design.

The information shown in Table 19.3 is a quick guide to identifying how to deal with specific situations in the workstation design process. The data is segregated according to the health and safety factors that were found to be relevant to the reviewed authors. These factors include organizational conditions that are not only linked with workers' well-being but with the company's bottom line.

When trying to assess how the healthy workplace elements [introduced by Parker and Griffin (2014) earlier in this chapter] can be incorporated into workplace design, several ideas were found from some of the authors above. Table 19.4 shows a summary.

Table 19.3 Health and safety proposals per workstation design stages

Health and safety category	Original process design stage	Process redesign stage
Measure indicator	To be determined at a later stage	Productivity increase (Boulila et al. 2018; Onawumi et al. 2016) Quality increase (Boulila et al. 2018) Occupational absenteeism reduction (Colim et al. 2020) Employee turnover rate (Colim et al. 2020)
Effectiveness measure	To be determined at a later stage	Score or evaluation on individual workstation redesign (Boulila et al. 2018; Onawumi et al. 2016)
Effectiveness level	High effectiveness at controlling risk for safety and health at the workplace since any anticipated hazards could be eliminated, substituted or engineered (Lingard et al. 2019)	Lower effectiveness at control of risk for safety and health at the workplace since a limited part of the process could be substituted or engineered; high usage of administrative and personal protective equipment and regulations compliance (Lingard et al. 2019)
Rules and regulations	Rules and regulations should be observed by process design (Lingard et al. 2019; Nafuna 2019)	Rules and regulations can be overpassed by workstation redesign
Training and education courses	Identified and implemented at an early stage. Reduced later on with careful selection of experts and team members	Essential for the proper use of equipment, look for actual hazards and anticipate them. Evaluate risk assessment, highlight hazards and risk levels (García-Gómez et al. 2020; Nafuna 2019; Nurwani and Minghat 2018; Tadesse et al. 2016)
People involvement and empowerment	Identified and implemented at an early stage. Reduced later on with careful selection of experts and team members	Workers involvement and empowerment is crucial. Promote usage of small projects, Kaizen, six sigma, and lean actions (Nafuna 2019; Sorensen et al. 2018)

Source Own creation

As Table 19.4 shows, the literature reviewed recommends considering physical elements in the early stages of the design process [during the initial design stage] since these factors are usually objective and easy to predict. Biomechanical elements should be revised during the process redesign stage as they may lead to musculoskeletal disorders from , and poorly designed tasks in which force, movement posture were

Table 19.4 Health workplace elements at design stages

Health workplace element	Hazardous aspects included	Authors	Design stage	Redesign stage
Physical elements	Chemical exposure, noise, lighting, heat and cold, environmental humidity, particles in the air, environmental gases	Al Horr et al. (2016), Cheung et al. (2016), Cox et al. (2005), De Araújo Vieira et al. (2016), Hawes et al. (2012), Kralikova et al. (2019), ILO (2017a), Leccese et al. (2017), Lingard et al. (2019), Martín Leal and Rojas Sánchez (2014), Neffa (2015), Parker and Griffin (2014), Parsons (2000), Pinosova et al. (2015), Sammarco et al. (2012) and Schonfeld and Chang (2017)	X	
Biomechanical elements	Force, movement, posture, vibration	Allahyari et al. (2016), Castillo Martínez (2007), Chávez et al. (2017), Colim et al. (2020), Del Rio Vilas et al. (2013), Gerr et al. (2014a, b), IMSS (2020), Inyang et al. (2012), Khanzode et al. (2012), Lin and Chan (2007), Neffa (2015), Onawumi et al. (2016), Vinay et al. (2012) and Vos et al. (2015)		X
Cognitive elements	Information load, processing load, complexity, duration of the task	Castillo Martínez (2007), Darvishi et al. (2016), Deeney and O'Sullivan (2017), Goh et al. (2015), Gujski et al. (2017), Haghshenas et al. (2018), Heidarimoghadam et al. (2019), Hoedt et al. (2017), Thorvald et al. (2019), Hu et al. (2017), ILO (2017b), ILO and WHO (1986), Körner et al. (2019), Martin (1983), Moreno-Jiménez and Báez-León (2010), Parker and Griffin (2014), Rubio-Valdehita et al. (2017), Stenfors et al. (2013), Vargas et al. (2012), Wallace and Chen (2005) and Wang et al. (2020)	X	X

(continued)

Table 19.4 (continued)

Health workplace element	Hazardous aspects included	Authors	Design stage	Redesign stage
Psychosocial elements	Equipment, task design, workload and work pace, work schedule, organizational culture, function in the organization, professional development, decision making and autonomy, interpersonal relationships at work, home–work interface	BSI (2011), Charria et al. (2011), Cox et al. (2000, 2005), Gil-Monte (2012), Grote (2014), Hollnagel (2014), ILO (2016), Jespersen et al. (2016), Leka et al. (2015), Leka and Jain (2016), Milczarek et al. (2007), Moreno-Jiménez and Báez-León (2010), Rathnayaka et al. (2014) and Rosário et al. (2016)		X

Source Own creation

inadequately set due to the evolving nature of the manufacturing process which is when operators gain more experienced, work instructions and methods change. Even machines and equipment are changed or customized. In line with these ideas, authors that consider psychosocial factors include the corrective approach by evaluating the organizational environment and adjusting when and where needed. Authors that refer to cognitive elements of workstation design recommend both the design and redesign approaches since technological changes associated with a higher mental workload can be predicted, and the workforce can be trained before the change is introduced. However, it should be checked and reviewed frequently mistakes due to distraction or absentmindedness to eliminate the root cause.

19.2.3.3 The Regulatory Approach to Workplace Design

There are two main ways to control workers' safety, health, and well-being conditions in the workplace. The first is done through the laws imposed by each country, and the second through the regulations developed by the industry. This section will address both.

The World Health Organization [WHO], aware that worker health is a fundamental requirement for productivity and economic growth, relinquishes responsibility of the creation of rules to ensure health and safety in every workplace to its member countries so that each of them can create their own rules through their legislative bodies (WHO 2007).

In the specific case of Mexico, this function is done through the Secretariat of Labor and Social Welfare [STPS for its Spanish acronym, which stands for *Secretaría del Trabajo y Previsión Social*], which regulates everything related to work safety and health at a national level through standards and regulations created under the Federal Labor Law (Cámara de Diputados 2014). From an operational point of view, social security is the responsibility of federal or local entities or branches, as well as decentralized organizations, such as the Mexican Social Security Institute [IMSS for its Spanish acronym which stands for *Instituto Mexicano del Seguro Social*], which is the basic instrument of social security, established as a national public service under the Social Security Law. This law, among other things, establishes the definitions for work-related risks, accidents, and illness, as well as the rights and obligations of employees and employers in that context (Cámara de Diputados 2018).

Since the year 2000, the Secretariat of Economy through the General Directorate of Standards issued Standard NMX-SAST-001-IMNC Occupational Health and Safety Management Systems [currently in its 2008 version], which specifies the requirements for Occupational Health and Safety Management Systems [OH&S], allowing organizations to control risks and improve their OH&S performance. However, this standard does not specify the OH&S performance criterion or provide detailed specifications to design a management system (Instituto Mexicano de Normalización y Certificación 2008).

On November 13, 2014, the Federal Occupational Health and Safety Regulations were published in the Official Federal Gazette (STPS 2014). These regulations establish the conditions to prevent risks and guarantee the workers' right to perform their activities in environments that ensure their life and health. Among the considerations of the new regulations, the obligation of employers to protect their employees from psychosocial and ergonomic hazards stands out. On October 23, 2018, the Official Mexican Standard NOM-035-STPS-2018 Occupational Psychosocial Risk Factors—Identification, analysis, and prevention, came into effect. The objective of the standard is to establish the elements to identify, analyze, and prevent psychosocial risk factors, as well as for the promotion of a favorable organizational environment in the workplace (STPS 2018a). Finally, on November 23, 2018, the Official Mexican Standard NOM-036-1-STPS-2018 Occupational Ergonomic Risk Factors—Identification, analysis, prevention, and control came into effect. Part 1: Manual handling of loads. Its objective is to establish the elements to identify, analyze, prevent, and control ergonomic risk factors in the workplace arising from the manual handling of loads to avoid affecting workers' health. Obligations for both employers and employees are established in this standard, and prevention, training, and development aspects (STPS 2018b).

In terms of the industry, it is vital to highlight the International Organization for Standardization [ISO], which is dedicated to creating regulations or standards to ensure the quality, safety, and efficiency of products and services. These are the ISO standards. ISO is currently present in 165 countries [including Mexico]. It is a non-government and independent organization with more than 23,000 ISO International Standards covering almost all aspects of technology and manufacturing, from food technology and safety to health and agriculture (ISO—About Us, n.d.). Among these, the primary standards that refer to occupational health and safety are grouped into the following categories: (1) management systems, (2) ergonomics, and (3) anthropometry and biomechanics.

In the first group, management systems, the standards that consider worker health and safety conditions are:

(a) ISO 9001:2015 Quality management systems—Requirements. This standard specifies requirements to establish a quality management system in public and private organizations, regardless of size or business activity. Its clause 7.1.4 Environment for the operation of processes mentions that the organization must determine, provide, and maintain a suitable work environment for the operation of its processes and achieve compliance of the products and services. This suitable environment can be achieved through a combination of factors: (a) social, (b) psychological, and (c) physical.

(b) ISO 45001:2018 Occupational health and safety management systems—Requirements with guidance for use. This standard specifies requirements for an occupational health and safety [OH&S] management system and gives guidance for its use to enable organizations to provide safe and healthy workplaces by preventing work-related injury and ill health, as well as by proactively improving its OH&S performance. However, it does not specify OH&S

performance criteria or provides detailed specifications to design a management system. It also does not address product safety, property damage, or environmental impacts beyond worker risks and other relevant aspects.

The regulations included in the second group are presented in Table 19.5. It can be seen that said standards offer general guidelines to approach the factors proposed by ISO 9001:2005 and strengthen the necessary actions of the health and safety management system that can be established based on ISO's application 45001:2018.

Finally, Table 19.6 presents the last group of standards that can be applied in the workplace. These standards go more in-depth into the study of the physical factors established by an earlier version of the ISO 9001 standard, the ISO 9001:2005, particularly for the performance of different tasks by workers.

19.3 New Trends in Manufacturing Workplace Design

19.3.1 Review Methodology

The cases presented in this section are the result of a recent literature review in which articles that present innovative ways to solve workplace design problems associated with manufacturing processes were identified. The review was carried out between August 20 and September 16 of 2020. Searches were done with the help of the meta

Table 19.5 ISO standards regarding ergonomics

Standard	Description
ISO 6385:2016 Ergonomic principles in the design of work systems	Establishes the fundamental principles of ergonomics as basic guidelines for the design of work systems and defines relevant basic terms
ISO 10075-1:2017 Ergonomic principles related to mental workload Part 1: General issues and concepts, terms, and definitions	Defines terms in the field of mental workload, covering mental stress and mental strain, and short- and long-term, positive and negative consequences of mental strain. It also specifies the relations between these concepts involved
ISO 10075-2:1996 Ergonomic principles related to mental workload Part 2: Design principles	Gives guidance on the design of work systems, including task and equipment and design of the workplace, as well as working conditions. Relates to the adequate design of work and use of human capacities
ISO 10075-3:2004 Ergonomic principles related to mental workload Part 3 Principles and requirements concerning methods for measuring and assessing mental workload	Establishes principles and requirements for the measurement and assessment of mental workload and specifies the requirements for measurement instruments (ISO 10075-3:2007 2007)

Source Own creation from information retrieved from ISO Web site (n.d.)

Table 19.6 ISO standards regarding anthropometry and biomechanics

Standard	Description
ISO 11226:2000 Ergonomics—Evaluation of static working postures	Establishes ergonomic recommendations for design or redesign of workplaces in relation to static working postures without any or only with minimal external force exertion, while taking into account body angles and time aspects
ISO 11228-1:2003 Ergonomics—Manual handling—Part 1: Lifting and carrying	Specifies recommended limits for manual lifting and carrying while taking into account, respectively, the intensity, the frequency, and the duration of the task. ISO 11228:2003 provides guidance to assess and expose recommendations that guarantee reasonable protection. It applies to manual handling of objects with a mass of 3 kg or more
ISO 11228-2:2007 Ergonomics—Manual handling—Part 2: Pushing and pulling	Gives the recommended limits for whole-body pushing and pulling. Like ISO 11228-1:2003, it provides guidance to assess and expose recommendations that guarantee reasonable protection
ISO 11228-3:2007 Ergonomics—Manual handling—Part 3: Handling of low loads at high frequency	Establishes ergonomic recommendations for repetitive work tasks involving the manual handling of low loads at high frequency. Like ISO 11228-1:2003, it provides guidance to assess and expose recommendations that guarantee reasonable protection

Source Own creation from information retrieved from ISO Web site (n.d.)

searcher of the Universidad Autónoma de Baja California library, which provides access to different databases such as EBSC Host, Elsevier. Emerald, IEEE/IET Electronic Library [IEL], Scopus, Springer and Wiley, among others, as well as Google Scholar. The inclusion criteria used were: academic peer-reviewed articles published from 2017 to date, which present approaches, methodologies, or application of tools for workplace design considering the well-being and safety of workers with a preventive or corrective approach. Articles that only relate improvements to process optimization and cost reduction without considering worker well-being are discarded.

The first selection produced 127 articles based on title and abstract. This group of studies was subsequently reviewed, considering the contents of the introduction of each and the results reported by the authors, applying the established exclusion criteria. Once the selection process was complete, a total of 22 studies remained, which contained specific information about workplace design considering worker well-being. These remaining studies were categorized.

19.3.2 New Trends

As we have already seen, according to the literature review, there are several innovative techniques being currently used to design or redesign a workplace. The articles identified in this section pertain to several approaches which have a holistic viewpoint to workplace design problems, all of them considering the human factor as a key element. The description of proposed improvements will be approached from three perspectives: use of technology, human factors and health and safety.

19.3.2.1 Technology's Helping Hand

The development of digital technology has proven to have great potential to support the workplace design process. As such, technological development has enabled the use of new tools to improve the design process by allowing the workplace design to be tested before it is implemented. An example of such technique is computer-integrated simulations based on virtual prototypes and digital human models [DHMs] (Peruzzini et al. 2019). These simulations compare different computer-integrated setups to support human-centered manufacturing workstations and then define a protocol analysis to support workstation design by analyzing both physical and cognitive aspects. This process allows for validating different design alternatives and optimizing the workstation design before creation.

Another example is using digital twins of real stations to minimize the time needed to develop and design a new work setup [i.e., a new assembly line]. The aim is to avoid the possibility of realizing that an existing worksite has ergonomic problems and correcting the design errors during the design phase instead of during the production phase. This approach makes it possible to achieve significant advantages in terms of the cost of correcting design errors and in terms of time to market, which will be significantly reduced (Caputo et al. 2019).

However, once the production process is already set and running, digital tools can also improve the current design. One way to achieve this is by assessing the workers' ergonomic performance and perceived comfort via the use of an eye-tracking device and a wearable biosensor, which would provide proper knowledge about the human asset of the factory that could be integrated with the knowledge derived from machine data collection, and then used to simulate the human–machine interaction [HMI]. These virtual prototypes can be used to avoid bottlenecks on the shop floor, optimize the workflows, and improve workstation design and layout (Peruzzini et al. 2020).

Another example is using a 3D computerized assessment model for local risk evaluation of work-related musculoskeletal disorders [WMSD], based on real-time and motion history volumes. The visual display of the WMSD risk level for each body segment is defined by color-coding at points surrounding an avatar's segment, representing an actual user. The values associated with an increased risk of WMSDs can be identified and iterated quickly to determine the optimal posture. Designers can share this knowledge by recording the user's postural interactions, defined by mapping geometric comfort data and WMSD risk-level categories (Eldar and Fisher-Gewirtzman 2019).

A different approach consists of using immersive virtual reality for layout redesign. Since workplace optimization is based on observational methods and software simulations which may not be insightful, and full-size prototypes signify high costs and implementation time, a system that allows the tracking of multiple users virtually performing assembly tasks is desirable. An option can be found in the use of an Automatic Virtual Environment [CAVE] system, where the virtual factory operators can be tracked along with the visualization of key performance indicators [KPIs] (e.g., completion time, traveled distance, and ergonomics) to support decision making by production engineers (Michalos et al. 2018).

Considering the relational perspective, when a company implements processes-oriented setups that may bring many advantages on an organizational level, shop floor workers can feel less able to maintain close contact with others and exchange awareness cues, and this diminished social connectedness can have consequences at the personal and organizational level. The implementation of information technology [IT] solutions within the workstations or as part of the equipment control systems can help to re-establish the connections between employees that may have dwindled after a company's reorganization. A positive impact on the employees' social connectedness can be achieved by allowing for social appraisal and by improving their sense of sharing and involvement when including IT communication tools on the workplace design (Richter et al. 2020). Table 19.7 summarizes the reviewed technologies.

One of the setbacks of the use of technology for the workplace design process is the lack of common references and structured protocols for the assessment of worker experience in industrial practices in an effective and predictive way. As a result, designers are poorly supported in the application of digital technologies. Thus, structured analysis protocols are needed in order to objectify and measure the experience of workers and the development of a mixed reality [MR] to shorten design time and improve the design's overall quality (Peruzzini et al. 2018). Another problem is that, due to new technological advances, the availability of a plethora of visualization techniques to develop ergonomic models can make it difficult to determine the most appropriate technique to convey maximum possible understanding (Eldar and Fisher-Gewirtzman 2019). Finally, there still are possibilities to harness new capabilities in cognitive and behavioral knowledge of future workstation–operator interaction as a principal component of the human–machine system in the Industry 4.0 era. These new configurations should allow an adaptive ongoing interaction that aims to improve operator performance, safety, well-being, and satisfaction, as well as production measures (Cohen et al. 2018). Along with the use and implementation of new digital tools and instruments for workplace evaluation, methodologies, guides, and protocols must be developed in order to facilitate the implementation of such technologies.

Table 19.7 Digital technology used on workplace design

Technology	Workplace design stage	Characteristics	Authors
Simulation and use of digital human models	Before workstation is implemented	Considers physical and cognitive aspects to test multiple scenarios before implementation	Peruzzini et al. (2019)
Digital twin of real workstation	Before workstation is implemented	Identifies ergonomic problems and errors of existing workstations to be eliminated in new design	Caputo et al. (2019)
Eye-tracking device and wearable biosensor	Implemented and running workstation	Assess workers' ergonomic performance and perceived comfort and, through simulation, optimizes the workflows and improves workstation design and layout	Peruzzini et al. (2020)
3D computerized assessment model	Implemented and running workstation	Evaluates local risk work-related musculoskeletal disorders in employees using avatars	Eldar and Fisher-Gewirtzman (2019)
Automatic Virtual Environment [CAVE] system	Implemented and running workstation	Tracks workers within an immersive virtual reality system for layout redesign	Michalos et al. (2018)
IT communication tools	Implemented and running workstation	Creates IT communication tools that can help keep social interaction on the shop floor	Richter et al. (2020)

Source Own creation

19.3.2.2 The Human Factor and Workplace Design

As manufacturing sites become cyber-physical systems [CPSs] in many areas, their effectiveness will be highly dependent on the extent to which they are designed for humans. Therefore, an integrated system design methodology which includes the human factors at an early stage is required. One proposal uses a hypothetical model to show the interdependencies between human-oriented work design and the resulting job performance in regards to cyber-physical production systems. By using an interdisciplinary approach consisting of research and findings from human factors, ergonomics, human–machine interaction, and work psychology in connection with engineering goals, the resulting model could provide exemplary work design actions for cyber-physical production systems (Stern and Becker 2017).

From the workplace ergonomics and human factors [E/HF] standpoint, it is recommended that E/HF expertise should be involved in early and appropriate phases of the workplace design process to leverage user needs and requirements to constrain the proposed design solution. A systems-theoretical framework suggests that the design decisions can be organized into five perspectives from which the design problem can be viewed holistically. This framework is intended to facilitate many types of design, including the design of work systems that allow all stakeholders to converge around design decisions that ensure that the work system is optimized to human characteristics and the activity to be performed (Bligård and Berlin 2019).

Another design framework based on the function–behavior–structure considers human factors earlier in the design process by analyzing the behavioral interactions between workers and manufacturing equipment driven from the system structure. Then, relevant human and health-related factors to these interactions are integrated into the analysis. These interactions are modeled and simulated to assess the system design using specific productivity and working condition indicators, such as fatigue, workload, stress, accidents, or productivity (Dantan et al. 2019).

The analysis of human factors, which strongly affect time and quality of manufacturing processes, is crucial for satisfying people involved in the manufacturing process and making them safe, preventing diseases, errors, and excessive workload. This analysis can be performed through a structured procedure that automatically extracts data from virtual ergonomic analysis made by digital manufacturing tools, and measures a set of indicators to validly assess manufacturing ergonomics. The expected result is a rapid and objective assessment, independent from the experience of the user, which can be executed during workspace design (Grandi et al. 2019).

Moreover, from the Industry 4.0 perspective, grounded on the integration of key technologies and CPSs (Henning et al. 2013), a profound modification in the manufacturing sector is expected. Work will change and different skills will be needed (Wang et al. 2018). The problem is how to govern this evolution and purposely guide the process of integrating people within CPSs in order to move toward the desired scenario. One way to achieve this is to use a methodology that supports the design and assessment of different work configurations, jointly considering the uniqueness of human labor and the characteristics of cyber-physical production within a comprehensive framework. The method should cover ordinary production as well as irregular scenarios, such as failure detection or maintenance intervention, particularly interesting for human work (Fantini et al. 2020).

From the collaborative perspective of workplace design, one technique that favors this approach is job crafting. Job crafting is defined as an activity performed by the employee that aims to change cognitive, task, and/or relational boundaries to shape, mold, or redesign their own job, and is capable of altering the meaning of work itself (Llorente-Alonso and Topa 2019; Wrzesniewski and Dutton 2001). Job crafting can also be positively associated with work attachment, as indicated by psychological ownership of the job and affective organizational commitment. The positive association between job crafting and work attachment can be stronger when employees experience tough times at work [i.e., low-quality leader–member dynamics that result from being rated as a poor performer or experiencing job insecurity] (Wang et al. 2018).

One way to implement this technique considers the effect of job features on job crafting as a developmental intervention. In the manufacturing sector, it has been found that four relational job features [supportive supervision, external pressure, valued social position, and contact with others] play a main role in accounting for job crafting. Similarly, in the IT sector, opportunities for skill acquisition and utilization, as well as career outlook have been identified as relevant to job crafting practice. Organizations can then refocus their developmental intervention with a goal, and train employees to craft their jobs in a way that fits them and the organization better (Latika 2018).

Research indicates that the greater the opportunity to participate in decision making, the greater the person's satisfaction and sense of self-esteem. Long-term autonomy for performing tasks is beneficial for mental health and worker's productivity (ILO 2016). Autonomy at work consists in the possibility of being an actor; lead one's professional life and have the possibility to progress and make a career; intervene in the production of goods or services with the capacity to participate in decision making, to use and develop professional skills; and keep learning (Neffa 2015). Autonomy seems to be one of the most salient work design features, but organizations need to be aware of the different autonomy dimensions and the different effects on employee attitudes and work outcomes. It is therefore crucial that based on the desired outcomes, firms have a more finely tuned understanding of the different employee autonomy dimensions so as to apply them in a more targeted way (Theurer et al. 2018). Table 19.8 summarizes the proposed methodologies.

19.3.2.3 The Health and Safety Approach to Workplace Design

Design can significantly enhance workplace safety and health when integrated into the project from conception rather than added as a stand-alone aspect during the execution phase. This approach is known as Prevention through Design [PtD] and focuses on eliminating risks posed to workers in facilities by work methods and operations, processes, equipment, tools, products, new technologies, and the organization of work, during the design phase by relying on engineering controls rather than behavior interventions (NIOSH 2013). Ezisi and Issa (2019) describe a case study aimed at developing a method to facilitate the implementation of the Prevention through Design approach and apply it to a pump station. The method used—in part—failure mode and effects analysis and involved tasking experts with analyzing the project's design documents to identify potential occupational health and safety failures that could occur throughout construction. The project's construction documents were also analyzed to determine actual, design-related occupational health and safety failures throughout construction. The application of the method identified 42 potential failure modes in the design, 38% of which were deemed high risk. A total of 18 failures was detected throughout construction. Of these, 89% were

Table 19.8 Methodologies that include human factors on workplace design

Methodology	Workplace design stage	Characteristics	Authors
Human-oriented work design and CPSs	Early workstation design stage	Uses a hypothetical model to show interdependencies between human-oriented work design and resulting job performance in regards to CPSs, applying an interdisciplinary approach	Stern and Becker (2017)
Systems-theoretical framework based on E/HF	Early workstation design stage	Organizes design decisions into five perspectives from which the design problem can be viewed holistically, to facilitate many types of design	Bligård and Berlin (2019)
Framework based on function–behavior–structure	Early workstation design stage	Considers human factors by analyzing the behavioral interactions between workers and manufacturing equipment to assess the system design	Dantan et al. (2019)
Automatic virtual ergonomic analysis	Implemented and running workstation	Automatically extracts data from virtual ergonomic analysis and measures a set of indicators to assess manufacturing ergonomics, so a rapid and objective assessment, independent from the experience of the user, is obtained	Grandi et al. (2019)
Different work configurations on CPSs	Implemented and running workstation	Designs and assess different work configurations, considering the uniqueness of human labor and the characteristics of CPSs, covering ordinary production as well as irregular scenarios	Fantini et al. (2020)

(continued)

Table 19.8 (continued)

Methodology	Workplace design stage	Characteristics	Authors
Job crafting and job features	Implemented and running workstation	Identify the relevant job features for the workforce and train employees to craft their jobs in a way that fits them and the organization better	Latika (2018)
Autonomy dimensions	Implemented and running workstation	Identify the relevant autonomy dimensions and the different effects on employee attitudes and work outcomes, so firms can apply them in a more targeted way	Theurer et al. (2018)

Source Own creation

predicted using failure mode and effects analysis and thus deemed preventable by design, indicating the potential effectiveness of the method.

Oakman and Macdonald (2019) consider that workplace risk management largely fails to address risk arising from psychosocial hazards, does not allow sufficient participation by workers, and often fails to control risk the source. To address these deficiencies, they introduce A Participative Hazard Identification and Risk Management [APHIRM] toolkit which has been formulated following a framework developed by the World Health Organization and implementation science principles. The toolkit is intended to be used by workplace managers and consultants responsible for occupational health and safety, with active participation from workers. APHIRM comprises a set of online tools that include automated data analysis and reporting modules, and procedures to guide users through the five stages of the conventional risk management cycle. It can assess both hazard and risk levels for groups of people doing a particular job, focusing on the job overall rather than only on tasks deemed to be hazardous. The resultant risk control interventions are customized to address the main physical and psychosocial hazards identified for the target job, and repetitions of the risk management cycle allow for ongoing evaluation of outcomes in terms of hazard and risk levels.

One factor that has recently emerged is the aging society in many developed countries and how it affects the ergonomic workplace design. Concerning the health of [aged] workers, it is crucial to reduce such risks. Otto et al. (2017) investigate the workplace design for order pickers that manually collect items from the shelves of a warehouse. Specifically, the authors treat storage assignment [the placement of products in shelves of different height] and zoning [partitioning of the storage space into areas assigned to separate pickers] in the fast pick area of a warehouse. The area unifies the most fast-moving items in a compact space, so that workers are relieved from unproductive travel, but face extraordinary ergonomic risks due to the frequent repetition of picking operations. A combined ergonomic storage assignment and zoning problem was defined to minimize the maximum ergonomic burden in all workers. This problem is formalized, and two construction heuristics and a tabu search procedure are proposed. Results show that neglecting ergonomic aspects and only focusing on picking performance leads to much higher ergonomic risks faced by the aging workforce.

Another factor that has sparked a growing interest in manufacturing processes design is the identification of burnout syndrome risk factors since it represents one of the main problems in the area of health at work. In her paper, Jiménez (2019) starts from new theoretical models that approach work design from an extended perspective [motivational, social, and contextual characteristics] to establish its relationship with burnout and learn to what extent the different dimensions that constitute burnout can be explained by different job characteristics. Using an ex post facto design, bivariate correlations and multiple regressions were applied to data collected from manufacturing workers. The results obtained show that dimensions of the motivational, social, and contextual characteristics of work design have a significant negative effect on burnout. For instance, it was found that the ergonomic aspects negatively explain depersonalization and burnout syndrome. Also, working conditions present a negative correlation with emotional tiredness and burnout syndrome.

One novel concept for health promotion through the workplace environment is to adopt a "salutogenic" perspective of health for workplace design, which more explicitly focuses on factors that support human health and well-being, as opposed to factors which cause disease. In its most general meaning, salutogenesis refers to a scholarly orientation focusing attention on the study of the origins of health and assets for health, versus the origins of disease and risk factors. Central to this study is that life experiences help shape one's sense of coherence. A strong sense of coherence helps one mobilize resources to cope with stressors and manage tension successfully (Mittelmark and Bauer 2017). According to Roskams and Haynes (2020), the workplace environment can be seen as a composite of pathogenic "demands" and salutogenic "resources". They identify environmental resources which might strengthen the three components of an employee's "sense of coherence" [comprehensibility, manageability and meaningfulness], an individual orientation associated with more positive health outcomes. Comprehensibility can be supported by effectively implementing a clear set of rules governing the use of the workplace. Manageability can be supported through biophilic design solutions, and through design which supports social cohesion and physical activity. Meaningfulness can be supported by recognizing the importance of personal identity expression and through design which reinforces the employees' sense of purpose.

Finally, Knapčíková et al. (2019) state that, despite the implementation of all available measures to increase safety, health, and awareness by adhering to the organization's occupational safety and health policy, undesirable situations leading to occupational accidents still occur. A system of measures that includes legislative, economic, social, organizational, technical, health, and educational indicators needs to be developed and implemented to create a safe work environment. The authors propose an educational measure to be used when working with the Tecnomatix

Plant Simulation software during the teaching process and thus, evaluate theory and practice. Table 19.9 summarizes the reviewed proposals.

Table 19.9 Workplace design and health and safety considerations

Approach	Workplace design stage	Characteristics	Authors
Prevention through design	Early workstation design stage	Use of failure mode and effects analysis, involve tasking experts and the project's construction documents to determine actual, design-related occupational health and safety failures throughout construction	Ezisi and Issa (2019)
Workplace risk management	Implemented and running workstation	Toolkit to perform risk control interventions customized to address the main physical and psychosocial hazards identified for the target job	Oakman and Macdonald (2019)
Ergonomics	Implemented and running workstation	Ergonomic redesign considering the limitations of an aging workforce	Otto et al. (2017)
Evaluation of burnout risk factors	Implemented and running workstation	Identify the relevant burnout risk factors to improve working conditions of workers	Jiménez (2019)
Salutogenic approach	Both stages	Consider the sense of coherence in supporting higher levels of physical and mental health, so as to provide truly "healthy" workplaces	Roskams and Haynes (2020)
Workforce education	Implemented and running workstation	System that measures legislative, economic, social, organizational, technical, health, and educational indicators to create a safe work environment	Knapčíková et al. (2019)

Source Own creation

19.4 Conclusion

19.4.1 Discussion

The aim of this chapter was to present new trends on workplace design. The literature relevant to the topic showed that there is a tendency to consider the design of workplaces in a more holistic way, taking into account a design concept that not only considers the characteristics of the job, but also takes into account the way workers link to their work, their colleagues and bosses and their work environment as a whole. However, in order to present the information obtained more clearly, the revised studies were classified into three categories, namely: technology, human factor and health and safety.

From a technology point of view, new trends focus on the best use of technologies already available, which have not necessarily been used in industrial environments. These tools can be used for the identification of existing problems that need to be corrected in the current process or in the implementation of new production lines.

One of the main advantages of digital technology is the ability to apply simulation techniques, which facilitate the identification of critical errors and allow the designer to test different configurations without affecting normal factory operation. Another advantage is being able to obtain valid real-time information about the activities that workers are carrying out, as well as their interaction with machinery, tools, and their co-workers. This allows for a more sophisticated level of knowledge about the actual operation of the system under evaluation.

However, there are disadvantages. By having access to a wide variety of technology tools, it is difficult to determine the one that best suits the needs of the design team. In addition, it may be that there are no protocols for the use and application of such tools. These methods must be developed by the design team, which can delay both data collection and analysis, and thus the development and implementation of the design.

In relation to the inclusion of the human factor into the design process, the literature mainly considers new methodological approaches to identify the labor factors that workers consider relevant to not only perform their work properly, but to have a good quality of life in the workplace. The authors agree that the measures considered for application in workplace redesign should benefit the company as well. This group of studies is where the use of techniques or methods linked to the relational and proactive approaches was identified, as, in both cases, their common factor when describing them is the human factor. An interesting feature of this category was the need for the use of technology for data collection and analysis, associated with the proposed methodologies.

With regard to the design approach that considers safety and health, it is important to note that some of the techniques described are considered emerging problems associated with an aging workforce and the characterization of new health risks such as burnout syndrome. On the other hand, it highlights the importance of health and safety legislation and the knowledge that workers must have about their legal responsibility in implementation and monitoring. Finally, it is important to highlight the amount of information that needs to be processed to ensure the safety of a job site and the support that technology can provide in these cases.

19.4.2 Conclusion

Manufacturing processes can be designed or redesigned depending on the product's life cycle. Original process design corresponds to the early introduction stage (Matuszek et al. 2020) looking at benefits for a new product introduction (Kuka 2018). Process re-design is related to increased benefits from the growing, maturing, and declining periods of the life cycle of the product (Caple et al. 2019; Jacquemin et al. 2012; Kuka 2018). Both stages of design or re-design of a workstation have different main goals, improvement focus, impact, methods, and techniques as shown in Table 19.2.

Health and safety have a positive correlation with workstation design and redesign according to several authors. The relationship with workstation design stages is shown in Table 19.3.

Some authors indicate that regulatory norms and regulations have less attention where legal duty is barely fulfilled and confusing because some still need a fine-tune (Al Horr et al. 2016; Cheung et al. 2016; Lingard et al. 2019). Others state that rules and regulations should be observed by process design (Lingard et al. 2019; Nafuna 2019) but are sometimes forgotten during the workstation redesign. Thus, a section with regulations and norms related to worker well-being is also presented.

Since there are too many hazards and risks at the workplace, a systematic approach for workstation design or redesign is crucial. Based on the suggestions of Parker and Griffin (2014), physical, biomechanical, cognitive, and psychosocial elements are defined as Health workplace elements since they could allow making decisions and taking actions in alignment with a wide range of hazards and risks, encouraging to eliminate or minimize them, thus promoting a healthy and safe workplace where work design optimizes human performance, productivity, and job satisfaction.

Psychosocial elements in the workplace are shown as a novelty. The correlation with psychosocial aspects in the workplace is supported by many authors where engaged employees protect their work from harm with safety behavior and self-protection (Chen et al. 2017; Dollard and Karasek 2010; Dollard et al. 2017; Kim et al. 2017).

The literature review in search of new trends provided information about emerging approaches to workplace design. An increasingly holistic approach that tries to consider the health and well-being of workers, as well as protect the interests of the organization was identified. This happens not only as part of the continuous improvement of the process but also as an approach that should be considered from the beginning and happens from a person-centered point of view.

On the other hand, an increasingly broader use of technology to identify problems in real time was observed. These tools include simulation and virtual reality technologies that could help in workplace redesign as well as to design new workstations.

Finally, a concern for emerging issues such as the aging of the economically active population and the characterization of the burnout syndrome within manufacturing systems was also identified.

References

Abdullah MZ, Othman AK, Ahmad MF, Justine M (2015) The mediating role of work-related musculoskeletal disorders on the link between psychosocial factors and absenteeism among administrative workers. Soc Work Public Health 30(1):64–74. https://doi.org/10.1080/19371918.2014.938393

Aghilinejad M, Ehsani AA, Talebi A, Koohpayehzadeh J, Dehghan N (2016) Ergonomic risk factors and musculoskeletal symptoms in surgeons with three types of surgery: open, laparoscopic, and microsurgery. Med J Islam Repub Iran 30(1):1–5

Al Horr Y, Arif M, Kaushik A, Mazroei A, Katafygiotou M, Elsarrag E (2016) Occupant productivity and office indoor environment quality: a review of the literature. Build Environ 105:369–389. https://doi.org/10.1016/j.buildenv.2016.06.001

Allahyari T, Mortazavi N, Khalkhali HR, Sanjari MA (2016) Shoulder girdle muscle activity and fatigue in traditional and improved design carpet weaving workstations. Int J Occup Med Environ Health 29(2):345–354. https://doi.org/10.13075/ijomeh.1896.00589

Andersen AL, El Maraghy H, El Maraghy W, Brunoe TD, Nielsen K (2018) A participatory systems design methodology for changeable manufacturing systems. Int J Prod Res 56(8):2769–2787

Barber LK, Santuzzi AM (2014) Please respond ASAP: workplace telepressure and employee recovery. J Occup Health Psychol 20(2). https://doi.org/10.1037/a0038278

Barrios León M, Illada R (2013) Valoración del desgaste laboral como riesgo psicosocial. Ing Ind 12(1):69–76

Beheshti MH, Tajpuor A, Jari A, Samadi S, Borhani Jebeli M, Rahmanzadeh H (2018) Evaluation of ergonomic risk factors for musculoskeletal disorders among kitchen workers. Arch Occup Health 2(2):128–135

Blakstad SH, Hatling M, Bygdås AL (2009) The knowledge workplace–searching for data on use of open plan offices. In: Proceedings EFMC 2009 research symposium, June 2009, pp 16–17

Bligård LO, Berlin C (2019) ACD3 as a framework for design of ergonomic workplaces. Work 62(1):5–12. https://doi.org/10.3233/WOR-182836

Boulila A, Ayadi M, Mrabet K (2018) Ergonomics study and analysis of workstations in Tunisian mechanical manufacturing. Hum Factors Ergon Manuf Serv Ind 28(4):166–185. https://doi.org/10.1002/hfm.20732

BSI (2011) PAS 1010:2011 Guidance on the management of psychosocial risks in the workplace

Butt J, Jedi S (2020) Redesign of an in-market conveyor system for manufacturing cost reduction and design efficiency using DFMA methodology. Designs 4(1):6. https://doi.org/10.3390/designs4010006

Cámara de Diputados del H. Congreso de la Unión (2014) Ley Orgánica de la Administración Pública Federal. In: Cámara de Diputados del H. Congreso de la Unión (ed). Diario Oficial de la Federación

Cámara de Diputados del H. Congreso de la Unión (2018) Ley del Seguro Social. Diario Oficial de la Federación

Caple D, Austin N, Begg F, Fitzgerald C, Hurst G, McLean C, Rigb T (2019) Chapter 34.3: health and safety in design in the OHS body of knowledge. Retrieved 17 Sept 2020 from https://www.ohsbok.org.au/chapter-34-3-health-and-safety-in-design/#1548328017774-594b1d39-e14f

Caputo F, Greco A, Fera M, Macchiaroli R (2019) Digital twins to enhance the integration of ergonomics in the workplace design. Int J Ind Ergon 71:20–31. https://doi.org/10.1016/j.ergon.2019.02.001

Castillo Martínez JA (2007) Elementos cognitivos para el análisis ergonómico del trabajo. Editorial Universidad del Rosario

Chanchai W, Songkham W, Ketsomporn P, Sappakitchanchai P, Siriwong W, Robson MG (2016) The impact of an ergonomics intervention on psychosocial factors and musculoskeletal symptoms among Thai hospital orderlies. Int J Environ Res Public Health 13(5). https://doi.org/10.3390/ijerph13050464

Charria VH, Sarsosa KV, Arenas F (2011) Factores de riesgo psicosocial laboral: métodos e instrumentos de evaluación. Rev Fac Nac Salud Pública 29(4):380–391

Chávez SAF, Páramo MS, Ortiz MOP, Velasco MDLAA (2017) Enfermedades músculo-esqueléticas por agentes ergonómicos en trabajadores afiliados al Instituto Mexicano del Seguro Social, México/Musculoskeletal diseases caused by ergonomic agents on workers affiliated to the Mexican Institute of Social Security, Mexico. Rev Int Hum Méd 6(1). https://doi.org/10.37467/gka-revmedica.v6.1466

Chen Y, McCabe B, Hyatt D (2017) Impact of individual resilience and safety climate on safety performance and psychological stress of construction workers: a case study of the Ontario construction industry. J Saf Res 61:167–176. https://doi.org/10.1016/j.jsr.2017.02.014

Cheung SS, Lee JKW, Oksa J (2016) Thermal stress, human performance, and physical employment standards. Appl Physiol Nutr Metab 41:S148–S164. https://doi.org/10.1139/apnm-2015-0518

Cohen Y, Golan M, Singer G, Faccio M (2018) Workstation–operator interaction in 4.0 era: WOI 4.0. IFAC-PapersOnLine 51(11):399–404. https://doi.org/10.1016/j.ifacol.2018.08.327

Colim A, Faria C, Braga AC, Sousa N, Rocha L, Carneiro P, Costa N, Arezes P (2020) Towards an ergonomic assessment framework for industrial assembly workstations—a case study. Appl Sci 10(9):3048. https://doi.org/10.3390/app10093048

Cox T, Griffiths A, Rial-Gonzalez E (2000) Research on work related stress. European Agency for Safety and Health at Work, Office for Official Publications of the European Communities, Luxembourg

Cox T, Griffiths A, Rial-González E (2005) Investigación sobre el estrés relacionado con el trabajo. Agencia Europea para la Seguridad y la Salud en el Trabajo

Daniels K, Gedikli C, Watson D, Semkina A, Vaughn O (2017) Job design, employment practices and well-being: a systematic review of intervention studies. Ergonomics 60(9):1177–1196. https://doi.org/10.1080/00140139.2017.1303085

Dantan JY, El Mouayni I, Sadeghi L, Siadat A, Etienne A (2019) Human factors integration in manufacturing systems design using function–behavior–structure framework and behaviour simulations. CIRP Ann 68(1):125–128. https://doi.org/10.1016/j.cirp.2019.04.040

Darvishi E, Maleki A, Giahi O, Akbarzadeh A (2016) Subjective mental workload and its correlation with musculoskeletal disorders in bank staff. J Manipulative Physiol Ther 39(6):420–426. https://doi.org/10.1016/j.jmpt.2016.05.003

De Araújo Vieira EM, Bueno de Silva L, Lopes de Souza E (2016) The influence of the workplace indoor environmental quality on the incidence of psychological and physical symptoms in intensive care units. Build Environ 109:12–24. https://doi.org/10.1016/j.buildenv.2016.09.007

Deeney C, O'Sullivan LW (2017) Effects of cognitive loading and force on upper trapezius fatigue. Occup Med 67(9):678–683. https://doi.org/10.1093/occmed/kqx157

Del Rio Vilas D, Longo F, Monteil NR (2013) A general framework for the manufacturing workstation design optimization: a combined ergonomic and operational approach. Simulation 89(3):306–329. https://doi.org/10.1177/0037549712462862

Dollard MF, Karasek RA (2010) Building psychosocial safety climate. Contemp Occup Health Psychol Glob Perspect Res Pract 1:208–233

Dollard MF, Dormann C, Tuckey MR, Escartín J (2017) Psychosocial safety climate (PSC) and enacted PSC for workplace bullying and psychological health problem reduction. Eur J Work Organ Psychol 26(6):844–857. https://doi.org/10.1080/1359432X.2017.1380626

Dorst K (2011) The core of 'design thinking' and its application. Des Stud 32(6):521–532. https://doi.org/10.1016/j.destud.2011.07.006
Driscoll TR, Harrison JE, Bradley C, Newson RS (2008) The role of design issues in work-related fatal injury in Australia. J Saf Res 39(2):209–214. https://doi.org/10.1016/j.jsr.2008.02.024
Eldar R, Fisher-Gewirtzman D (2019) Ergonomic design visualization mapping-developing an assistive model for design activities. Int J Ind Ergon 74:102859. https://doi.org/10.1016/j.ergon.2019.102859
El-Nounu A, Popov A, Ratchev S (2018) Redesign methodology for mechanical assembly. Res Eng Des 29(1):107–122. https://doi.org/10.1007/s00163-017-0255-6
Ezisi U, Issa MH (2019) Case study application of prevention through design to enhance workplace safety and health in Manitoba heavy construction projects. Can J Civ Eng 46(2):124–133. https://doi.org/10.1139/cjce-2017-0454
Fantini P, Pinzone M, Taisch M (2020) Placing the operator at the centre of Industry 4.0 design: modelling and assessing human activities within cyber-physical systems. Comput Ind Eng 139:105058. https://doi.org/10.1016/j.cie.2018.01.025
Foldspang L, Mark M, Hjorth LR, Langholz-Carstensen C, Poulsen OM, Johansson U, Rants LL (2014) Working environment and productivity: a register-based analysis of Nordic enterprises. Nordic Council of Ministers, Denmark
Freeney Y, Fellenz MR (2013) Work engagement, job design and the role of the social context at work: exploring antecedents from a relational perspective. Hum Relat 66(11):1427–1445
García-Gómez F, González-Gaya C, Rosales-Prieto VF (2020) An approach to health and safety assessment in industrial parks. Sustainability 12(9):3646. https://doi.org/10.3390/su12093646
Gerr F, Fethke NB, Anton D, Merlino L, Rosecrance J, Marcus M, Jones MP (2014a) A prospective study of musculoskeletal outcomes among manufacturing workers: II. Effects of psychosocial stress and work organization factors. Hum Factors 56(1):178–190. https://doi.org/10.1177/0018720813487201
Gerr F, Fethke NB, Merlino L, Anton D, Rosecrance J, Jones MP, Meyers AR (2014b) A prospective study of musculoskeletal outcomes among manufacturing workers: I. Effects of physical risk factors. Hum Factors 56(1):112–130. https://doi.org/10.1177/0018720813491114
Gil-Monte PR (2012) Riesgos psicosociales en el trabajo y salud ocupacional. Rev Peru Med Exp Salud Publica 29(2):237–241
Goh J, Pfeffer J, Zenios SA, Rajpal S (2015). Workplace stressors & health outcomes: health policy for the workplace. Behav Sci Policy 1(1):43–52. https://doi.org/10.1353/bsp.2015.0001
Gonçalves MT, Salonitis K (2017) Lean assessment tool for workstation design of assembly lines. Paper presented at the 27th CIRP design 2017, United Kingdom, Europe
Granath JA, Hinnerson J, Lindahl G (2005) Case study: Orebro University Hospital. The O-building. Usability of workplaces. Report on case studies. CiB Task Group, 51
Grandi F, Peruzzini M, Zanni L, Pellicciari M (2019) An automatic procedure based on virtual ergonomic analysis to promote human-centric manufacturing. Procedia Manuf 38:488–496. https://doi.org/10.1016/j.promfg.2020.01.062
Grant AM, Parker SK (2009) Redesigning work design theories: the rise of relational and proactive perspectives. Acad Manag Ann 3(1):317–375
Grewal S (2011) Manufacturing process design and costing: an integrated approach. Springer, London
Grote G (2014) Adding a strategic edge to human factors/ergonomics: principles for the management of uncertainty as cornerstones for system design. Appl Ergon 45(1):33–39. https://doi.org/10.1016/j.apergo.2013.03.020
Gujski M, Pinkas J, Juńczyk T, Pawełczak-Barszczowska A, Raczkiewicz D, Owoc A, Bojar I (2017) Stress at the place of work and cognitive functions among women performing intellectual work during peri- and post-menopausal period. Int J Occup Med Environ Health 30(6):943–961. https://doi.org/10.13075/ijomeh.1896.01119

Haghshenas B, Habibi E, Haji Esmaeil Hajar F, Ghanbary Sartang A, van Wijk L, Khakkar S (2018) The association between musculoskeletal disorders with mental workload and occupational fatigue in the office staff of a communication service company in Tehran, Iran, in 2017. J Occup Health Epidemiol 7(1):20–29

Hawes BK, Brunyé TT, Mahoney CR, Sullivan JM, Aall CD (2012) Effects of four workplace lighting technologies on perception, cognition and affective state. Int J Ind Ergon 42(1):122–128. https://doi.org/10.1016/j.ergon.2011.09.004

Heidarimoghadam R, Saidnia H, Joudaki J, Mohammadi Y, Babamiri M (2019) Does mental workload can lead to musculoskeletal disorders in healthcare office workers? Suggest and investigate a path. Cogent Psychol 6(1):1664205. https://doi.org/10.1080/23311908.2019.1664205

Henning K, Wolfgang W, Johannes H (2013) Recommendations for implementing the strategic initiative INDUSTRIE 4.0. Final report of the Industrie 4:82

Hoedt S, Claeys A, Van Landeghem H, Cottyn J (2017) The evaluation of an elementary virtual training system for manual assembly. Int J Prod Res 55(24):7496–7508. https://doi.org/10.1080/00207543.2017.1374572

Hollnagel E (2014) Human factors/ergonomics as a systems discipline? "The human use of human beings" revisited. Appl Ergon 45(1):40–44. https://doi.org/10.1016/j.apergo.2013.03.024

Hossain MD, Aftab A, Al Imam MH, Mahmud I, Chowdhury IA, Kabir RI, Sarker M (2018) Prevalence of work related musculoskeletal disorders (WMSDs) and ergonomic risk assessment among readymade garment workers of Bangladesh: a cross sectional study. PLoS ONE 13(7):e0200122. https://doi.org/10.1371/journal.pone.0200122

Hu Q, Schaufeli WB, Taris TW (2017) How are changes in exposure to job demands and job resources related to burnout and engagement? A longitudinal study among Chinese nurses and police officers. Stress Health 33(5):631–644. https://doi.org/10.1002/smi.2750

ILO (2016) Estrés en el trabajo: un reto colectivo. In: Organización Internacional del Trabajo (ed). Organización Internacional del Trabajo

ILO (2017a) Occupational safety and health in the oil and gas industry in selected sub-Saharan African countries

ILO (2017b) Psychosocial risks, stress and violence in the world of work. Geneva

ILO and WHO (1986) Psychosocial factors at work. Recognition and control, vol 56. International Labor Organization

IMSS (2020) La fisioterapia, uno de los pilares de la rehabilitación en el IMSS. Boletín de Prensa No. 620. Retrieved 15 Sept 2020 from https://www.imss.gob.mx/prensa/archivo/202009/620

Instituto Mexicano de Normalización y Certificación AC (2008) Sistema de Gestión de Seguridad y Salud en el Trabajo—Requisitos NMX-SAST001-IMNC-2008 In: IMNC

Inyang N, Al-Hussein M, El-Rich M, Al-Jibouri S (2012) Ergonomic analysis and the need for its integration for planning and assessing construction tasks. J Constr Eng Manag 138(12):1370–1376

ISO (n.d.) Standards. Retrieved 12 Sept 2020 from https://www.iso.org/standards.html

Iuga A, Popa V, Popa L (2017) Industrial product life cycle stages and lifecycle eco-design. In International conference on advanced manufacturing engineering and technologies, June 2017. Springer, Cham, pp 365–374

Jacquemin L, Pontalier PY, Sablayrolles C (2012) Life cycle assessment (LCA) applied to the process industry: a review. Int J Life Cycle Assess 17(8):1028–1041. https://doi.org/10.1007/s11367-012-0432-9

Jespersen AH, Hohnen P, Hasle P (2016) Internal audits of psychosocial risks at workplaces with certified OHS management systems. Saf Sci 84:201–209. https://doi.org/10.1016/j.ssci.2015.12.013

Jiménez EMH (2019) Work design and burnout: a study of their relationship. Rev Psicol Organ Trab 19(4):744–754. https://doi.org/10.17652/rpot/2019.4.17502

Khanzode VV, Maiti J, Ray PK (2012) Occupational injury and accident research: a comprehensive review. Saf Sci 50(5):1355–1367. https://doi.org/10.1016/j.ssci.2011.12.015

Kim KW, Park SJ, Lim HS, Cho HH (2017) Safety climate and occupational stress according to occupational accidents experience and employment type in shipbuilding industry of Korea. Saf Health Work 8(3):290–295. https://doi.org/10.1016/j.shaw.2017.08.002

Knapčíková L, Behúnová A, Behún M (2019) Workplace optimization in the manufacturing enterprise supported by digitalization software. TEM J 8(4):1259. https://doi.org/10.18421/TEM84-22

Körner U, Müller-Thur K, Lunau T, Dragano N, Angerer P, Buchner A (2019) Perceived stress in human–machine interaction in modern manufacturing environments—results of a qualitative interview study. Stress Health 35(2):187–199. https://doi.org/10.1002/smi.2853

Kovács G, Kot S (2017) Facility layout redesign for efficiency improvement and cost reduction. J Appl Math Comput Mech 16(1):63–74. https://doi.org/10.17512/jamcm.2017.1.06

Kralikova R, Dzunova L, Pinosova M, Wessely E, Koblasa F (2019) Man-machine-environment system analyses and impact of environment factors to productivity and health of employees. Ann DAAAM Proc 30

Kuka MGK (2018) Product development and management strategies. In: Product lifecycle management: terminology and applications, 11

Latika K (2018) The effect of job features on job crafting as a developmental intervention. J Contemp Manag Res 12(2):36–63

Leccese F, Salvadori G, Montagnani C, Ciconi A, Rocca M (2017) Lighting assessment of ergonomic workstation for radio diagnostic reporting. Int J Ind Ergon 57:42–54. https://doi.org/10.1016/j.ergon.2016.11.005

Leka S, Jain A (2016) International initiatives to tackle psychosocial risks and promote mental health in the workplace: is there a good balance in policy and practice? In: Psychosocial factors at work in the Asia Pacific. Springer, Cham, pp 23–43

Leka S, Van Wassenhove W, Jain A (2015) Is psychosocial risk prevention possible? Deconstructing common presumptions. Saf Sci 71:61–67. https://doi.org/10.1016/j.ssci.2014.03.014

Lin RT, Chan CC (2007) Effectiveness of workstation design on reducing musculoskeletal risk factors and symptoms among semiconductor fabrication room workers. Int J Ind Ergon 37(1):35–42. https://doi.org/10.1016/j.ergon.2006.09.015

Lingard H, Leifels K, Rahnama S, Fletcher H, Harley J (2019) Applying the hierarchy of control to occupational health risks in construction: barriers to effective decision-making

Llorente-Alonso M, Topa G (2019) Individual crafting, collaborative crafting, and job satisfaction: the mediator role of engagement. J Work Organ Psychol 35(3):217–226. https://doi.org/10.5093/jwop2019a23

Maakip I, Keegel T, Oakman J (2016) Prevalence and predictors for musculoskeletal discomfort in Malaysian office workers: investigating explanatory factors for a developing country. Appl Ergon 53:252–257. https://doi.org/10.1016/j.apergo.2015.10.008

Machac J, Steiner F, Tupa J (2017) Product life cycle risk management. In: Risk management treatise for engineering practitioners. IntechOpen

Magnusson Hanson LL, Chungkham HS, Åkerstedt T, Westerlund H (2014) The role of sleep disturbances in the longitudinal relationship between psychosocial working conditions, measured by work demands and support, and depression. Sleep 37(12):1977–1985. https://doi.org/10.5665/sleep.4254

Martin M (1983) Cognitive failure: everyday and laboratory performance. Bull Psychon Soc 21(2):97–100

Martín Leal S, Rojas Sánchez GA (2014) Exposición a ruido en la fábrica de Materiales Higiénico Sanitarios de Sancti Spíritus. Gac Méd Espirit 16(1):34–40

Matuszek J, Seneta T, Moczała A (2020) Assessment of the design for manufacturability using fuzzy logic. Appl Sci 10(11):3935. https://doi.org/10.3390/app10113935

Michalos G, Karvouniari A, Dimitropoulos N, Togias T, Makris S (2018) Workplace analysis and design using virtual reality techniques. CIRP Ann 67(1):141–144. https://doi.org/10.1016/j.cirp.2018.04.120

Milczarek M, Brun E, Houtman I, Goudswaard A, Evers M, Bovenkamp M et al (2007) Expert forecast on emerging psychosocial risks related to occupational safety and health. European Agency for Safety and Health at Work

Mittelmark MB, Bauer GF (2017) The meanings of salutogenesis. The handbook of salutogenesis 7–13

Molleman E, Slomp J (2001) The impact of technological innovations on work design in a cellular manufacturing environment. New Technol Work Employ 16(3):152–163. https://doi.org/10.1111/1468-005X.00085

Moreno-Jiménez B, Báez-León C (2010) Factores y riesgos psicosociales, formas, consecuencias, medidas y buenas prácticas. Universidad Autónoma de Madrid

Mourtzis D (2016) Challenges and future perspectives for the life cycle of manufacturing networks in the mass customisation era. Logist Res 9(1):2. https://doi.org/10.1007/s12159-015-0129-0

Mourtzis D (2020) Simulation in the design and operation of manufacturing systems: state of the art and new trends. Int J Prod Res 58(7):1927–1949. https://doi.org/10.1080/00207543.2019.1636321

Nafuna BA (2019) Occupational safety and health hazards in work places: a case study of Sugar Corporation of Uganda Limited. Doctoral dissertation, Makerere University

Neffa JC (2015) Los riesgos psicosociales en el trabajo: contribución a su estudio. CEIL-CONICET, Buenos Aires

NIOSH (2013) Prevention through design. Center for Disease Control and Prevention. Available from https://www.cdc.gov/niosh/topics/ptd/

Nurwani NE, Minghat AD (2018) Knowledge about safety practices in workshop among student in Industrial Training Institute. J Adv Res Occup Saf Health 2(1):7–12

Oakman J, Macdonald W (2019) The APHIRM toolkit: an evidence-based system for workplace MSD risk management. BMC Musculoskelet Disord 20(1):504. https://doi.org/10.1186/s12891-019-2828-1

Oldham GR, Fried Y (2016) Job design research and theory: past, present and future. Organ Behav Hum Decis Process 136:20–35. https://doi.org/10.1016/j.obhdp.2016.05.002

Onawumi AS, Dunmade IS, Ajayi OO, Adebiyi KA, Omotosho AO (2016) Investigation of work related health hazards prevalent among metal fabrication workers in Nigeria. Eur J Eng Technol 4(4)

Otto A, Boysen N, Scholl A, Walter R (2017) Ergonomic workplace design in the fast pick area. OR Spectrum 39(4):945–975. https://doi.org/10.1007/s00291-017-0479-x

Parker SK (2014) Beyond motivation: job and work design for development, health, ambidexterity, and more. Annu Rev Psychol 65. https://doi.org/10.1146/annurev-psych-010213-115208

Parker SK (2015) Does the evidence and theory support the good work design principles? Safe Work Australia, Canberra

Parker SK, Griffin MA (2014) Principles and evidence for good work through effective design. Report commissioned by Comcare (RFQ) 13/373 to inform the Safe Work Australia Members Collaborative Project 'Good Work Through Effective Design'

Parsons KC (2000) Environmental ergonomics: a review of principles, methods and models. Appl Ergon 31(6):581–594. https://doi.org/10.1016/S0003-6870(00)00044-2

Peruzzini M, Grandi F, Pellicciari M (2018) How to analyse the workers' experience in integrated product-process design. J Ind Inf Integr 12:31–46. https://doi.org/10.1016/j.jii.2018.06.002

Peruzzini M, Pellicciari M, Gadaleta M (2019) A comparative study on computer-integrated set-ups to design human-centred manufacturing systems. Robot Comput-Integr Manuf 55:265–278. https://doi.org/10.1016/j.rcim.2018.03.009

Peruzzini M, Grandi F, Pellicciari M (2020) Exploring the potential of Operator 4.0 interface and monitoring. Comput Ind Eng 139:105600

Pilemalm S (2018) Participatory design in emerging civic engagement initiatives in the new public sector: applying PD concepts in resource-scarce organizations. ACM Trans Comput-Hum Interact (TOCHI) 25(1):1–26

Pinosova M, Andrejiova M, Kralikova R, Hricova B, Lumnitezer E, Wessely E (2015) Assessment of hearing impairment risk from the point of view of long-term exposure to noise in working environment. In: DAAAM international scientific book, pp 345–358

Rathnayaka S, Khan F, Amyotte P (2014) Risk-based process plant design considering inherent safety. Saf Sci 70:438–464. https://doi.org/10.1016/j.ssci.2014.06.004

Richter A, Leyer M, Steinhüser M (2020) Workers united: digitally enhancing social connectedness on the shop floor. Int J Inf Manage 52:102101. https://doi.org/10.1016/j.ijinfomgt.2020.102101

Rosário S, Fonseca JA, Nienhaus A, da Costa JT (2016) Standardized assessment of psychosocial factors and their influence on medically confirmed health outcomes in workers: a systematic review. J Occup Med Toxicol 11(1):19. https://doi.org/10.1186/s12995-016-0106-9

Roskams M, Haynes B (2020) Salutogenic workplace design: a conceptual framework for supporting sense of coherence through environmental resources. J Corp Real Estate 22(2):139–153. https://doi.org/10.1108/JCRE-01-2019-0001

Rubio-Valdehita S, López-Núñez MI, López-Higes R, Díaz-Ramiro EM (2017) Development of the CarMen-Q questionnaire for mental workload assessment. Psicothema 29(4):570–576. https://doi.org/10.7334/psicothema2017.151

Sammarco JJ, Pollard JP, Porter WL, Dempsey PG, Moore CT (2012) The effect of cap lamp lighting on postural control and stability. Int J Ind Ergon 42(4):377–383. https://doi.org/10.1016/j.ergon.2012.04.001

Schabracq MJ (2003) What an organisation can do about its employees' well-being and health: an overview. In: Schabracq MJ, Winnubst JAM, Cooper CL (eds) Handbook of work & health psychology, 2nd edn. Wiley, West Sussex, p 585

Schonfeld IS, Chang CH (2017) Occupational health psychology. Springer Publishing Company

Solarzano A (n.d.) Product life cycle and strategies. Retrieved from https://asolorzano125.wordpress.com

Sorensen G, Sparer E, Williams JA, Gundersen D, Boden LI, Dennerlein JT et al (2018) Measuring best practices for workplace safety, health and wellbeing: the workplace integrated safety and health assessment. J Occup Environ Med 60(5):430. https://doi.org/10.1097/JOM.0000000000001286

Stenfors CU, Hanson LM, Oxenstierna G, Theorell T, Nilsson LG (2013) Psychosocial working conditions and cognitive complaints among Swedish employees. PLoS ONE 8(4):e60637. https://doi.org/10.1371/journal.pone.0060637

Stern H, Becker T (2017) Development of a model for the integration of human factors in cyber-physical production systems. Procedia Manuf 9:151–158. https://doi.org/10.1016/j.promfg.2017.04.030

STPS (2014) Reglamento Federal de Seguridad y Salud en el Trabajo, 13 Nov 2014. Diario Oficial de la Federación, México, p 39

STPS (2018a) Norma Oficial Mexicana NOM-035-STPS-2018 Factores de riesgo psicosocial en el trabajo—Identificación, análisis y prevención, 23 Oct 2018. Diario Oficial de la Federación, México, pp 85–128

STPS (2018b) Norma Oficial Mexicana NOM-036-STPS-2018 Factores de riesgo ergonómico en el Trabajo-Identificación, análisis, prevención y control. Parte 1: Manejo manual de cargas, 23 Nov 2018. Diario Oficial de la Federación, México, pp 20–69

Tadesse S, Kelaye T, Assefa Y (2016) Utilization of personal protective equipment and associated factors among textile factory workers at Hawassa Town, Southern Ethiopia. J Occup Med Toxicol 11(1):6. https://doi.org/10.1186/s12995-016-0096-7

Theurer CP, Tumasjan A, Welpe IM (2018) Contextual work design and employee innovative work behavior: when does autonomy matter? PLoS ONE 13(10):e0204089. https://doi.org/10.1371/journal.pone.0204089

Thorvald P, Lindblom J, Andreasson R (2019) On the development of a method for cognitive load assessment in manufacturing. Robot Comput-Integr Manuf 59:252–266. https://doi.org/10.1016/j.rcim.2019.04.012

Unda S, Uribe F, Jurado S, García M, Tovalín H, Juárez A (2016) Elaboración de una escala para valorar los factores de riesgo psicosocial en el trabajo de profesores universitarios. J Work Organ Psychol 32(2):67–74. https://doi.org/10.1016/j.rpto.2016.04.004

Vandergrift JL, Gold JE, Hanlon A, Punnett L (2011) Physical and psychosocial ergonomic risk factors for low back pain in automobile manufacturing workers. Occup Environ Med 69(1):29–34. https://doi.org/10.1136/oem.2010.061770

Vargas JDP, Ríos MF, Vielma RGR (2012) Diseño del trabajo y satisfacción con la vida. Rev Venez Gerenc 17(59):466–481

Vaughn RC (2010) Introducción a la ingeniería industrial, 2a. edn. Reverte

Vinay D, Kwatra S, Sharma S, Kaur N (2012) Ergonomic implementation and work station design for quilt manufacturing unit. Indian J Occup Environ Med 16(2):79–83. https://doi.org/10.4103/0019-5278.107081

Vos T, Barber RM, Bell B, Bertozzi-Villa A, Biryukov S, Bolliger I, Duan L (2015) Global, regional, and national incidence, prevalence, and years lived with disability for 301 acute and chronic diseases and injuries in 188 countries, 1990–2013: a systematic analysis for the Global Burden of Disease Study 2013. Lancet 386(9995):743–800

Wallace JC, Chen G (2005) Development and validation of a work-specific measure of cognitive failure: implications for occupational safety. J Occup Organ Psychol 78(4):615–632. https://doi.org/10.1348/096317905X37442

Wang HJ, Demerouti E, Blanc PL, Lu CQ (2018) Crafting a job in 'tough times': when being proactive is positively related to work attachment. J Occup Organ Psychol 91(3):569–590. https://doi.org/10.1111/joop.12218

Wang B, Liu Y, Parker SK (2020) How does the use of information communication technology affects individuals: a work design perspective. Acad Manag Ann 14(2):695–725. https://doi.org/10.5465/annals.2018.0127

Waschull S, Bokhorst JAC, Molleman E, Wortmann JC (2019) Work design in future industrial production: transforming towards cyber-physical systems. Comput Ind Eng 139. https://doi.org/10.1016/j.cie.2019.01.053

Weber A (2005) Lean workstations: organized for productivity. Assembly 48(2):40. Retrieved from https://www.assemblymag.com/articles/84001-lean-workstations-organized-for-productivity

WHO (2007) Salud de los trabajadores: plan de acción mundial. Organización Mundial de la Salud

Widanarko B, Legg S, Devereux J, Stevenson M (2014) The combined effect of physical, psychosocial/organisational and/or environmental risk factors on the presence of work-related musculoskeletal symptoms and its consequences. Appl Ergon 45(6):1610–1621. https://doi.org/10.1016/j.apergo.2014.05.018

Wrzesniewski A, Dutton JE (2001) Crafting a job: revisioning employees as active crafters of their work. Acad Manag Rev 26(2):179–201. https://doi.org/10.5465/amr.2001.4378011

Chapter 20
Ergonomic Study of Construction Workers in Odisha (India): A Case Study in Construction Sites

Debesh Mishra and Suchismita Satapathy

Abstract Work-related productivity falls for the most part on account of labor, which is the most significant factor. Since labor has to know how the work is to be done and issues can be decreased by taking usual breaks, avoiding multitasks, minimizing intrusions during work. Moreover, improper design of workplaces causes most of the occupational accidents as well as injuries in production sites that are labor-intensives. A number of factors adversely affect the workers reducing their productivity such as compelling efforts, repetitive-tasks, static and awkward postures, tools, and materials. Thus, adequate safety management is required to be undertaken and proper postures need to be followed by the workers. Therefore, an attempt was carried out in the present study to identify the potential risk factors for the construction site workers through ergonomic interventions, such that through proper education, training, and ergonomic design, the hazards and dis-comfortableness can be reduced to a minimum level.

Keywords Construction · Musculoskeletal disorders · Ergonomics · REBA · QEC · Odisha

20.1 Introduction

The present construction projects have become more challenging and complex. Construction workers require a wide assortment of aptitudes to be able to accomplish venture objectives inside specific time period, spending plan, and determinations. These are more work concentrated as well as labor-intensive in comparison to other manufacturing industries. The intricacy of undertakings frequently expects laborers to go past the unnatural physical breaking points, or perform redundant assignments for a longer duration. After some time, such continued physical interest on laborers' bodies may cause health problems and substantial injuries. And the injuries related to works of this sort are known as Musculoskeletal Disorders (MSDs). Other than

D. Mishra · S. Satapathy (✉)
KIIT Deemed to be University, Bhubaneswar, Odisha, India

J. L. Garcia Alcaraz et al. (eds.), *Trends in Industrial Engineering Applications to Manufacturing Process*, https://doi.org/10.1007/978-3-030-71579-3_20

the unfavorable physical ramifications, MSDs can likewise prompt significant financial misfortunes. Various rules and guidelines + have been established by different health and safety organizations as remedial measures for the identification of the risks associated with certain tasks performances. Such endeavors go for ergonomic structure of the working environment to coordinate physical employments with laborers' regular abilities. One of the most hazardous industries is the construction field in which laborers are presented to an assortment of risks in site. In the form of new hazards such as physical, chemical, biological, and ergonomic hazards are arising on a regular basis in sites during the performance of different tasks. These kinds of hazards may result in different work-related injuries as well as illness. Ergonomics in working environment typically exist inside three variables; undertakings which laborers ought to perform, laborers who are playing out the errand, and working environment condition. A portion of the factor which influences the musculoskeletal framework incorporates working in the same position for long time, staying at work longer than required, conveying, lifting, or moving substantial materials and so forth, working in clumsy stance, insufficient the preparing, performing same assignments again and again, mental-pressure, work-fulfillment, poor work rehearses and so on. This may bring about injuries to lower and upper extremities of the laborers body. Thus, an attempt was carried out in the present study to identify the potential risk factors for the construction site workers through ergonomic interventions, so as to reduce the hazards and dis-comfortableness to a minimum level. The novelty of this research lies in the ergonomic-assessment and possible recommendations by a group of experts in order to avoid the potential ergonomic risk factors, which has not been carried out so far in the constructional sectors of Odisha in India.

20.2 Literature Review

The assessment of work tasks is involved in ergonomics for recognizing the ergonomic risk factors, such that suitable engineering and job practices can be undertaken so as to control or decrease the recognized hazardous factors. Typically, ergonomic changes are made to acquire better fit among the work or task demands and the labor's capabilities (Woodside 1997), instead of physical forcing of the labors' body to fit the work (Koningsveld and Van der Molen 1997). Injuries' tracking allows untimely detection of injury risks, and hence can bring about more effectual prevention (Welch and Hunting 2003). There are many tasks where the workers perform manual materials-handling tasks on ground surfaces that are not completely flat in various work environments, such as in agriculture, constructions, and maritime workplaces, etc. The lifting technique or strategy may be influenced by the sloped ground surfaces, adopted by the lifter. Which-in-turn alters the spinal bio-mechanical loading.

Shin and Mirka (2004) have revealed in a study that slope-angle of ground influences the lifting kinetics and kinematics, and for that reason it requires due attention while the evaluation of risks needs to be done for lower back injuries under such working conditions. Work related musculoskeletal disorders have been revealed as a widespread occupational health problem for construction workers (Stattin and Järvholm 2005), which is approximately 16% higher than other industrial workers (CPWR 2013). higher physical demands is the major cause of musculoskeletal disorders among construction workers (Schneider 2001), such as lifting heavy loads, repetitive actions, in addition to awkward working postures like bending and twisting, going down on knees, working with arms above the height of shoulders (CPWR 2013; Jaffar et al. 2011). The reduction of the risk of musculoskeletal disorders among the construction workers may be achieved through ergonomic solutions (CPWR 2013; van der Molen et al. 2005).

In most cases, ergonomists have the opinion that for construction workers, the reduction in the physical workload should be facilitated by the use of ergonomic measures (Dale et al. 2012; Kramer et al. 2010). Albers and Hudock (2007) have conducted a study of iron-workers for evaluating their risk of development in hand and back injuries from hand-tying reinforcing steel bar, by measuring wrist and forearm movements with gonio-meters as well as videotaping, and analyzed trunk postures. The power tying tools were revealed as an effective intervention for the prevention of work-related musculoskeletal disorders. Mitropoulos and Namboodiri (2011) have used the "Task Demand Assessment (TDA)" technique to measure the safety risk in construction activities on two different operations, such as a roofing activity and a concrete paving process, and then analyzed the affect on the potential for accidents by variations in operation parameters. Memarian and Mitropoulos (2013) have analyzed a large masonry company, by considering three perspective incidents such as the nature of events i.e. falls, overexertion, etc., the production activity performed by the injured worker, and the incident in relation to the position of injured workers in the crew. It was found that the accident rate for laborers were significantly higher than the masons. Aminbakhsh et al. (2013) have made an attempt to prioritize the safety risks in construction projects, based on the theory of "Cost of Safety" model and the "Analytic Hierarchy Process (AHP)". Muhammad et al. (2013) have carried out a study for construction works in Chittagong (Bangladesh) during June 2012 to November 2012 to observe the status of practicing safety regulations and ergonomic principles.

The lower back pain was seen to constitute 25% of total work-related musculoskeletal symptoms as a result of manual handling of heavy materials, while elbow and wrist injuries constituted 19.8% because of cutting operations. From the study, it was found that a total of nineteen construction workers reported of consulting physicians for their musculoskeletal injuries, and 5.5% of total workers for lower back injuries. The leading causes of work-related musculoskeletal symptoms were identified as awkward posture and working in same positions for longer periods. Yuan and Buchholz (2014) have used a bio-mechanical model incorporated approach for drywall installers to investigate the effects of positions along with sizes of drywall on physical demands. The results showed that for a single person, it would be physically difficult and even impossible for lifting heavier and bigger drywall sheets. Thus, it was recommended to have sound engineering such as lifting tables, and administrative solutions like teamwork with two-person to handle massive drywall sheets.

Julitta et al. (2015) have evaluated the use of ergonomic measures associated with musculoskeletal disorders among construction workers. The response rate was obtained as 63% i.e. 713 out of 1130. The worker's proportion for vertical transport in using ergonomic measures was increased to 34%, while no change was reported for horizontal transport and for positioning of materials. A decrease of 28% in the proportion of workers who reported of shoulder complaints was obtained. Moreover, no relationship was found between the use of ergonomic measures and MSDs. While 83% of the workers indicated to have adequate knowledge of ergonomic measures. Yuan and Buvens (2015) have evaluated the ergonomic risks associated with building or erection of scaffold for one of the local construction company by the use of "Rapid Entire Body Assessment (REBA)" tool for the estimation of risks of entire body disorders and injuries. Building or erection of scaffold requires carrying/lifting of weighty and bulk materials, uncomfortable working postures like reaching and holding overheads in addition to kneeling on the scaffolds, and repetition in movements i.e. hammering of cup-locks.

The musculoskeletal injuries and disorders are because of the exposure to those hazards, particularly to the shoulder and back for scaffold builders. It was further recommended to install scaffold hoist pulley systems or other hoist supporting systems, followed by training to all field personnel on ergonomics, proper work-rest scheduling along with workplace elongating programs. Ray et al. (2015) made a bio-mechanical evaluation of several manual material tasks involving carrying and lifting of materials like normal along with "Reinforcement Concrete Cement" bricks and heavy jack pipes, at a construction site in India through direct observations with video-graphs. Further, a motion analysis system called "Ariel Performance Analysis System (APAS)" was used for each task's work cycles to find out linear and angular accelerations for body joints as well as segments along with all three coordinates. The forces in compression for the construction workers at L5/S1 disc were found beyond the threshold value of 3.4 KN. Nath (2017) has used time-motion study for collecting data via smart phone sensors, and analyzed the data for health along with productivity assessment in the architecture, engineering, and construction (AEC) domains. The calculated risk levels for eleven of the thirteen postures were found as identical to true values. Nath et al. (2017) have used built-in smart phone sensors for monitoring worker's body postures and for identifying potential work-related ergonomic risks. It was seen by smart phone sensors data that measurements of trunk and shoulder flexion of workers were very close to corresponding measurements through observations.

Kumar and Maheswari (2017) have suggested a design concept called "Design for Construction Safety (DfCS)", for construction projects related ergonomic factors, in order to improve health and safety of workers by reducing physical factors, eliminating avoidable movements, and also reducing the reimbursement cost of the workers. It was also revealed that several industries in different ways have successfully implemented design concepts for ergonomic solutions through procurement of new tools, modification in existing equipment and creation of obstacles free workplaces.

Mohan (2018) has identified nine ergonomic risk factors in construction occupations from field data of 589 ergonomic injuries such as: (a) frequent/heavy lifting, (b) fixed/awkward body postures, (c) pushing, pulling and carrying heavy loads, (d) hand tools and equipment, (e) work methods, (f) forced, repetitive, or prolonged actions, (g) noises, (h) vibration of whole body, and (i) personnel relations, respectively. Ozkaya et al. (2018) have used "Ovako Working Posture Analysis System (OWAS)", to find the workloads as well as the risk of work-related musculoskeletal disorders, by considering the unloading and loading functions of workers, frequencies of postures, and operational times. Kratzenstein et al. (2019) have investigated the shoulder muscles' neuro-muscular responses at dissimilar attachment heights during arm movement of a carrying system. A hip belt was used to overcome muscle injuries, and height adjustments relieving the shoulder muscles. It was observed that increase in attachment height helped in relieving the muscles.

The studies on internal as well as external stresses that act on human-body are dealt in ergonomics, and thus, the use of ergonomic principles help in reducing the fatigue levels experienced by the human-body during the performances of various tasks. Abdul-Tharim et al. (2011) have suggested enhancement of ergonomics implementations in the workplaces through better communication as well as management controls, which need to be followed by suitable ergonomics-design, organized education, and training. Moreover, different studies have been conducted in order to assess the postures of the workers in a variety of building-construction tasks. For instance, in order to carry out the postural analysis during construction activities, Purnomo and Apsari (2016) have used REBA method; Kulkarni and Devalkar (2018) have used RULA and REBA as ergonomic tools; Zengin and Asal (2020) have used three ergonomic risk-assessment methods such as REBA, OWAS, and QEC; and so on.

20.2.1 Analytical Methods for Ergonomic Analysis

McAtamney and Corlett (1993) have developed an ergonomic evaluation method such as "Rapid-Upper-Limb Assessment (RULA)" method, which is a survey method for analyzing work-related musculoskeletal upper-limb disorder. A quick assessment is made in this method for the upper-limb postures in the vein of neck, trunk, muscle functions,and external forces experienced by the body. Steven and Arun (1995) have developed the "Strain Index" method for the analysis of the work-related musculoskeletal disorders of the working tasks upper-limb, which is based on the biomechanics, physiology as well as epidemiology principles. The variables determined in this method include the exertion intensity and duration, number of exertions per minute, the hand-wrist segment's postures, speed of tasks performances, as well as the task duration during the working day. Occhipinti (1998) has developed the "Occupational Repetitive Action (OCRA)" method that analyses the work-related musculoskeletal disorders of the workers' upper-limb, and particularly preferred for redesigning or thorough analysis of the workplace ergonomics. Hignett and McAtamney (2000) have developed the "Rapid-Entire-Body Assessment (REBA)" sharing the same principles as RULA pertaining to the risk factors evaluation procedures, and selected assessment of body parts'.

For whole-body assessment, REBA is considered to be a better tool in comparison to RULA that focuses more on the upper-body evaluations. Further, on the basis of physical observations, the "Quick Exposure Check (QEC)" methods has been developed by Geoffrey et al. (2005) which helps to create action categories by risk parameters like postures, duration, frequencies in addition to external forces. The most commonly used two assessment methods for ergonomic evaluation are "Rapid-Upper-Limb Assessment (RULA)" and "Rapid-Entire-Body Assessment (REBA)", while "Microsoft-Kinect" and "Inertial Measurement Units (IMU)" are the well-liked devices (Tee et al. 2017). For instance, the RULA method has been used to assess the ergonomic risks in computer professionals (Sharan and Ajeesh 2012), while the REBA method was used for the assessment of the workers in a book-selling company (Lasota 2014). Khandan et al. (2018) have used "Novel-Ergonomic-Postural-Assessment (NERPA)", RULA, and REBA methods to assess the working postures of 455 employees of operational-units of four-companies such as drug-producers, dairy, printing and publishing houses, and drinks-producers, during 2014 in Iran. Significant risk-levels were found between NERPA and RULA, while it was not-significant with NERPA and REBA output. The RULA, REBA, and QEC analysis were performed by considering eight different vendors selling a variety of eatable-products at the railway-stations as well as in the trains (Khan and Deb 2019). Both RULA and REBA methods were used for evaluating the postures during manual-feeding of a wood-chipper by an operator (Cremasco et al. 2019).

Moreover, a significant correlation was found between the QEC and REBA methods for the identification of risk-oriented jobs and for the determination of the potential-risks for work-related musculoskeletal-disorders (Motamedzade et al. 2011). Both QEC and REBA methods were used by considering 82 workers engaged in different activities in an "anodizing and aluminum profiles producing-industry" in Tehran (Iran) (Nadri et al. 2013). The QEC method has been reported to be a better method among the methods such as REBA, OWAS, and QEC for the ergonomic risk-assessments in the building-construction (Zengin and Asal 2020). Thus, for the present analysis, the ergonomic methods such as REBA and QEC were used to assess the working postures during various manual tasks in construction sites of Odisha in India.

20.3 Research Methodology

20.3.1 Materials and Methods

In this study, a construction site was selected in the district of Khorda in Odisha (India) where some construction activities were going on by a group of laborers. By using video recording as well as still photography, different actions and postures of the workers engaged in different construction activities were recorded and captured for subsequent analysis of the obtained data through ergonomic risk assessment tools like REBA and QEC.

20.3.2 Ergonomic Risk Assessment

For postural assessment by the use of ergonomic methods, the most important issue lies in the determination of the desired postures. In order to achieve this purpose, the task analysis of all activities was performed in this study by dividing all construction-related jobs into tasks and accordingly, the duration of each task was measured. Then, by considering the assigned-time to each task, the postural assessment of each task performed by the workers with the selection of worst-postures was carried out through the ergonomic risk assessment tools (Karwowski and Marras 1998; Mishra and Satapathy 2019a, b; Motamedzade et al. 2011; Qutubuddin et al. 2013). Moreover, with the use of "ErgoFellow 3.0" software, the REBA tool was used for analyzing the postures taken by workers during construction works. While the QEC tool was utilized for assessing the workers' in carrying some loads on head based on a predefined set of questionnaires for both worker as well as observer. The steps followed in this study are as illustrated in Fig. 20.1.

20.4 Results

During a construction project work, different workers were found performing various tasks with a variety of postures. From their still photography, personal interactions and type of action performances, the REBA scores along with the necessary recommendations were obtained as follows.

When the wood-related works were performed during the project as shown in Fig. 20.2, then different information based on the postures of worker as well as performing actions were considered as input variables for the ergonomic REBA analysis. In the analysis for neck, trunk, and legs (Fig. 20.3), the neck was found at more than 20° and it was twisted and side bending also. For trunk, it was between 20 and 40° and it was twisted and side bending also. Likewise, the legs were supporting the laborer and were found bent at more than 60°. The operating load by the laborer

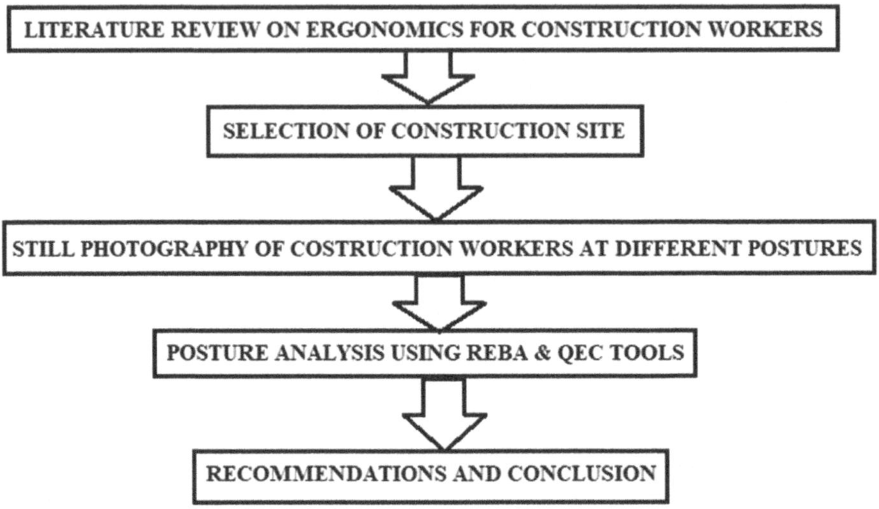

Fig. 20.1 Steps followed

Fig. 20.2 Worker doing wood works

was between 5 and 10 kg (Fig. 20.4), and the upper arm was at 20–45° and abducted. The lower arm was between 0 and 60°, while both the wrists were found at more than 15° up as well as down and bent from mid-line or twisted (Fig. 20.5). Moreover, the coupling observed during this activity was good (Fig. 20.6), and repeated small change actions with more than 4 s per minute were observed (Fig. 20.7). Based on this information, the obtained REBA score was 12 with the corresponding recommendation as "Very high risk, implement change" as shown in Fig. 20.8, which indicated that the task was at very high risk, and thus implementation of changes were necessary.

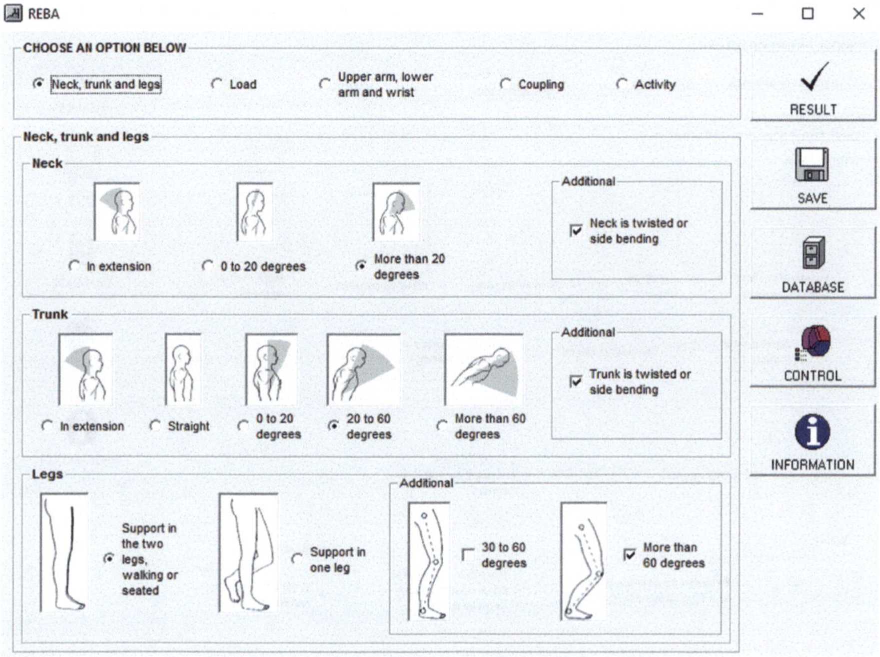

Fig. 20.3 REBA input for neck, trunk and legs for worker doing wood works

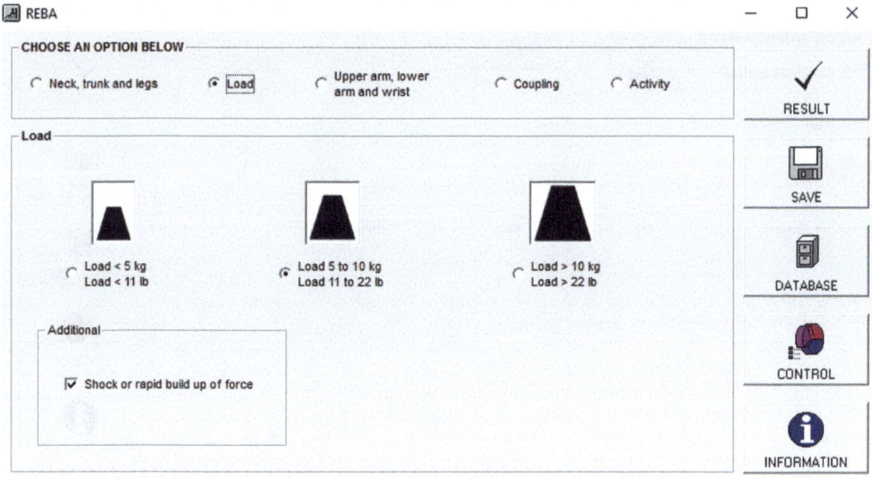

Fig. 20.4 REBA input for load during wood works

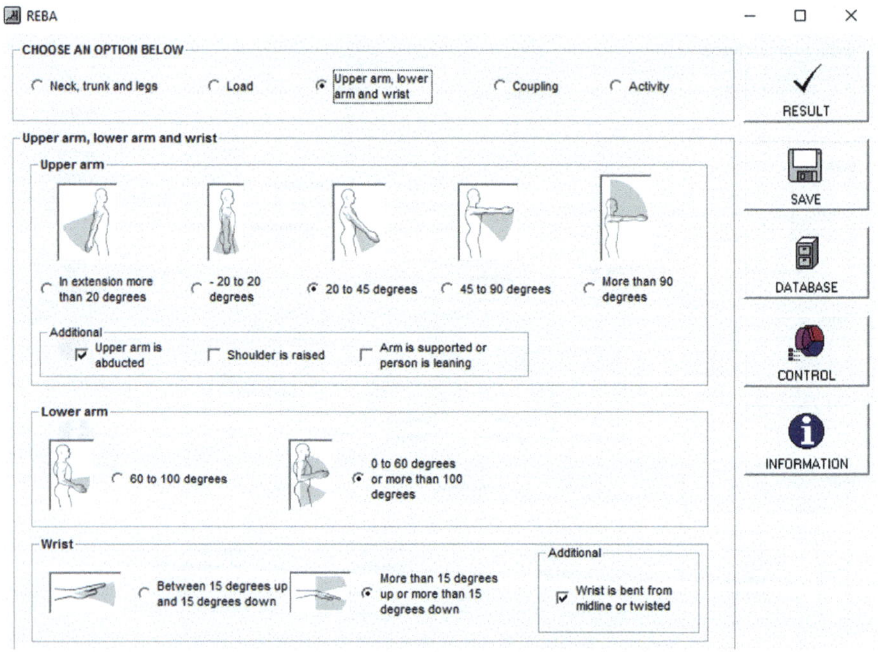

Fig. 20.5 REBA input for upper arms, lower arms and wrists during wood works

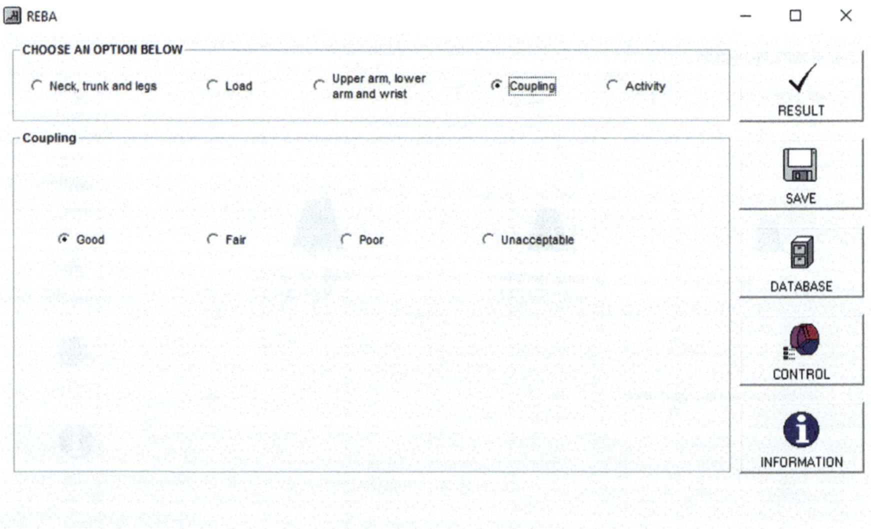

Fig. 20.6 REBA input for type of coupling during wood works

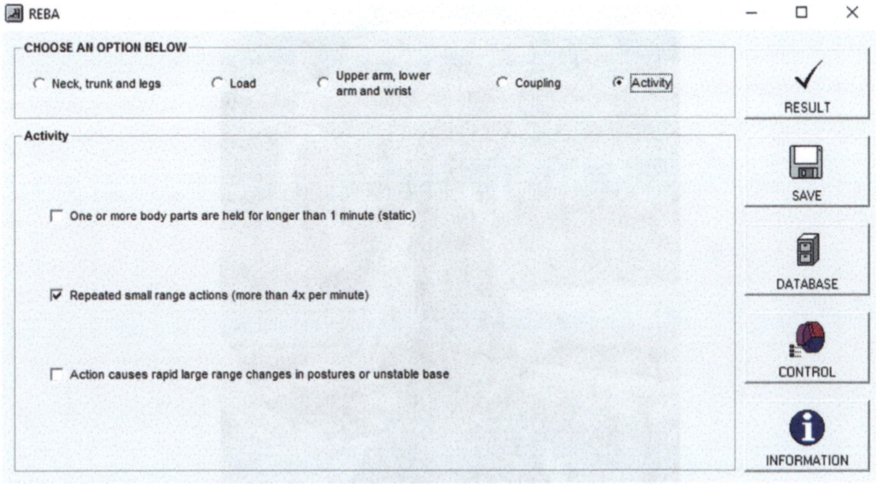

Fig. 20.7 REBA input for type of activity during wood works

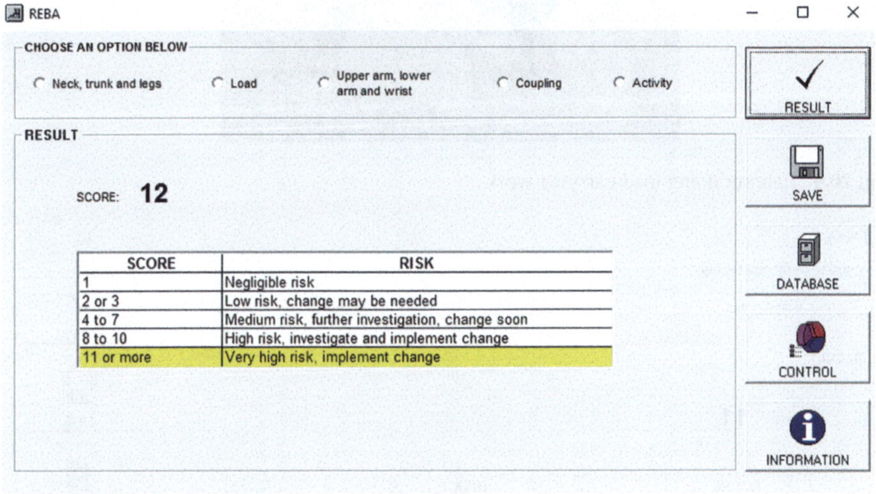

Fig. 20.8 REBA score for worker doing woodwork

In the process of executing manually load carrying tasks such as carrying bricks or sands as shown in Fig. 20.9, the corresponding REBA score obtained was 11 with the recommendation as "very high risk, implement change" (Fig. 20.10).

For the QEC analysis, from the observer's assessment point of view as shown in Fig. 20.11 the parts of body considered include back, shoulder or arm, wrist or hand, and neck, respectively. While performing such load-carrying task on head, the back of laborer was found to be moderately twisted as well as side bent. Further, the

Fig. 20.9 Laborer doing load-carrying work

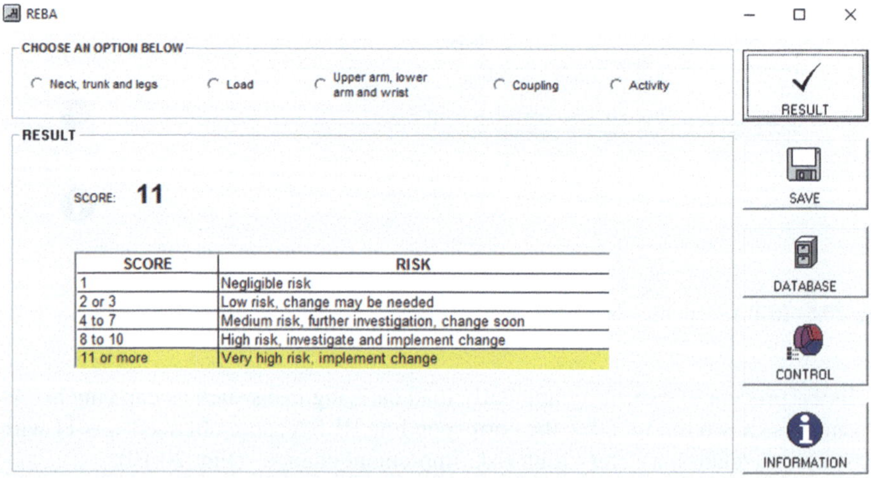

Fig. 20.10 REBA score for laborer doing load-carrying work

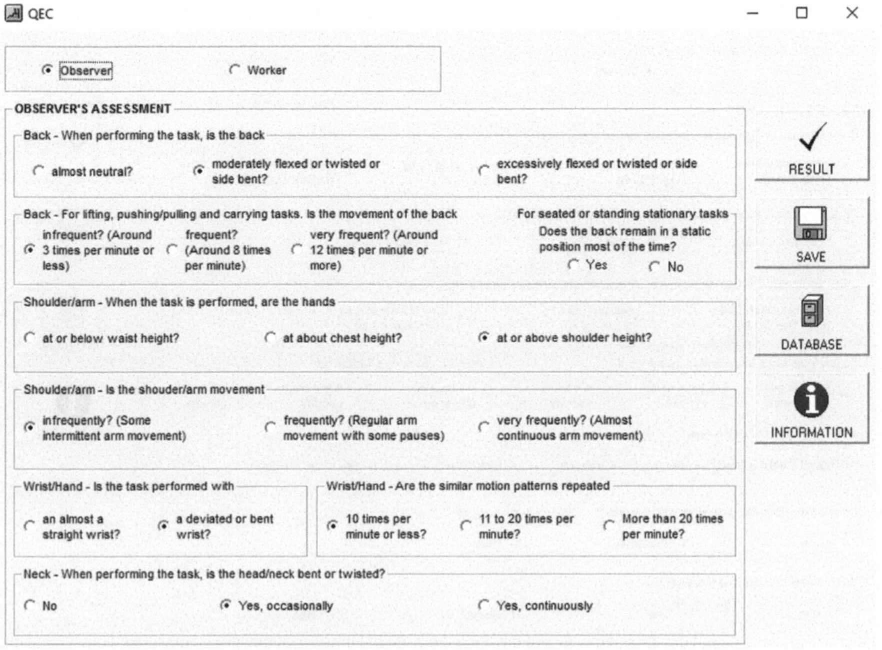

Fig. 20.11 QEC input from observer's point of view during load-carrying works

movement of the back was infrequent occurring less than 3 times per minute. The hands were found above shoulder height, and infrequent with some intermittent arms movements found for the shoulder as well as arms of laborer. The wrist was found to be bent and deviated, with motion pattern repeating less than 10 times. Further, the head as well as neck was found to be bent and twisted occasionally. Moreover, from the worker's assessment point of view as shown in Fig. 20.12 the maximum load handled by laborer during such task was reported to be moderate between 6 and 10 kg. Furthermore, the average time spent per day for such a task was reported as 2–4 h, and the maximum force level exerted by one hand was high, i.e., more than 4 kg. The vibration experienced during the task was reported to be nil, and no vehicle was reported of driven during such work. The visual demand for this work was high and sometimes difficulty of performing such task with medium stressful level was revealed by laborer. The result sheet of QEC output based on this information, and the obtained scores for different parameters along with their usual interpretation was as illustrated in Fig. 20.13 and Table 20.1. It was observed for laborer that the worse situations existed for the wrist or hand and neck, respectively.

During the execution of pillar construction works using the iron and steel bars as shown in Fig. 20.14, the corresponding REBA score obtained was 10 suggesting "high risk, investigate and implement change" (Fig. 20.15). Thus, these types of postures are not acceptable and need to be changed by taking appropriate suitable measures.

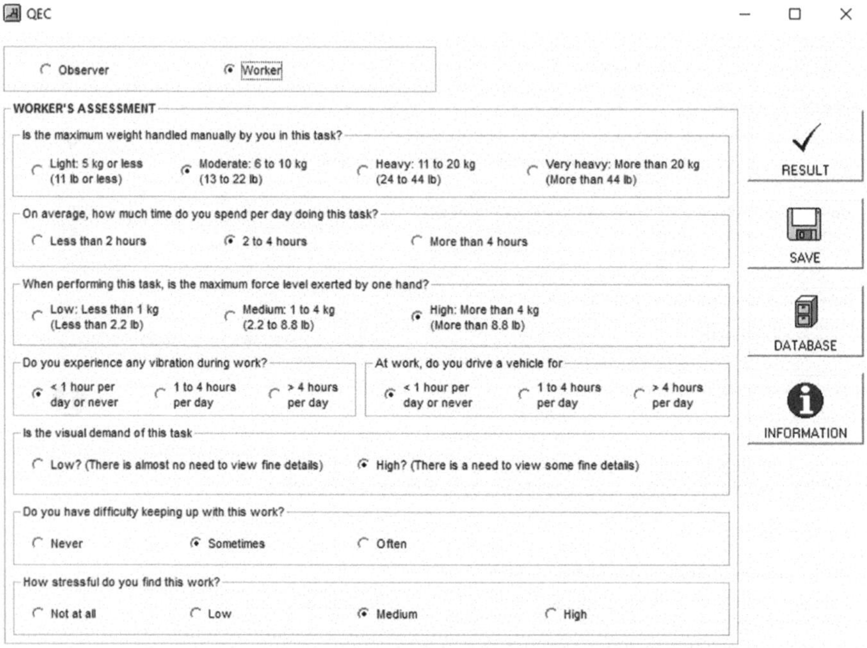

Fig. 20.12 QEC input from worker's point of view during load-carrying works

While performing the water supply works in the construction site as shown in Fig. 20.16, the corresponding REBA score obtained was 6 suggesting "medium risk, further investigations and changes soon" (Fig. 20.17). Thus, this type of postures is at medium level of risks, requiring further investigations and changes are also to be made sooner by suitable measures.

20.5 Discussion

Construction sectors have been considered the most hazardous as compared to agricultural, mining, and manufacturing sectors. Different hazards are emerging day by day as physical, chemical, biological as well as ergonomic hazards while performing various tasks. Particularly in construction sectors of Odisha in India, these hazards not only affect the health and safety of workers, but also the firm's productivity. The construction projects related hazards are due to tools, equipment, materials, and work environment condition. Ergonomic risk can be physical, environmental, or physiological factors inside the work environment that can hurt the musculoskeletal arrangement of the laborers. The components which influence the musculoskeletal systems incorporate redundant work, unbalanced positions, static postures, contact pressure, vibration and abrupt power, and so on. This is a result of poor structure of

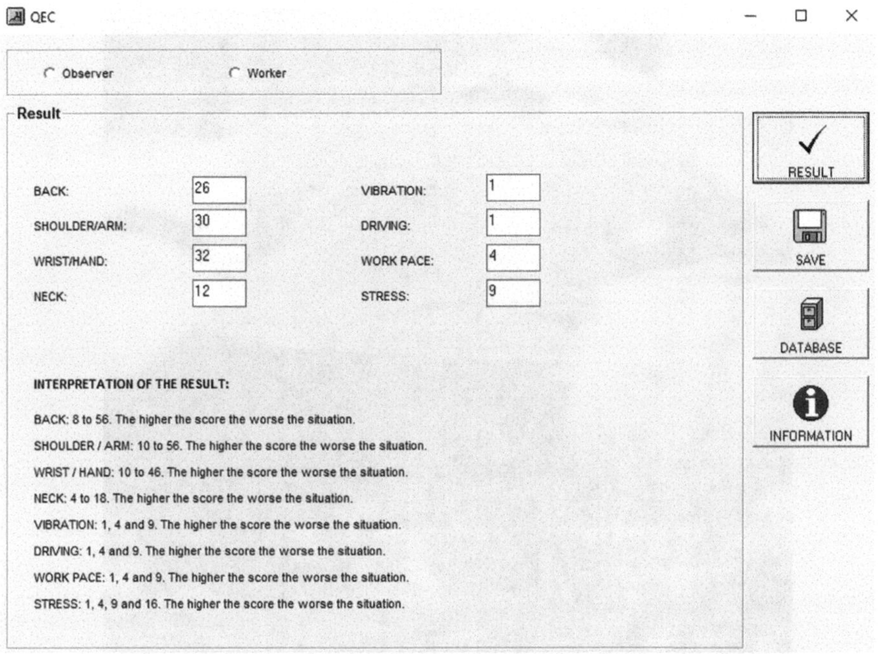

Fig. 20.13 Result sheet of QEC output

Table 20.1 Interpretation of QEC scores

Parameters	QEC scores	Interpretation of scores
Back	26	Moderate level of situation
Shoulder or arm	30	Moderate level of situation
Wrist or hand	32	Worse situation
Neck	12	Worse situation
Vibration	1	Lower risk level situation
Driving	1	Lower risk level situation
Work pace	4	Moderate level of situation
Stress	9	Moderate level of situation

devices, materials, equipment as well as the working environment of the construction sectors.

As shown in Fig. 20.2, it was observed that the workers or laborers performing the wood-related works tend to bend the neck at more than 20° with twisting as well as and side bending also. Further, the trunk was found to be bent in between 20 and 40° twisting as well as and side bending. Further, the worker was found to perform the activities in awkward squatting postures and based on the REBA score, this posture was at very higher risk requiring changes. Moreover, these findings were supported by Kurnianto and Mulyono (2014), who have performed research on

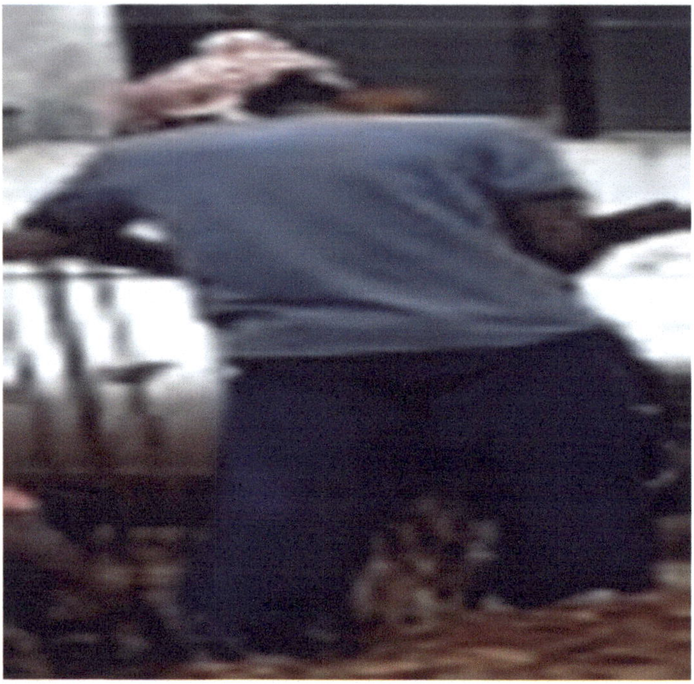

Fig. 20.14 Worker doing iron and steel work

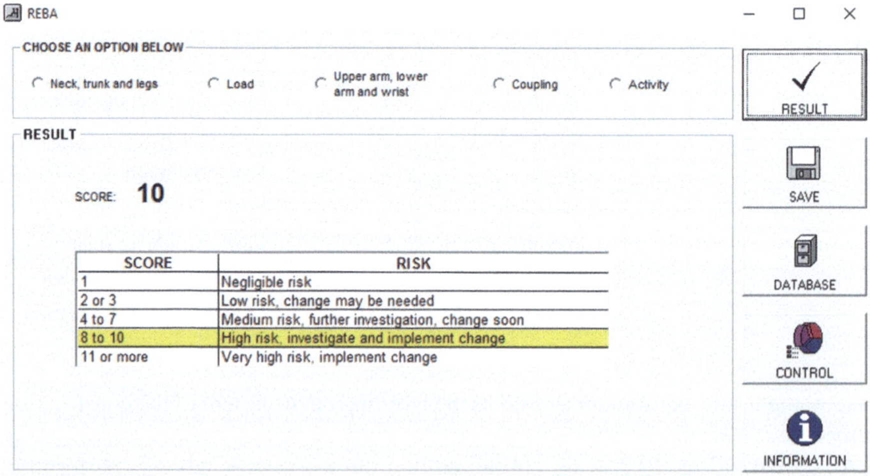

Fig. 20.15 REBA score for worker doing iron and steel work

Fig. 20.16 Worker doing water supply work

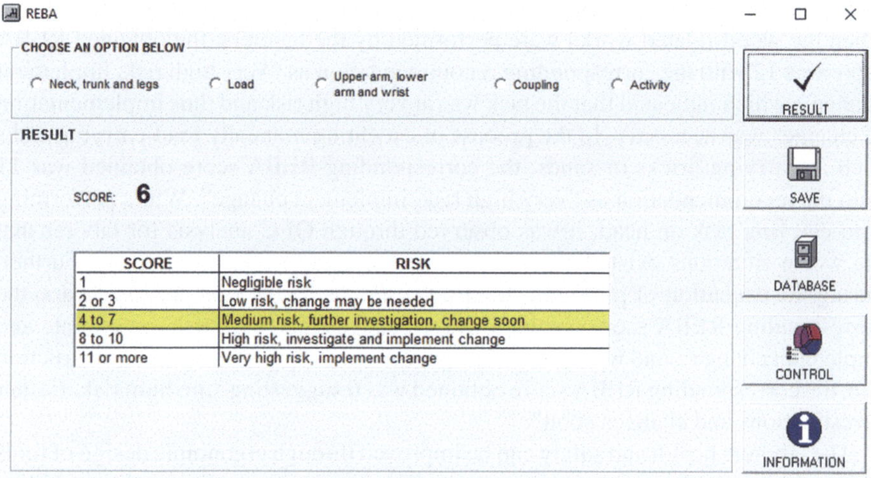

Fig. 20.17 REBA score for worker doing water supply work

welders in squatting postures and found 92.31% complaining about disorders from bending and squatting, which subsequently result in musculoskeletal disorders in the waist. Similarly, during the execution of pillar construction works using the iron and steel bars (Fig. 20.14), and the water supply works in the construction site as shown in Fig. 20.16, the postures observed were unacceptable resulting in musculoskeletal disorders. The findings of these postures recommend further changes based on the respective REBA scores, which was supported by the findings of Sundari (2011) that defined the causes of musculoskeletal disorders were because of working in a bending position. It may be noted that proper height levels of workstation should be maintained, either at similar elbow height or at 0–15 cm above the height of elbows, in order to avoid the musculoskeletal disorders (Dul and Weerdmeester 1993). Too much bending postures may affect the workers' hip and trunk, and result in more stresses on legs. Moreover, the discomfort feeling was more in the "Low-back" of workers during the construction of the buildings (Zengin and Asal 2020).

During the execution of the manually load-carrying tasks such as carrying bricks or sands as shown in Fig. 20.9, based on the REBA score, the recommended action was "very high risk, implement change". Further, based on QEC analysis, worse situations were obtained for the wrist or hand and neck. Thus, by the utilization of REBA as well as QEC as the ergonomic assessment tools, the postures related risks were almost similar (Moussavi-Najarkola and Mirzaei 2012; Motamedzade et al. 2011).

20.6 Conclusions and Recommendations

In this study for the constructional sectors of Odisha in India, it was found that when the wood-related works were performed by the laborers, the obtained REBA score was 12 with the corresponding recommendation as "Very high risk, implement change", which indicated that the task was at very high risk and thus implementation of changes was necessary. In the process of executing manually load-carrying tasks such as carrying bricks or sands, the corresponding REBA score obtained was 11 with the recommendation as "very high risk, implement change". While performing load-carrying task on head, it was observed through QEC analysis for laborer that the worse situations existed for the wrist or hand and neck, respectively. Further, during the execution of pillar construction works using the iron and steel bars, the corresponding REBA score obtained was 10 suggesting "high risk, investigate and implement change", and while performing the water supply works in the construction site, the corresponding REBA score obtained was 6 suggesting "medium risk, further investigations and changes soon".

The laborers health and safety can be improved through ergonomic design of tools and equipment in the construction projects. The upside of actualizing plan and ideas in starting phase of ergonomics is, laborers can realize how to work or deal with the undertaking they perform before they reach and can choose what sort of well-being safety measures ought to be taken while carrying out that responsibility. In order to reduce the laborer damages and diseases, work assignments ought to be structured out in the underlying phase of the venture for ergonomic elements to improve the laborers safety.

Further, by consulting with domain-experts, the following recommendations and suggestions were obtained for the potential developments in the construction sectors:

- Requirement of re-designing in activities.
- In order to reduce unnecessary trunk-movements, the necessary platforms need to be elevated.
- For materials transportation activities, trolleys or conveyor-belts need to be used for minimizing the lifting-actions and external loads on workers' body.

- Platforms must be provided for the workers to keep their load carriages or buckets for loading the excavated materials, without the need for much bending.
- Appropriate height levels should be maintained for carrying out the necessary tasks to reduce avoidable upper-limb movements.
- All the necessary materials should be kept in reach of the workers to reduce unnecessary movements.

The smaller sample-size was a limitation of this study; thus, similar studies with larger and different populations may be conducted for better interpretations.

References

Abdul-Tharim AH, Jaffar N, Lop NS, Mohd-Kamar IF (2011) Ergonomic risk controls in construction industry—a literature review. In: Proceedings of 2nd international building control conference. Procedia Eng 20:80–88

Albers JT, Hudock SD (2007) Biomechanical assessment of three rebar tying techniques. Int J Occup Saf Ergon 13(3):279–289

Aminbakhsh S, Gunduz M, Sonmez R (2013) Safety risk assessment using analytic hierarchy process (AHP) during planning and budgeting of construction projects. J Saf Res 46:99–105. https://doi.org/10.1016/j.jsr.2013.05.003

Boschman JS, Frings-Dresen MHW, van der Molen HF (2015) Use of ergonomic measures related to musculoskeletal complaints among construction workers: a 2-year follow-up study. Saf Health Work 6:90–96. https://doi.org/10.1016/j.shaw.2014.12.003

Center for Construction Research and Training (CPWR) (2013) The construction chart book, 5th edn. CPWR, Silver Spring, 34 p. Available from https://www.cpwr.com/publications/chart-book-hazards-and-exposures

Cremasco MM, Giustetto A, Caffaro F, Colantoni A, Cavallo E, Grigolato S (2019) Risk assessment for musculoskeletal disorders in forestry: a comparison between RULA and REBA in the manual feeding of a wood-chipper. Int J Environ Res Public Health 16:793. https://doi.org/10.3390/ijerph16050793

Dale AM, Jaegers L, Buchholz B, Welch L, Evanoff BA (2012) Using process evaluation to determine effectiveness of participatory ergonomics training interventions in construction. Work 41:3824–3826

Dul J, Werdmeester BA (1993) Ergonomics for beginner a quick reference guide. Taylor & Francis, Washington

Geoffrey D, Woods V, Buckle P (2005) Further development of the usability and validity of the quick exposure check. Prepared by University of Surrey for the Health and Safety Executive

Hignett S, McAtamney L (2000) Rapid entire body assessment (REBA). Appl Ergon 31(2):201–205

Jaffar N, Abdul-Tharim AH, Mohd-Kamar IF, Lop NS (2011) A literature review of ergonomics risk factors in construction industry. Procedia Eng 20:89–97

Karwowski W, Marras WS (1998) The occupational ergonomics handbook. CRC Press, USA

Khan IA, Deb RK (2019) Postural analysis through RULA, REBA and QEC of vendors selling edible items at railway stations and in the trains. Int J Eng Adv Technol 9(1):7269–7277. https://doi.org/10.35940/ijeat.A9878.109119

Khandan M, Vosoughi Sh, Poursadeghiyan M, Azizi F, Ahounbar E, Koohpaei A (2018) Ergonomic assessment of posture risk factors among Iranian workers: an alternative to conventional methods. Iran Rehabil J 16(1):11–16. https://doi.org/10.29252/NRIP.IRJ.16.1.11

Koningsveld P, Van der Molen F (1997) History and future of ergonomics in building and construction. Ergonomics 40:1025–1034

Kramer DM, Bigelow PL, Carlan N, Wells RP, Garritano E, Vi P, Plawinski M (2010) Searching for needles in a haystack: identifying innovations to prevent MSDs in the construction sector. Appl Ergon 41:577–584
Kratzenstein S, Wanstrath M, Behrenbruch K (2019) Height adjustments on backpack-carrying systems and muscle activity. Appl Ergon 74:172–176
Kulkarni VS, Devalkar RV (2018) Postural analysis of building construction workers using ergonomics. Int J Constr Manag. https://doi.org/10.1080/15623599.2018.1452096
Kumar CV, Maheswari JU (2017) A concept note to prevent construction workers from ergonomic threats. J Ergon Res 1:1
Kurnianto RY, Mulyono (2014) Gambaran Postur Kerja dan Resiko Terjadinya Muskuloskeletal Pada Pekerja Bagian Welding di Area Workshop Bay 4.2 PT. Alstom Power Energy Systems Indonesia. Indones J Occup Saf Health Environ 1(1):61–72
Lasota AM (2014) A REBA-based analysis of packers workload: a case study. Log Forum 10:1
McAtamney L, Corlett EN (1993) RULA: a survey method for the investigation of world-related upper limb disorders. Appl Ergon 24(2):91–99
Memarian B, Mitropoulos P (2013) Accidents in masonry construction: the contribution of production activities to accidents, and the effect on different worker groups. Saf Sci 59:179–186
Mishra D, Satapathy S (2019a) Ergonomic risk assessment of farmers in Odisha (India). Int J Syst Assur Eng Manag 10(5):1121–1132. https://doi.org/10.1007/s13198-019-00842-5
Mishra D, Satapathy S (2019b) An integrated MCDM and ergonomic approach for agricultural sectors of Odisha in India: a critical analysis for farming sustainability. In: Advanced multi-criteria decision making for addressing complex sustainability issues. IGI Global, USA, pp 181–221. https://doi.org/10.4018/978-1-5225-8579-4.ch009
Mitropoulos P, Namboodiri M (2011) New method for measuring the safety risk of construction activities: task demand assessment. J Constr Eng Manag 137(1):30–38
Mohan SB (2018) Identifying and controlling ergonomic risk factors in construction. J Ergon 8:235. https://doi.org/10.4172/2165-7556.1000235
Motamedzade M, Ashuri MR, Golmohammadi R, Mahjub H (2011) Comparison of ergonomic risk assessment outputs from rapid entire body assessment and quick exposure check in an Engine Oil Company. J Res Health Sci 11(1):26–32
Moussavi-Najarkola SA, Mirzaei R (2012) Assessment of musculoskeletal loads of electric factory workers by rapid entire body assessment. Health Scope 1(2):71–79
Muhammad S, Amran MIUA, Kabir MH, Hossain MM, Kader MA (2013) A study on the safety and ergonomics for construction works in Chittagong. Int J Sci Eng Res 4(11)
Nadri H, Fasih F, Nadri F, Nadri A (2013) Comparison of ergonomic risk assessment results from Quick Exposure Check and Rapid Entire Body Assessment in an anodizing industry of Tehran, Iran. JOHE 2(4):195–202
Nath ND (2017) Construction ergonomic risk and productivity assessment using mobile technology and machine learning. MSU graduate theses, 3157. Retrieved on 23 Aug 2019, from https://bearworks.missouristate.edu/theses/3157
Nath ND, Akhavian R, Behzadan AH (2017) Ergonomic analysis of construction worker's body postures using wearable mobile sensors. Appl Ergon 62:107–117. https://doi.org/10.1016/j.apergo.2017.02.007
Occhipinti E (1998) OCRA: a concise index for the assessment of exposure to repetitive movements of the upper limbs. Ergonomics 41(9):1290–1311
Ozkaya K, Polat O, Kalinkara V (2018) Physical workload assessment of furniture industry workers by using Owas method. Ergon Open J 11(1)
Purnomo H, Apsari AE (2016) REBA analysis for construction workers in Indonesia. J Built Environ Technol Eng 1:104–110. ISSN: 0128-1003
Qutubuddin SM, Hebbal SS, Kumar ACS (2013) An ergonomic study of work related musculoskeletal disorder risks in Indian Saw Mills. J Mech Civ Eng 7(5):7–13
Ray PK, Parida R, Sarkar S (2015) Ergonomic analysis of construction jobs in India: a biomechanical modelling approach. In: 6th international conference on applied human factors and ergonomics

(AHFE 2015) and the affiliated conferences, AHFE 2015. Procedia Manuf 3:4606–4612. https://doi.org/10.1016/j.promfg.2015.07.542

Schneider SP (2001) Musculoskeletal injuries in construction: a review of the literature. Appl Occup Environ Hyg 16:1056–1064

Sharan D, Ajeesh P (2012) Correlation of ergonomic risk factors with RULA in IT professionals from India. Work 41(Supplement 1):512–515

Shin G, Mirka G (2004) The effects of a sloped ground surface on trunk kinematics and L5/S1 moment during lifting. Ergonomics 47(6):646–659

Stattin M, Järvholm B (2005) Occupation, work environment, and disability pension: a prospective study of construction workers. Scand J Public Health 33:84–90

Steven MJ, Arun G (1995) The strain index: a proposed method to analyze jobs for risk of distal upper extremity disorders. Am Ind Hyg Assoc J 56(5):443–458

Sundari KL (2011) Sikap Kerja yang Menimbulkan Keluhan Muskuloskeletal dan Meningkatkan Beban Kerja Pada Tukang Bentuk Keramik. J Ilmiah Tekn Ind 10(1)

Tee KS, Low E, Saim H, Zakaria WNW, Khialdin SBM, Isa H, Awad MI, Chin Fhong S (2017) A study on the ergonomic assessment in the Workplace. AIP Conf Proc 1883:020034. https://doi.org/10.1063/1.5002052

van der Molen HF, Sluiter JK, Hulshof CT, Vink P, Frings-Dresen MH (2005) Effectiveness of measures and implementation strategies in reducing physical work demands due to manual handling at work. Scand J Work Environ Health 31:75–87

Welch LS, Hunting K (2003) Injury surveillance in construction: what is an "injury", anyway? Am J Ind Med 44(2):191–196

Woodside G (1997) Environmental, health, and safety portable handbook. McGraw Hill, pp 175–178

Yuan L, Buchholz B (2014) The effects of position and size of drywall on the physical demands for installers. Proc Hum Factors Ergon Soc Annu Meet 58(1):1612–1616

Yuan L, Buvens M (2015) Ergonomic evaluation of scaffold building. Procedia Manuf 3:4338–4341

Zengin MA, Asal Ö (2020) Evaluation of employee postures in building construction with different ergonomic risk assessment methods. J Fac Eng Archit Gazi Univ 35(3):1615–1630. https://doi.org/10.17341/gazimmfd.548028

Chapter 21
Are Productivity and Quality in Electronics Manufacturing Industry Affected by Human Factors? A Quantitative Analysis Using Statistical Tools

Berónica Botello-Lara, Margarita Gil-Samaniego Ramos, Juan Ceballos-Corral, and Arturo Sinue Ontiveros-Zepeda

Abstract Productivity and quality in electronic manufacturing companies with a high level of automation are key performance indicators; customer complaints represent the most tangible examples that affect those indicators as well as recurring events. The aim of this chapter is to investigate if human factors have an incidence in the loss of productivity and quality in manufacturing companies. To achieve this, a case study was considered using different approaches, both for quantitative and qualitative analysis. For the first, statistical tools were applied, and for the later, cognitive ergonomics evaluations were performed. Productivity and quality historical data from the production department were collected and evaluated statistically, and interviews were conducted with related workers for assessing their suitability for the job assigned. Results showed that psychological factors (lack of compliance with rules/not following procedures) and socio-technical factors (inadequate and/or incomplete skills development) combined with ergonomically deficient process design are among the most important human and technical factors affecting customer's satisfaction and key performance indicators of manufacturing companies.

Keywords Productivity · Quality · Cognitive ergonomics · Human factors · Manufacturing

---

B. Botello-Lara · M. G.-S. Ramos (✉) · J. Ceballos-Corral
Department of Industrial Engineering, Universidad Autónoma de Baja California, Blvd. Benito Juárez S/N, 21280 Mexicali, Baja California, México
e-mail: margarita.gil.samaniego.ramos@uabc.edu.mx

A. S. Ontiveros-Zepeda
Department of Industrial Engineering, Universidad Autónoma de Baja California, Blvd. Universidad 1, 21460 San Fernando, Tecate, Baja California, Mexico

J. L. Garcia Alcaraz et al. (eds.), *Trends in Industrial Engineering Applications to Manufacturing Process*, https://doi.org/10.1007/978-3-030-71579-3_22

21.1 Introduction

The high levels of production in conjunction with the aim of obtaining the best manufacturing costs have led companies to systematic changes in the manufacturing processes, all based on the philosophy and structure of continuous improvement. Indicators such as quality are affected by the use of equipment with a high technological level which demand a high level of knowledge and cognitive capacity in the personnel that operates it, for the maintenance personnel that repairs it, for engineering that designs the processes; hence the proposal to create synergy through the use of lean manufacturing methodologies, cognitive ergonomics and socio-technical systems. Other strategies such as motivation, communication, ergonomics, and automation have a direct influence on the individual performance of workers, because the greater the awareness of human factors and their safety, the higher the expected level of productivity (Karanikas et al. 2017).

Information processing theory is one of the most widely used models in human error research and is perhaps the most useful cognitive error model for industrial applications. It states that people perceive information via their senses, interpret this information, and make decisions concerning its meaning and relevance based on their previous understanding and current interpretation (Van Vuuren et al. 1997). On their part, Manchi et al. (2013) say that the emerging model of cognition provides at least partial model of cognitive mechanism to understand the way human thinking works and that the most effective way to deal with error due to human behavior and unpredictable environment is by safety culture and favorable system design. Finally, Gordon et al. (2005) developed a human factors incident investigation tool (HFIT) based on the dominant psychological theories of accident causation, which has the potential to improve the quality of human factors incident data.

Therefore, in this chapter, we present an investigation based in a case study, seeking to find if the root causes of the loss of productivity and quality in a manufacturing company were due to the human factor, the work process design, or a combination of both. The chapter is organized as follows. First, we made an extensive literature review looking for previous research on how human error is present in many manufacturing environments and affects production; we also searched for the importance of process design and automation for preventing errors in manufacturing and how socio-technical systems contributes to consider the human factor in the process of work design. In the section of methodology, a case study was described as well as the statistical tools used. We also explained the tests made to obtain the psychological profile of the workers to find if this factor was relevant to productivity and quality. In the next section, results and discussion are presented, and finally, the conclusions are portrayed.

21.2 Literature Review

21.2.1 Human Error in Manufacturing

Misoperations are those activities documented in the manufacturing process that are omitted and generate damages to the product and even put at risk the physical and/or mental integrity of the personnel involved as well as the environment; these are frequently attributed to the human factor, specifically at the operator level and no further analysis is made. But if these assumptions are not addressed correctly, they can decrease the productivity levels of the companies. Human error is sometimes described as being one of the following: an incorrect decision, an improperly performed action, or an improper lack of action (inaction). The types of problems caused by human factors are often unavoidable (Rothblum et al. 2002). Under certain situations, human beings will always make mistakes and there is a limit to what can be done to modify the behavior itself. As they are inevitable, the approach to error management focuses on reducing the possibility of their occurrence and minimizing the impact.

A good system should not allow people to make mistakes easily, this may sound obvious, but all too commonly system design is carried out in the absence of feedback from its potential users, which increases the chance that the users will not be able to interact correctly with the system (POST 2001). Hence, the importance is of making an exhaustive analysis of the error causal factors of the human–machine interaction in order to minimize failures between their interactions that led to misoperations. In this chapter, a framework was developed for aiding in the prevention of misoperations related to the human–machine system interface and thus improving productivity and quality. Its contribution relies on a procedure to incorporate the human factor in the design phase of the process, along with its technical features.

Mclaughlin (2013) indicates that the improvement of manufacturing processes and productivity is linked to the culture of the organization which is in its commitment to the best manufacturing practices that include: lean manufacturing, continuous improvement, and customer service strategy to define the competitive advantage of the company and its permanence in the market. In addition, considering human factors and operator satisfaction with the work environment is key to the continuous improvement of quality, so the evaluation of performance and feedback of the worker should be a standard procedure in companies to determine the level of individual fulfillment.

In their study of methods for locating operators in manufacturing cells, Haraguchi et al. (2014) suggested that productivity and quality can be improved by controlling operations and focusing on employee morale, which is feasible through a systematic approach in technical and human aspects and/or their interaction. Many other authors (not mentioned here because of space limitations) also found that the human factor is as important as the technological factor in maintaining or increasing productivity in a manufacturing environment. Also, Toledo-Rosillo (2013) highlights that when designing a new work method, the creation of efficient teams should be considered through the appropriate combination of individual personality traits and placing people in work activities that are compatible with their profile.

Systematic error is formed from a limited number of primitive errors that had their origin in the fundamental characteristics of human cognition (psychological biases) that are considered as the nature of the processing of human information itself and place human performance above artificial intelligence and considers it irreplaceable.

21.2.2 Importance of Process Design in Productivity and Quality

Technological, economic, organizational, and human factors affect the work behavior and well-being of people as part of a production system. Applying ergonomic knowledge in the light of practical experience in the design of a work system is intended to satisfy human requirements.

Assessing the risk of the man–machine interface in the design of a process is an essential part of the quality and process control systems as indicated by risk management section of the standard ISO 9001: 2015. The technique "Analysis of the mode and effect of the failure" (FMEA) developed by Carlson (2012) can be the guide for the development of a complete set of actions that will reduce the risk associated with the production system through the Risk Priority Number (RPN), which is a numerical classification of the risk of each mode or cause of potential failure. This number is computed by multiplying the indexes severity, detectability, and occurrence (Aguilar-Otero et al. 2010). The severity index is the variation in the severity of the failure mode. Occurrence index is the probability that a specific cause will result in a failure mode. Finally, detectability is a numerical value that classifies the probability of finding the fault before the product reaches the customer.

On the other hand, Wells et al. (2016) state that quality is considered a critical indicator to be competitive and stay in the market and to consider human factors to reduce to the minimum level the errors that staff can generate is part of the strategy of any quality management system. These "quality risk factors" (QRF) are divided into three design stages: (1) Product design: in which the assembly task characteristics are determined; (2) process design: which sets the stage for the performance of assembly defining flows, task division, and materials supply strategies; and (3) workstation design: defining the dimensional layouts that will define the operators postures while performing the assembly tasks. If during the design of the process the human factor is not considered for risk assessment, it will result in a weak process, which will propitiate conditions for the staff to omit actions or make wrong decisions. Also, their cognitive process will be affected at the moment of the execution of their activity and, depending on how complex it is, can put at risk large amounts of production, the equipment, or systems and even their physical integrity and that of others.

A failure to embed human factors across all stages of the design lifecycle of future technologies will be catastrophic, leading to inefficient, unrewarding, and unsafe human work (Salmon and Read 2018). However, as Sobhani et al. (2015) indicate, human factors (ergonomics) studies have mostly focused on occupational health and safety aspects without accounting for operational elements of the system and that they are infrequently able to systematically integrate causes and effects of employee performance variation.

Finally, Bargelis et al. (2014) emphasize that the integrated design of a new product and its components together with manufacturing process at the early development stage applying concurrent engineering and innovative models and tools would be the way of achieving required quality and reliability tasks of the end product.

21.2.3 Automation and Process Design

Human factors, especially the cognitive aspects, are often misunderstood and neglected in system design. These factors are growing in importance as systems become automated, more complex, and more interconnected to other systems, making the role of the human operator more critical than ever before (Sheridan and Parasuraman 2005). In manufacturing processes, automation is often applied to ensure process reliability and to improve quality. Developing automation solutions for processes with strong interconnection between humans and processes discloses certain challenges, especially in the complexity of the resulting automation solution. In recent research, it is shown that the tendency to integrated human-automation-systems (HAS) has increased. Concepts of fusing human and automation capabilities to achieve an optimal benefit of both domains are necessary (Langer and Söffker 2015). For their part, Anderson and Raiken (1988) state that the theory of socio-technical systems proposes the interaction between technological aspects and human resources, although physical factors such as lighting, noise levels, and working posture must also be taken into account.

Complex productive processes, in parallel with the increase in the use of equipment or systems with a high level of automation, require that the cognitive level of the people interacting with them to be higher. For this reason, the integration of the human factor is recommended from the design phase to consider all those specific human skills required for a correct execution and increase in the productive levels. The challenge of integrating human factors (HF) during the design process is to improve the usability and reliability of the product, without affecting the functional specifications (Scaravetti 2015), and it can be made during the stage of capturing requirements and specifications. Although this recommendation is more focused on the human–machine interface, Takano et al. (1994) say that it applies to any development stage. Another perspective is provided by them when proposing to give more importance to human factors in dynamic environments because of their critical reliability when making decisions in their interaction with equipment or critical operational conditions with the objective of revealing underlying mechanisms to an action that was judged as human error. The interactions between the cognitive performance of the operator and the situations of risk for the production process presented must be fully investigated along with their specific execution failure scenario. When human error is repetitive, it is considered to be propitiated by cognitive activities and it is latent, while serial operations are carried out and have their origin in primitive forms called psychological biases (Takano and Reason 1999), which are by nature the processing of human cognition.

21.2.4 Work Design Through Socio-Technical Systems

The design of work through the use of socio-technical systems is based on the premise that an organization or a work unit is a combination of social and technical parts, and this is because both must interact together to perform the tasks to produce physical and social or psychological results. This ensures a better quality of working life as well as the job satisfaction of the team members. The well-being of the employee should be seen as a desired output (Ulhøi and Jorgensen 2010). Finally, the central point of the proposal for the application of socio-technical systems is to integrate the soft factors (human) within the hard factors (technical) and achieve their adequate interdependence.

Within the work design activity from the perspective of the socio-technical system, there is a proposal to moderate the effects of the characteristics of the tasks or activities, an essential approach for the project to reduce defects/customer complaints since its correct application helps to moderate the relationships that occur between the inputs, the process, and results of the group or the interaction of all of them (Leung and Wang 2015).

The fundamental principles of the mission of the socio-technical systems that contribute to an effective work design can be summarized as follows (Appelbaum and Hester 1997):

1. Overall productivity is directly related to the accurate analysis of the social development system and technical needs and requirements.
2. A precise analysis of the social aspect and technical needs generally leads to work on job designs with the following characteristics:

    - Minimum specifications of rules.
    - Control of variation in the process.
    - Multi-skilled employees.
    - Boundary location, based on one or more of the following three criteria: technology, territory, or time.
    - Information flow.
    - Support with consistency in job design.
    - Human values and design.

If companies want results above the standard, these can be achieved through the joint optimization of individual and organizational needs, in which it is essential to view the human system as an asset and that technologies are support tools to make their work more effective and safe.

21.3 Methodology

As part of the investigation process, we used the case study methodology, which according to Martínez (2006), is a research strategy aimed at understanding the dynamics present in unique contexts that allows studying phenomena from multiple perspectives and not from the influence of a single variable. A real case was selected from an electronic manufacturing company with high-tech automatized equipment, in order to analyze the possible influence of human factors on their productivity and quality, due to recurrent complaints from internal and external customers about the reception of defective products. There is a generalized assumption that in companies with "cutting-edge technology and high level of development," there is a very small chance to generate human errors of operation, since the personnel is very well trained in the interaction with their equipment and systems. However, it can be noted that they still generate errors that trigger conditions impacting productivity and quality that lead to consider human performance as the key factor that generated the problem. One of the objectives of this chapter was to detect which factors were influencing these incidents.

Figure 21.1 shows the behavior of misoperations in the company's case study during the fiscal year of 2018 through the different processes.

The bar chart shown in Fig. 21.1 indicates that Process 1 had the highest occurrence of misoperations with 51.6%, so it was selected to be investigated for this case of study.

The methodology used was mainly focused to find if several human factors had a significative influence in productivity and quality, from a statistical approach. For the analysis, the software Minitab was employed. We had access to historical data kept by the production department, concerning levels of rejections due to defective products, as well as the technical profile of the personnel in charge of the specific operation. The statistical analyses performed were mainly correlation and design of experiments. For the psychological profile of the workers, interviews were performed.

The investigation was divided into the five stages described below.

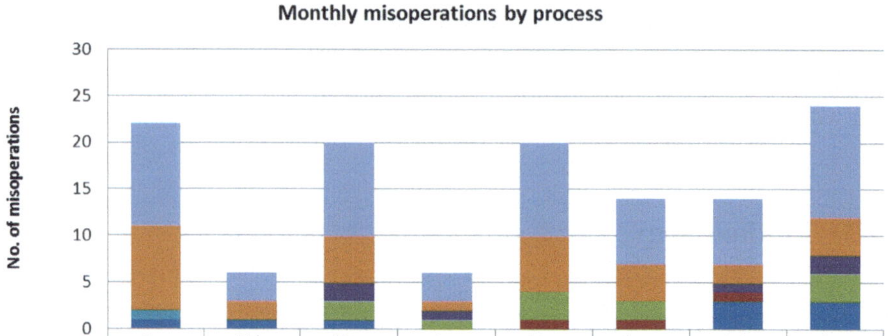

	Jan	Feb	Mar	Apr	May	June	July	Aug
Total	11	3	10	3	10	7	7	12
Process 1	9	2	5	1	6	4	2	4
Process 2	1	0	0	0	0	0	0	0
Process 3	0	0	2	1	0	0	1	2
Process 4	0	0	2	1	3	2	0	3
Process 5	0	0	0	0	1	1	1	0
Process 6	1	1	1	0	0	0	3	3

Fig. 21.1 Monthly behavior of misoperations in the fiscal year 2018

21.3.1 Stage 1: Analysis of Correlation Between the Level of Rejected Lots and Worker Seniority

This stage was focused to find if a statistical correlation exists between the number of rejected lots with the human factor seniority, which is related to the experience of the operator due to his time working in the company. Correlation is a measure of linear association between two variables; and its intensity is measured by the Pearson correlation coefficient, whose values are in the range of -1 to 1. For a two-tailed test of correlation, the corresponding hypothesis is shown in Eqs. (21.1) and (21.2):

$$\mathrm{H}_0\colon \rho = 0 \tag{21.1}$$

$$\mathrm{H}_1\colon \rho \neq 0 \tag{21.2}$$

where ρ is the correlation between a pair of variables.

To validate this hypothesis, a correlation diagram was made to evaluate if there is a relationship between the work profile against the process quality audit level (PQA's) generated during quarters 1 and 2 of fiscal year 2018.

21.3.2 Stage 2: Identification of Psychological Profiles

In stage 2, a psychometric test to the operational personnel was applied in order to identify if their psychological profiles were compatible with the job descriptions. Knowing if the company had the appropriate operational personnel for the tasks to be performed implied the analysis of the type of profile to determine if it influenced the execution of the activity. For this purpose, the questionnaire type survey "Cleaver" was used. The Cleaver test is an administrative technique designed to be used by the managers of a company. It is aimed at obtaining information from people and positions to improve the management of human resources, that is, to determine the behavior required for the position and compare this with the characteristics of the employee in their daily behavior. It is useful also to discover the worker's possible limitations, as well as his internal motivations and those that may influence to improve their identification with the position they occupy (Montano et al. 2019).

21.3.3 Stage 3: Design of Experiments

In stage 3, several statistical design of experiments (DOE) analyses were applied following the methodology proposed by Tanko (2007), aimed to find if different human factors significantly influenced productivity. This stage was divided into four steps, that is, four DOEs were performed, each one of them seeking to find specific relationships. DOE is a methodology for systematically applying statistics to experimentation. It consists of a series of tests in which purposeful changes are made to the input variables (factors) of a product or process, so that one may observe and identify the reasons for these changes in the output response. DOE provides a quick and cost-effective method to understand and optimize products and processes. Tanko et al. (2007) proposed a model to simplify the application of design of experiments that include a series of activities that must be completed before continuing to the next step. The complete methodology uses some brief guidelines to explain and guide users through the different activities necessary to complete a DOE project, as summarized in Fig. 21.2.

The activities indicated by the model are:

1. Define: Select a team, formulate problem, state relevant background, choose response, state objective.
2. Measure: Identify factors, classify factors, validate measurement systems, choose strategies for nuisance factors, choose ranges and levels, state actual process knowledge.
3. Pre-analyze: Characterize the factors, define characteristics needed for the design, choose experimental design, select levels.
4. Experiment: Outline experiment, evaluate trial runs, perform the experiment, and recollect data.

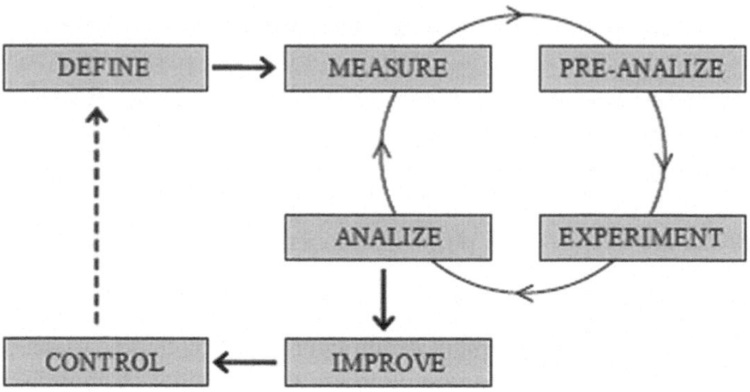

Fig. 21.2 Model of the methodology for applying DOE (Tanko 2007)

5. Analyze: Determine factor effects, determine significant effects, model building, optimization, evaluate new experiments.
6. Improve: Confirming testing, draw conclusions and do recommendations, implement new conditions.
7. Control: Implement controls, validate results.

The four steps, i.e., the four DOEs, of this stage were carried out as follows.

Step 3.1. Here, DOE 1 was performed to determine if the rank and classification of the operating technicians are significative factors in the generation of defects attributed to product handling. A repeatability and reproducibility (R&R) analysis was applied to technical operators to validate that the faulty products segregation process was reliable.

The procedure proposed by Tanko (2014) was as followed:

1. Define: Determine if the classification and rank of the operating technician influences the level of segregation of defects attributed to handling.
2. Measure: Identify the factors:

   - Factor 1: Classification of operating personnel. Levels: operating technician I, operating technician II.
   - Factor 2: Rank of operating personnel. Levels: Talent Mark 1 (TM1) or inspector 1 and Talent Mark 2 (TM2), or inspector 2.
   - Noise factor: Skill level of operating personnel.

3. Validate measurement systems: A R&R analysis was applied to technical operators to validate that the faulty products segregation process is reliable. This analysis is valid for all the DOEs performed in this investigation.

1. Choose strategies for noise factors
   - Select operating personnel randomly to participate in the DOE.
   - Select operational personnel who are certified in the assigned operation and who will participate in the DOE to ensure the same level of competencies.
2. Establish ranges and levels
   The levels are two, the evaluation applies to technical operator level one and two: and the ranges are the classification TM1, TM2.
3. Use prior knowledge of the process to predict the effects of factors and their interactions. Historical information on the personnel selected for the DOE was used to identify the level of segregation by management prior to the current experimentation.
4. Pre-analysis
   a. Characterize the factors
      The levels of the considered factors will consider the use of the operator classification factor TM1, TM2, since it is considered that the range in which the operator technician was classified influences adherence to procedures and consequently a higher quality index and/or less segregation of defects attributed to handling.
   b. Define the characteristics required for the design
      The development of the DOE to be applied consists of the following activities:
      - Selection of operational personnel "operator technician level one and two" and their classification TM1, TM2; randomly.
      - Identify production batches that are assigned to pre-selected operator technicians and identify parts per million (PPMs) found associated with handling coding.
      - Create and analysis of variance (ANOVA) to identify and validate if there is a significant difference in the classification of operational personnel which is associated with their classification.
   c. Choose the experimental design according to the previously defined characteristics.
      The DOE selected to apply is the 2×2 factorial design (two factors with two levels each and nine replications).
   d. Select levels
5. Define the experiment
   (a) Factors and levels to be considered in the DOE are determined.
   (b) It is considered that the classification and rank of the technical operator influence the level of segregation, and the objective is to validate the hypothesis.
   (c) The parts per million of rejected parts (PPMs) obtained during the defined period will be the response variable.

(d) Select operating personnel at random who meets the conditions of the selected factors.
(e) Obtain information on batches processed by the personnel to be evaluated and the segregation of failures by handling in terms of PPMs, and the period considered was the second quarter of 2016.

6. The hypotheses are (two for the main effects and one for the interaction):

   (a) H_0 = There is no effect due to the rank factor.
   H_1 = There is an effect due to the range factor.
   (b) H_0 = There is no effect due to the classification factor.
   H_1 = There is an effect due to the classification factor.
   (c) H_0: There is no interaction between the rank and classification factors.
   H_1: There is an interaction between the rank and classification factors.

7. An analysis of variance (ANOVA) was performed to analyze the experiment data and draw conclusions on its influence in the level of product rejection.

Step 3.2. In this step, DOE 2 was applied to determine if the classification of technical operator versus equipment platform (or type of processing equipment) where the material was processed influenced the product quality.

1. Define: Determine if the process equipment platform and classification of operator technician affects the level of defects attributed to handling.
2. Measure: Identify the factors:

   - Factor 1: Equipment platform. Levels: (1) 8028, (2) Maxum, (3) ASM, (4) Icon, and (5) Procu platforms.
   - Factor 2: Classification of operator technician. Levels: (1) operator technician 1 (OT1) and (2) operator technician 2 (OT2).

3. Validate measurement systems: A R&R analysis was applied to technical operators to validate that the faulty products segregation process is reliable (see DOE 1). The procedure proposed by Tanko (2014) was followed for all DOEs performed, as explained in detail in DOE 1.
   In this step, the hypotheses are as follows:

   (a) H_0 = There is no effect due to the equipment platform factor.
   H_1 = There is an effect due to the equipment platform factor.
   (b) H_0 = There is no effect due to the classification of operator technician factor.
   H_1 = There is an effect due to the classification of operator technician factor
   (c) H_0: There is an interaction between the equipment platform and type of operator.
   H_1: There is no interaction between the equipment platform and type of operator.

Step 3.3. Similarly, DOE 3 was implemented to reveal if classification of technical operator versus area where the product is processed (assembly process) affected the level of rejection due to quality flaws.

1. Define: Type of experimental design: 2 × 2 factorial design.
2. Measure: Identify factors:
   - Factor 1: Assembly process, levels: dice sticking area (DA) and wiring area (WA).
   - Factor 2: Classification of operator technician, levels: OT1 and OT2.
3. Validate measurement systems: The R&R analysis was applied to technical operators to validate that the faulty products segregation process is reliable which is valid for all the DOEs performed, as explained in DOE 1. The procedure proposed by Tanko (2014) was followed for all DOEs performed, as explained in detail in DOE.
   In this step, the hypotheses are as follows:
   (a) H_0: There is no effect of the assembly process factor on the level of rejections.
   H_1: There is an effect of the assembly process factor on the level of rejections.
   (b) H_0: There is no effect of the type of operator technician on the level of rejections.
   H_1: There is an effect of the type of operator technician on the level of rejections.
   (c) H_0: There is no interaction between the assembly process and type of operator technician.
   H_1: There is an interaction between the assembly process and type of operator technician.

Step 3.4. Finally, DOE 4 was carried out to find out if profile compatibility with inspector position, lighting in wafer inspection area, and inspector's level of knowledge affected the level of quality (wrong rejection).

1. Define: Determine if profile compatibility, area lighting, and level of knowledge of the inspector influence the level of rejection of inspected wafers.
2. Measure:
   (a) Identify the factors:
      - Factor 1: Profile rating. Levels: (low: 0–49; high: 50–100%).
      - Factor 2: Level of knowledge. Levels: (low: 0–85%; high: 86–100%).
      - Factor 3: Lighting levels. Levels: (low: 560; high: 1980 lm).

   Table 21.1 shows the factors and levels considered for DOE 4 (three factors, two levels each).

Table 21.1 Factors and levels to consider in DOE 4

Knowledge (%)	Lighting (lm)	Compatibility profile
0–85	580–590	0–49
86–100	1000–1980	50–100

(b) Choose strategies for noise factors

- Select inspection personnel at random to participate in the DOE.
- Select inspection personnel who are certified in the assigned operation and who will participate in the DOE to ensure the same level of competence.

(c) Establish ranges and levels
There are three factors with two levels each, the evaluation applies to compatibility profile (low and high), lighting (low and high), and the level of knowledge (low and high).
Use prior knowledge of the process to predict the effects of factors and their interactions. The supervisors suspected that the lighting level affected the accuracy of the inspection process and its rejection level. Historical information from personnel selected for DOE 4 about the level of knowledge of the inspection process as well as the compatibility profile was used to identify if those factors affected the level of rejection of the parts.

3. Pre-analysis and characterization of the factors: As said before, the treatments considered for DOE were the lighting factors (560–1980 lm), profile compatibility (low and high), and knowledge level (low and high).
4. Define the characteristics required for the design.
The development of the DOE to be applied consisted of the following activities:

- Selection of profiles "compatibility with position," "level of knowledge" and lighting in the inspection area.
- Identify wafers rejected by inspector and identify percentage of rejection.

Table 21.2 shows the information considered for DOE 4, non-replicated three factors with two levels each.
For this step, the proposed hypotheses are as follows:

(a) H_0: There is no effect due to the knowledge factor.
H_1: There is an effect due to the knowledge factor.
(b) H_0: There is no effect due to the profile compatibility factor.
H_1: There is an effect due to the profile compatibility factor.
(c) H_0: There is no effect due to the lighting factor.
H_1: There is an effect due to the lighting factor.
(d) H_0: There is no interaction between the knowledge factor and profile compatibility.
H_1: There is an interaction between the knowledge factor and profile compatibility.

Table 21.2 Data for DOE 4

Factors			Response variable
Compatibility	Knowledge	Lighting (lm)	Level of rejection
49	100	560	4.06
49	85	560	3.64
100	85	560	2.58
100	85	1980	1.66
49	85	1980	3.63
100	85	1980	2.82
49	100	1980	11.78
100	100	560	4.1

(e) H_0: There is no interaction between the knowledge factor and lighting.
H_1: There is an interaction between the knowledge factor and lighting.

(f) H_0: There is no interaction between profile and lighting compatibility.
H_1: There is an interaction between profile and lighting compatibility.

(g) H_0: There is no interaction between the knowledge factor, profile compatibility, and lighting.
H_1: There is an interaction between the knowledge factor, profile compatibility, and lighting.

21.3.4 Stage 4: Analysis of Case Study

In stage 4, a case study was analyzed. It was related to customer complaints or voice of customer (VOC's) related to bad operations (misoperations) that led to faulty products to be sent to the customer.

21.3.5 Stage 5: Proposals of Cognitive Ergonomics Approach

In stage 5 of the investigation, a proposal was made for the inclusion of cognitive ergonomics in the product/process/system design phase. It includes the generation of a FMEA and determination of the RPN index, both for the design of the technical aspects of the process and also for taking into account the phases of the worker's cognitive process (detection of the problem, assessment of the situation, working memory, response planning, long-term memory, implementation of responses).

21.4 Results and Discussion

21.4.1 Analysis of Correlation Between the Level of Rejected Lots and Worker Seniority

As the first data collection and analysis activity, the statistical correlation was carried out, which is shown in Fig. 21.3. It provided the value of the correlation coefficient between the seniority of the operator versus the generation of batches with more of 100 failures in defects related to the following processes: under die, moved die, and excess epoxy. The result obtained corresponds to a Pearson's correlation coefficient of −0.0265, which indicates a poor and negative correlation between seniority and the level of rejection. From the outset, the indicator "worker's seniority" has no direct relationship with the loss of productivity (generation of defects).

21.4.2 Identification of Psychological Profiles

Results obtained from the Cleaver psychometric tests showed that 60% of the personnel meet the required profile. However, this is not conclusive to determine that there is a relationship between the profile and the incidence of non-adherence to procedures. The results are conclusive as the data is extremely scattered and the correlation coefficient is 0.377, as seen on Fig. 21.4.

On the other hand, the result of the determination coefficient R^2 (14.2%) does not explain the variability of the regression equation. R^2 is the percentage of variation of

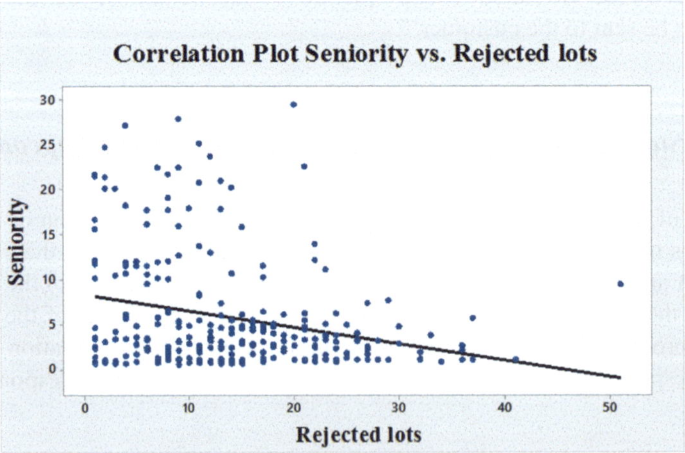

Fig. 21.3 Correlation plot of batches with more than 100 failures and the worker's seniority, in years

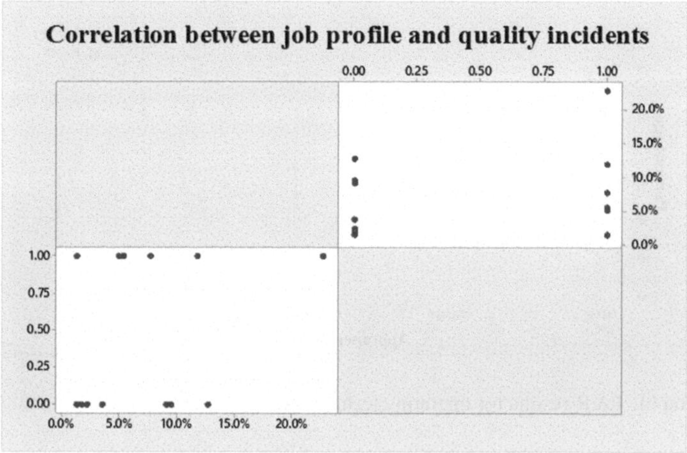

Fig. 21.4 Correlation matrix between work profile and quality incidents

the response variable that is explained by its relationship with one or more predictor variables. In general, the higher R^2, the better the model will fit the data. R^2 is always between 0 and 100%. The *p*-value was greater than $\alpha = 0.05$, so the null hypothesis is accepted, and there is no significative correlation between both variables.

The following are the results of the correlation and regression analysis as returned by the software Minitab®:

- Correlation variables: Job profile, process quality audit level (PQA)
- Pearson correlation coefficient (Job profile and PQA) = 0.377
- *p*-value = 0.150
- Regression analysis: Findings for audits of processes versus job profile
- The regression equation is: PQA = 0.1745 + 3.208 * Job profile
- $S = 0.479415$, $R^2 = 14.2\%$, adjusted $R^2 = 8.1\%$.

Results returned by Minitab®, statistically validated that the data is dispersed, without a clear pattern (see Fig. 21.4) and statistically affirmed that the work profile and quality incidents (PQA) are not related as R^2 was 14.2%. Also, the *p*-value was 0.150, greater than a (0.05), which indicates that the correlation coefficient is equal to 0, and no correlation exists between those two variables.

Table 21.3 shows the results of the evaluation of the level of effectiveness in the inspection process of technical operators. It was in the range of 77.3%. The level of

Table 21.3 R&R analysis results

General die attach process				Rejections		
Operator	No. of pieces inspected	Correct measurement	%	Over rejection	Escape	% over rejection
Various	1650	1275	77.30	267	108	71.20

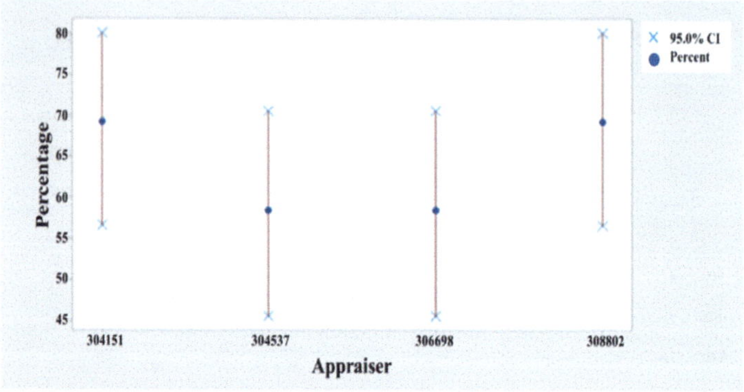

Fig. 21.5 Graphic R&R results for operating technician

over rejection was 71.2%, which means that when the staff is in doubt whether the product is acceptable, rejects it.

Similarly, Fig. 21.5 graphically shows the results of the evaluation of the level of effectiveness in the inspection process of technical operators compared to each other. We can conclude that there is no variation between their means, and therefore, they have the same level of effectiveness in the inspection process.

To analyze the variables defined in the operationalization process, four experimental designs were performed to validate the relationship between several human factors and the level of lot rejections, which impacted in the productivity and quality of the analyzed processes.

21.4.3 Design of Experiments Results

21.4.3.1 DOE 1. Classification Versus Rank

In DOE 1 and according to the results of the ANOVA shown in Table 21.4, all the null hypotheses were accepted. These mean that statistically there is no evidence that the treatments influence the response variable.

The *p*-value results obtained in the ANOVA of Table 21.4 indicate that they are greater than the significance value, (0.05), and therefore the null hypotheses are accepted, concluding that there is no statistical evidence that the treatments influence the response variable.

Figure 21.6 shows the graphical results of the model's assumption verification of design of experiments 1.

(a) Assumptions verification

- Normality: The normality assumption is met since the data is aligned to a straight line.

Table 21.4 ANOVA of DOE 1

Source	DF	Adj SS	Adj MS	F-value	p-value
Model	3	243.22	81.074	0.88	0.463
Linear	2	143.22	71.611	0.78	0.469
Rank	1	136.11	136.111	1.47	0.234
Classification	1	7.11	7.111	0.08	0.783
Two-way interactions	1	100	100	1.08	0.306
Rank * position	1	100	100	1.08	0.306
Error	32	2954	92.313		
Total	35	3197.22			

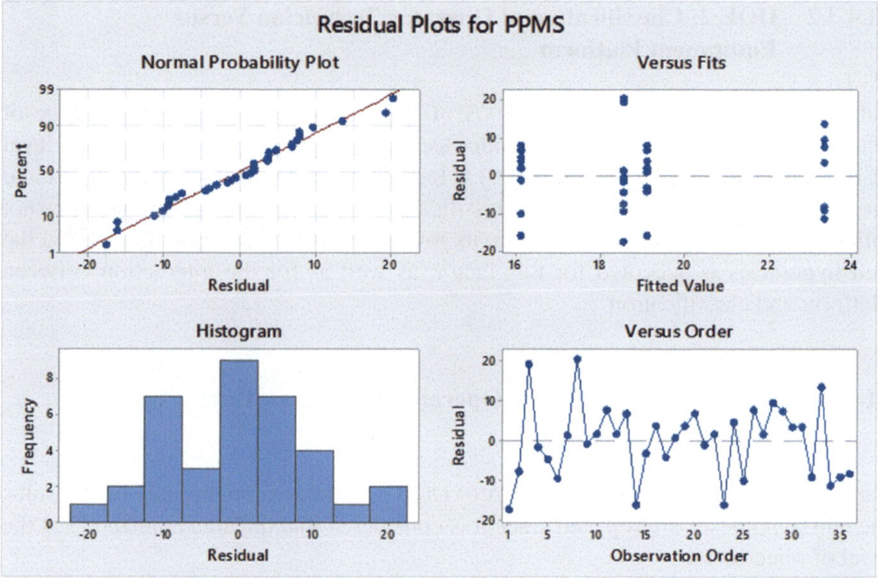

Fig. 21.6 Graph of DOE 1 residuals

- Constant variance: The data are distributed without any clear pattern, so it is considered that the treatments have the same variance.
- Independence: The data behaves randomly, and it does not present a defined pattern.

Due to the results of the ANOVA of DOE 1, it was concluded that the rank and classification of the operating technician do not influence the level of segregation by handling.

Table 21.5 ANOVA of DOE 2

Source	DF	Adj SS	Adj MS	F-value	p-value
Model	11	1127.5	1025.0	3.00	0.036
Linear	6	10,350.8	1725.1	5.04	0.008
Platform	5	8909.3	136.111	5.21	0.009
Classification	1	1441.5	1441.5	4.21	0.06
Two-way interactions	5	924.5	184.9	0.54	0.742
Platform * class	5	924.5	184.9	0.54	0.742
Error	12	4104.0	342.0		
Total	23	15,379.3			

21.4.3.2 DOE 2: Classification of Operator Technician Versus Equipment Platform

Table 21.5 corresponds to the ANOVA of DOE 2. Results show that the type of equipment platform used by the operator technician affects significantly to the level of rejections, since the p-value (0.009) is less than the significance level (0.05) and the null hypothesis is rejected. The classification of the operator technician does not influence the level of rejections, since its p-value is 0.06, greater than Alfa, and the null hypothesis is accepted for this factor, as well as for the interaction between platform and classification.

21.4.3.3 DOE 3 Classification of Operator Technician Versus Assembly Process

Table 21.6 corresponds to the ANOVA of DOE 3, and based on the statistical results, the null hypotheses are rejected since it is considered that the area contributes at the level of rejected lots.

Table 21.6 ANOVA of DOE 3 as returned by Minitab 17

Source	DF	Adj SS	Adj MS	F-value	p-value
Model	3	5866.2	1955.39	9.66	0.000
Linear	2	5829.4	2914.71	14.39	0.000
Classification	1	65.3	65.33	0.32	0.0.573
Area	1	5764.1	5764.08	28.47	0.000
Two-way interactions	1	36.8	36.8	0.18	0.672
Class * area	1	36.8	36.8	0.18	0.672
Error	44	8909.5	202.49		
Total	47	14,775.7			

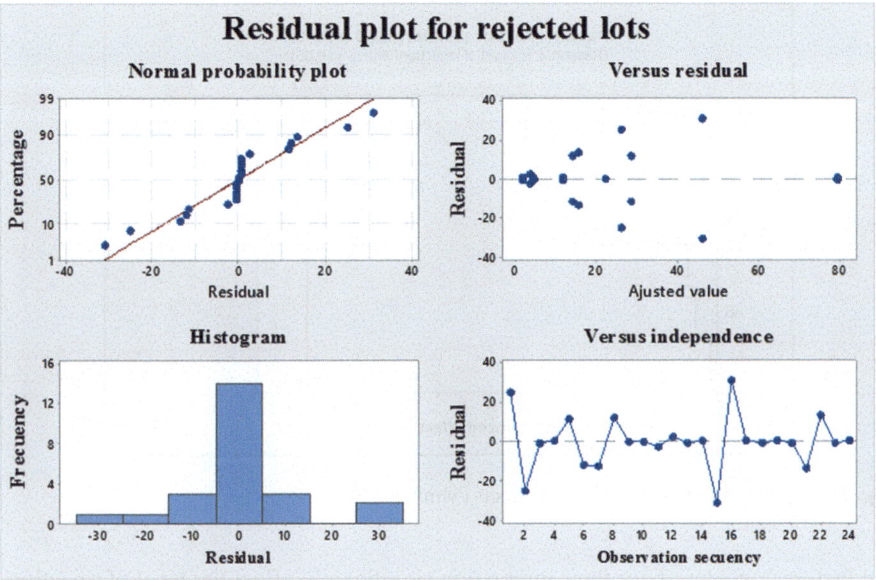

Fig. 21.7 Graph of DOE 3 for model assumptions verification

The interpretation of Fig. 21.7 is as follows:

1. Normality: The assumption of normality is fulfilled since the data is aligned to a straight line.
2. Constant variance: The data is distributed without any clear pattern, so it is considered that the treatments have the same variance.
3. Independence: The data behave randomly; it does not present a defined pattern.

21.4.3.4 DOE 4. Profile Compatibility Versus Lighting Level Versus Knowledge Level

Table 21.7 shows the ANOVA results as returned by the software Minitab 17, where the *p*-value indicates that the null hypothesis are accepted, concluding that neither of

Table 21.7 ANOVA of the DOE 4 experiment as returned by Minitab 17

Source	DF	SS sec	Adj MS	*F*-value	*p*-value
Main effects	3	45.9441	1955.39	22.12	0.155
Two-interactions of (No.) factors	3	22.4319	2914.71	11.11	0.216
Residual error	1	0.6728	0.6728		
Pure error	1	0.6728	0.6728		
Total	7	69.0488			

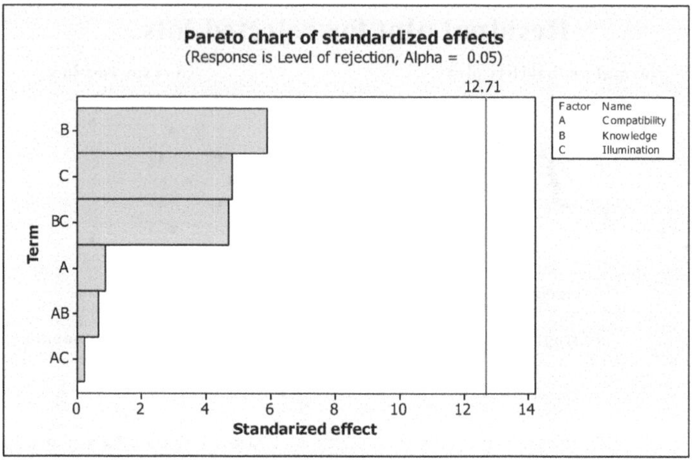

Fig. 21.8 Pareto chart of standardized effects (Minitab 17)

the factors considered nor their interaction significantly affect the level of rejection, at a level of significance of 5%.

Results shown in the Pareto chart of the effects in Fig. 21.8 determine the magnitude and importance of an effect. The diagram shows the absolute value of the effects and draws a reference line on the graph. Any effect that extends beyond this baseline is potentially significant. If there is no error term as when is a single replica experiment, the Lenth's method is used to draw the line and displays the non-standardized. According to this results and those of the ANOVA, the null hypotheses are accepted, and thus, statistically there is no evidence that the factors influence the response variable.

Statistical model assumptions verification is shown in Fig. 21.9:

- Normality: The assumption of normality is fulfilled since the data is aligned to a straight line.
- Constant variance: The data are distributed without any clear pattern, so it is considered that the treatments have the same variance.
- Independence: The data behaves randomly and does not present a defined pattern.

21.4.3.5 Summary of DOE's Results

Several DOEs were performed to investigate if different factors statistically influenced the level of parts rejection, thus affecting productivity and quality of the company. The human factors analyzed were training level of the operational personnel (classification and rank of operator technician), job profile of the workers, their knowledge level, and the profile compatibility with the task assigned to the operator. Technical factors considered were the kind of technological platform of the

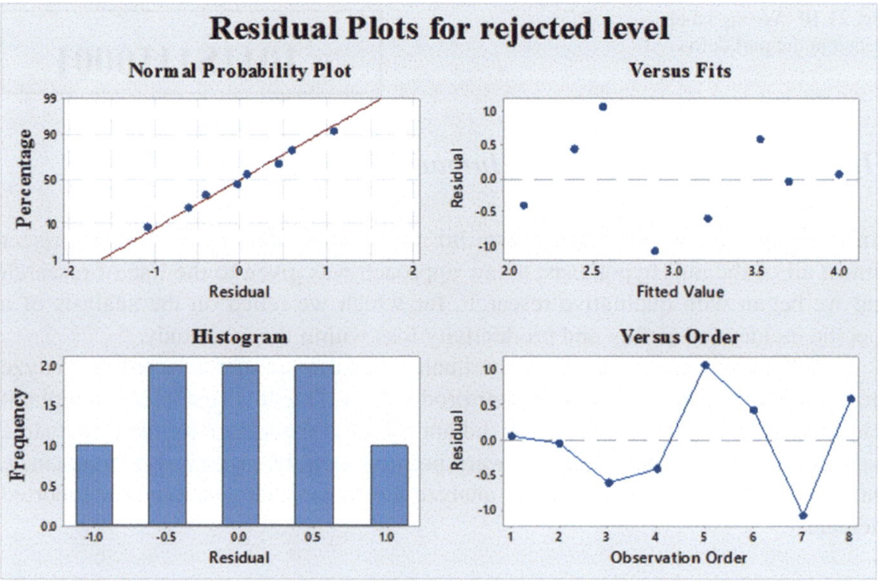

Fig. 21.9 Graphical results of the verification of assumptions of DOE 4

automated equipment used by the workers, the process area in the production floor, and the lighting of the work area.

1. DOE 1 considered two factors: "level of the operator technician" (OT1 and OT2) versus rank (TM1-TM2). Results are conclusive, and these factors do not contribute to the level of rejections for handling and to carry out further analysis.
2. DOE 2 results gives us the acceptance of the null hypotheses, and we conclude that the different equipment platforms used in the evaluated manufacturing process influence the response variable, but the classification of the operator technician does not affect the level of rejection.
3. In DOE 3, we again used a factorial design 2 × 2, considering as factors the operator's classification and two process areas, gluing the dice and wiring. Results show significant effects of the type of process area in the response variable. Since the null hypotheses are rejected, the process area is one factor that contributes to the level of rejected lots.
4. DOE 4 were carried out in which three factors were considered: "profile compatibility," "lighting," and "level of knowledge" to determine if they influence the level of rejection in the wafer inspection process. Results were consistent with those obtained in the first DOE when rejecting the null hypotheses and accepting the alternative hypotheses in which it was concluded that none of the factors considered influence the level of rejection. Given these results, our line of research must take a new approach since it was found that the operating personnel (technical operators and inspectors) are not factors that affect productivity and quality in the manufacturing process.

Fig. 21.10 Wrong label placed in the part delivered

1011S1110001

21.4.4 Further Analysis of Human Factors

Since the quantitative research phase did not obtain statistical evidence to reject almost all of the null hypothesis, a new approach was given to the line of research and we began with qualitative research, for which we relied on the analysis of a specific incident of quality and productivity loss within the case study.

A notification was received by the quality department and selected to analyze the influence of human factors on their productivity. Due to a customer's complaint related to the receipt of products that did not meet the specifications, the responsible staff and a multidisciplinary group were notified: quality, processes, maintenance, and manufacturing engineers met to analyze the causes that generated the reported incident.

21.4.4.1 Analysis of the Incident "Mislabeled Part"

The problem was that the warehouseman physically delivered the item 1011**S**1110001 instead of item 1011**N**1110001, whose label was written *upside down* on the product, as exemplified in Fig. 21.10. Although visually we can only distinguish between the two pieces a difference of one character at each (**S** and **N**), this functionally represents that the final product does not fulfill the functions for which it was created.

This error produced faulty units that affected the customer's order. Here, the human factor associated with the malfunction was due to ergonomic conditions not favorable for a correct process because the procedure indicated to make a visual validation of the characters in the product, which were physically reversed. The storekeeper had to read them from right to left, whereas the comparison should have been made simultaneously with a normally written serial number (left to right). Furthermore, the serial number to be identified was written backwards in relation to the operator's eyes. Those instructions were impossible to achieve correctly due to how the human brain processes information. Here, it is important to emphasize that the dimensions of the characters require the use of a magnifying glass.

21.4.4.2 3L5W Analysis for Incorrectly Labeled Parts

To start finding the root cause of the problem, the lean sigma tool 5W3L was used. It is a method for problem solving that gets to some hard to identify causes and gives the opportunity to see issues that have a leveraging effect on the overall process. When properly deployed, it takes to the root of the problem because it is a simple process of logical connections. It consists on a series of questions of the type of "Why did this happen?" repeating the process about five times. Afterward, it will typically come to a root cause.

In the problem analyzed, the questionnaire was divided in three parts: occurrence, detectability, and systematic. Due to space limitations, the questionnaire was not included here.

21.4.4.3 Actions Implemented

The actions implemented to attack the latent causes that led to the mislabeled part event were the following:

(1) In the occurrence part:

  (a) For parts warehouse: The physical validation procedure in the parts warehouse was modified, in which the use of an optical recognition camera was documented, which communicates with an internal application, and only if the data matches, the system will allow the product to be released to the next process.

  (b) For the manufacturing area: Character recognition was activated in the equipment to read the physical identification of the part and print the label of the reading carried out.

(2) In the detection part:

  (a) For the cell leader: After printing, it will use this label to validate in the flow management system through the scan, and if the data is correct, you will be able to continue, and if it is not, it will restrict you and you will not be able to continue until someone takes corrective actions.

  (b) For the production inspector: For the production inspector, the use of the label printed by the equipment in the product validation module was defined through its scanning, this facilitates the process and eliminates the probability of error or dependence on the inspector as the applications are the ones that make the comparison and release the product once the veracity of the data has been confirmed.

(3) In the systematic part:

  (a) Figure 21.11 shows the design flow of the original process that considers the technical aspects with their corresponding calculations of the Risk Priority Number (RPN) values used for the introduction of the new product, of which the customer reported a defective part, a condition generated by incorrect process labeling (identification). It is important to note that the higher the RPN value obtained, the greater the risk for the product/process to fail.

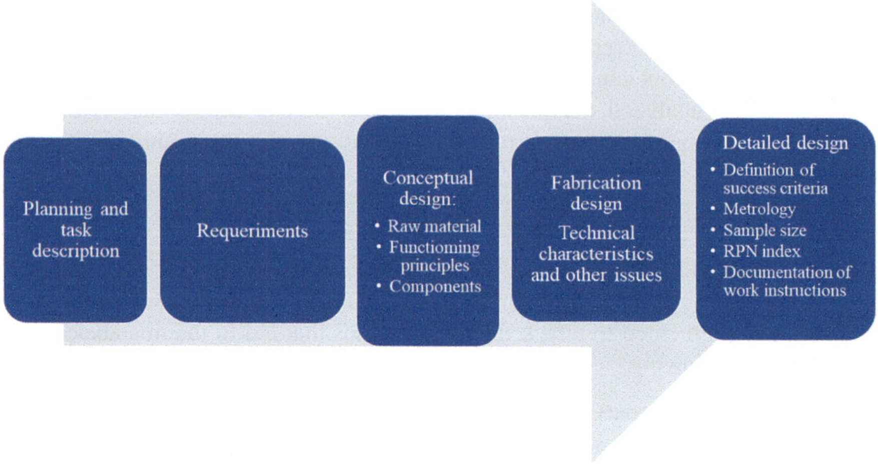

Fig. 21.11 Initial design process flow for 12-inch item (original)

Because the conditions that led to this incident were born in the design of the process, the following actions were implemented to attack the root cause.

- Application of the failure mode and effect analysis (FMEA) for evaluating the process design, through the determination of the RPN index for each success criteria and the complement of the conceptualization of Fig. 21.12, including cognitive ergonomics. Here, the function "Technical characteristics" was the initial stage of the introduction of the new product (NPI), the 12-inch article. Our proposal is the inclusion of the "cognitive ergonomics" evaluation during the FMEA of the new product design in order to detect which phases of this function are the most critical for this new element and to create and implement actions to

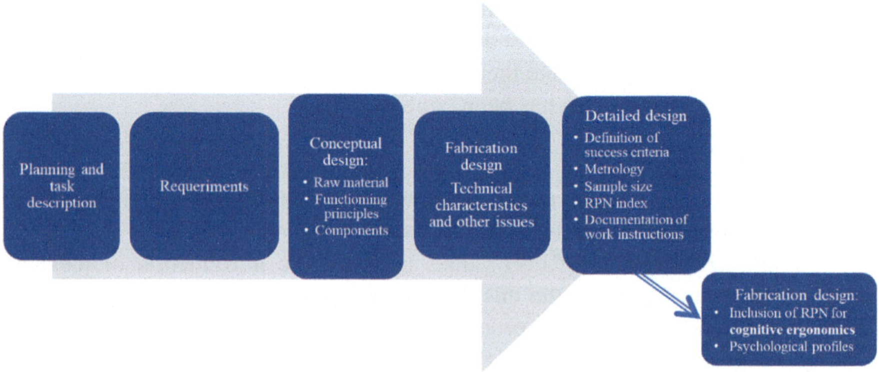

Fig. 21.12 Proposal of the design process flow for the introduction of the new product, considering and including the evaluation of the human factor

avoid repetition of incorrect operations and improve operator performance, as a reminder of the importance to note that the higher the RPN value, the more critical the failure mode.
- Development of competencies of project engineers for the introduction of new products in the discipline of design and manufacturing execution.
- Involvement of project engineers and introduction of new products in the process and use of the same manufacturing flow management applications to increase the level of knowledge in the procedures that they document and that the personnel will then execute.
- Review and compare the new design against the current execution process, visiting the production floor to ensure that the designed process is functional and in accordance with the documented work procedures. If this does not apply with the new product introduction, make the necessary modifications.

21.4.5 Proposed Framework for Designing Work Methods that Incorporate the Human Factor

Although the initial focus was on the level of execution of the operator's activities, the results obtained from the various analysis activities carried out throughout the present research work showed that the operation error events were derived from designs of incomplete and inadequate process, where the inclusion of cognitive ergonomics was omitted during the generation stage of the risk analysis (RPN) and incorrect documentation of the work instructions was derived from this. Too often system design is carried out without input from potential users, and without feedback, critical process activities remain unfinished. Keeping the operators involved will provide benefits such as process optimization, error reduction, and satisfaction with the correct execution of their work; this is the primary responsibility of product/process engineers "to develop automated systems less vulnerable to human error."

Figure 21.12 is an update of Fig. 21.11 during the definition of the new product process in the design conceptualization stage, and the evaluation of cognitive ergonomics (human factors) is included as a key activity that helps to optimize the process and mainly the elimination of incorrect operations during the execution phase. This proposal aims to redefine the criteria of success in the FMEA that is carried out in each new design and to be used by those responsible for the development and introduction of processes and new products or as part of the philosophy of continuous improvement of the companies.

Table 21.8 is the FMEA with its determination of the proposed RPN index and the complement of the conceptualization of Fig. 21.12. Here, the function "technical characteristics" was the initial stage of the introduction of the new product (NPI) 12 in the article. As a complement, our proposal is the inclusion of the "cognitive ergonomics" evaluation during the FMEA of the new product design in order to detect which phases of this function are most critical for this new element and to create and implement actions to avoid repetition of incorrect operations and improve

Table 21.8 Short FMEA example, RPN index determination for NPI 12 inch part

Function	Failure mode	Effect	Severity	Occurrence	Detection	Initial RPN
Technical characteristics "NPI for 12 in. product"	Air particles 591	Breaking risk	4	8	33	96
	Broken piece	Functional failure	99	4	7	252
	Labeling	Wrong consumption	8	6	8	384
	Line type hair	Functional failure	10	4	1	40
	Strings	Lifting problems	2	6	1	12
	Adherence	Breaking risk	6	3	4	72
Cognitive ergonomics phases "NPI for 12in"	Problem detection	No problem detection	9	6	4	216
	Situation evaluation	Incomplete or erroneous	8	6	7	196
	Working memory	Insufficient	7	7	4	196
	Planned response	Out of time/inadequate	7	6	5	210
	Long-range memory	Insufficient	7	6	4	168
	Immediate response	Incomplete or erroneous	6	8	4	192

operator performance, as a reminder it is important to note that the higher the RPN value, the more critical the failure mode.

21.5 Conclusions

The focus of this research project was to determine if human factors contribute to the loss of productivity and quality, both key indicators in manufacturing companies. To validate this approach, different statistical analysis was performed in a multinational manufacturing company with highly automated processes. Preliminary results were inconclusive about the influence of the human factors considered. A further analysis was made within the same company, where a real case study was selected to analyze the causes that led to a client's complaint after receiving a non-conforming product. For this, a methodological process was followed, using a series of problem-solving tools.

The different statistical analysis were performed to find if age, seniority, level of training or position (operator or inspector), job profile, knowledge, among other human factors influenced the effectiveness of lots rejections, and thus the levels of productivity and quality. Results showed that none of the factors analyzed influence the response variable, that is, the level of lots rejected. Also, the results obtained from the application of the cleaver psychometric test applied to the personnel were not conclusive to determine if the work profile influences quality incidents. The recommendations are as follows:

1. In the process of hiring personnel, it is recommended to apply the survey to the candidate and document it for future reference and based on the previously defined profile "high stability and high compliance," select those who meet the necessary characteristics to carry out their activities properly.
2. In the case of incidents of error in which it is determined that the human factor was the highest contributor, apply the psychometric test again to verify if there was a change in your profile from the moment you were hired to the date of the incident, and based on these results, evaluate cognitive process, consider the change of activity if applicable, or give feedback.

Although the initial approach was focused at the operator level, the results showed that his event was derived from a deficient process design, which omitted a critical manufacturing activity for the administration of the product flow, which required a high cognitive level.

Also, a risk analysis of the human factor conditions on the process was not carried out. Within the literature review, some authors make statements about misoperations, mostly focused on the execution process but not in the phase of development of products. A more profound and precise analysis of the problem took us to study the structural levels of the manufacturing process as it is the design, development, and introduction of new products; for not considering a critical activity for the process which was already documented at the level of manufacture production and therefore was not considered in the measurement of risk analysis.

Error events were derived from designs of incomplete and inadequate process, where the inclusion of cognitive ergonomics was omitted during the generation stage of the risk analysis and its indicator "Risk Priority Number" (RPN), and incorrect documentation of the work instructions was derived from this. Too often system design is carried out without input from potential users, and without their feedback, critical process activities remain unfinished. Keeping the operators involved will provide benefits such as process optimization, error reduction, and satisfaction for the correct execution of their work; this is the primary responsibility of product/process engineers to develop automated systems less vulnerable to human error.

After two years of the improvement, activities to eliminate wrong material labeling identification were implemented, no more issues had been presented, so we can determine that the root cause was correctly defined and the measures to avoid recurrence were successfully implemented.

21.5.1 Key Points

- In manufacturing process execution, the 3L5W is an effective tool of continuous improvement to solve issues related to customer complaints.
- Use socio-technical systems perspective in order to eliminate misoperations that have their origin in the conditions related to human factors.
- The design phase of the introduction of new products must include the human factors involved in the activity evaluated, the FMEA with the RPN index calculation will give you the priorities of intervention to attack the failure modes related to the most critical human factors and define corrective actions that eliminate and/or decrease incidents of misoperations.

References

Aguilar-Otero J, Torres-Arcique R, Magaña-Jiménez D (2010) Analisis de modos de falla, efectos y criticidad (AMFEC) para la planeación del mantenimiento empleando criterios de riesgo y confiabilidad. Tecnol Cienc Educ 25(1):15–26. https://www.redalyc.org/articulo.oa?id=482/48215094003

Anderson SJ, Raiken N (1988) Human factors isues in the aceptance of workgroup productivity tools: Work in Progress at Applied Voice Technology, IPCC '88 Conference Record 'On the Edge: A Pacific Rim Conference on Professional Technical Communication'. https://doi.org/10.1109/IPCC.1988.24029

Appelbaum SH, Hester AJ (1997) Socio-technical systems theory: an intervention strategy for organizational development. Socio-technical systems theory as a diagnostic tool for examining underutilization of wiki technology. Manag Decis J Knowl Manag 35(1):452–463

Bargelis A, Čikotiene D, Ramonas Z (2014) Impact of human factors and errors for product quality and reliability in the integrated approach of product and process design, maintenance and production. Mechanika 20(1):92–98

Carlson CS (2012) Failure mode and effects analysis (FMEA), Understanding the fundamental definitions and concepts of FMEAs. Wiley

Gordon R, Flin R, Mearns K (2005) Designing and evaluating a human factors investigation tool (HFIT) for accident analysis. Saf Sci 43(3):147–171. https://doi.org/10.1016/j.ssci.2005.02.002

Haraguchi H, Kaihara T, Fuji N, Nonaka T (2014) A study on operator allocation method considering the process of skill proficiency in cell manufacturing systems. In: Proceedings of the SICE annual conference, pp 1400–1406

Karanikas N, Melis DJ, Kourousis KI (2017) The balance between safety and productivity and its relationship with human factors and safety awareness and communication in aircraft manufacturing. Saf Health Work 9:257–264

Langer M, Söffker D (2015) Event-discrete formal representation of a semi-automated manufacturing process as framework for human guidance and assistance concepts: analysis and application. Int J Prod Res 53(8):2321–2341

Leung K, Wang J (2015) Social processes and team creativity in multicultural teams: A socio-technical framework. J Organ Behav 36(7):1008–1025. https://doi.org/10.1002/job.2021

Manchi GB, Gowda S, Hanspal JS (2013) Study on cognitive approach to human error and its application to reduce the accidents at workplace. IJEAT 2(6):236–242

Martínez PC (2006) El método de estudio de caso: Estrategia metodológica de la investigación científica. Pensam Gestión Rev Div Cienc Adm Univ Norte 20:165–193

Mclaughlin P (2013) Manufacturing best practice and UK productivity. Future of manufacturing project: evidence paper 21

Montano EM, Luevanos A, Sanchez CV (2019) Análisis de la evaluación psicométrica como un proceso de la gestión de talentos en la organización. U.A. Coahuila. https://riico.org/wp-content/uploads/2019/03/1.14-Ana%CC%81lisis-de-la-evaluacio%CC%81n-psicome%CC%81trica-como-un-proceso-de-la-gestio%CC%81n-de-talentos-en-la-organizacio%CC%81n.pdf

POST (2001) Managing human error. Parliam Off Sci Technol 156(156):1–8

Rothblum A, Wheal D, Withington S, Shappell SA, Wiegmann DA, Boehm W, Chaderjian M (2002) Improving incident investigation through inclusion of human factors. United States Department of Transportation—publications & papers 32, p 150. https://digitalcommons.unl.edu/usdot/32

Salmon PM, Read GJM (2018) Using principles from the past to solve the problems of the future: human factors and sociotechnical systems thinking in the design of future work. Hum Factors Ergon Manuf 28(6):277–280

Scaravetti D, Montaignier E (2009) Early human factor involvement in product design processes. In: Proceedings of the international conference on engineering design, ICED'09, Stanford University, CA, USA

Sheridan BTB, Parasuraman R (2005) Human-automation interaction. Rev Hum Factors Ergon 1(1):89–129. https://doi.org/10.1518/155723405783703082

Sobhani A, Wahab MIM, Neumann WP (2015) An innovative modeling method to evaluate human factor effects on the performance of manufacturing systems. In: Proceedings of the IEEE international conference on industrial engineering and engineering management, pp 130–134

Takano K, Reason J (1999) Psychological biases affecting human cognitive performance in dynamic operational environments. J Nucl Sci Technol 36(11):1041–1051

Takano K, Sawayanagi K, Kabetani T (1994) System for analyzing and evaluating human-related nuclear power plant incidents: development of remedy-oriented analysis and evaluation procedure. J Nucl Sci Technol 31(9):894–913

Tanko M, Viles, LI, Alvarez MJ (2007) Manufacturing Industries Need Design of Experiments (DoE). Procedings of the Worls Congress on Enginering 207 Vol II WCE 207, London, U. K

Toledo-Rosillo HG (2013) Test de selección basado en la personalidad del candidato. Gestiopolis

Ulhøi F, Jorgensen F (2010) Linking humanity with performability through social-technical systems theory. Int J Performability Eng 6(1)

Van Vuuren W, Shea CE, van der Schaaf TW (1997) The development of an incident analysis tool for the medical field. In: EUT—BDK report. Dept. of Industrial Engineering and Management Science, vol 85. Technische Universiteit Eindhoven

Wells WP, Wells W, Fahd K (2016) Human factors in production system. IFAC-PapersOnLine 49(12):1721–1724

Index

A
Absenteeism, 79
Absorption, 316
Accident causation, 530
Action performance, 513
Adaptability, 376
Additive manufacturing, 259
Additive manufacturing system, 160
Adequate amount, 410
Administration function, 297
Administration process, 301
Advanced manufacturing technology, 136
Air pressures, 471
Algorithms, 61
Alternative preferences, 232
Alternatives for waste management, 230
AMPL IDE, 408
Analysis data, 372
Analysis of correlation, 535
Analytical models, 159
Analytic hierarchy process, 221
Anecacao, 398
Analysis of variance (ANOVA), 411, 539
Anthropometry and biomechanics, 487
APHITM, 493
Applicability of the standard, 270
Arcs of Integration, 35
Arrival rate, 162
AS-IS model, 60
Aurora profile, 349
AutoCAD, 415
Automated Layout Design Program (ALDEP), 394
Automated systems, 249
Automation, 533
Automatized equipment, 535
Autonomy dimensions, 492
Available area, 93
Available budget, 192

B
Back pain, 472
Balanced workloads, 468
Barriers of standard implementation, 280
Baseline scenario, 175
Beneficiaries, 292, 293
Benefits, 125
Benefits of standard implementation, 279
Bidirectional communication, 252
Big data, 258
Billing process, 340
Biomechanical elements, 470
Biomechanical loading, 508
Bottleneck, 340
Brainstorming, 375
Breaking risk, 556
Budget, 192
Burn out, 313, 473
Business strategic initiatives, 252
Business strategy, 328

C
Cacao company, 397
Capability study, 383
Capacity, 68
Case study, 166, 187, 319
Categories, 9
Categories of waste management, 228
Cause-effects, 343
CAVE, 487
Certificates, 265
Certified organizations, 276

J. L. Garcia Alcaraz et al. (eds.), *Trends in Industrial Engineering Applications to Manufacturing Process*, https://doi.org/10.1007/978-3-030-71579-3

Challenges, 367
Chocolate, 389
Chocolate molds, 411
Chocolate production, 398
Circuit boards, 344
Circular economy, 10
Classification, 9, 546
Class of membership, 50
Client requirements, 295
Closed-cycle glass supply chain, 225
Closed loop supply chain, 215
Closeness rating, 396
Coefficient of variation, 171
Coercive pressures, 13
Cognitive elements, 473
Cognitive ergonomics, 543
Coherence, 289
Collected data, 346
Collectors, 227
Combination of levels, 43
Commercial organizations, 292
Common defects, 342
Communication, 91, 170
Company production, 400
Compatibility profile, 541
Competitive advantage, 150
Competitive priorities, 434, 444
Competitive status, 360
Components analysis, 62
Computerized Facility Aided Design (COFAD), 394
Computerized Relationship Layout Planning (CORELAP), 394
Concatenation of values, 306
Conditional probability, 145
Confrontation matrix, 207
Congestion levels, 177
CONOCER, 371
Constant variance, 550
Constraints, 67
Construction, 91
Construction workers, 509
Consumers, 157
Container, 378
Containment plan, 380
Contextual barriers, 16
Continuous improvement of processes, 272
Control recipe model, 250
$COP, 446
Corporation strategies, 328
Correlation matrix, 545
Correlation variables, 545
Corridors, 90
Corroboration of hypothesis, 325
Cost-effective method, 537
Cost function, 161
Cost of transport, 70
Cost saving, 70
Countries, 7
Coverage level, 190
Covered demand, 192
Critical requirements, 469
Critical success factors, 277
Cultural analysis, 368
Cultural aspects, 369
Cultural barriers, 16
Cultural values, 369
Culture, 305
Customer demand, 140
Customer-facing, 49
Customer level, 48–50
Customer requirements, 283
Customers, 42
Customer service, 531
Customer's order, 201
Cyber-Physical Systems (CPSs), 490, 491
Cycle time, 157

D

Damaged T3, 352
Data collection, 188
Data screening, 143
Decentralization and vertical integration, 257
Decision criteria, 367
Decision factors, 381
Decision-makers, 192, 441
Decision making, 452
Decision-making autonomy, 476
Decision-making process, 228
Decision variables, 186
Decrease incidents, 558
Dedication, 324
Dedication of the employee, 316
Defects, 342
Degree of integration, 43
Delivery performance, 141
Delivery time, 73
Demand, 168
Demand patterns, 400
Demand centers, 218
Dependence relationship, 442
Dependent variables, 148, 324
Descriptive statistics, 376
Design alternatives, 396

Design and installation, 77
Design For Assembly (DFA), 476
Design for construction safety, 510
Design of alternatives, 405
Design of Experiments (DOE), 537
Develop a solution, 375
Diagnosis, 52
Dimensional reductions, 62
Direct effects, 150
Direct flow supply chain, 219
Discrete choice model, 183
Discriminant analysis, 47
Disengagement, 313
Distance, 107
Distance to power, 374
Distribution, 64, 65
Distribution center, 68
Distribution plan, 90
DMAIC, 337
DMAIC approach, 360
DMAIC implementation, 337
DMAIC methodology, 372
Draft committee, 271

E

Ease of use, 255
Eco-design, 18
Economic analysis, 414
Economic evaluation, 69
Economic growth, 200
Economic performance, 21
Economic recovery, 216
Ecosystemic imbalance, 3
Ecosystems, 466
Ecuadorian aromas, 397
Effectiveness measure, 481
Effective performance, 304, 474
Effect size, 148
Efficiency, 125
Efficiency indices, 147
Efficient operation, 208
E/HF, 491
Elbows, 523
Electronic boards, 343
Elements of purpose, 295
Empirical work, 328
Employer branding, 312, 314
Employer branding scale, 321
Empowerment, 320
Enablers, 280
Engagement of People and Improvement (EPI), 281
Engineering terms, 379
Equipment and tooling, 467
Enterprise Information Systems (EIS), 247
Enterprise Resource Planning (ERP), 479
Environmental certification, 268
Environmental impacts, 23
Environmental management system, 276
Environmental performance, 21
Environmental signals, 441
Environment monitoring, 308
Equipment, 96, 117
Ergonomic, 508
Ergonomic analysis, 511
Ergonomic hazardous, 470
Ergonomic principles, 486
Ergonomic risk, 511, 520
Exclusion restriction, 441
Existence of a organization, 293
Experiment, 118
Experimental design, 412
Experimentation with the model, 397
Expert choice software, 225
Expert guidance, 270
Expert knowledge, 467
Exploratory Factor Analysis (EFA), 279
External factors, 17
External integration, 38

F

Facilities, 81, 392
Facility design, 395
Facility Location Problem (FLP), 182, 184, 391
Facility planning, 391
Factor analysis, 40
Factorial design, 541, 551
Factors and levels, 541
Failures, 378
Failure Mode & Effects Analysis (FMEA), 478
Family, 121
Family of products, 392
Farmers market, 183
Fast integration, 367
Financial investment, 218
Finished goods, 98
Finished product, 119, 404
Fishery sector, 199
Fitness test, 355
Fixtures, 98
Flatness, 356
Flexible parameters, 358

Flexibility, 115, 136
Flowchart, 117
Flow matrix, 163, 403
Flow variability, 166
Food dessert, 183
Food production, 400
Food insecurity, 193
Food supply chain, 182
Formulation process, 437
Framework developed, 439
Framework validity, 443
Frequencies, 43
Functional area, 300
Functional failure, 556
Functional strategy map, 207
Functions, 296
Functions for developing, 297

G
Gamma distribution, 170
General direction of standardization, 265
Genetic information, 291, 307
Genetic knowledge, 300
Genetics of a organization, 289
Glass container, 226
Global competition, 36
Global cycle time, 175
Global confectionery market, 392
Global economy, 275
Globalized market, 365
Good Manufacturing Practices (GMP), 416
Goodness of fit, 324
Google maps application, 190
Green distribution, 19
Green human resources, 20
Green information systems, 20
Green manufacturing, 19
Green purchasing, 19
Green supply chain management, 3, 223
GR&R, 380
GSCM literature, 5
GSCM practices, 18

H
Harmonious organization climate, 306
Hazard Analysis And Critical Control Points (HACCP), 416
Hierarchical structure, 231
Highest performance, 436
High level structure, 274
High probability, 446
High quality products, 208
Hiring personnel, 557
Historical information, 539
Holistic, 87
Holistic framework, 5
Holistic way, 497
Horizontal organizations, 368
Human-Automation-Systems (HAS), 533
Human factors, 466, 529, 552
Human Factors Incident investigation Tool (HFIT), 530
Human–Machine Interaction (HMI), 488
Human resources, 115, 209
Human resources management, 282
Human response, 470
Human values, 534
Hybrid methods, 222
Hydromassage tubs, 78
Hypotheses, 144
Hypotheses test, 540
Hypothesis, 320

I
IEC 61300, 349
Identity matrix, 163
Implementation, 11, 110
Implementation and validation process, 121
Implementation plan, 378
Implement change, 517
Improve, 337, 538
Improved performance, 338
Improved process, 357
Improved sigma level, 338
Improvement actions, 355
Improvement focus, 479
Improvement of supply chain management, 200
IMSS, 484
Inadequate lighting, 471
Inclusion relationship, 442
Independence, 549
Indian food, 339
Indirect collection, 235
Indirect effects, 148
Individualism, 370
Individual level, 318
Industrial activities, 283
Industrial association, 253
Industrial automation, 258
Industrial engineering, 87, 105
Industry 4.0, 245
Industry average, 440
Inefficient inspection, 353

Influential authors, 8
Influential factors, 12
Information Management System (IMS), 247
Information technology, 141
Initial state, 233
Innovation and growth, 316
Inspection profile, 352
Installation, 95
Institutional vision, 303
Integration factors, 43
Integration of the supply chain, 35
Inter-arrival times, 171
Interconnectivity, 246
Internal and external customers, 202
Internal drivers, 15
Internal integration, 38
Internal logistic, 113
International standard, 272
Internet of Things (IoT), 246
Interdependencies among skills, 470
Interpersonal acceptance, 317
Interpretative Structural Modelling (ISM), 10
Introjection, 302
Inventory systems, 93
Inward facing, 39
Ishikawa diagram, 343
ISO 9001: 2015, 532
ISO 9001 standard, 263
ISO 10075-1, 486
ISO 10075-3, 487
ISO 11226, 487
ISO 11228-1, 487
ISO 11228-2, 487
ISO 9001:2015, 486
IT Communication, 490
Iterative process, 97

J
JIT philosophy, 139
Job features, 494
Job profile, 545
Joint technical coordination group, 274
Joint venture decisions, 202
Journals, 7

K
Kaizen, 83
Kaizen event, 113
Kanbans, 82, 135
Khorda, 513
Knowledge, 542
Knowledge transfer, 453
Key Performance Indicators (KPIs), 489

L
Latent variables, 146, 323
Layout planning, 77
Leadership, 278
Leadership engagement, 329
Leadership process, 301
Lead time, 107
Lean manufacturing, 82, 105, 336
Lean sigma, 368, 385, 552
Lean tool, 123
Legal criteria, 232
Level process map, 204
Life cycle of the product, 477
Linear association, 536
Linear programming model, 185
Localization, 85
Local staff, 385
Location, 118
Location of landfills, 222
Logic distribution, 119
Logistics, 203
Long-term investment, 437
Long-term reliability, 377
Lot rejections, 546
Lower arms, 516
Lower back pain, 509
Lower risk level, 521
3L5W analysis, 552

M
Machine learning, 61
Macro logistic processes, 203
Management system requirements, 275
Manual manipulation, 348
Manufacturing, 59
Manufacturing and service systems, 104
Manufacturing companies, 340, 438
Manufacturing enterprise solutions, 250
Manufacturing equipment, 135
Manufacturing Execution Systems (MES), 245
Manufacturing execution taxonomy, 254
Manufacturing factors, 254
Manufacturing levers, 438
Manufacturing operations, 3, 369
Manufacturing operations management, 250
Manufacturing process, 477, 498
Manufacturing systems, 336

Manufacturing technology, 360
Maquiladora industry, 134
Market, 150
Marketing strategy, 315
Marketing times, 151
Market probability, 189
Material handling, 82
Material Handling Cost (MHC), 393, 408
Material Handling System (MHS), 389, 392
Material transferred, 403
Mathematical models, 109
Maturity index, 452
Maturity model, 207
Maximum capacity, 400
Measure pollution, 357
Measures, 23
Medical device, 341
Medium risk, 520
Mental process, 302
Meta-analysis, 371
Methodologies, 52
Mexican institute of standardization, 266
Mexican manufacturing, 343
Mexican manufacturing policy, 299
Mimetic pressures, 15
Minimum cost flow, 161
Minitab, 548
Miscalibrated equipment, 353
Misoperations, 531
Mission, 294
Mixed-integer linear programming, 185
Mixed Integer Programming (MIP), 405
Mix of products, 137
Mobile Facility Routing Problem (MFRP), 391
Model, 225
Model constraints, 187
Model development, 411
Model elements, 186
Model parameters, 186
Motivation, 101
Multi-criteria model, 233
Multi-criteria path, 97
Multi-criteria techniques, 220
Multicultural environment, 385
Multi-level functional hierarchy, 251
Municipal solid waste, 222
Musculoskeletal disorders, 507

N
Nanostores, 194
National Institute for Occupational Safety and Health (NIOSH), 492
Neck, 521
Negative environment impact, 219
Net Present Value (NPV), 415
Networks, 117
Nonconformities, 477
Normality, 549
Normality test, 375
Normalized score, 43
Normative pressures, 14
NP-Hard, 393

O
Objective function, 65, 186, 407
Occupational health hazards, 4
Odisha, 524
Open network, 166
Operationalization, 39, 308
Operationalization matrix, 210
Operational perspectives, 469
Operations, 176
Operations research, 61
Operations strategy, 436
Operations times, 109
Operators, 121
Optical fibers, 346
Optimization model, 184
Optimization strategies, 107
Optimization tools, 59
Organic products, 402
Organizational assessment, 202
Organizational culture, 84, 314, 366
Organizational goals, 85
Organizational purpose, 290
Organizational strategy, 23
Organizational structure, 443
Organization goals, 282
Organization management, 269
Organization policy, 298
Organizations, 265
Organization sustainability, 298
Organization vision, 299
Original fixture, 352
OS formulation, 444
Overseas, 366

P
Packaging, 227
Packaging process, 106
Packing method, 341
Papers, 7
Parameters of the model, 66

Pareto chart, 344, 373, 550
Partial Least Squares (PLS), 133
Participants, 319
Parts Per Million (PPM), 539
Pearson's correlation, 545
Percentage of utilization, 118
Performance measures, 116, 174
Perimetral areas, 401
Personal Protective Equipment (PPE), 480
Personnel behavior, 307
Personnel needs, 95
Personnel performance, 291
Physical tasks, 473
Pilot tests, 96, 412
Planning, 344
Planning horizon, 191
Policies, 296
Policies testing, 191
Poor integrated, 40
POS improvement, 442
Positive economic performance, 22
Possible solutions, 347
Postural assessment, 513
Practices, 11
Practitioners, 454
Pressures/drivers, 12
Printing process, 384
Prioritization of criteria, 231
Prioritization of criteria and subcriteria, 231
PRISMA, 371
Probability, 151, 163
Probability distributions, 165, 171
Problem-solving, 466
Problem statement, 209
Process approach, 278, 300
Process design, 532
Process flexibility, 36
Process flow, 554
Process flowchart, 91
Processing costs, 73
Process mapping, 86
Process time, 115
Procurement, 212
Product demand, 128
Production, 64, 78, 106
Production areas, 112
Production batches, 402
Production cost, 479
Production cycle, 216
Production elements, 88
Production facility, 402
Production factors, 290
Production flow, 121
Production levels, 399
Production lines, 80, 345
Production/Operations System (POS), 435
Production process, 81
Production runs, 468
Production scenario, 407
Production systems, 94, 103
Production transfer, 99
Productivity, 80
Productivity and quality, 550
Product packing, 203
Product quality, 136
Professional development, 476
Profitability, 414
Project budget, 211
Project definition, 71
Project main opportunities, 72
Project main threats, 72
Project objectives, 71
Proposed street market locations, 194, 195
Proposed technology, 382
Proximity, 47
Psychological profiles, 544
Psychometric test, 537
Psychosocial hazards, 475
Pull systems, 139
Pull systems tools, 135
Pulping area, 404
Punctuation factorial, 44–46
Purchase management, 205
Purchase order management process, 206
Purchase process automatization, 205
Purchasing process, 211
Purpose of a organization, 292
Push/pull view, 201
P-values, 147, 383

Q

QEC analysis, 517
QEC output, 519
Quadratic Assignment Problem (QAP), 393
Quality assurance standard, 267
Quality characteristics, 384
Quality department, 552
Quality Function Deployment (QFD), 437, 478
Quality management, 256
Quality Management Principles (QMP), 268
Quality Management System (QMS), 264, 304
Quality standard, 276
Quality tool, 256

Quantitative and qualitative approaches, 101
Quantitative values, 373
Questionnaires, 142, 322
Queueing networks, 159
Quick exposure check, 512

R
R&R analysis, 538
Rapid Entire Body Assessment (REBA), 510
Raw materials, 88
Real lead times, 130
REBA input, 515
REBA score, 517
Receiving zone, 403
Recovery period, 69
Recycling, 229
Recycling of post-consumer waste, 216
Re-design, 498
Re-designing in activities, 524
Redistribution, 121
Regional case study, 187
Regional culture, 366
Regulatory approach, 484
Regulatory norms, 498
Rejection rate, 358
Relationships, 56, 144
Relative frequencies, 56
Reliability analysis, 323
Remanufacturing, 218
Removable particles, 349
Research and development, 390
Research methodologies, 9
Resources optimization, 104
Responsibility, 294
Responsiveness to Market (RtM), 140
Responsive Supply Chain Strategy (RSCS), 137
Return Over Investment (ROI), 416
Return sources, 221
Reverse supply chain, 218
Risk based thinking, 270
Risk-control, 480
Risk factors, 485, 496
Risk level, 489
Risk Priority Number (RPN), 555
Rolled Throughput Yield (RTY), 338
Root Cause Analysis (RCA), 345, 379
Root causes, 342
Route distance, 118
RPN index, 543
Rules and regulations, 481
Running workstation, 496

S
Sample, 146
Saturations, 41
Scenarios, 151
Secretaría del Trabajo y Previsión Social (STPS), 484
Senior management, 299
Sensitivity analysis, 144, 233
Servant leadership, 317
Servant leadership scale, 322
Service times, 157, 171
Sets, 65
Shell and tube, 339
Shipping warehouse, 90
Shoulder, 521
Signal level, 356
Simulated lead times, 130
Simulation, 63, 103, 490
Simulation model, 109, 169, 411
Single-Minute Exchange of Dies (SMED), 445
Social behavior, 368
Social criteria, 232
Social performance, 23
Socio-technical factors, 529
Socio-technical systems, 534
Solid waste management, 215
Solution, 176
Solving time, 376
Sophisticated level, 497
Space requirements, 404
Spaces, 117
Spaghetti diagram, 86
Spray painting, 339
Staff attitudes, 370
Standard deviation, 164
Standardization, 358
Standardized regression, 325
Standardized solutions, 465
Standard requirements, 272
State of mind, 313
Statistical analysis, 556
Statistical methods, 189
Statistical software, 281
Steel work, 522
Storage spaces, 399
Strategic level, 453
Strategic matrix, 204
Strategic role, 439
Strategic supply chains, 37
Strategy development, 436
Strategy formulation, 444
Strategy implementation, 448

Stratification technique, 374
Street market locations, 194, 195
Street markets, 181
Structural Equation Modeling (SEM), 133, 283, 311
Subcriteria identification, 229
Success factors, 13
Successful implementation, 386
Suitability, 345
Supplier-facing, 48
Supplier level, 48–50
Suppliers, 37, 42, 157
Supplier selection, 223
Supply chain, 158, 174
Supply chain management, 134
Supply chain strategy, 61
Supply management, 199
Supply nodes, 167
Survey, 9, 142, 263
Survey for improvement, 270
Sustainability, 3, 235, 303
Sustainable development goals, 217
Sustainable performance, 6
Sustainable performance dimensions, 23
Synergy, 305
Systematic Layout Planning (SLP), 84
Systematization of processes, 100
System integration, 258
Systems audit administration, 276
Systems evolution, 248

T
Tasks, 100
Team level, 318
Technical characteristics, 555
Technical Committee (TC), 265
Technical reinforcement, 445
Technicians' visibility, 352
Technician vs equipment, 548
Technological advances, 465
Technological factors, 255
Technology device, 474
Technology tools, 497
Temporal evolution, 248
Thermoforming, 341
TO-BE model, 60
Total effects, 148
Total Productive Maintenance (TPM), 445
Training and education, 481
Transparency in manufacturing processes, 256
Transport, 176
Transportation activities, 99
Transportation costs, 79
Transportation methods, 407
Transport costs, 168
Traveled distance, 93
Triple Bottom Line (TBL), 21
Trunk and legs, 513
T3, 346
T3 transformer, 355
Tukey's procedure, 413
Two-sample test, 376
Type of chocolate, 398
Types of failures, 474
Types of postures, 519
Types of supply chains, 219

U
Upper arms, 516
Usability, 469
Utilization, 121

V
Validation, 110, 143
Validity of scales, 322
Validity strategy, 443
Value position, 315
Value Stream Map (VSM), 108
Variability, 158, 163, 173
Variables, 63
Variety of postures, 513
Verein Deutscher Ingenieure (VDI), 253
Very integrated, 40
Vibration, 519
Vigor, 319
Vigor model, 324
Virtual ergonomics, 493

W
Warehouse, 88
Waste, 336
Wastewater, 224
Water supply, 523
Weights of variables, 324
Whys analysis, 376
Wood works, 515
Work activities, 473
Work design, 468, 534
Work engagement, 311, 325
Work engagement scale, 321
Worker seniority, 544
Workflow diagram, 359

Workforce, 107
Work in process, 173
Work intensity, 472
Workplace, 312, 373
Workplace design, 470
Workplace environment, 495
Workplace violence, 475
Work-Related Musculoskeletal Disorders (WMSD), 488
Work schedule, 476
Workstation design, 496
Workstations, 88, 113, 169, 467
World-class, 385
World class manufacturing, 440
World Health Organization (WHO), 484
Wrong inspection, 353
Wrong material, 557